Food and Beverage Cost Control

Food and Beverage Cost Control

EIGHTH EDITION

DAVID K. HAYES
LEA R. DOPSON

WILEY

VICE PRESIDENT	Amanda Miller
SENIOR EDITORIAL DIRECTOR	Justin Jeffryes
EXECUTIVE EDITOR	Todd Green
ASSISTANT EDITOR	Kelly Gomez
SENIOR MANAGING EDITOR	Monica Chandra Sekar
CONTENT REFINEMENT SPECIALIST	Philo Antonie
COVER PHOTO	© MarFot/stock.adobe.com

This book was typeset in 10pt/12pt SabonLTStd by Straive™.

Published by John Wiley & Sons, Inc., Hoboken, New Jersey.

EPUB ISBN: 9781394365968

The inside back cover will contain printing identification and country of origin if omitted from this page.

In addition, if the ISBN on the cover differs from the ISBN on this page, the one on the cover is correct.

Library of Congress Cataloging-in-Publication Data Applied for

ISBN: 9781394365951

Cover Design: Wiley
Cover Image: © MarFot/stock.adobe.com

DEDICATION

This edition of *Food and Beverage Cost Control* is dedicated to the memory of Dr. Lea R. Dopson.

I had the honor of knowing Lea from a number of different perspectives over the course of her life and her professional career.

The first was as my student. I initially met Lea when I was the Head of the Restaurant, Hotel, and Institutional Management (RHM) department at Texas Tech university. Lea enrolled in the program, and it was very clear from the beginning that she would be an outstanding student. She graduated from the Bachelor's degree program with honors, and immediately enrolled in, and completed, the RHIM Master's degree program. Of the thousands of students I have had the pleasure to teach over the course of my 45+-year career in hospitality education, Lea was undoubtedly my very best.

I also knew Lea as a most talented educator. On graduation with her Master's degree, Lea took a prestigious management position with one of the major hotel chains in Texas. We stayed in touch and, when I was appointed associate Dean of the Conrad Hilton College of Hotel Administration at the University of Houston (U of H), I persuaded Lea to move from Dallas to Houston and serve as that college's Director of Student Services.

She did an amazing job.

As planned, she also enrolled in and completed her doctorate at U of H. She then immediately secured a faculty position at Cal Poly Pomona, which houses one of California's most prestigious hospitality programs.

Her talent as an educator did not go unnoticed. That led to her being appointed as the head of the hospitality program at the University of North Texas (UNT). Her leadership of that program was remarkable, and it quickly became known among hospitality faculty and students as an outstanding and innovative program.

It was not a surprise when, a few years later, the administration of Cal Poly Pomona reached out to Lea to ask her to return and take over the administration of their entire hospitality school. She accepted their offer and served as the Dean of that school until her untimely passing.

To say that Lea was admired by her colleagues as an educator and administrator would be a tremendous understatement. She was elected by the membership to serve a number of key roles with the International Council of Hotel and Institutional Educators (ICHRIE), including her election as that organization's President in 2022.

As well, she was awarded ICHRIE's prestigious Howard B Meek Award. The Howard B. Meek Award is the highest individual recognition a member of ICHRIE may receive. It recognizes a member's lifetime contributions and outstanding service both to hospitality education and to ICHRIE. This award serves as a living memorial to the late Howard B. Meek, a pioneer in American hospitality education, the first Dean of the School of Hotel Administration at Cornell University, and an Executive Vice President of ICHRIE.

It is important to note that the Howard B. Meek Award is *not* given to the most esteemed hospitality educator in the United States; rather it is awarded to the most esteemed hospitality educator in the world. That was Lea.

Lea was a gifted author. In 2001 I asked her to be my co-author on the second edition of *Food and Beverage Cost Control*, and she agreed. Subsequently, I had the honor and pleasure of working with Lea on six editions of that text, as well as the first and second editions of the John Wiley published *Managerial Accounting for the Hospitality Industry*.

In all cases Lea was consistently dedicated to ensuring that each edition was easy to read, easy to understand, and easy to remember. By doing so she won the admiration of both those instructors who utilize the texts in their classrooms and the students who benefited from them. She was an outstanding writing partner.

Finally, and most importantly to me, Lea was my closest professional friend. During our 35+ years friendship we shared our professional and personal challenges and successes, and were unwaveringly fierce supporters of each other's life choices and experiences.

For me, one of the highlights of each year was dinner with Lea at the Carson's for Ribs restaurant, where we would meet during the National Restaurant Association annual show in Chicago. For 30+ years we never missed our Saturday night standing dinner date, and it is a tradition that, in her memory, I continue to uphold.

With Lea's passing the hospitality industry has lost an outstanding educator, a consummate professional, a talented author, and a most beautiful person. I continue to miss her dearly and hope she would be pleased with my efforts to continue her legacy via the publication of this new edition of *Food and Beverage Cost Control*.

David K. Hayes, Ph.D.

Holt, MI

CONTENTS

PREFACE

The publication of the first edition of *Food and Beverage Cost Control* fulfilled original lead author Professor Jack Miller's vision for a college-level text that would provide students and practicing managers with the essential tools needed to effectively manage costs in foodservice operations.

For the second through seventh editions, Dr. David Hayes and Dr. Lea R. Dopson collaborated in the preparation of each revision. After Dr. Dopson's untimely passing, Dr. Hayes developed this eighth edition with significant assistance from consulting author Joshua D. Hayes, Ph.D., building upon Dr. Dopson's steadfast dedication to the book's content that mirrored Professor Miller's vision.

The author hopes that the study of cost management creates in readers the same interest and excitement for the topic that the author has experienced. If so, he will have been successful in his efforts to be true to this text's original vision of creating an outstanding learning tool that prepares its readers to be successful managers in the exciting hospitality industry.

It has been said that there are three kinds of managers: those who know what has happened in the past, those who know what is happening now, and those who know what will happen in the future. Clearly, the manager who possesses all three traits is best prepared to manage effectively and efficiently. This text will give the reader the tools required to maintain sales and cost histories (the past), develop systems for monitoring current activities (the present), and learn the techniques required to anticipate and forecast what is to come (the future).

Today's professional foodservice managers face myriad complex challenges. As was true in the past, the tools and information they need to properly address these challenges must be well understood if they are to be readily applied. Students will find this edition easy to read. Instructors will find the information in it easy to present. Practicing managers will find the information in it easy to apply. The author is convinced that is exactly what Professors Miller and Dopson would have wanted for the eighth edition of *Food and Beverage Cost Control*, and he is honored to play his part in fulfilling their vision.

TO THE STUDENT

This book will provide you with all of the cost-control-related information and tools you will need to achieve success levels that will match your highest career goals.

If you work hard and do your best, you will master all of the information in this text. When you do, you will have gained an invaluable set of management skills that will enhance both your knowledge of the hospitality management industry and your professional self-confidence.

TO THE INSTRUCTOR

Experienced managers know that effective cost control in a foodservice operation is built upon a variety of systems that depend on the skillful use of mathematics. In many cases, however, hospitality students may be unsure of their mathematical abilities. Because that is so, any textbook addressing cost control must possess two essential math-related traits.

First, it must be accurate. The author is grateful to the number of editors and reviewers who helped to ensure that the mathematical formulas and examples presented in this edition are as error-free as humanly possible. He is grateful, as well, to the number of students and instructors who, in editions 1 through 7, provided feedback that helped ensure all mathematical examples and end-of-chapter questions and answers retained in this edition are clearly presented and accurate.

Secondly, the mathematical concepts included in the text must be presented clearly. In a heightened effort to address this key concern, this edition retains the popular "Here's How It's Done" feature included in each chapter. This feature was retained in direct response to instructors' desire that their students have step-by-step explanations and illustrations of some of the text's more challenging math concepts. In this unique feature, students are shown, using real work setting examples, how the math concepts presented in the chapter are applied and their results evaluated.

Instructors utilizing this text as their primary classroom teaching tool can expect to find detailed discussions of the most important new cost-control-related issues their students will face upon graduation. This is so because the author's combined years of teaching hospitality management at the undergraduate, graduate, and continuing professional education levels have helped enormously to shape this textbook's original content and this new revision. Specifically, the following became special areas of emphasis as this new edition was developed:

Simplification of Presentation. The readers of our text have always been a primary focus, and the author is pleased that, again and again, creative graphics and clearly written narrative help to enhance the book's reader-friendliness and, as a result, present complex ideas in easily understandable ways. Special care was taken in this edition to review each paragraph and sentence to ensure that the content they contain was presented in the clearest possible manner.

Recognition of Industry Change and Evolving Cost Control Concerns.

One of the most significant cost-control-related issues facing foodservice operators today is the continued move by guests to off-premise dining.

Whether it is increased use of their drive-thru, carryout, or off-site deliveries, many foodservice operations are adjusting to the fact that their guests increasingly choose to consume the products they purchase in off-premise locations, rather than on-site.

The impact of this consumer movement affects everything from the size of new facilities being constructed (because operations doing a significant amount of off-premise business can be smaller), to issues related to the cost of packaging and delivering products at their peak quality levels.

To address the myriad issues related to this changing consumer preference, this text includes a new feature: "The Move to Off-Premise Dining." This feature, which is included in each chapter, details how the move to off-premise dining impacts many of an operator's cost control efforts.

For example, in Chapter 5 (Monitoring Food and Beverage Product Costs), "The Move to Off-Premise Dining" boxed feature addresses the fact that foodservice operators doing a significant amount of off-premise takeout or delivery sales must make a decision about how to account for the cost of the packaging materials used for these off-premise items.

Historically, the Uniform System of Accounts for Restaurants (USAR) has recommended that the cost of paper and packaging used for off-premise food and beverage sales be listed as a "Direct Operating Expense," rather than a cost of food or beverage sales.

Increasingly, however, some foodservice operators are electing to include required packaging items as part of their "Cost of Goods Sold" (COGS) expense. The view of these operators is that packaging that is directly related to the cost of producing a menu item, such as boxes or wrappers, should be listed separately, but included in COGS. Thus, for example, for these operators the cost of the box used when selling a carryout pizza would be considered part of their foodservice operations' COGS.

Items such as paper or plastic bags, foam containers, napkins, utensils, plastic cups, lids, straws, and sauce packets can often cost as much as 3 to 5 percent of an operation's total sales. Thus, whether the decision is made to record these expenses as a COGS or as a Direct Operating Expense, it is important to understand how the control of these costs affect an operation's profitability.

The above example is just one of many off-premise-related issues that are addressed in detail in this edition, ensuring that readers understand the full cost-related ramifications of this significant consumer trend.

TO MANAGERS

While *Food and Beverage Cost Control* has always been produced in a textbook format, it has also consistently been an invaluable tool for the practicing manager. The easy, step-by-step approach used to estimate future customer counts (Chapter 2) and apply measures of labor productivity (Chapter 7) are just two examples of its very practical application. The formulas used to calculate edible portion product yields (Chapter 5) and the information utilized to properly establish prices for menu items (Chapter 6) are two more such examples.

From information needed to convert standardized recipes from the US system of weights and measures to the metric system (Chapter 3), to tips for calculating and analyzing variances on profit and loss statements (Chapter 9), to recognizing the difference between a POS system's X and Z reports (Chapter 11), managers responsible for the operation of smaller to higher-volume foodservice units will find that the information in this book is vitally important and easily applicable to their operations.

Effective foodservice managers are skilled problem solvers. The information found in the eighth edition of *Food and Beverage Cost Control* is designed especially to provide professional problem solvers with the tools they need to manage efficient and highly profitable foodservice operations.

NEW IN THE EIGHTH EDITION

Eighth edition readers will be pleased to find major enhancements both in the text's content and in its structure.

NEW CONTENT

One of the ongoing strengths of *Food and Beverage Cost Control* has been the author's commitment to continually and carefully monitoring the field of food and beverage cost control to identify changes that must be made to ensure that the book presents the most up-to-date and accurate information available. Significant content additions to this edition include the following:

- ✓ The inclusion of one or more "The Move to Off-Premise Dining" boxed features per chapter specifically addressing the significant cost-control-related challenges of this continually growing customer preference.
- ✓ The insertion of a time/temperature control for safety foods definition (formerly potentially hazardous food) as a food that requires time/temperature control for safety to limit pathogenic microorganism growth or toxin formation.
- ✓ The updating of all relevant information related to the release of the 2022 (previously 2017) Food Code. The US Food and Drug Administration (FDA) publishes the FDA Food Code, which serves as a model for state and local food safety regulations.
- ✓ The inclusion of a section providing tips for the responsible service of "to-go" alcohol sales.
- ✓ The addition of a section addressing how "to-go" packing materials, disposable service ware, and pre-packaged condiments impact total product costs.
- ✓ A section addressing the pros and cons of providing QR codes (only) for guest viewing of menus.
- ✓ A discussion of dynamic pricing and its potential use in foodservice operations.
- ✓ A discussion of the need for, and value of, providing easy access to online employment applications, as well as the importance of utilizing the E-Verify system when hiring new employees.
- ✓ An examination of situations in which employee scheduling can be improved using cloud-based services and smart-device applications.
- ✓ The addition of detailed definitions of non-cash operating expenses (e.g., depreciation and amortization).
- ✓ The addition of a new section addressing the importance of monitoring user-generated content review sites and detailed recommendations for responding to negative guest reviews that are posted online.
- ✓ The addition of a new figure detailing the individual entries on a POS system-generated guest receipt, including detailed definitions of transaction numbers, reference IDs, and authorization IDs.
- ✓ An examination of the increasing use of digital wallets (e.g., Apple Pay, Google Pay, PayPal, Venmo, and the like) as well as a discussion of the controversy surrounding the move to noncash-only payment forms.
- ✓ A detailed explanation of the legality of imposing service fees on guests' debit and credit card purchases.

RETAINED IN THE EIGHTH EDITION

Much of the popularity of this text has been due to the quality of the elements and features developed for it in prior editions. In this eighth edition, the author was pleased to update and retain from the previous edition the following key text elements:

OVERVIEW: Each chapter begins with a brief overview of what the chapter contains. The overview focuses on why students will benefit from learning the information presented in the chapter. Thus, this element directly informs readers about what is to be presented in the chapter and why it is important to know it.

CHAPTER OUTLINE: One-tier outlines are presented at the beginning of each chapter to inform readers about the specific topics to be addressed. This helpful feature also makes it easier to find specific material contained in the chapter.

LEARNING OUTCOMES: Students want to know how the information they learn will be useful to them in their careers. This feature specifically identifies what readers will know and what they will know how to do when they have mastered the material in the chapter.

HERE'S HOW IT'S DONE: This math "help" feature was originally developed and introduced in the sixth edition. Its introduction was extremely well-received by students and instructors. This feature is inserted, where applicable in a chapter, to assist students with the arithmetic required to understand the cost-control-related concept being presented. Created to provide easy-to-follow instructions and a step-by-step numerical example, this popular feature has been retained and expanded to include even more mathematical illustrations in this new edition.

GREEN AND GROWING: More than ever, students and customers alike recognize that environmental consciousness is as important at work as it is at home. As a result, hospitality professionals are increasingly adopting "green" practices and policies that aid the planet as well as their own bottom lines. In this feature, students become familiar with the "whys" and the "hows" of responsibly growing their businesses by implementing earth-friendly business practices specific to the hospitality industry.

CONSIDER THE COSTS: One of the most exciting things about learning any new skill is the ability to directly apply what has been learned to situations the learners will actually encounter. To give students an opportunity to do just that, "Consider the Costs" micro case studies have been developed to present students with common cost-control-related challenges they will likely encounter at work. Each case study poses questions that allow readers to apply information learned in the chapter to these real-world work situations and problems. Instructors will also find these micro case studies are fun for their students to read and discuss in class.

FUN ON THE WEB!: This important feature of the text adds to student learning by integrating the use of the Internet to the study of cost control. This feature provides Web-based resources that can help managers more effectively do their jobs.

TECHNOLOGY TOOLS: These updated listings of real-life application examples demonstrate to students that they can utilize advanced smart device applications, sophisticated wired and wireless communication tools, and much more to help manage costs and improve operating efficiencies. While not all managers will use all of the tools suggested, it is important for students to understand the rapidly expanding technology-based resources available to them today.

APPLY WHAT YOU HAVE LEARNED: This innovative pedagogical feature allows students to draw on their own problem-solving skills, ideas, and opinions using the concepts explored within each chapter. Challenging and realistic, yet purposely brief, these industry-specific scenarios provide excellent starting points for class discussions or, if the instructor prefers, outstanding written homework assignments.

KEY TERMS AND CONCEPTS: Students often need help in identifying key terms and concepts that should be mastered after reading a section of a book. These are listed at the conclusion of each chapter and in the order in which they appeared in the chapter to make finding them easier. This new edition contains over 300 industry-specific terms.

TEST YOUR SKILLS: This popular feature has been retained and expanded. As was true in previous editions, predesigned Microsoft Excel spreadsheets are employed in most of the questions to allow students to practice problem solving. Doing so enhances the instructor's ability to evaluate student mastery of cost control concepts and student skill in understanding and using spreadsheets.

Excel spreadsheets with answers to all Test Your Skills are available in the instructor's website at www.wiley.com/go/dopson/foodandbeveragecostcontrol8e. Student versions (without answers) of the Excel spreadsheets are available at the book website at wiley.com.

As well, the author has again included an appendix: *Frequently Used Formulas for Managing Costs*. It is available on the instructor's website as well as the student companion site. This feature allows readers to look up mathematical formulas for many of the important computations presented in the text.

COMPANION WEBSITE

To help instructors effectively manage their time and to enhance student learning opportunities, several significant educational tools have been developed specifically for this text.

INSTRUCTOR'S MATERIALS

A password-protected online Instructor's Companion Site, available at www.wiley.com/go/dopson/foodandbeveragecostcontrol8e, has been meticulously developed and classroom tested for this text. The manual includes the following, each of which is presented in a stand-alone format:

- ***Lecture Outlines*** for each chapter.
- ***Power Point Presentations*** for each chapter: These easy-to-read teaching aids are excellent tools for instructors presenting their lectures in class or online.
- Suggested answers to each chapter's "***Consider the Costs***" micro case studies.
- Suggested answers for "***Apply What You Have Learned***" questions for each chapter.

- Suggested answers to each chapter's "***For Your Consideration***" conceptual queries.
- Correct answers to all chapter-ending "***Test Your Skills***" problems. Instructors will be able to access answers and formulas to the "Test Your Skills" spreadsheet exercises at the end of each chapter.
- A ***Test Bank*** including 25 multiple choice (4-alternative) and 10 True and False (2-alternative) exam questions developed for each chapter. The author recognizes the importance that instructors place on well-designed exam questions. I am convinced the questions developed for this text are among the very best in all of hospitality education.

The Test Bank for this text has been specifically formatted for Respondus, an easy-to-use software for creating and managing exams that can be printed or published directly to Blackboard and other eLearning systems. Instructors who adopt *Food and Beverage Cost Control, Eighth Edition* can download the Test Bank for free.

ACKNOWLEDGMENTS

The first seven editions of this text have been very popular. As a result, this book continues to be one of the market leaders among hospitality cost control texts. This success has stemmed in large part from the testing of its concepts and materials in classes at Purdue University, Texas Tech University, the University of North Texas, the University of Houston, California State Polytechnic University at Pomona, and Lansing Community College, as well as from those original St. Louis Community College students who received their instruction under Professor Jack Miller.

We are also extremely grateful to the myriad professionals in institutional, commercial, and hotel foodservice operations with whom we consulted and who so freely gave of their time and advice in this endeavor.

This edition could not have been produced without the assistance of a great many colleagues, friends, and family who supported my efforts. As always, heartfelt gratitude goes to those who have been so supportive of me throughout my writing career: Peggy, Scott, and Trishaunna, I appreciate all of you!

A particular shout-out goes to Mike and Judy Kavanagh of Biloxi, Mississippi, who continue, through their thoughts and deeds, to reinforce to me that "hospitality" is best understood as an active verb, and not merely an abstract noun.

Finally, a special thank you goes to Joshua D. Hayes, Ph.D., who agreed to serve as a Consulting Author for this edition.

As always, I am also deeply grateful to Judy Howarth and all of the staff at John Wiley, as well as freelance copy-editor Lesley Montford, for their intellect, patience, and faithfulness in helping to produce this eighth and most comprehensive edition of *Food and Beverage Cost Control.*

David K. Hayes, Ph.D.
Holt, MI

ABOUT THE COMPANION WEBSITE

This book is accompanied by a companion website.

www.wiley.com/go/dopson/foodandbeveragecostcontrol8e

To access the website materials, click on the link under Related Resources on the book's page on wiley.com. You can also find this by searching www.wiley.com for the ISBN [**9781394365951**].

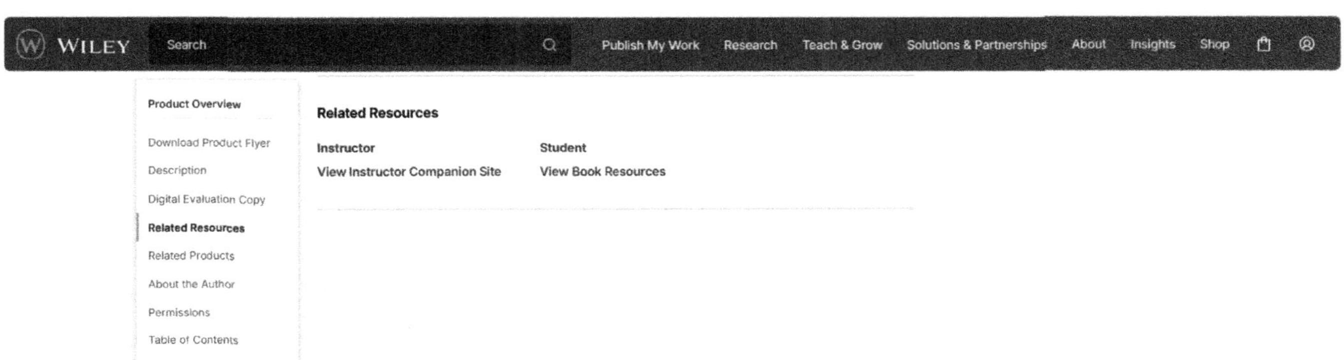

This website includes:

- Test Bank
- Respndus Comp. Test Bank
- Power Points
- Lecture Outline
- Apply What You Have Learned
- Consider the Costs
- For Your Consideration
- TYS For Instructors
- TYS For Students
- Appendix B

CHAPTER 1

Managing Revenue and Expense

OVERVIEW

This chapter presents the relationship between a foodservice business's revenue, expenses, and profit. As a professional foodservice manager, you must understand the relationship that exists between monitoring and controlling these three areas and the resulting success of your operation. In addition, the chapter presents the mathematical foundation you must know to report your operating results and express them as a percentage of your revenue or budget. This process is a best-practice standard within the hospitality industry.

Chapter Outline

- Professional Foodservice Manager
- Profit: The Reward for Service
- Getting Started
- Understanding the Income (Profit and Loss) Statement
- Understanding the Budget
- Technology Tools
- Apply What You Have Learned
- For Your Consideration
- Key Terms and Concepts
- Test Your Skills

LEARNING OUTCOMES

At the conclusion of this chapter, you will be able to:

- Apply the formula used to determine business profits.
- Express business expenses and profits as a percentage of revenue.
- Compare actual operating results with budgeted operating results.

PROFESSIONAL FOODSERVICE MANAGER

To be a successful foodservice manager, you must be a talented individual. Consider, for a moment, your role in the operation of a profitable foodservice facility. As a foodservice manager, you are both a manufacturer and a retailer. A professional foodservice manager is unique because all of the functions of a product's sale, from menu development to guest service, are in the hands of the same individual. As a manager, you are in charge of securing raw materials, producing a product, and selling it—all under the same roof. Few other managers are required to have the breadth of skills that effective foodservice operators must have. Because foodservice operators are in the service sector of business, many aspects of management are more challenging for them than for their manufacturing or retailing management counterparts.

Consider also that a foodservice manager is one of the few types of managers who actually have contact with the ultimate consumer. This is not true for the managers of a cell phone factory or an automobile production line. Those individuals produce a product, but they do not sell it to the person who will actually use it. In a like manner, furniture and clothing store managers will sell products to those who use them, but they have had no role in actually producing the products they sell. The face-to-face guest contact in the hospitality industry requires that you assume the responsibility of standing behind your own work and the work of your staff, in a one-on-one situation with the ultimate consumer, or end user, of your products and services. This is true even if many of your guests choose **off-premise dining**, which are those meals that are ordered from a restaurant but eaten elsewhere. These meals can be delivered, picked up in your operation, sold via drive-thru, or picked up curbside.

The management task checklist in Figure 1.1 shows some of the areas in which foodservice, manufacturing, and retailing managers differ in their responsibilities.

In addition to your role as a food factory supervisor, you must serve as a cost control manager, because if you fail to perform this vital role, your business will perform poorly or may even cease to exist. Foodservice management provides the opportunity for creativity in a variety of settings. The control of revenue and expense is just one more area in which an effective foodservice operator can excel. In fact, in most areas of foodservice, excellence in operation is measured in terms of a manager's ability to produce and deliver quality products in a way that ensures an appropriate operating profit for the owners of the business.

The Move to Off-Premise Dining

A solid argument could be made to state that the most significant change in foodservice operations over the past decade relates to guests' accelerating move to off-premise dining. According to the National Restaurant Association's State of the Restaurant Industry 2024 report, of the surveyed restaurants, 43 percent of full-service restaurants and 55 percent of limited-service restaurants said off-premise accounted for more business than it did five years ago.[1]

The move to off-premise dining in the foodservice had already begun due to increased numbers of quick-service restaurants providing drive-thru service and the emergence of popular delivery apps such DoorDash, Uber Eats, and Grubhub, but that trend exploded during the COVID-19 pandemic.

Even after the pandemic ended, the trend toward greater numbers of guests increasingly preferring to buy from a foodservice operator but consume the purchased products elsewhere expanded. The trend toward off-premise buying is projected to continue, increasing from $3.7 billion in 2023 to an estimated $90.3 billion by 2032.[2]

It is easy to see that the trend toward off-premise dining will continue to have a direct impact on how foodservice operators control their off-premise dining-related operating costs because the move to increased levels of off-premise dining provides foodservice operators with both cost-related benefits and challenges, and these will be examined throughout this book.

[1] https://www.restaurantdive.com/news/how-off-premise-diner-demand-is-evolving-what-the-data-shows/708564. Retrieved 2/21/2025.

[2] https://trycake.com/blogs/news/28-most-influential-online-ordering-stats-to-know-in-2024#:~:text=In%20 2023%2C%20online%20ordering%20accounted,be%20 using%20food%20delivery%20services. Retrieved 2/30/2025.

FIGURE 1.1 Management Task Checklist

Task	Foodservice Manager	Manufacturing Manager	Retail Manager
1. Secure raw materials	Yes	Yes	No
2. Manufacture product	Yes	Yes	No
3. Market to end user	Yes	No	Yes
4. Sell to end user	Yes	No	Yes
5. Reconcile problems with end user	Yes	No	Yes

PROFIT: THE REWARD FOR SERVICE

In the foodservice industry, a manager's primary responsibility is to deliver quality products and services to guests at a price mutually agreeable to both parties. In addition, the quality must be such that buyers of the product or service feel that excellent value was received for the money they spent. When they do, a business will prosper. If, however, management focuses more on reducing costs than providing value to guests, problems will inevitably occur.

It is important to remember that serving guests causes businesses to incur costs. It is wrong to think that "low" costs are good and "high" costs are bad. A restaurant with $5 million in sales per year will have higher costs than the same-size restaurant achieving only $500,000 in sales per year. The reason is quite clear. The amount of products, labor, and equipment needed to sell $5 million worth of food and beverages is greater than those required to sell only $500,000. Remember, if there are fewer guests, there are likely to be lower costs, but fewer sales and less profit as well! Because that is true, a business will suffer if management attempts to reduce costs with no regard for the impact on the balance between managing costs and maintaining high levels of guest satisfaction. In addition, efforts to reduce costs that result in unsafe physical conditions for guests or employees are never wise. Although some short-term savings may result, the expense of a lawsuit resulting from a guest or employee injury can be very high. Managers who, for example, neglect to spend the money to shovel and salt a snowy restaurant entrance area may find that they spend thousands of dollars more defending themselves in a lawsuit brought by an individual who slipped and fell on the ice than they would have spent clearing the snowy walkway.

For an effective manager, the question to be considered is not whether costs are high or low. The question is whether costs are *too high* or *too low*, given the value a business seeks to create for its guests. Managers can eliminate nearly all costs by closing their operations' doors. Obviously, however, when you close the doors to nearly all expenses, you also close the doors to sales and, more importantly, to profits. Expenses, then, must be incurred, but managed in a way that allows the operation to achieve its desired profit levels.

It is especially important for you to understand profits. Some people assume that if a business purchases an item for $1.00 and sells it for $4.00, the profit generated is $3.00. In fact, this is not true. As a business operator, you must realize that the difference between what you have paid for the goods you sell and the price at which you sell them does not represent your profit. Instead, all expenses, including advertising, the building that houses your operation, management salaries, and the labor required to generate the sale, just to name but a few, are among the many expenses that must be subtracted from your income *before* you can determine your profits accurately.

Every foodservice operator must know and understand well the profit formula given below:

Revenue − Expenses = Profit

Thus, when you manage your facility, you will receive **revenue**—the money you take in from selling your products—and you will incur **expenses**—the cost of everything required to operate the business and to generate your revenue. **Profit** is the amount of money that remains after all expenses have been paid. Because doing so is common in the industry, in this book the authors will use the following terms interchangeably: revenues and sales; expenses and costs.

The profit formula holds true even for managers in the nonprofit sectors of foodservice such as schools, hospitals, military bases, and businesses providing meals to their workers. For example, consider the situation of Hector Bentevina. Hector is the foodservice manager at the headquarters of a large high-tech corporation that employs many office workers. Hector supplies the foodservice to a large group of these workers, each of whom is employed by the corporation that owns the facility Hector manages. In this situation, Hector's employer may not consider generating profits from its foodservice operation as a primary goal. That is so because, in most **business dining** situations, meals are provided to the company's employees either as a no-cost (to the employee) benefit or at a greatly subsidized price. In most cases, however, even nonprofit operations will be concerned about their levels of revenue and their operating expenses.

Figure 1.2 shows the flow of business for the typical foodservice operation. Note that, near the end of the flow process, profit dollars are to be taken out, or management will be in the position of simply trading equal amounts of cash for cash.

In your own operation, if you find that revenue is consistently less than your expenses, with no reserve for the future, you will also find that there is no money for new equipment, needed facility maintenance may not be performed, and employee raises (including your own) may be few and far between. In addition, your facility will eventually become outdated due to a lack of funds for remodeling and upgrading. The fact is that all foodservice operations must be properly managed if they are to generate revenue in excess of expenses.

THE MANAGEMENT PROCESS

While there are a number of ways to view the management process, one good way is to consider the five functions that are essential for the effective management of any organization. These are listed in Figure 1.3.

- *Planning:* This initial step in the management process addresses the creation of goals and objectives. This is the first step in the management process because it identifies precisely what the organization wants to achieve through its efforts.
- *Organizing:* After its objectives have been identified, an organization must ensure that it has the funding, staff, equipment, and raw

FIGURE 1.2 Foodservice Business Flowchart

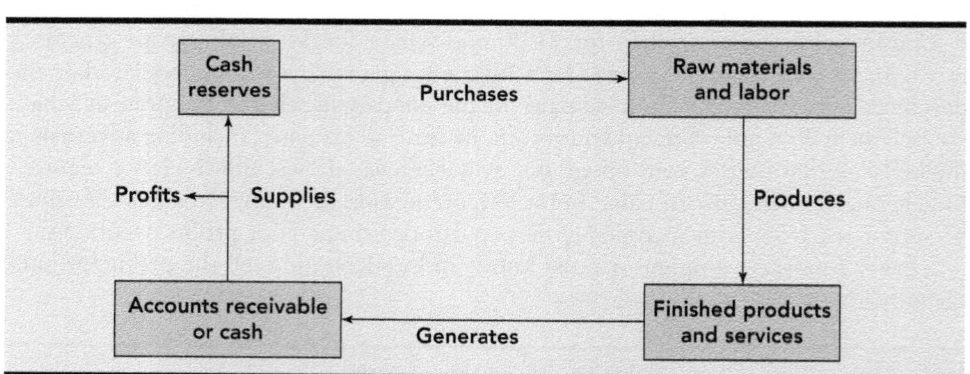

FIGURE 1.3 The Management Process

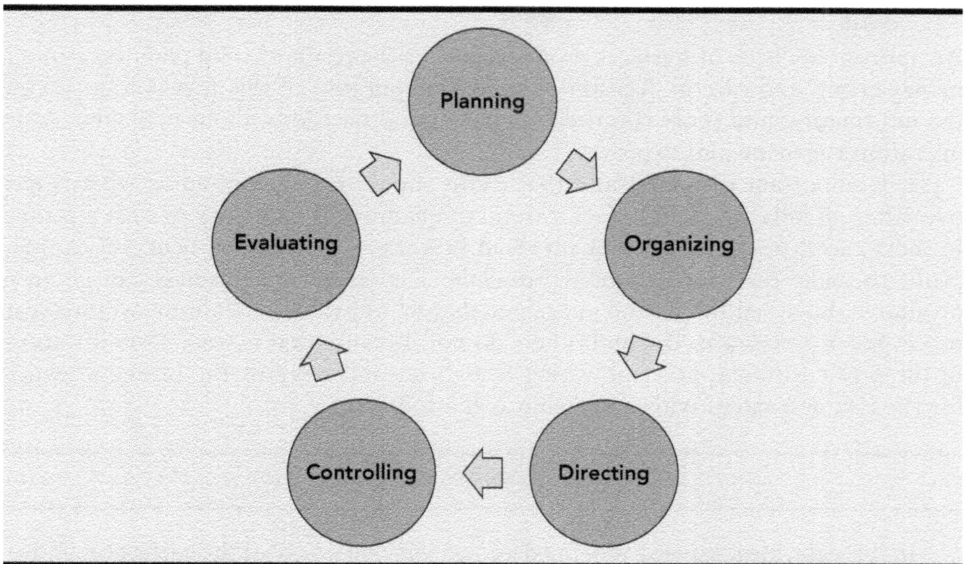

materials it will need to achieve its objectives. These assets must then be arranged (organized) in a way that optimizes the organization's ability to achieve its objectives.

- *Directing:* This important management function addresses the task of telling, and showing, all staff members exactly what is expected of them. When given clear directions, all staff members will know the important roles they will play in helping the organization achieve its objectives.
- *Controlling:* By continually assessing the work processes and procedures they have put place, managers can better identify situations that could prevent the organization from meeting its objectives. In the foodservice industry, important processes that must be controlled include those activities related to purchasing, receiving, storing, preparing, and serving menu items.
- *Evaluating:* This final management activity requires that an organization assess its current performance and compare the performance to planned performance. If significant differences are found to exist, the organization must determine the reason for the differences and then either change its objectives or change the methods used to achieve them.

Although this book addresses each of the five key management processes, its primary focus is on control. Specifically, it addresses the important things food and beverage managers must know, and do, to properly control their costs. Essentially, the proper control of costs in a foodservice organization requires managers to do the following:

- Establish performance standards based on organizational objectives.
- Measure and report actual performance.
- Compare actual results to desired (planned for) results.
- Take corrective actions as needed.

GETTING STARTED

An appropriate level of business profits is always the result of solid planning, sound management, and careful decision making. The purpose of this text is to give you the information and tools you need to make good decisions about managing your operation's revenue and expenses.

It is important to understand that profit should not be viewed as what is left over after all bills are paid. In fact, careful preplanning is necessary to earn a profit. In most cases, investors will not invest in businesses that do not generate enough profit to make their investment worthwhile. The restaurant business can be very profitable; however, there is no guarantee that an individual restaurant will make a profit. Some restaurants do, and others do not. Because that is true, a modification of the profit formula, and one which recognizes and rewards the business owner for the risk associated with ownership is the following:

$$\text{Revenue} - \text{Desired profit} = \text{Ideal expense}$$

In this case, **ideal expense** is defined as management's view of the correct or appropriate amount of expense necessary to generate a given level of sales. **Desired profit** is defined as the profit that the owner wants to achieve at that level of revenue. This formula clearly places profit as a reward for providing service, not as a leftover. When foodservice managers deliver quality and value to their guests, anticipated revenue levels and desired profits can be achieved. Desired profit and ideal expense levels are not, however, easily achieved. It takes a talented foodservice operator to consistently make good decisions that will maximize revenue while holding expenses to the ideal or appropriate amount. This book will teach you how to make those good decisions.

REVENUE

Revenue dollars are the result of units sold to customers. Units may consist of individual menu items, lunches, dinners, drinks, or any other item produced by your operation. Revenue varies with both the number of guests coming to your business and the amount of money spent by each guest. You can increase revenue by increasing the number of guests you serve. You can do this by adding additional seating or drive-thru windows, promoting guest pickup of items sold, extending operating hours, and building additional foodservice units. Suggestive selling by service staff, creative menu pricing techniques, and discounts for very large purchases are examples of efforts to increase the amount of money each guest spends.

Green and Growing!

Good food and service will attract foodservice customers. So will other important factors customers care about, including location, unique décor, and, increasingly, how "green" an operation is perceived to be. Green is the term used to describe those foodservice operations that incorporate environmentally conscious activities into the design, construction, and operation of their businesses. These activities can be related to packaging and shipping materials reduction, energy conservation, or sustainable development, a term used to describe a variety of Earth-friendly practices and policies as "development that meets the needs of the present population without compromising the ability of future generations to meet their own needs."

The positive benefits that accrue when businesses incorporate green activities are significant and are increasing. Managers of green operations help protect the environment. For example, did you know that every ton of 100 percent post-consumer waste recycled paper saves 12 trees, 1,976 pounds of greenhouse gases, and

390 gallons of oil? Green operating is also gaining in popularity because more and more guests seek out and frequent green restaurants simply because they are committed to preserving the environment.

The Green Restaurant Association℠ (GRA) is a nonprofit, national environmental organization founded to help restaurants and their customers become more "green" (environmentally sustainable) in ways that are convenient and cost-effective. The GRA's agenda includes issues related to the following:

Research
Environmental consulting
Education
Public relations and marketing
Community organizing and consumer activism

To learn more about this increasingly high-profile group, visit the GRA's website, dinegreen.com. To learn more about how your foodservice operation can increase profits by implementing sustainable activities, watch for the Green and Growing feature in each upcoming chapter of this book.

UN Brundtland Commission, "Report of the World Commission on Environment and Development: Our Common Future," 42nd session, Development and International Cooperation: Environment, August 4, 1987, Chapter 2 opener.

Management's primary task is to take the steps necessary to attract new and repeat guests to the foodservice operation. This is true because the profit formula begins with sales made to guests. Experienced foodservice operators know that adding guests and selling more to each guest are extremely effective ways of increasing overall profitability, but *only* if effective cost management systems are also in place.

The focus of this text is on managing and controlling costs, not on generating additional revenue. While the two topics are related, they are very different. Marketing efforts, restaurant design, site selection, employee training, and proper food preparation methods are good examples of factors that directly impact an operation's ability to increase sales levels. Effective expense control cannot solve the problems caused by inadequate revenue resulting from ineffective marketing, inferior product quality, or poor service levels. Effective cost control when coupled with management's aggressive attitude toward meeting and exceeding guests' expectations, however, can result in outstanding revenue and profit performance.

EXPENSES

Expenses are used to generate revenue. But managers must carefully control their expenses. There are four major foodservice expense categories that you must learn to control:

1. Food costs
2. Beverage costs
3. Labor costs
4. Other expenses

FOOD COSTS

Food costs are the costs associated with actually producing the menu items sold to guests. They include the expense of meats, dairy products, fruits, vegetables, and other categories of food items. When calculating their food costs, some managers include the cost of minor paper and plastic items, such as the paper wrappers used to wrap sandwiches, as well as the cups and straws used when serving drinks. Other managers record these expenses in their "Other expenses" category. In most cases, food costs will make up the largest or second-largest expense category of expense you must learn to manage.

BEVERAGE COSTS

Beverage costs are those expenses related to the sale of alcoholic beverages. It is common practice in the hospitality industry to consider beverage costs of a nonalcoholic nature as an expense in the food cost category, not the beverage category. Thus, milk, tea, coffee, waters, carbonated beverages, and other nonalcoholic beverage items are *not* generally considered a beverage cost.

Alcoholic beverages accounted for in the beverage cost category include beer, wine, and liquor. This cost category may also include the expense of other drink ingredients such as cherries, lemons, olives, limes, and mixers (such as carbonated beverages), as well as the fruit and vegetable juices needed to produce alcoholic drinks. It may also include the cost of miscellaneous service-related items such as stir sticks, straws, cocktail napkins, and coasters.

LABOR COSTS

Labor costs include the cost of all staff (nonmanagement) employees, as well as management employees needed to run a business. This expense category includes wages, salaries, the cost of any benefits you provide your workers, and the amount of taxes you are required to pay when you have employees on your payroll.

While labor costs is the term most managers use to refer to the expense required to staff a foodservice operation, experienced managers know that "team costs" would be an equally descriptive term. This is so because a manager's cost control success depends upon the efforts of the entire food operations team. Every employee in a food and beverage operation, from the executive chef to dishwashers, plays an important cost-related role, and it is the job of the unit manager to create a team in which all workers understand their roles. When managers are successful leaders, and are able to create such teams, their cost control efforts are enhanced. The importance of effective team management is easy to understand when you recognize that, in many foodservice operations, labor costs are actually the operation's highest cost, or they are second only to food and beverage costs in the total number of dollars spent.

Consider the Costs

"I'm feeling pretty good about our cost management efforts," said Rachel. "Our labor cost is higher than our food cost."

"I'm pleased with our efforts too," said Jules. "Our food cost is higher than our labor cost."

"That's great, Julie," said Joseph. "I just calculated our monthly costs, and our food and labor expenses are just about equal. Sounds like we are all doing well!"

Rachel, Jules, and Joseph had all attended hospitality school together. Each had taken a job in the same large city, so they often got together over coffee to talk about their businesses and their jobs. One manages Chez Paul's, a fine dining French-style restaurant known for impeccable service. Another manages Fuby's, a family-style cafeteria known for its tasty, home-style cooking, and one had taken a job with Gardinos, a national restaurant chain that offered mid-priced Italian cuisine in a beautiful Tuscan-style decor.

1. Which foodservice operation do you think Rachel manages? Why?
2. Which foodservice operation do you think Jules manages? Why?
3. Which foodservice operation do you think Joseph manages? Why?

FUN ON THE WEB!

The foodservice industry is large and continues to grow. The trade association supporting over 500,000 foodservice operations is the National Restaurant Association (NRA).

Enter "National Restaurant Association" in your favorite browser to visit their website.

When you arrive at the site, enter "State of the Industry" in the search bar to see the association's current revenue projections for the restaurant industry, which had annual sales of $1.75 trillion in 2027.

OTHER EXPENSES

Other expenses are composed of all of those expenses not included in food, beverage, or labor costs. Examples include business insurance, utilities, rent, and items such as linens, china, glassware, kitchen knives, and pots and pans.

Although this expense category is sometimes incorrectly referred to by some as "minor expenses," your ability to successfully control this expense area will be critical to the overall profitability of your foodservice operation.

Good managers must learn to understand, control, and manage their expenses. Consider the case of Tabreshia Larson, the food and beverage director of the 200-room Renaud Hotel, located in a college town and built near an interstate highway exit. Tabreshia has just received her end-of-the-year operating reports for the current year. She is interested in comparing these results to those of the prior year. The numbers she received are shown in Figure 1.4.

Tabreshia is concerned about her operation, but she is not sure if she should be. Revenue is higher than last year, so she feels her guests must like the products and services they receive. In fact, repeat business from corporate meetings and special-events meals is increasing. Her profit is greater than last year also, but Tabreshia has the uneasy feeling that things are not going as well as they could. The kitchen appears to run smoothly. The production staff, however, often run out of needed items, and there sometimes seems to be a large amount of leftover food that must be thrown away. Also, at times, there seem to be too many employees on the property and not enough work for them to do; at other times, there seems to be too few employees, and her guests have to wait too long to get served. Tabreshia also feels that employee theft may be occurring, but she certainly doesn't have the time to watch every storage area within her operation. She would really like to get a handle on the problems (if there are any), but how and where should she start?

The answer for Tabreshia, and for you, if you want to develop a serious expense control system, is very simple. You start with the basic mathematics skills that you must have to properly analyze your revenue and expenses. The mathematics required, and used in this text, are not hard. They consist of only addition, subtraction, multiplication, and division. These tools will be sufficient to build a cost control system that will help you professionally manage the expenses you incur.

To see why managers must be able to analyze their businesses, consider what it would mean to you if a fellow foodservice manager told you that yesterday he

FIGURE 1.4 Renaud Hotel Operating Results

	This Year	Last Year
Revenue	$1,106,040	$850,100
Expenses	$1,017,557	$773,591
Profit	$88,483	$76,509

spent $1,000 on food. Obviously, it means very little unless you know more about their operation. Should he have spent $1,000 yesterday? Was that too much? Too little? Was it a "good" day or a "bad" day? These questions raise a challenging problem. How can you properly compare your revenue or expenses today with those of yesterday, or your own foodservice unit with another, so that you can see how well you are doing?

The answer to that question becomes even more complex because we know that the value of dollars changes over time. For example, a restaurant that generated revenue of $1,000 per day in 1954 would be very different from that same restaurant with daily revenue of $1,000 today because the value of the dollar today is quite different from what it was in 1954. Generally, inflation causes the purchasing power of the dollar today to be less than that of a dollar from a previous time period. Inflation can make it challenging to answer the simple question, "Am I doing as well today as I was doing 5 years ago?"

Alternatively, consider the problem of an individual responsible for the management of several foodservice units. They own two food trucks that sell tacos on either side of a large city. One food truck uses $500 worth of food products each day: the other uses $600 worth of food products each day. Are both units being efficiently operated? Does the second truck use an additional $100 worth of food each day because it serves more customers or because it is less efficient in utilizing its food?

The answer to all of the preceding questions, and many more, can be determined if we use percentages to relate the expenses of an operation to the revenue it generates. Percentage calculations are important for at least two major reasons. First and foremost, percentages are the most common tools used to evaluate costs in the foodservice industry. Therefore, knowledge of what a percentage is and how it is calculated is vital. Second, as a manager in the foodservice industry, you will be evaluated primarily on your ability to compute, analyze, and control these percentages.

Although it is true that many basic management tools and apps such as Microsoft Excel, Apple Numbers, Google Sheets, Apache OpenOffice Calc, and others can compute percentages for you, it is important that you understand what the percentages mean and how they should be interpreted. Percentage calculations are used extensively in this text and are a cornerstone of any effective cost control system.

PERCENTAGE REVIEW

Understanding percentages and how they are mathematically computed is essential for all foodservice managers. The following review may be helpful for some readers. If you already thoroughly understand the percentage concept, you may skip this section and the *Computing Percentages* section and proceed directly to the *Using Percentages* section of this chapter.

Percent (%) means "out of each hundred." Thus, 10 percent would mean 10 out of each 100. If we asked how many guests would buy blueberry pie on a given day, and the answer is 10 percent, then 10 people out of each 100 we serve will select blueberry pie. If 52 percent of your employees are female, then 52 out of each 100 employees are female. If 15 percent of your employees will receive a raise this month, then 15 out of each 100 employees will get their raise. There are three basic ways to express a percentage:

1. Common form
2. Fractional form
3. Decimal form

Figure 1.5 shows these three forms, or ways, of writing a percentage.

FIGURE 1.5 Forms of Expressing Percent

	Percentage		
Form	1%	10%	100%
Common	1%	10%	100%
Fraction	1/100	10/100	100/100
Decimal	0.01	0.10	1.00

COMMON FORM

In its common form, the % sign is used to express the percentage. If we say 10 percent, then we mean "10 out of each 100" and no further explanation is necessary. If we say 50 percent, then we mean "50 out of each 100" and no further explanation is necessary. In the common form, the percentage is equivalent to the same amount expressed in either the fraction or the decimal form.

FRACTION FORM

In fraction form, the percentage is expressed as the part, or a portion of 100. Thus, 10 percent is written as 10 over 100 (10/100). Similarly, 50 percent is written as 50 over 100 (50/100). When using the fraction form, the "part" is the numerator and is always placed on top, while the "whole" is the denominator and it is always placed on the bottom.

Despite the fact that it looks different, when writing 10 percent, or any other percent, use of the fraction form is simply another way of expressing the relationship between, in this example, the part (10) and the whole (100).

DECIMAL FORM

A decimal is a number developed directly from the counting system we use. It is based on the fact that we count to 10, then start over again. In other words, each of our major units—10s, 100s, 1,000s, and so on—is based on the use of 10s, and each number can easily be divided by 10.

Unlike the common or fraction form, the decimal form of expressing a percentage uses the decimal point (.) to present the percent relationship. Thus, 10 percent is expressed as 0.10 in decimal form. 50 percent is expressed as 0.50. When utilizing the decimal form, the numbers to the right of the decimal point express the percentage.

Each of these three methods of expressing percentages is used by professionals throughout the foodservice industry. To be successful, you must develop a clear understanding of how a percentage is computed and when it is properly used. Once you know that, you can express the percentage in any form that is required or that is useful to you.

COMPUTING PERCENTAGES

To determine what percentage one number is of another number, you divide the number that is the part by the number that is the whole. Usually, but not always, this means dividing the smaller number by the larger number. For example, assume that 840 guests were served during a banquet at your hotel and that 420 of them asked for coffee with their meal. To find what percentage of your guests ordered coffee, divide the part of the group ordered coffee (420) by the size of the whole group (840).

The process looks like the following:

$$\frac{\text{Part}}{\text{Whole}} = \text{Percent or } \frac{420}{840} = 0.50$$

Recall that a percentage can be expressed in three ways. Thus, 50 percent (common form), 50/100 (fraction form), and 0.50 (decimal form) all represent the proportion of people at the banquet who ordered coffee.

Some new foodservice managers have difficulty computing percentage figures. That's because sometimes it is easy to forget which number goes on the top and which number goes on the bottom. In general, if you attempt to compute a percentage and get a whole number (a number larger than 1.0), either a mistake has been made or revenue is extremely low and/or costs are extremely high!

Some people also become confused when converting from one form of percentage to another. If that is a problem for you, remember the following conversion rules:

1. To convert from common form to decimal form, move the decimal two places to the left and drop the percent sign, that is, 50.00% = 0.50.
2. To convert from decimal form to common form, move the decimal two places to the right and add the percent sign, that is, 0.40 = 40.00%.

In a restaurant, the "whole" is most often a revenue (sales) figure. Expenses and profits are the "parts," which are usually expressed in terms of a percentage.

It is interesting to note that, in the United States, the same system in use for our numbers is in use for our money. Each dime contains 10 pennies; each dollar contains 10 dimes, and so on. Thus, in discussions of money, it is true that a percentage refers to "cents out of each dollar" as well as "out of each 100 dollars." When we say 10 percent of a dollar, we mean 10 cents, or "10 cents out of each dollar." Likewise, 25 percent of a dollar represents 25 cents, 50 percent of a dollar represents 50 cents, and 100 percent of a dollar represents $1.00.

Sometimes, when using percentage to express the relationship between portions of a dollar and the whole dollar, we can find that the part is indeed larger than the whole. Figure 1.6 demonstrates the three possible outcomes that can occur when computing a percentage. Great care must always be taken when computing percentages, so that the percentage arrived at is of help to you in your work and does not represent an error in mathematics. In turn, the mathematical errors could cause you to make poor foodservice decisions.

FIGURE 1.6 Percentage Computations

Possibilities	Examples	Results
Part is smaller than the whole	$\frac{61}{100} = 61\%$	Always less than 100%
Part is equal to the whole	$\frac{35}{35} = 100\%$	Always equals 100%
Part is larger than the whole	$\frac{125}{50} = 250\%$	Always greater than 100%

HERE'S HOW IT'S DONE 1.1

When a manager calculates a food expense percentage, labor expense percentage, or any other expense percentage, the mathematical result is usually a number less than 1. In most cases, the mathematical formula you will use to calculate a percentage will result in the percentage being expressed in decimal form.

For example, if the cost of food required to make a menu item is \$3.60 and the menu item sells for \$12.00, the percentage that represents the cost of food is calculated as follows:

$$\frac{\$3.60 \text{ Cost of food}}{\$12.00 \text{ Selling price}} = 0.30$$

In this example, the menu item's food cost is expressed in decimal form (0.30).

When a percentage is expressed in decimal form, the numbers to the right of the decimal represent the size of the percentage. To convert the percentage from decimal form to common form, simply multiply the decimal form amount by 100 and add the percent sign.

In this example, the conversion from decimal form percentage to common form percentage would be:

$$0.30 \times 100 = 30\%$$

In this example, the menu item's food cost percentage, when expressed in common form, is 30 percent.

If the percentage in this example is to be expressed in fraction form, the part (30) is the numerator and would be placed on top of the whole (100). 100 is the whole and thus is the denominator. The denominator is always placed on the bottom. In this example, the result would be expressed as 30/100.

USING PERCENTAGES

The ability to calculate a percentage is important because percentages are useful tools. To illustrate, consider a restaurant that you are operating. Imagine that your revenue for a week is \$4,800. Expenses for that same week are \$3,600. Given these facts and the information presented earlier in this chapter, your profit formula for the week would be as follows:

$$\text{Revenue} - \text{Expenses} = \text{Profit}$$

or

$$\$4,800 - \$3,600 = \$1,200$$

If you had planned for a \$1,500 profit for the week, you would have been short of your profit goal by \$300. Using the alternative profit formula that identifies desired profit and presented earlier, you would find the following:

$$\text{Revenue} - \text{Desired profit} = \text{Ideal expense}$$

or

$$\$4,800 - \$1,500 = \$3,300$$

Note that your actual expense in this example (\$3,600) exceeded your ideal expense of \$3,300 and, as a result, too little profit was achieved.

These numbers can also be expressed in terms of percentages. If we want to know what percentage of our revenue went to pay for our expenses, we would compute it as follows:

$$\frac{\text{Expense}}{\text{Revenue}} = \text{Expense \%}$$

or

$$\frac{\$3{,}600}{\$4{,}800} = 0.75 \text{ or } 75\,\%$$

In this example, expenses are 75 percent of revenue. Another way to state this relationship is to say that each dollar of revenue costs 75 cents to produce. As a result, each revenue dollar taken in results in 25 cents profit:

$$\$1.00 \text{ revenue} - \$0.75 \text{ expense} = \$0.25 \text{ profit}$$

As long as your business's expenses are smaller than its revenues, some profit will be generated, even if it is not as much as you had planned.

Managers can calculate their expense percentages, and they can also calculate their profit percentages. You can compute profit percentage using the following formula:

$$\frac{\text{Profit}}{\text{Revenue}} = \text{Profit \%}$$

In our example

$$\frac{\$1{,}200 \text{ profit}}{\$4{,}800 \text{ revenue}} = 0.25 \text{ or } 25\,\% \text{ profit}$$

You can calculate the desired (planned) profit percentage by using the following formula.

$$\frac{\text{Desired profit}}{\text{Revenue}} = \text{Profit percentage}$$

In this example,

$$\frac{\$1{,}500 \text{ desired profit}}{\$4{,}800 \text{ revenue}} = 0.3125 \text{ or } 31.25\,\% \text{ desired profit}$$

In simple terms, we had planned to make 31.25 percent profit but instead made only a 25 percent profit. Excess costs account for the difference. If these costs could be identified and then corrected, we could perhaps achieve the desired profit percentage in the future.

Expenses expressed as cost percentages are important to manage and most foodservice operators compute many cost percentages, not just one. As you have learned, the major cost divisions used in foodservice are:

1. Food cost
2. Beverage cost
3. Labor cost
4. Other expenses

Because these are the major cost categories, in foodservice operations, a modified profit formula can be developed as follows:

Revenue − (Food cost + Beverage cost + Labor cost + Other expenses) = Profit

Put in another format, the equation for the profit formula would look like the following:

	Revenue (100 %)
Minus (−)	Food Cost %
Minus (−)	Beverage Cost %
Minus (−)	Labor Cost %
Minus (−)	Other Expenses %
Equals (=)	Profit %

This expression of the profit formula makes sense because it clearly shows that managers start with 100 percent of their revenue, subtract their expense percentages, and the amount that remains represents the operation's profit percentage. Regardless of the form in which percentages are reported, professional foodservice managers carefully evaluate their revenue and expenses, and they use percentages to do so.

UNDERSTANDING THE INCOME (PROFIT AND LOSS) STATEMENT

Foodservice operations generate revenue and incur expenses whenever they are open for business. Periodically, these operations will want to report their income and expense activity. To see how this is typically done, consider Figure 1.7, a summary of the revenue, expenses, and profits generated by Dan's Steakhouse.

All of Dan's expenses and profits can be computed as percentages by using the revenue figure of $800,000 as the whole and the combined food and beverage cost, labor cost, other expenses, and profit representing the parts as follows:

$$\frac{\text{Food and beverage cost}}{\text{Revenue}} = \text{Food and beverage cost percentage.}$$

or

$$\frac{\$300{,}000}{\$800{,}000} = 37.5\%.$$

FIGURE 1.7 Dan's Steakhouse

Revenue		**$800,000**
Expenses		
Food and beverage cost	$300,000	
Labor cost	$350,000	
Other expenses	$50,000	
Total expenses		$700,000
Profit		**$100,000**

All of Dan's expenses and profits can be computed as percentages by using the revenue figure of $400,000 as the whole and the combined food and beverage cost, labor cost, other expenses, and profit representing the parts as follows:

$$\frac{\text{Food and beverage cost}}{\text{Revenue}} = \text{Food and beverage cost \%}$$

or

$$\frac{\$300{,}000}{\$800{,}000} = 37.50\ \%$$

$$\frac{\text{Labor cost}}{\text{Revenue}} = \text{Labor cost \%}$$

or

$$\frac{\$350{,}000}{\$800{,}000} = 43.75\ \%$$

$$\frac{\text{Other expenses}}{\text{Revenue}} = \text{Other expenses \%}$$

or

$$\frac{\$50{,}000}{\$800{,}000} = 6.25\ \%$$

$$\frac{\text{Total expense}}{\text{Revenue}} = \text{Total expense \%}$$

or

$$\frac{\$700{,}000}{\$800{,}000} = 87.50\ \%$$

$$\frac{\text{Profit}}{\text{Revenue}} = \text{Profit \%}$$

or

$$\frac{\$100{,}000}{\$800{,}000} = 12.50\ \%$$

The accounting tool managers use to report their operations' revenue, expenses, and profit for a specific time period is called the **statement of income and expense**. The statement of income and expense is also commonly known as the **Income Statement** or the **profit and loss statement**, which is very often shortened by foodservice managers to the **P&L**.

The P&L lists all of an operations' revenue, food and beverage cost, labor cost, and other expenses. The P&L also identifies profits by using the profit formula. Remember that the profit formula is:

$$\text{Revenue} - \text{Expenses} = \text{Profit}$$

Figure 1.8 is a very simplified P&L statement for Dan's Steakhouse (note: detailed information on properly preparing and analyzing an income statement will be addressed in Chapter 9: Analyzing Results Using the Income Statement). Notice the similarity of Figure 1.8 to Figure 1.7. In Figure 1.8, expenses and profits are expressed in terms of both dollar amounts and percentages of revenue.

Yet another way of looking at Dan's simplified P&L is shown in Figure 1.9. The pieces of the pie represent Dan's cost and profit categories. Costs and profit total 100 percent, which is equal to Dan's total revenues. Put another way, out of every sales dollar Dan generates, 100 percent is designated as either costs or profit.

Dan knows from the P&L that revenues represent 100 percent of the total dollars available to cover his expenses and provide for a profit. In this example, combined food and beverage cost is 37.50 percent, and labor cost percentage equals 43.75 percent. Other expenses percentage equals 6.25 percent, and the total expense percent is 87.50 percent (37.50 + 43.75 + 6.25 = 87.50 percent). The steakhouse profit equals 12.50 percent. Thus, for each dollar in revenue, Dan earns a profit of 12.50 cents.

FIGURE 1.8 Dan's Steakhouse P&L

Revenue		$800,000	100%
Food and beverage cost	$300,000		37.5%
Labor cost	$350,000		43.75%
Other expenses	$50,000		6.25%
Total Expenses		$700,000	87.5%
Profit		$100,000	12.5%

FIGURE 1.9 Dan's Steakhouse Costs and Profit as a Percentage of Revenues

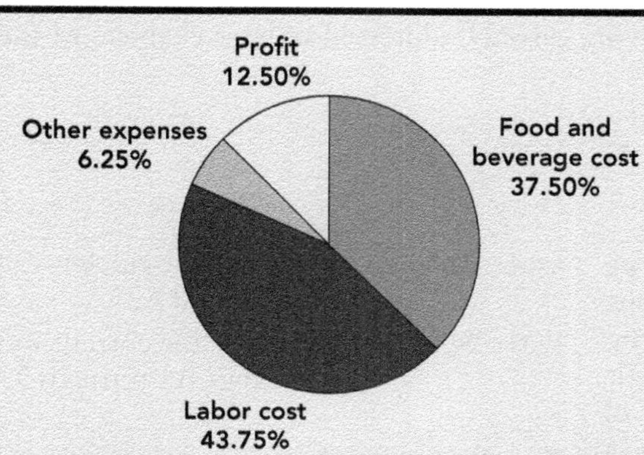

In restaurants that serve alcohol, food revenues and beverage revenues are most often reported separately. Likewise, in these types of operations, food costs and beverage costs are most often separated into two categories in the P&L. This is done so that food costs can easily be compared to food revenues and beverage costs can be easily compared to beverage revenues. This is helpful when, for example, one manager is responsible for controlling food costs in the restaurant and another manager is responsible for controlling beverage costs in the bar.

The P&L is an important management tool because it indicates the efficiency and profitability of a business. It is essential that this financial report is accurate and easily understood. Because so many individuals and groups are interested in a facility's performance, it is also important that the P&L and other financial statements are prepared in a manner that is consistent with other facilities. If, for example, you own two Italian restaurants, it would be very confusing if the units' two managers used different methods for preparing and reporting their P&Ls. You, your investors, accountants, governmental taxing entities, and your creditors will all be interested in your operational results, and unless you report and account for these in a manner that is consistent and that can be easily understood, confusion is likely to result.

To avoid such confusion, a **Uniform System of Accounts for Restaurants (USAR)** is used to report financial results in many foodservice units. This system was created to ensure uniform reporting of financial results. A uniform system of accounts exists for restaurants, another for hotels, and another for clubs. These uniform accounting systems are continually reviewed and periodically revised. This text was prepared using reporting principles contained in the most recent (eighth) edition of the USAR. Important specific recommendations of the USAR will be addressed in detail in the appropriate portions of this text.

The use of the USAR when producing a P&L is not mandatory, but it is highly recommended. This is because the primary purpose of preparing a P&L is to clearly identify revenue, expenses, and profits for a specific time period, and good managers want their operations' financial records to accurately reflect their efforts.

As a manager, your individual efforts will greatly influence your operation's profitability. Good managers want to provide excellent value to their guests, which will cause guests to return. When they do, sales will increase. In addition, good managers know how to analyze, manage, and control their costs. When costs are controlled well, an operation's expenses are held to the amounts that were preplanned by its manager. The final result is the desired profit level.

The best managers influence the success of their units and their own employees. The results for them personally are promotions, added responsibilities, and salary increases. If you wish to succeed in the hospitality industry, it is important to remember that your performance will be evaluated primarily on your ability to achieve the profit levels your operation has planned for.

In addition to your own efforts, many factors influence profit dollars and profit percentage, and you must be aware, and in control, of all of them. All of the factors that make up professional food and beverage cost control and, as a result, will impact your profits are directly addressed in later chapters of this text.

HERE'S HOW IT'S DONE 1.2

Managers can use spreadsheet programs to speed up their work when calculating percentages or performing other calculations needed to assess their cost control efforts. Microsoft Excel, Apple Numbers, Google Sheets, and Apache OpenOffice Calc are examples of software programs that allow managers to enter text, a numeric value, or a formula to help meet their unique information needs.

Essentially, a spreadsheet program presents users with easily identifiable cells designated by a column letter and a row number. In the following example, cell A1 refers to the 1st cell in column A. Cell D5 refers to the 5th cell in column D.

	A	B	C	E	D
1					
2					
3					
4					
5					

Using a spreadsheet is easy. To do so, first click on the cell in which you wish to enter a number or formula. Then, for example, when using the Excel spreadsheet program, a manager would enter the = sign to begin. The following are examples of how managers would use Microsoft Excel spreadsheets to perform basic adding, subtracting, multiplying, and dividing tasks.

For Adding

Enter: 15+20+35	Adds the entered numbers 15, 20, and 35 in the selected cell
Enter: A1+A2+A3	Adds the current values in cells A1, A2, and A3
Enter: Sum(A1:A5)	Adds the current values in cell range A1 through A5

For Subtracting

Enter: 35–15	Subtracts 15 from 35 in the selected cell
Enter: A1–A2	Subtracts the current value in cell A2, from the current value in cell A1

For Multiplying

Enter: = 15*10	Multiplies 15 times 10 in the selected cell
Enter: = A1*A2	Multiplies the current value in cell A1 times the current value in cell A2
Enter: = A1*10	Multiplies the current value in cell A1 times 10

For Dividing

Enter: 15/10	Divides 15 by 10 in the selected cell
Enter: A1/A2	Divides the current value in cell A1 by the current value in cell A2
Enter: A1/10	Divides the current value in cell A1 by 10

From simple formulas to more complex, multifunction formulas, the use of computer-based software and apps currently available in the industry can help make a manager's cost control efforts faster and more accurate.

FUN ON THE WEB!

For restaurant managers, learning is an ongoing process. Fortunately, there are a large number of sources of important information readily available to managers who seek to better understand trends in their industry. To examine some of these resources, enter "restaurant industry financial results" in your favorite search engine and review current data about how the restaurant industry is performing.

UNDERSTANDING THE BUDGET

Some foodservice managers do not generate revenue on a daily basis. Consider, for example, the individual who manages the foodservice operation at Camp Eureka, a children's summer camp. In this case, parents pay a fixed fee to cover their children's housing, activities, and meals for a set period of time. The foodservice manager, in this situation, will be just one of several camp managers who must share this revenue. If too many dollars are spent on providing housing or play activities, too few dollars may be available to provide an adequate quantity or quality of meals. On the other hand, if too many dollars are spent on providing foodservice, there may not be enough left to cover expenses needed in other areas of the camp. In a case like this, as in many other cases, foodservice operators must prepare a budget.

A **budget** is simply an estimate of projected revenue, expense, and profit. In some hospitality companies, the budget is known as the forecast, or the **plan**, referring to the fact that the budget details the operation's estimated, or planned for, revenue and expense for a given accounting period. An **accounting period** is any specific hour, day, week, month, or other specified period of time in which an operator wishes to report and analyze an operation's revenue and expenses.

All effective managers, whether in the commercial (for-profit) or nonprofit sector, should use budgets. Budgeting is simply planning for an operation's revenue, expenses, and profit. If these items are planned for, you can determine how close your actual performance is to your plan or budget.

To illustrate the importance of a budget using the Eureka summer camp example, assume that the following information is known:

1. Number of campers to be served each day is 180.
2. Number of meals served to each camper per day is 3.
3. Length of each camper's stay is 7 days.

With 180 campers eating 3 meals each day for 7 days, 3,780 meals will be served (180 campers × 3 meals per day × 7 days = 3,780 meals).

Generally, in a case such as the summer camp, the foodservice manager is given a dollar amount that represents the allowed expense for each meal to be served. For example, if $2.85 per meal is the amount budgeted for this foodservice manager, the total revenue budget would equal $10,733 ($2.85 per meal × 3,780 meals = $6,993).

From this figure, we can begin to develop an expense budget. In this case, we would be interested in the amount of expenses budgeted and the amount actually spent on those expenses. Equally important, we are interested in the percentage of the budget we actually used, a concept known as comparing **performance to budget**.

A simple example may help to explain the idea of budget and performance to budget. Assume that a child has $2.00 per day to spend on candy. On Monday morning, the child's parents give the child $2.00 for each day of the week, or $14.00 total ($2.00 × 7 days = $14.00).

If the child spends exactly $2.00 per day, they will be able to buy candy all week. If, however, too much is spent on any one day, there may not be any money left at the end of the week. To ensure a full week of candy eating, a good candy purchasing budget could be created, such as the one shown in Figure 1.10.

To prepare this budget, the "% of Total" column is computed by dividing $2.00 (the part) by $14.00 (the whole). Notice that we can determine the percentage of total that should have been spent by any given day, that is, each day equals 14.28 percent, or 1/7 of the total.

This very same logic applies to a foodservice operation. Managers determine how much of their total budgets can be spent in any specific time period. Figure 1.11 represents commonly used budget periods and their accompanying proportional amounts.

In the foodservice industry, the use of monthly budgets is very popular. Some foodservice operations, however, recognize that different months include different numbers of days. As a result, many operators are changing from one-month budget periods to budget periods of 28 days. The **28-day-period** method divides a year into 13 equal periods of 28 days each. Therefore, each period has four Mondays, four Tuesdays, four Wednesdays, and so on. This helps the manager compare performance from one period to the next without having to compensate for extra days in any one month or unequal numbers of specific weekdays (e.g., unequal numbers of Mondays, Saturdays, or Sundays in the same period).

The disadvantage of the 28-day period approach is that managers can no longer talk about the month of March because if a budget began on the first of the year,

FIGURE 1.10 Candy Purchases

Weekday	Budgeted Amount	% of Total
Monday	$2.00	14.28%
Tuesday	$2.00	14.28%
Wednesday	$2.00	14.28%
Thursday	$2.00	14.28%
Friday	$2.00	14.28%
Saturday	$2.00	14.28%
Sunday	$2.00	14.28%
Total	$14.00	100.00%

FIGURE 1.11 Common Foodservice Budget Periods

Budget Period	Portion	% of Total
One week	One day	1/7 or 14.28%
Two-week period	One day	1/14 or 7.14%
	One week	1/2 or 50.0%
One month	One week	1/4 or 25.0%
28 days	One day	1/28 or 3.6%
30 days	One day	1/30 or 3.3%
31 days	One day	1/31 or 3.2%
Six months	One month	1/6 or 16.7%
One year (365 days)	One day	1/365 or 0.3%
	One week	1/52 or 1.9%
	One month	1/12 or 8.3%

Period 3 would occur during part of February and part of March. Although using the 28-day-period approach takes a while to get used to, it is often an effective way to measure performance and plan from period to period. Managers are interested in regularly comparing their actual performance to their budgeted performance and detailed information about this process is addressed in this text in Chapter 10: Planning for Profit.

Many operators use the concept of **significant variation** to determine whether a cost control problem exists. In this case, a significant variation is any variation in expected costs that management feels is an area of concern. This variation can be caused by costs that were either higher or lower than the amount originally budgeted or planned for.

In the foodservice industry, the essence of good cost control is identifying actual costs, comparing them to planned costs, and then taking corrective action when necessary. When you manage a foodservice operation and you find that significant variations from your planned results occur, you must do the following:

1. Identify the problem.
2. Determine the cause.
3. Take corrective action.

It is crucial to know the nature of the problem you have if you are to be an effective problem solver. Management's attention must be focused on the proper area of concern. Why? Remember that expenses create revenue, thus it is not your goal to eliminate expense. In fact, managers who focus too much on eliminating expense, instead of building revenue, often find that their expenses are completely eliminated when they are forced to close their operation's doors permanently because guests did not feel they received good value for the money spent at that restaurant! Control and management of revenue and expenses are important. Elimination of either is not desired.

As you have learned, revenue and expenses directly impact profit. Your important role as a hospitality manager is to analyze, manage, and control your costs so that you achieve planned results. It can be done, and it can be fun.

The remainder of this text discusses how you can best manage and account for foodservice revenue and expenses. With a good understanding of the relationship among revenue, expenses, and profit, and your ability to analyze these areas using percentages, you are ready to begin the cost control and cost management process.

Technology Tools

Most hospitality managers would agree that an accurate and timely income statement (P&L) is an invaluable aid to their management efforts. There are a variety of software programs on the market that can be used to develop this statement for you. You simply fill in the revenue and expense portions of the program, and a P&L is produced. Variations include programs that compare your actual results to budgeted figures or forecasts, to prior-month performance, or to prior-year performance. In addition, P&Ls can be produced for any time period, including days, weeks, months, quarters, or years. Most income statement programs will have a budgeting feature and the ability to maintain historical sales and cost records. Some of these have been developed specifically for restaurants, but cost-effective generic products are also available.

A second issue, and one that must be kept foremost in mind, is that of information accessibility. For example, an executive chef would certainly need to have information readily available on food cost; however, it may not be wise to allow servers or cooks access to payroll information. While labor expense certainly affects costs, individual worker pay should be shared only with those who need to know. Thus, as you examine (in this chapter and others) the cost control technology tools available to you, keep in mind that not all information should be accessible to all parties, and that security of your cost and customer information can be just as critical as its accuracy.

FUN ON THE WEB!

Intuit is the company that produces the popular QuickBooks line of accounting software. QuickBooks can help you create a monthly income statement (P&L) and do much more. To view QuickBooks product offerings, go to the Intuit company website and review the features available in their newest versions of QuickBooks for small businesses.

Cost Control Around the World

Food and beverage cost control is truly a topic of international importance. Consider that there are now over 42,000 McDonald's restaurants operating in 110 countries. The operations serve over 70 million customers per day. While there are over 13,000 McDonald's located in the United States, over twice that number of McDonald stores are located outside the United States.

Although McDonald's is among the best known, many other hospitality companies operate in the international market and the number of those doing so increases each year. Burger King, KFC, Wendy's, Hilton, Hyatt, Marriott, Aramark, Subway, Dunkin Donuts, Baskin Robbins, Pizza Hut, Taco Bell, and Rainforest Cafe are just a few examples of the increasingly large number of US-based companies expanding their international presence. As a result, more and more foodservice professionals working in restaurants, hotels, and contract foodservice businesses are being given cost control-related assignments in their companies' rapidly expanding international divisions.

Apply What You Have Learned

Jennifer Caratini has recently accepted the job as the foodservice director for Techmar Industries, a corporation with 1,000 employees. As their foodservice director, Jennifer's role is to operate a company cafeteria, serving 800 to 900 meals per day, as well as an executive dining room, serving 100 to 200 meals per day. All of the meals are provided free of charge to the employees of Techmar. One of Jennifer's first jobs is to prepare a budget for next year's operations.

1. In addition to the cost of food products and foodservice employees, what other expenses will Techmar incur by providing free meals to its employees?
2. Since employees do not pay for their food directly, what will Jennifer likely use as the "revenue" portion of her budget? How do you think this number should be determined?
3. In addition to her know-how as a foodservice director, what skills will Jennifer likely need as she interacts with the executives at Techmar who must approve her operating budget?

For Your Consideration

1. Effective cost control requires foodservice managers to know the specific revenues achieved, and expenses incurred, in their own operations. What will be the likely result if they do not know this key information?
2. In most cases, a foodservice manager's performance will be evaluated in terms of the manager's ability to achieve budgeted, or planned for, financial results. Why do you think that is so?

Key Terms and Concepts

The following are terms and concepts addressed in the chapter that are important for you as a manager. To help you review, define the terms below:

Off-premise dining
Revenue
Expenses
Profit
Business dining
Ideal expense
Desired profit
Food costs
Beverage costs
Labor costs
Other expenses
Percentages
Statement of income and expense
Income statement
Profit and loss statement
P&L
Uniform System of Accounts for Restaurants (USAR)
Budget
Plan
Accounting period
Performance to budget
28-day-period
Significant variation

Test Your Skills

You may download the Excel spreadsheets for the Test Your Skills exercises from the student companion website at http://www.wiley.com/go/dopson/foodandbeveragecostcontrol8e.

Complete the exercises by placing your answers in the shaded boxes and answering the questions as indicated.

1. Last month Lani's Pizza Parlor generated $77,000 in revenues. Her total expenses last month were $64,000. How much profit did she make?

Lani's Pizza Parlor	
Revenue	
Total expenses	
Profit	

2. At the conclusion of her first month of operating Val's Donut Shop, Val computed the following revenue and expense figures:

Week	Revenue	Expense	Profit/Loss
1	$935.50	$771.80	
2	$1,177.60	$571.46	
3	$1,461.80	$933.33	
4	$1,545.11	$1,510.20	
Month			
To Receive $1,400 Profit for the Month			
Month			

Prepare both weekly and monthly profit formulas so that Val has a good idea about her current profit situation. Also, given her sales for the month, tell her how much her ideal expense should have been to realize her desired profit of $1,400.

3. Last year Isabella's sub shop generated $455,500 in revenues. Her food cost for the year was $143,500, her labor cost was $122,400, and her other costs were $81,750.

   Complete the spreadsheet Isabela needs to calculate the amount of profit she made last year. How much profit did she make? What was her profit percentage? Calculate your percentage answer to two decimal places.

Isabella's Sub Shop: Last Year Results		
	$	%
Revenue	455,500	100
Food cost		
Labor cost		
Other costs		
Total costs		
Profit		

4. Xanthi operates Zaytoon's Mediterranean restaurant featuring lamb, chicken, Greek salads, falafel, moussaka, spanakopita, paella, tabbouleh, tzatziki, and hummus. Xanthis's operation offers on-premise dining, carryout, and delivery service. She carefully tracks her sales in each of these three service categories.

   Help Xanthi evaluate her sales this year compared to last year by completing her sales comparison graph, then answer the questions that follow. Calculate your percentage answers to two decimal places.

Zaytoon's Sales Comparison						
Sales	Sales Last Year	% of Sales	Sales This Year	% of Sales	$ Sales Change	% Change
Dine-in	$825,000		$915,500			10.97
Carry-out	$115,000		$152,750			
Delivery	$72,500		$115,000			
Total Sales	$1,012,500		$1,183,250			

   Which of Xanthis' sales categories increased by the greatest amount of money this year compared to last year? Which sales category increased by the greatest percentage amount this year compared to last year?

5. Su Chan manages a popular restaurant called the Bungalow. Her P&L for the month of March is as follows:

The Bungalow's March P&L		
Revenue	$100,000.00	100.0%
F&B expense	$34,000.00	34.0%
Labor expense	$40,000.00	40.0%
Other expenses	$21,000.00	21.0%
Total expenses	$95,000.00	95.0%
Profit	$5,000.00	5.0%

Su has a meeting with the owner of the Bungalow next week, so she decided to create a pie chart showing the percentage of her costs in relation to her total sales (see the following diagram).

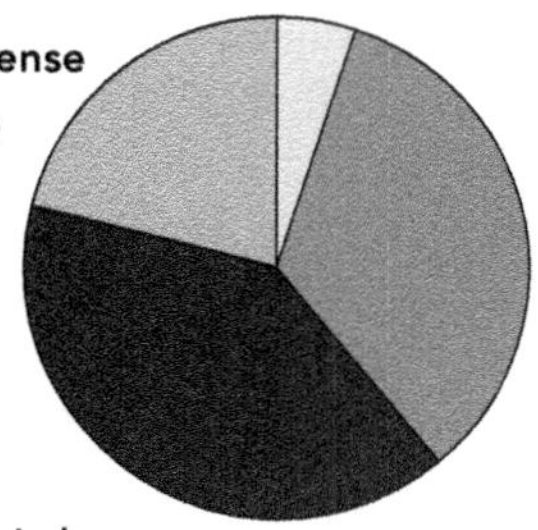

At the meeting with the owner, Su is asked to change the information on the pie chart to reflect the next month's projections. The owner suggests that April revenues and costs should be as follows:

April revenues = $120,000
Food and beverage expense = $44,000
Labor and other expenses remain constant.

Using these numbers, is the owner's profit percentage going to be higher or lower than that in March? By how much?

After looking at the owner's projections, she thinks it might be too difficult (and not so good for her guests) if she cannot increase labor costs along with sales. She proposes a compromise and tells the owner that if he will agree to increase labor costs, she will try to decrease other expenses. So, Su proposes the following:

April revenues = $120,000
Food and beverage expense = $44,000
Labor expense = $50,000
Other expense = $19,000

Using these numbers, is the owner's profit percentage going to be higher or lower than that in March? By how much?

Which set of projections has more reasonable goals?

Note: If you are using the Excel spreadsheets from this book's student resources site, the changes you make to the numbers should be reflected in the pie charts as well.

6. The dining room at the Roadrock Inn is extremely popular. Terry Ray, the food and beverage director, is pleased to see that his revenue is higher than last year's. Of course, expenses are higher also. Express Terry's expenses and profit as a percentage of total revenue, for both this year and last year (fill in all empty blanks).

	This Year	%	Last Year	%
Revenue	$963,971		$875,421	
F&B expense	$367,069		$347,168	
Labor expense	$338,090		$316,151	
Other expenses	$147,896		$142,068	
Total expenses				
Profit				

How is Terry doing in managing his expenses this year compared to last year? How do changes in revenue affect his performance?

7. Pamela Cantu operates a school foodservice department in a small, rural community. She feeds approximately 1,000 students per day in three different locations. She receives an average of $1.20 in revenues per meal. Her budget, set at the beginning of the school year by the superintendent, is developed in such a way that a small amount is to be reserved for future equipment purchases and dining-room renovation. These funds are available, however, only if Pamela meets her budget. She hopes to use this year's reserve (profit) to buy a $5,000 refrigerated salad bar for the high school. Because it is the midpoint of her school year, help her determine her performance to budget thus far (fill in all blanks).

Item	Budget	Actual	% of Budget
Meals served	300,000	$149,800	
Revenue			
Food expense	$170,000	$84,961	
Labor expense	$125,000	$63,752	
Other expense	$60,000	$31,460	
Total expenses	$355,000		
Reserve	$5,000		

Assuming that the year is 50 percent completed, and Pamela continues doing what she is doing, is she likely to meet the reserve requirement and thus be able to purchase the salad bar by the end of the year? If not, what changes should she make over the next 6 months to ensure that she will have the $5,000 in reserve?

8. Sam Guild operates a dining room reserved for doctors in a large hospital in the Northeast. Sam's boss has given Sam a target of a 33 percent food cost but has indicated that the target may be adjusted. Currently, the doctors' meals sell for $16.00. Sam knows he currently can spend $5.33 for the food required to produce each meal. Fill out the chart below to help Sam find out how much he will be able to spend on each meal at various food cost percentage levels if his boss adjusts his target.

Meal Selling Price	Food Cost %	Amount That Can Be Spent for Food
$16.00	20%	
$16.00	25%	
$16.00	30%	
$16.00	33%	$5.33
$16.00	35%	
$16.00	40%	

How will the doctors' meals likely be affected if the target cost percentage is reduced? What if the target cost percentage is increased?

9. Dawne Juan is the food and beverage director for a mid-size hotel in a beach destination area. The general manager of the hotel has given Dawne a target of 10 percent profit for this year. Dawne's staff is predominantly composed of her beach buddies. Although she is good at controlling most of

her costs, she has a hard time telling her friends to go home when business slows down and she needs to reduce her staff. If she doesn't make her profit goal, her general manager will likely reprimand her, and she could possibly lose her job. Express Dawne's expenses and profit as a percentage of total revenue, both this year and last year, to determine if she met her profit goal.

	This Year	%	Last Year	%
Revenue	$1,448,956		$1,094,276	
F&B expense	$463,666			35%
Labor expense	$652,030			40%
Other expenses		15%	$186,027	
Total expenses				
Profit				8%

Was Dawne effective at controlling all of her expenses? Did she meet the profit goal set by the general manager? If not, what could Dawne do in the future to help her make her target profit?

10. Lee Ray operates the Champs Steak and Seafood restaurant. Last month, Lee budgeted $3,500 for food, in the specific categories listed below. It was a busy month, but Lee thought he did a good job managing his costs. Imagine his surprise when, at the end of the month, Lee calculated his actual expenses and entered them in the chart below. He found he was way over budget! Calculate Lee's % of budget in each category listed on the chart, as well as the total.

	Budget	Actual	% of Budget
Meat and poultry	$1,500	$1,675	
Seafood	$1,200	$1,550	
Fruit and vegetables	$350	$370	
Dairy products	$200	$210	
Groceries	$250	$270	
Total			

By how much money was Lee over his total budget? In which categories did Lee's costs vary more than 10 percent from the amount he had originally budgeted? Lee said it was a busy month. Would the number of customers he served affect his actual costs? What would you recommend he do next to further analyze the reasons for his restaurant's budget performance last month?

11. Daudi owns and manages a restaurant featuring Middle Eastern cuisine. His operating results for this year are listed below. For next year, Daudi expects that his revenue will increase 5 percent. He also expects that the percentage of revenue he spends on food will remain unchanged, but that employee raises and rising health-care costs will mean he will spend 10 percent more for the cost of labor next year than he spent this year. Because of new cost control measures he plans to implement, Daudi expects the total amount that he will spend for other expenses next year will be unchanged from this year.

 Help Daudi prepare a budget for next year that will show the amount of revenue, expense, and profit his operation will likely experience. Show each amount in dollars and as a percentage of revenue. Should Daudi's profits next year be greater or lesser than this year? By how much?

	This Year Actual	Percentage	Next Year Budget	Next Year Percent
Revenue	$1,650,000	100%		
Cost of food		35%		
Cost of labor		30%		
Other expenses		20%		
Total expenses		85%		
Profit	$247,500	15%		

12. Richard Shaul owns and operates four Rapid Richard's food trucks in a large city. He identifies the trucks as Northside, Southside, Eastside, and Westside because those are the general locations in which each truck is assigned to operate.

Richard is developing simplified profit and loss statements and a combined profit and loss statement that summarizes last year's financial performance for his four trucks. Help Richard complete the financial summaries and then answer the questions that follow.

	Northside	%	Southside	%	Eastside	%	Westside	%	All Units	%
Revenue	$320,000	100.0%		100.0%	$295,000	100.0%	$375,000	100.0%	$1,415,000	100.0%
Food expense	$93,750		$128,750		$85,500		$106,250		$414,250	29.3%
Labor expense	$96,250		$128,500		$102,750		$129,800			
Other expenses	$67,500	21.1%	$87,500		$72,750	24.7%	$71,750	19.1%	$299,500	21.2%
Total expenses						88.5%			$1,171,050	
Profit	$62,500	19.5%	$80,250			11.5%				

Which of Richard's trucks achieved the highest amount of sales last year? Which truck achieved the lowest food expense percentage? Which truck achieved the lowest labor expense percentage? Which truck achieved the highest dollar amount of profit? Which truck(s) achieved a higher than average profit percentage?

CHAPTER 2

Creating Sales Forecasts

OVERVIEW

This chapter presents the methods and procedures you must learn to create accurate records of what you have sold in the past as well as forecasts of how much business you will do in the future. This includes predicting the total amount of revenue you will generate, the number of guests you will serve, and the amount of money each guest will spend. Knowledge of these management techniques will provide the information you must have to analyze your historical sales and use your findings to better serve your future guests.

Chapter Outline

LEARNING OUTCOMES

At the conclusion of this chapter, you will be able to:

- Develop a procedure to record current sales.
- Compute percentage increases or decreases in sales over time.
- Develop a procedure to predict future sales.

IMPORTANCE OF FORECASTING SALES

When they open their facility's doors at the beginning of the day, the questions most foodservice managers must ask themselves are very simple: "How many guests will I serve today?—And what will they buy?"

The answers to these questions are critical because the guests you will serve provide the revenue to pay your operating expenses and create a profit. Simply put, if too few guests are served, total revenue may not be enough to equal your costs, even if these costs are well managed. In addition, decisions regarding the type and quantity of food and beverage products to purchase for resale to guests depend on knowing the number of guests who will be coming to buy those products.

The labor required to serve guests is also determined based on a manager's "best guess" of the projected number of customers to be served, as well as what these guests will buy. In an ongoing operation, it is often true that future sales estimates, or projected sales, will be heavily based on sales history. This is true because what has happened in the past in your operation is often a good predictor of what will happen in the future. Those managers who can best predict the future are those who are most prepared to control how it will affect them.

In the hospitality industry, sales are important, and we have many ways of counting or defining sales. In the simplest case, sales can be defined as the dollar (or other currency) amount of revenue collected during a predetermined time period. The time period may be defined as an hour, a shift, a day, a week, a month, or a year. When used this way, sales and revenue are considered interchangeable terms.

When you predict the future number of guests you will serve and the revenue they will generate in a given time period, you have created a **sales forecast**. You can calculate your actual sales for a current time period in several ways. Most foodservice managers utilize a computerized system called a **point of sales (POS) system**, which, among other tasks, is programmed to record sales and payment information.

Alternatively, manually produced guest checks or head counts are other methods foodservice managers could use to help identify how many sales were completed. Today, however, even the smallest of foodservice operations should take advantage of the speed and accuracy provided by modern POS systems when recording their operations' sales.

FUN ON THE WEB!

A large number of companies sell hospitality industry-specific POS systems used to record sales information and do much more. Some of the most popular companies include the following:

Square
Toast
Clover
Aloha
Touch Bistro
Lightspeed

To review their product offerings, enter the company name followed by "POS" into your favorite search engine.

It is important to remember that a distinction is made in the hospitality industry between actual sales (revenue) and **sales volume**. Sales volume is the number of units sold by a foodservice operation. Consider Manny, a bagel shop manager, whose

Monday bagel business consisted of $3,000 in sales (revenue) because he sold 1,000 bagels (sales volume) for $3.00 each. It is important for Manny to know how much revenue was taken in, so he can evaluate the expenses required to generate his revenue and the number of units that have been sold. With this information, he can be better prepared to serve additional guests the next day.

In many areas of the hospitality industry, such as in hospitals, retirement centers and college and university residence halls, it is customary that no cash actually changes hands during a particular meal period. However, the manager of such a facility still created sales during that meal period and would be interested in sales volume or how much food was actually consumed by the residents or students on any particular day. This is critical information because all foodservice managers must be prepared to answer the questions, "How many individuals did I serve today, and how many should I expect tomorrow?" In some cases, a food and beverage operation may be a blend of cash and noncash sales.

Consider Tonya Brown, a hospital foodservice director. It is very likely that Tonya will be involved in serving both cash-paying guests (in an open-to-the-public cafeteria) and noncash patients (with tray-line-assembled meals that are then delivered to patients' rooms). In addition, meals for hospital employees may be made as cash sales, but at a reduced or subsidized rate. Clearly, Tonya's operation will create sales each day, and it will be important for her and her staff to know, as accurately as possible, how many of each type of guest she will serve and the menu items these guests will be served.

An understanding of anticipated sales, in terms of revenue dollars, guest counts, or both, will help you have the right number of staff members working with the right amounts of products available to them, at the right time. In this way, you can begin to effectively manage your costs. In addition to the importance of accurate sales records for purchasing and staffing, sales records are valuable to an operator developing labor standards to improve efficiency.

Consider, for example, managing a large restaurant with 400 seats. If an individual server can serve 25 guests at lunch, you would need 400/25, or 16, servers per lunch shift if all your seats were filled. If management keeps no accurate sales histories and makes no sales forecasts, too few or too many servers might be scheduled to work.

With accurate sales records, a sales history can be developed for each foodservice outlet you operate, and better decisions can be made with regard to planning for each unit's operation. Figure 2.1 lists some of the advantages that you gain when you can accurately predict the number of people you will serve in any future time period.

FIGURE 2.1 Advantages of Accurate Sales Forecasts

- Accurate revenue estimates
- Improved ability to predict expenses
- Greater efficiency in scheduling needed workers
- Greater efficiency in scheduling menu item production schedules
- Better accuracy in purchasing the correct amount of food for immediate use
- Improved ability to maintain proper levels of perishable and nonperishable food inventories
- Improved budgeting ability
- Lower selling prices for guests because of increased operational efficiencies
- Increased dollars available for current facility maintenance and future growth
- Increased profit levels and stockholder (or owner) value

SALES HISTORIES

A **sales history** is the systematic recording of all sales achieved during a predetermined time period. It is an accurate record of what and how much your operation has sold. Before you can develop a sales history, however, you must think about the definition of sales that is most helpful to you and your understanding of how your own operation functions.

The simplest type of sales history records revenue only. The sales history format used by Rae's Restaurant and shown in Figure 2.2 is a typical one for an operation recording its sales revenue on a daily and weekly basis.

Notice that, if you managed Rae's Restaurant, you would determine daily sales either from your POS system or from manually adding the information recorded on paper guest checks. You then transfer that number on a daily basis to your sales history by entering the amount of your daily sales in the column titled Daily Sales.

Sales to date is the cumulative total of sales reported in the unit. **Sales to date** is the number you will get when you add today's daily sales to the sales of all prior days in the **reporting period**—the time period for which sales records are being maintained.

To illustrate, sales to date on Tuesday, January 2, is computed by adding Tuesday's sales to those of Monday (the prior day) to arrive at a sales to date total of $1,826.27 ($851.90 + $974.37 = $1,826.27). As a result, the Sales to Date column is a running total of the sales achieved by Rae's Restaurant for the week.

Should Rae's manager prefer it, the reporting period could be defined in time blocks other than 1 week. Common alternatives are meal periods (breakfast, lunch, dinner, and so forth), days, weeks, 2-week periods, 4-week (28-day) periods, months, quarters (3-month periods), or any other unit of time that helps managers better understand their business. Most modern POS systems allow you to choose the specific reporting period of most interest to you.

In some foodservice operations, you will not have the ability to consider your sales in terms of revenue generated. Figure 2.3 is the type of sales history you can use when no cash sales are typically reported. In this case, the manager of the Eureka summer camp is interested in recording sales based on serving periods rather than an alternative time frame, such as a 24-hour (1-day) period. This approach is often used in such settings as all-inclusive hotels and resorts, extended care facilities, retirement homes, college residence halls, correctional facilities, military bases, hospitals, summer camps, or any other situation where knowledge of the number of actual guests served during a given period is critical for planning purposes.

FIGURE 2.2 Sales History (Revenue)

Rae's Restaurant			
Sales Period	**Date**	**Daily Sales**	**Sales to Date**
Monday	1/1	$851.90	$851.90
Tuesday	1/2	$974.37	$1,826.27
Wednesday	1/3	$1,004.22	$2,830.49
Thursday	1/4	$976.01	$3,806.50
Friday	1/5	$856.54	$4,663.04
Saturday	1/6	$1,428.22	$6,091.26
Sunday	1/7	$1,241.70	$7,332.96
Week's Total			$7,332.96

FIGURE 2.3 Sales History (Guests)

Camp Eureka								
	Guests Served							
Serving Period	Mon	Tues	Wed	Thurs	Fri	Sat	Sun	Total
7:00–9:00 A.M.	121							
9:00–11:00 A.M.	40							
11:00–1:00 P.M.	131							
1:00–3:00 P.M.	11							
3:00–5:00 P.M.	42							
5:00–7:00 P.M.	161							
Total Served	506							

Given the data in Figure 2.3, the implications for the Tuesday staffing of food servers at the camp are evident. Fewer servers will likely be needed from 9:00 to 11:00 A.M. than from 7:00 to 9:00 A.M. The reason is obvious. On Monday, fewer campers ate between 9:00 and 11:00 A.M. (40) than between 7:00 and 9:00 A.M. (121). As a knowledgeable manager, if you were operating this camp, you could either reduce serving staff during the slower service period or move those workers to some other necessary task.

You might also decide not to produce as many menu items for consumption during the 9:00 to 11:00A.M. period. In that way you could make more efficient use of both your labor and food products. It is simply easier to manage well when you know the answer to the question "How many guests will I serve?"

Sales histories can be created to record revenue, guests served, or both. In all cases, however, it is important that you keep good records of how much you have sold in the past because doing so is one key to accurately predicting the amount of sales you will likely achieve in the future.

COMPUTING AVERAGES FOR SALES HISTORIES

In some cases, knowing the average number of revenue dollars generated in a past time period, or the average number of guests served in that period, may be a real benefit to you. This is because in many operations, future guest activity will often be very similar to past guest activity. As a result, using historical sales averages from your operation can be useful as you project future guest sales and counts.

The two major types of averages you are likely to encounter as a foodservice manager are as:

- Fixed average
- Rolling average

FIXED AVERAGE

A **fixed average** is an average for a specific (fixed) time period, for example the first 14 days of a given month. In this case, you would compute the average amount of sales or guest activity for this 14-day period. Note that this average is called fixed because the first 14 days of the month will always consist of the same days (the 1st through the 14th) as shown in Figure 2.4.

HERE'S HOW IT'S DONE 2.1

An **average**, which is also called an arithmetic "mean," is simply a value computed by identifying the number of items in a series, adding the quantities of each item in that series, and then dividing the total of the quantities by the number of items in the series. It's easy.

The first and most important thing to do when you calculate an average is to identify exactly how many items will be included in the series. The number of items in the series you are examining is important because it will be the denominator in the calculation of the average. The sum of the assigned values of these items will be the numerator.

For example, if a manager wants to calculate their operation's average electric bill last year, they must first determine that the number of items in the series will be 12: the number of months in last year and, as a result, the number of electric bills to be added.

Similarly, if a manager wants to calculate the average sales per day last week, the number of items in the series will be the number of days the restaurant operated last week. After the number of items in the series is identified, the quantity of each item in the series must be totaled.

To illustrate how managers calculate an average, assume that a manager buys three different knives. One knife costs $16.00, another costs $19.00, and the third costs $28.00. To calculate the average cost of a knife, the manager first recognizes that there are three quantities (knives purchased) in the series of items. Then the sum of the quantities (the cost of the three knives) is calculated.

In this example, the sum of the quantities equals $63.00 ($16.00 + $19.00 + $28.00 = $63.00).

Next, the manager divides the arrived at sum by 3, which is the number of items (knives) in this series:

$$\frac{\$63.00 \text{ sum of the knives' cost}}{3 \text{ knives purchased}} = \$21.00 \text{ average price per knife}$$

Thus, $21 is the average of $16 + $19 + $28. The number of items in the series is 3, that is, they are 16, 19, and 28.

The sum of the quantities in this case equals 63 (16 + 19 + 28 = 63). Thus, in this example, $63.00 ÷ 3 = $21.00; the average price paid by the manager for knives.

FIGURE 2.4 14-Day Fixed Average

Lauraina's Take-Out Coffee	
Day	**Daily Sales**
1	$350.00
2	$322.00
3	$388.00
4	$441.00
5	$419.00
6	$458.00
7	$452.00
8	$458.00
9	$410.00
10	$434.00
11	$476.00
12	$460.00
13	$418.00
14	$494.00
14-Day Total	$5,980.00

$$\frac{\$5{,}980}{14} = \$427.14 \text{ sales per day}$$

Figure 2.4 details the sales activity of Lauraina's Take-Out Coffee. The calculation of this average (total revenue/number of days) is fixed, or constant, because Lauraina's management has identified the 14 specific days used to make up the average. The number $427.14 may be very useful because it might, if management wishes, be used as a good predictor of the revenue volume that should be expected for the first 14 days of next month.

ROLLING AVERAGE

A **rolling average** is the average amount of sales or volume over a *changing* time period. A fixed average is computed using a specific or constant set of data, but a rolling average is computed using data that will change. To illustrate, consider the case of Ubalda Salas, who operates a sports bar. Ubalda is interested in knowing what the average revenue dollars were in her operation for each prior 7-day period.

In this case, the prior 7-day period changes, or rolls forward, by 1 day as each day passes. It is important to note that Ubalda could have been interested in her average daily revenue last week (fixed average), but she prefers to know her average sales for the last 7 days. This means that she will, at times, be using data from both last week and this week to compute the last 7-day average.

Using the sales data recorded in Figure 2.5, the 7-day rolling average for Ubalda's Sports Bar is computed as shown in Figure 2.6.

Note that each 7-day period is made up of a group of daily revenue numbers that changes over time. The first 7-day rolling average is computed by summing the first 7 days' revenue (revenue on days 1–7 = $28,280) and dividing that number by 7 to arrive at a 7-day rolling average of $4,040 ($28,280 revenue ÷ 7 days = $4,040 average).

Each day, Ubalda adds her daily revenue to that of the prior 7-day total and drops the day that is now 8 days past. This gives her the effect of continually "rolling" the most current 7 days' data forward. The rolling average, although more complex and time-consuming to calculate than a fixed average, can be extremely useful in recording data to help you make effective predictions about the sales levels you can expect in the future. This is true because, in many cases, rolling data are more current and, thus, more relevant than some fixed historical averages.

FIGURE 2.5 14-Day Sales Levels

Ubalda's Sports Bar	
Day	**Sales**
1	$3,500.00
2	$3,200.00
3	$3,900.00
4	$4,400.00
5	$4,200.00
6	$4,580.00
7	$4,500.00
8	$4,600.00
9	$4,100.00
10	$4,400.00
11	$4,700.00
12	$4,600.00
13	$4,180.00
14	$4,940.00

FIGURE 2.6 Seven-Day Rolling Average

Ubalda's Sports Bar								
Day	1–7	2–8	3–9	4–10	5–11	6–12	7–13	8–14
1	$3,500	—						
2	$3,200	$3,200	—					
3	$3,900	$3,900	$3,900	—				
4	$4,400	$4,400	$4,400	$4,400	—			
5	$4,200	$4,200	$4,200	$4,200	$4,200	—		
6	$4,580	$4,580	$4,580	$4,580	$4,580	$4,580	—	
7	$4,500	$4,500	$4,500	$4,500	$4,500	$4,500	$4,500	—
8		$4,600	$4,600	$4,600	$4,600	$4,600	$4,600	$4,600
9			$4,100	$4,100	$4,100	$4,100	$4,100	$4,100
10				$4,400	$4,400	$4,400	$4,400	$4,400
11					$4,700	$4,700	$4,700	$4,700
12						$4,600	$4,600	$4,600
13							$4,180	$4,180
14								44,940
Total	**$28,280**	**$29,380**	**$30,280**	**$30,780**	**$31,080**	**$31,480**	**$31,080**	**$31,520**
7-Day	**$4,040.00**	**$4,197.14**	**$4,325.71**	**$4,397.14**	**$4,440.00**	**$4,497.14**	**$4,440.00**	**$4,502.86**

To best predict future sales in operations you manage, you may choose to compute fixed averages for some time periods and rolling averages for others. For example, it may be helpful to know your average daily sales for the first 14 days of last month as well as your average sales for the most recently past 14 days. If, for example, these two numbers are very different, you will know whether the number of sales you can expect in the future is increasing or declining. Regardless of the type of average you feel is best for your operation, you should always document your sales history because it is from your sales history that you will be better able to predict future sales levels.

RECORDING REVENUE, GUEST COUNTS, OR BOTH

Some foodservice operations do not record revenue as the primary measure of their sales activity. For them, developing sales histories by recording the number of individuals they serve each day makes the most sense. Thus, **guest count,** the term used in the hospitality industry to indicate the number of people served, is recorded on a regular basis.

You may decide that your operation is best managed by tracking both generated revenue and guest counts. In fact, if you do decide to record both revenue and guest counts, you have the information you need to compute **average sales per guest,** a term sometimes referred to as "**check average.**"

Average sales per guest is determined by the following formula:

$$\frac{\text{Total sales}}{\text{Number of guests served}} = \text{Average sales per guest}$$

HERE'S HOW IT'S DONE 2.2

The average sales per guest (sometimes called the "check average") is one of a foodservice manager's most commonly utilized formulas. And it's not hard to calculate it. To compute an operation's check average, managers use these steps:

Step 1. Identify the amount of sales achieved in a specific time period.

Step 2. Identify the number of guests served in the same time period.

Step 3. Apply the average sales per guest formula.

For example, assume that a manager recorded that the amount of sales achieved last week was $18,750 (Step 1). Further, assume that 750 guests were served last week (Step 2). Applying the average sales per guest formula (Step 3), the manager would find that last week's check average was $25.00.

$$\frac{\$18{,}750 \text{ total sales last week}}{750 \text{ guest served last week}} = \$25.00 \text{ average sales per guest}$$

One reason why the ability to accurately calculate the average sales per guest, or check average, is so important is that managers are often evaluated on the success of their efforts to maintain, or to increase, the average amount spent by each guest visiting their operations.

FIGURE 2.7 Sales History

Sales Period	Date	Day	Sales	Guests Served	Average Sales per Guest
Monday	Jan 1	Monday	$2,365.00	190	$12.45
Tuesday	Jan 2	Tuesday	$3,750.00	314	$11.94
Two-Day Average			$3,057.50	252	$12.13

Consider the information in Figure 2.7, in which the manager of Brothers' Family Restaurant has decided to monitor and record the following:

1. Sales
2. Guests served
3. Average sales per guest

Most POS systems are programmed to report the amount of revenue generated in a selected time period, the number of guests served, and the average sales per guest. In the case of Brothers' Family Restaurant, Monday's revenue was $2,365, they served 190 guests, and thus the average sales per guest that day was $12.45 ($2,365/190 = $12.45). On Tuesday, the average sales per guest was $11.94 ($3,750/314 = $11.94).

To compute the 2-day revenue average, the Brothers' manager would add Monday's revenue and Tuesday's revenue and then divide the total by 2, yielding a 2-day revenue average of $3,057.50 [($2,365 + $3,750)/2 = $3,057.50].

In a similar manner, the 2-day average for the number of guests served is computed by adding the number of guests served on Monday to the number served on Tuesday and then dividing by 2, yielding a 2-day average of guests served of 252 [(190 + 314)/2 = 252].

It might be logical to think that the manager of Brothers' could compute the Monday and Tuesday combined average sales per guest by adding the averages from each day and then dividing by 2. It is important to understand that this would *not* be correct.

A formula consisting of Monday's average sales per guest plus Tuesday's average sales per guest divided by 2 [($12.45 + $11.94)/2] yields $12.20.

In fact, the actual 2-day average sales per guest is $12.13 [($2,365 + $3,750)/(190 + 314) = ($6,115/504) = $12.13].

FIGURE 2.8 Weighted Average

	Sales	Guests Served	Average Sales per Guest
Day 1	$100.00	20	$5.00
Day 2	$4,000.00	400	$10.00
Two-Day Average	$2,050.00	210	???

Although the difference of $0.07 might, at first glance, appear to be an inconsequential amount, assume that you are the president of a restaurant chain with 4,000 units worldwide. If each unit served 1,000 guests per day and you miscalculated average sales per guest by $0.07, your daily revenue calculation would be "off" by $280,000 per day [(4,000 units × 1,000 guests × $0.07) = $280,000]!

Returning to the Brothers' Family Restaurant example, the correct formula for computing the 2-day average sales per guest is:

$$\frac{\text{(Monday sales + Tuesday sales)}}{\text{(Monday guests + Tuesday guests)}} = \text{2-day average sales per guest}$$

or

$$(\$2{,}365 + \$3{,}750) / (190 + 314) = \$12.13$$

The correct computation in this example is a **weighted average**, that is, an average that weighs the number of guests served in different time periods with how much they spend in those same time periods.

To demonstrate further the importance of weighted averages, consider the data in Figure 2.8 and assume that you want to answer the question "What is the combined average sales per guest?"

From the data in Figure 2.8, it is easy to see that the 2-day average would not be $7.50 [($5.00 + $10.00)/2 = $7.50] because many more guests were served on the day the average sales per guest was $10.00 than on the day the average sales per guest was $5.00. With so many guests spending an average of $10.00, and so few spending an average of $5.00, the overall average should be much closer to $10.00 than to $5.00.

In fact, utilizing the average sales per guest formula, the correct weighted average sales per guest would be $9.76, as follows:

$$\frac{\text{(Day 1 sales + Day 2 sales)}}{\text{(Day 1 guests + Day 2 guests)}} = \text{2-day average sales per guest}$$

or

$$\frac{\$100 + \$4{,}000}{20 + 400 \text{ guests}} = \$9.76 \text{ per guest}$$

MAINTAINING SALES HISTORIES

Although a sales history may consist of revenue, number of guests served, and average sales per guest, depending upon the type of operation you manage, you may want to know even more detailed information about your sales. This may include information such as the number of guests served in a specific meal or time period (e.g., breakfast, lunch, or dinner), the method of meal delivery (e.g., drive-thru sales *vs.* carryout or dine-in sales), or the method used to order (online *vs.* on-site).

The important concept to remember is that you have the power to develop the sales history information that best suits your operation. That information should be updated at least daily, and a cumulative total for the appropriate time periods should also be maintained. In most cases, your sales histories should be kept for a period of at least 2 years. This allows you to have a good sense of what has happened to your business in the recent past.

Of course, if you are the manager of a new operation, or one that has recently undergone a major concept change, you may not have the advantage of reviewing meaningful sales histories because they simply do not exist. If you find yourself in such a situation, it is imperative that you begin to build and maintain your sales histories as soon as possible so you will, as quickly as possible, have good sales information on which to base your future managerial decisions.

SALES VARIANCES

After an accurate sales history system has been established, you may begin to see that your operation experiences some **sales variance** or differences from previous sales levels. These sales variances are normal and will give you an indication of whether your sales are increasing, declining, or staying the same. Because that information is so important to predicting future sales levels, many foodservice managers improve their sales history information by including sales variance as an additional component of the history.

Figure 2.9 details a portion of a sales history that has been modified to include a sales variance column, which allows the manager to see how sales are different from a prior period. In this case, the manager of Quick Wok wants to compare sales for the first 3 months of this year to sales for the first 3 months of last year, as recorded in that period's sales history report.

The variance in Figure 2.9 is determined by subtracting sales last year from sales this year. In January, the variance figure is obtained as follows:

Sales this year − Sales last year = Variance

or

$54,000 − $51,200 = $2,800

Thus, the manager of Quick Wok can see that the sales for the first quarter are greater than last year's first quarter. In fact, all 3 months in the first quarter of the year showed revenue increases over the prior year.

FIGURE 2.9 Sales History and Variance

Quick Wok			
Month	**Sales This Year**	**Sales Last Year**	**Variance**
January	$54,000	$51,200	$2,800
February	$57,500	$50,750	$6,750
March	$61,200	$57,500	$3,700
First-Quarter Total	$172,700	$159,450	$13,250

FIGURE 2.10 Sales History, Variance, and Percentage Variance

Quick Wok				
Month	Sales This Year	Sales Last Year	Variance	Percentage Variance (%)
January	$54,000	$51,200	$2,800	5.5%
February	$57,500	$50,750	$6,750	13.3%
March	$61,200	$57,500	$3,700	6.4%
First-Quarter Total	$172,700	$159,450	$13,250	8.3%

The total sales improvement for the first quarter was $13,250 ($172,700 − $159,450 = $13,250). The sales history and variance format used in Figure 2.9 lets a manager know the dollar value of revenue variance. But good managers want to know even more because simply knowing the dollar value of a variance has limitations.

To illustrate, consider two restaurant managers. One manager's restaurant had revenue of $1,000,000 last year. The second manager's restaurant generated one-half as much revenue, or $500,000. This year both managers had sales increases of $50,000. But it is clear that while both experienced a $50,000 sales increase, that increase represents a much greater proportional change in the second restaurant than in the first. Because that is true, effective managers are often interested in the **percentage variance,** or percentage change, in their sales in one time period when compared to a different time period.

Figure 2.10 shows how the sales history data at the Quick Wok can be expanded to include percentage variance as part of that operation's complete sales history.

Percentage variance is obtained by subtracting sales last year from sales this year and then dividing the resulting number by sales last year. Thus, in the month of January, the percentage variance is calculated as follows:

$$\frac{\text{Sales this year} - \text{Sales last year}}{\text{Sales last year}} = \text{Percentage variance}$$

or

$$\frac{\$54{,}000 - \$51{,}200}{\$51{,}200} = 0.055 \text{ (in common decimal form, 5.5\%)}$$

Note that the resulting decimal form percentage can be converted to the more frequently used common form explained in Chapter 1 by moving the decimal point two places to the right, or through multiplying it by 100.

Note that when a sales history includes a variance column, an alternative, and shorter, formula for computing the percentage variance for January is as follows:

$$\frac{\text{January variance}}{\text{January sales last year}} = \text{January percentage variance}$$

or

$$\frac{\$2{,}800}{\$51{,}200} = 0.055 \text{ (in common decimal form, 5.5\%)}$$

Yet another way to compute the sales percentage variance for January is to use a math shortcut, as follows:

$$\frac{\text{January sales this year}}{\text{January sales last year}} - 1 = \text{January percentage variance}$$

or

$$\frac{\$54{,}000}{\$51{,}200} - 1 = 0.055 \text{ (in common form, 5.5\%)}$$

Regardless of the mathematical method you use, calculating percentage variance is invaluable when you want to compare the operating results of two foodservice operations of different sizes.

Returning to our previous example of the two restaurant managers, each of whom achieved $50,000 revenue increases, by using the first percentage variance formula presented, you will see that the restaurant with higher sales increased its revenue by 5.0 percent [($1,050,000 − $1,000,000)/$1,000,000 = 0.05] or 5 percent in common form. The restaurant with less original revenue and the same $50,000 increase in sales, however, achieved a 10 percent increase [($550,000 − 500,000)/$500,000 = 0.10] or 10 percent in common form.

As your level of expertise increases, you will find additional areas in which knowing the percentage variance in revenue and in your expenses (addressed later in this text) will greatly assist you in your management decision making.

PREDICTING FUTURE SALES

It has been said that truly outstanding managers have an ability to see the future in regard to the revenue figures they will achieve and the number of guests they expect to serve. You, too, can learn to do this when you apply your knowledge of percentage variances to estimating your operation's future sales. Depending on the type of facility you manage, you may be interested in predicting, or forecasting, future revenues, guest counts, or average sales per guest levels. We will examine the procedures for all three of these in detail.

FORECASTING FUTURE REVENUES

Erica Tullstein is the manager of Rock's Pizza Pub on the campus of State College. Her guests consist of college students, most of whom come to the Rock to talk, surf the Web, text, eat, and study. Erica has done a good job maintaining sales histories in the 2 years she has managed the Rock. She records the revenue dollars she achieves on a daily basis, as well as the number of students frequenting the Rock. Revenue data for the last 3 months of the year are recorded in Figure 2.11.

FIGURE 2.11 Revenue History

Rock's Pizza Pub				
Month	Sales This Year	Sales Last Year	Variance	Percentage Variance (%)
October	$75,000	$72,500	$2,500	3.4%
November	$64,250	$60,000	$4,250	7.1%
December	$57,500	$50,500	$7,000	13.9%
Fourth-Quarter Total	$196,750	$183,000	$13,750	7.5%

FIGURE 2.12 First-Quarter Revenue Forecast

Rock's Pizza Pub				
Month	Sales Last Year	% Increase Estimate	Increase Amount	Revenue Forecast
January	$68,500	7.5%	$5,137.50	$73,637.50
February	$72,000	7.5%	$5,400.00	$77,400.00
March	$77,000	7.5%	$5,775.00	$82,775.00
First-Quarter Total	$217,500	7.5%	$16,312.50	$233,812.50

As can be seen in Figure 2.11, fourth-quarter revenues for Erica's operation have increased from the previous year. Of course, there could be a variety of reasons for this. Erica may have extended her hours of operation to attract more students. She may have increased the size of pizzas and held her prices constant, thus creating more value for her guests and thus attracting more customers. She may have raised the selling price of pizzas. Or perhaps a competing pizza parlor was closed during this time period. Using all of her knowledge of her own operation and her market, Erica would like to predict the sales level she will likely experience in the first 3 months of next year. This sales forecast will be most helpful as she plans for next year's anticipated expenses, optimized staffing levels, and anticipated profits.

The first question Erica must address is the amount her sales have actually increased. Revenue increases range from a low in October of 3.4 percent, to a high in December of 13.9 percent. The overall quarter (3-month) average of 7.5 percent is the figure that Erica elects to use as she predicts her sales revenue for the first quarter of the coming year. She feels it is neither too conservative, as would be the case if she used the October percentage increase, nor too aggressive, as would be the case if she used the December figure.

If Erica were to use the 7.5 percent average increase from the fourth quarter of last year to predict her revenues for the first quarter of this year, a revenue forecast could be developed for the first quarter of next year as shown in Figure 2.12.

The revenue forecast for this time period is determined by multiplying sales last year by the percentage increase estimate and then adding the percentage increase amount to sales last year.

For the month of January, the revenue forecast is calculated using the following formula:

Sales last year + (Sales last year × % increase estimate) = Revenue forecast

or

$68,500 + ($68,500 × 0.075) = $73,637.50

An alternative way to compute the revenue forecast is to use a math shortcut as follows:

Sales last year × (1 + % increase estimate) = Revenue forecast

or

$68,500 × (1 + 0.075) = $73,637.50

Erica is using the increases she has experienced in the past to predict increases she may experience in the future. Monthly revenue figures from last year's sales history plus percentage increase estimates based on those histories can give Erica a very good idea of the revenue levels she may achieve in January, February, and March of the coming year.

Cost Control Around the World

Experienced foodservice managers working in the United States know that Mother's Day is the busiest day of the year for restaurants. An American foodservice manager's revenue forecast for that day would reflect this fact. But not every country celebrates Mother's Day, and there are some who do celebrate it but choose to do so on a day other than the second Sunday in May. As a result, predicting future revenues when operating a restaurant in a foreign country requires managers to have a good understanding of that country's holidays, festivals, and special events. In many cases, these will be very different from those holidays celebrated or events held in the United States.

To cite just one example, consider that on April 12, 1952, the Organization of American States (OAS) and the United Nations Children's Fund (UNICEF) asked each country to declare a Children's Day ("Dia del Nino" in Spanish).

Bolivia selected April 12, the same day on which the Declaration was drafted, and Dia del Nino has been celebrated in Bolivia since 1955. Argentina celebrates Dia del Nino on the second Sunday in August. In both Ecuador and Nicaragua, this holiday is celebrated on June 1.

While it is not celebrated in the United States, Dia del Nino is one of South America's biggest holidays. On that day, it is common for parents to visit their children at school, give the children small presents, and then take their families out to eat to celebrate the day. Certainly, foodservice managers working in South America need to be aware of when this special day is celebrated and plan accordingly if they are to accurately estimate their revenue and be ready for the big Dia del Nino crowds!

FORECASTING FUTURE GUEST COUNTS

Using the same techniques employed in estimating increases in sales, you can estimate increases or decreases in the number of guests you will serve. Figure 2.13 shows how Erica, the manager of Rock's Pizza Pub, used last year's fourth-quarter guest counts to determine the variance and percentage variance for the guest counts she achieved in her facility in the fourth quarter of this year.

If Erica were to use the 6.1 percent average increase from the fourth quarter of last year to predict her guest counts for the first quarter of the coming year, a planning sheet could be developed as presented in Figure 2.14. It is important to

FIGURE 2.13 Guest Count History

Rock's Pizza Pub				
Month	Guests This Year	Guests Last Year	Variance	Percentage Variance (%)
October	14,200	13,700	+500	3.6%
November	15,250	14,500	+750	5.2%
December	16,900	15,500	+1,400	9.0%
Fourth-Quarter Total	46,350	43,700	+2,650	6.1%

FIGURE 2.14 First-Quarter Guest Count Forecast

Rock's Pizza Pub				
Month	Guests Last Year	% Increase Estimate	Guest Increase Estimate	Guest Count Forecast
January	12,620	6.1%	770	13,390
February	13,120	6.1%	800	13,920
March	13,241	6.1%	808	14,049
First-Quarter Total	38,981	6.1%	2,378	41,359

note that Erica is not required to use the same percentage increase estimate for each coming month. Any forecasted increase management feels is appropriate can be used to predict future sales.

The guest count forecast is determined by multiplying guest count last year by the percentage increase estimate this year and then adding the result to the guest count last year. In the month of January, the guest count forecast is calculated using the following formula:

Guest count last year + (Guest count last year × % Increase estimate) = Guest count forecast

or

12,260 + (12,620 × 0.061) = 13,390

This process can be simplified by using a math shortcut, as follows:

Guest last year × (1.00 + % increase estimate) = Guest count forecast

or

12,260 × (1.00 + 0.061) = 13,390

You may notice that in January the guest increase estimate is rounded from 769.82 to 770. This is because you cannot serve 0.82 people! Utilizing normal rounding procedures, you should choose and use the guest count forecast formula you feel most comfortable with.

Consider the Costs

"A 60-minute wait! You've got to be kidding!" said the guest.

"I'm sorry sir," replied Romy. "We'll seat your party as quickly as possible."

Romy was the dining room host at the Al-Amir restaurant. The Al-Amir featured Middle Eastern and North African Cuisine. Guests loved the Al-Amir's chicken shawarma, baba ghanoush, tabouleh, and kibbeh. As a result, the restaurant was often very busy. When that happened, the line of guests waiting to be seated got long and, sometimes, waiting customers got upset.

"Listen," replied the guest, "I understand when places are busy. It can take a while to serve everyone. But look, nearly half of your dining room is empty. The tables just need to be cleared and reset."

"Yes sir," replied Romy. "But the workers we do have are clearing and resetting the tables as fast as they can."

"Then you need more dining room help. We'll just come back another time," said the guest, as he left the restaurant along with his female dining companion, and two small children in hand.

"I'm really very sorry, sir," said Romy to the guest's back as he watched him leave. Romy thought to himself, "This happens way too often!"

Assume that the Al-Amir does not have an effective sales forecast system in place.

1. What will be the long-term impact on the revenue-generating ability of the restaurant if it consistently understaffs its dining room?
2. What would be the long-term impact on Al-Amir's staff?
3. Sometimes even the best sales forecasts can be inaccurate. What steps can managers take to ease the difficulties encountered when their sales forecasts prove to be incorrect?

FORECASTING FUTURE AVERAGE SALES PER GUEST

Recall that average sales per guest (check average) is simply the average amount of money each guest spends during a visit. The same formula is used to forecast an operation's future average sales per guest as was used in forecasting total revenue and guest counts. Using data taken from the sales history, the following formula is used:

Last year's average sales per guest
+ Estimated increase in sales per guest
= Sales per guest forecast

Alternatively, the manager of Rock's Pizza Pub could compute average sales per guest using the data collected from revenue forecasts (Figure 2.12) and combine that data with the guest count forecasts you created (Figure 2.14). If that is done, the revenue forecast data presented in Figure 2.15 would result.

An average sales per guest forecast is obtained by dividing the revenue forecast by the guest count forecast. Thus, in the month of January, the average sales per guest forecast is determined by the following formula:

$$\frac{\text{Revenue forecast}}{\text{Guest count forecast}} = \text{Average sales per guest forecast}$$

or

$$\frac{\$73{,}637.50}{13{,}390 \text{ guests}} = \$5.50 \text{ per guest}$$

Increasingly, sophisticated POS systems analyze not only how many guests were served on a given day but also provide data that helps managers better understand why that number was served. For example, historical weather patterns can be analyzed to discover if rainy days affect total sales volume (and thus perhaps indicate that a forecast adjustment should be made on a day on which rain is predicted). Weather, day of the month, and day of the week are just some of the many factors that may affect sales volume and should be carefully considered by sophisticated foodservice managers as they prepare their sales forecasts.

The Move to Off-Premise Dining

As part of their response to an increased levels of off-premise sales, some foodservice operators have found that it makes good sense to calculate and monitor separate check averages for their on-premise and off-premise diners. Doing so can often help them better predict their future revenues.

In some cases, on-premise diners generate significantly higher check averages than do off-premise diners. This can be due to the fact that on-premise dining provides opportunities for face-to-face suggestive selling of appetizers and desserts by service staff, the sale of alcoholic beverages, and the promotion of higher-priced off-menu specials.

In other cases, operators may find that their off-premise diners generate significantly higher average sales than do on-premise diners because, in many cases, off-premise diners are buying takeout for an entire family rather than for an individual.

Also, tech savvy operators can modify their online order menus to include features that allow guests who have selected a particular entrée to see specific suggestions about additional items to order. This could be the case, for example, when a simple click allows off-premise diners to add chips and queso or a side of guacamole to an off-premise enchilada dinner order.

It is important to note that sales histories, regardless of how well they have been developed and maintained, may not provide all of the information needed to accurately predict future sales. Your knowledge of potential price changes, new or lost competitors, facility renovations, and improved selling programs are additional factors that you may need to consider when predicting future sales. There is no question, however, that a sales history report is easily developed and will serve as

FIGURE 2.15 First-Quarter Average Sales per Guest Forecast

Rock's Pizza Pub			
Month	Revenue Forecast	Guest Count Forecast	Average Sales per Guest Forecast
January	$73,637.50	13,390	$5.50
February	$77,400.00	13,920	$5.56
March	$82,775.00	14,049	$5.89
First-Quarter Total	$233,812.50	41,359	$5.65

the cornerstone of other financial control systems you will design. Without accurate sales data, the control systems you implement, regardless of their sophistication, will perform poorly.

When added to your knowledge of the unique factors that impact your unit, properly maintained sales histories can help you answer two important control questions, namely, "How many people are coming tomorrow?" and "How much is each person likely to spend?" The judgment of management is critical in forecasting answers to these questions. Since you can now answer those questions, you are ready to develop systems that will allow you to prepare an efficient and cost-effective way of serving your guests.

You want to be ready to provide your arriving guests with quality food and beverage products, and enough staff to serve them properly. You have done your homework with regard to the number of individuals who may be coming to your operation. You also know how much they are likely to spend while they are there. Now you must prepare for their arrival. That is the topic addressed in the next chapter!

FUN ON THE WEB!

Maintaining a membership in a professional trade organization is a good way to stay up to date on new techniques for reporting and forecasting sales. One excellent way to do this is by attending the organization's annual meetings and trade shows. Visit the websites of these popular trade associations to find out when and where they will be holding their next trade show or convention.

Associations

National Restaurant Association
American Hotel & Lodging Association
Asian American Hotel Owner's Association
The Academy of Nutrition and Dietetics
Association of Nutrition & Foodservice Professionals
American Culinary Federation
Club Managers Association of America
School Nutrition Association
National Association for Catering and Events
National Association of College and University Food Services
Hospitality Financial and Technology Professionals
International Hotel and Restaurant Association

Green and Growing!

If your operation is "going green" and your customer counts are going up, don't be surprised. Customers are increasingly concerned about the quality of their environment. These customers also prefer to frequent foodservice operations that share their concerns and act on them. As a result, guest counts can be increased by undertaking green initiatives and then proudly letting customers know about what your operation is doing.

It is important to market your products and services directly to the rapidly growing market segment of educated, savvy customers who care about the food they eat and their impact on the world around them. Implementing sustainable practices that focus on conservation as well as utilizing organic, seasonal, and locally grown products can help build customer and employee loyalty, as well as boost profits.

Technology Tools

The hardware and software included in their POS systems are among the most important cost control tools available to foodservice managers. In this chapter, you learned that utilizing accurate sales histories when forecasting future sales is essential and your POS system can be instrumental in helping you with your forecasting efforts. But POS systems can supply managers with a massive amount of additional information. Modern POS systems can be property-based or cloud-based and can provide you with essential operational detail in a number of key areas. Among the most important of these areas are listed as follows:

1. Revenue generation
2. Product usage
3. Guest payment
4. Guest relations

Revenue generation: One of a POS system's most essential functions is that of accurately recording revenue. A POS may be programmed to do the following:

1. Track sales by day, week, month, and year
2. Track sales by time of day or meal period (e.g., breakfast, lunch, and dinner)
3. Record sales made by individual servers
4. Track sales by source of business (e.g., drive-thru *vs.* dine-in)
5. Calculate unit check averages
6. Maintain sales exception records (e.g., refunds, voids, coupons, and gift card usage)
7. Maintain rolling sales averages
8. Compare current sales to prior-period sales
9. Forecast future sales in predetermined increments:
   Monthly
   Weekly
   Daily
   By meal period
   Hourly
   Portions of an hour (e.g., 15 minutes, 30 minutes)
10. Maintain actual sales to forecasted sales variance reports

Product usage: A POS system can be used to give valuable information about an operation's individual menu item sales and product usage. Among other product-usage related information, a POS system may be programmed to do the following:

1. Maintain menu items sold histories
2. Calculate individual menu item profitability
3. Compute sales mix by menu item and by menu subcategory
4. Calculate expected inventory levels
5. Prepare purchase orders

Guest payment: The accurate recording and collecting of guest payments is essential to ensure that an operation receives all of the revenue it has earned, as well as to prevent possible employee theft or fraud. A modern POS can be programmed to do the following:

1. Maintain daily guest payment totals
2. Track cash *vs.* credit *vs.* virtual wallet sales
3. Allow for mobile or online guest payments
4. Maintain guest payment totals by server or shift
5. Maintain tip histories by server
6. Compute sales taxes due
7. Create cashier over and short reports

Guest relations: Today, customer management features are increasingly built into a POS system. These features help managers identify and monitor repeat guests, which then allows for the creation of targeted marketing programs. Among the most popular customer management features are those that:

1. Forecast reservation demand
2. Manage call-in, online, and mobile reservations
3. Automate advanced bookings up to 24 months in advance
4. Track no-shows and cancelations
5. Manage table usage to optimize seating availability
6. Maintain guest complaint records
7. Record total dollars spent by individual guests
8. Build a guest database for mailings and e-mailings
9. Track guest birthdays, favorite tables, and food preferences
10. Establish and maintain customer loyalty programs to reward repeat guests

POS systems can do even more! Additional POS features will be introduced to you as you read the remaining chapters of this book.

FUN ON THE WEB!

POS systems continually undergo development to meet the evolving needs of hospitality managers. To explore product offerings from a variety of progressive and innovative POS companies, enter "restaurant POS" and any one of the following terms into your favorite search engine:

Cloud computing
Tableside payment

Mobile phone-based ordering/payment
Online payment systems
QR code payment systems

When you arrive at a POS company's website, review the POS product offerings, paying especially close attention to the features related to the maintenance of sales histories.

Apply What You Have Learned

Pauline Cooper is a Registered Dietitian Nutritionist (RDN) and the foodservice director at Memorial Hospital. Increasingly, the hospital's marketing efforts have emphasized its skill in treating diabetic patients. As a result, the number of diabetic meals served by Pauline's staff has been increasing. As a professional member of the hospital's management team, Pauline has been asked to report on how the hospital's diabetic treatment marketing efforts have affected her area.

1. How important is it that Pauline have historical records of the type of meals served by her staff and not merely the total number of meals served? Why?
2. Assume that Pauline's per-meal cost has been increasing because diabetic meals are more expensive to produce than regular meals. Could Pauline use sales histories to estimate the financial impact of serving additional diabetic meals in the future? How?
3. What are other reasons managers in a foodservice operation might keep detailed records of meal types (i.e., vegetarian, low-sodium, gluten-free, and the like) served, as well as the total number of meals served?

For Your Consideration

1. Many of the food items purchased by a foodservice manager are highly perishable. Breads, pastries, dairy, produce, meats, and seafood items are common examples. How would these inventory items, and the entire operation, be affected if a manager's sales forecasts were highly inaccurate, or did not even exist?
2. Worker scheduling, in terms of both the number and the types of workers needed to properly service guests, is directly affected by the number of guests estimated to be served. How would the scheduling of workers be affected if a manager's sales forecasts were highly inaccurate, or did not even exist?

Key Terms and Concepts

The following are terms and concepts addressed in the chapter that are important for you as a manager. To help you review, define the terms below:

Sales forecast
Point of sales (POS) system
Sales volume
Sales history
Sales to date
Reporting period
Average (mean)
Fixed average
Rolling average
Guest count
Average sales per guest (check average)
Weighted average
Sales variance
Percentage variance

Test Your Skills

You may download the Excel spreadsheets for the Test Your Skills exercises from the student companion website at http://www.wiley.com/go/dopson/foodandbeveragecost-control8e. Complete the exercises by placing your answers in the shaded boxes and answering the questions as indicated.

1. Vanessa is the snack bar manager at the Silver Glen Country Club. Each day, Vanessa records the revenue generated and the number of guests served in the snack bar. Using the information below, calculate her total revenue, the total number of guests served, her average revenue per day, and the average number of guests served per day.

	Revenue	Guests Served
Monday	$480	45
Tuesday	$535	50
Wednesday	$595	60
Thursday	$395	40
Friday	$940	85
Saturday	$2,450	220
Sunday	$1,120	90
Week's Total		
Daily Average		

2. Laurie Fitsin owns a small sandwich shop called Laurie's Lunch Box. She has developed a sales history for the first week of March using total sales and guests served. Help Laurie calculate her average sales per guest for each day of the week and calculate her totals. Laurie has decided that she could take a shortcut and calculate the average sales per guest for the week by adding Monday through Sunday's average sales per guest and dividing by 7. Would this shortcut make a difference in her total average sales per guest for the week? If so, how much of a difference? Should she take this shortcut? Why or why not?

Laurie's Lunch Box				
Sales Period	**Date**	**Sales**	**Guests Served**	**Average Sales per Guest**
Monday	3/1	$1,248.44	200	
Tuesday	3/2	$1,686.25	360	
Wednesday	3/3	$1,700.00	350	
Thursday	3/4	$1,555.65	300	
Friday	3/5	$1,966.31	380	
Saturday	3/6	$2,134.65	400	
Sunday	3/7	$2,215.77	420	
Total				

3. Mike and Judy Kavanaugh own the Out and In Burgers restaurant. The operation provides on-premise dining where customers place their orders at the counter and the orders are then delivered to the guests' seats. It also offers its guests a drive-thru service. Mike and Judy separately record their counter-placed sales and their drive-thru sales. They also calculate a separate average sale (check average) for each service style.

   Using the chart below, calculate their daily average sale for both their counter sales and drive-thru sales. Also, calculate their operation's overall average sale for the entire week using data that you have combined from both revenue sources.

Out and In Burgers Weekly Sales				
	Sales		**Number of Sales**	**Average Sale**
Monday sales				
Counter sales	$1,845		85	$21.71
Drive-thru sales		$2,750	107	$25.70
Tuesday sales				
Counter sales	$2,150		89	
Drive-thru sales		$3,190	127	
Wednesday sales				
Counter sales	$1,975		91	
Drive-thru sales		$2,850	102	
Thursday sales				
Counter sales	$2,420		106	
Drive-thru sales		$3,725	141	
Friday sales				
Counter sales	$2,850		123	
Drive-thru sales		$4,225	162	
Saturday sales				
Counter sales	$3,200		140	
Drive-thru sales		$6,400	196	

(continued)

Out and In Burgers Weekly Sales				
	Sales		Number of Sales	Average Sale
Sunday sales				
Counter sales	$1,875		84	
Drive-thru sales		$2,375	83	
Weekly Total Sales By Service Style				
All Sales Week Total				

4. Kyle Marler operates a craft beer pub that has been open for just over 1 year. His sales have been steadily increasing. Kyle's sales histories indicate that, over the past 6 weeks, his Monday through Thursday sales have averaged a 5 percent per day increase when compared to his prior year's sales. His Friday, Saturday, and Sunday sales have averaged a 7 percent increase when compared to his prior year's sales. Using that information, and his last year's sales figures, help Kyle create his sales forecast for next week, then answer the questions that follow.

Kyle's Craft Beer Pub Sales Forecast		
	Last Year	Next Week's
Day of Week	Sales	Forecast
Monday	$1,135	$
Tuesday	$1,150	$
Wednesday	$1,270	$
Thursday	$1,225	$
Friday	$1,410	$
Saturday	$2,855	$
Sunday	$1,760	$
Total	$	$
Average	$	$

What were Kyle's total sales for the week last year? What is Kyle's total sales forecast for next week? What was Kyle's average sale per day for the week last year? What is Kyle's forecasted average sale per day for next week?

5. Peggy Richey operates Peggy's Pizza Place in southern California. She has maintained a sales history from January through June and wants to compare this year's sales with last year's sales. Calculate her sales variances and percentage variances for the first 6 months of the year.

Peggy's Pizza Place				
Month	Sales This Year	Sales Last Year	Variance	Percentage Variance
January	$37,902.73	$34,861.51		
February	$33,472.03	$31,485.60		
March	$36,692.98	$33,707.79		
April	$36,550.12	$32,557.85		
May	$36,990.12	$37,852.42		
June	$37,742.52	$37,256.36		
6-Month Total				

6. The Lopez brothers, Angelo, Antonio, and Isaiah, own the Lopez Cantina. Angelo is in charge of marketing, and he is developing his sales forecast for next year. Because of his marketing efforts, he predicts a 5 percent increase in his monthly guest counts. Angelo is not aware of any anticipated menu price increases and assumes, therefore, that his weighted check average will remain stable.

   A. Using last year's sales and guest counts, estimate Angelo's weighted check average (average sales per guest) for the year. (Spreadsheet hint: Use the ROUND function to two decimal places on the cell containing the weighted check average, cell D18, because it will be used in another formula in part B.)

Month	Sales Last Year	Guest Count Last Year	Check Average
January	$45,216	4,800	
February	$48,538	5,120	
March	$50,009	5,006	
April	$45,979	4,960	
May	$49,703	5,140	
June	$48,813	5,300	
July	$55,142	5,621	
August	$59,119	6,002	
September	$55,257	5,780	
October	$50,900	5,341	
November	$54,054	5,460	
December	$50,998	5,400	
Total			

   B. Using the weighted check average calculated in part A, determine Angelo's projected sales, assuming a 5 percent increase in guest counts. (Spreadsheet hint: Use the ROUND function to zero decimal places in the Guest

Count Forecast column, cells C23:C34. Use the SUM function for the total, cell C35. Otherwise, your answers will not be correct.)

Month	Guest Count Last Year	Guest Count Forecast	Weighted Check Average	Projected Sales
January	4,800			
February	5,120			
March	5,006			
April	4,960			
May	5,140			
June	5,300			
July	5,621			
August	6,002			
September	5,780			
October	5,341			
November	5,460			
December	5,400			
Total				

7. Donna Berger is a hotel food and beverage director at a 500-room hotel. Donna knows that as the number of rooms sold in the hotel increases, the number of guests she serves breakfast increases also. Based on historical records, Donna will serve breakfast to 55 percent of the hotel's registered guests. Help Donna plan for the number of breakfasts she will serve by completing the following chart:

Number of Guests in Hotel	Historical % of Guests Eating Breakfast	Estimated Number of Guests to Be Served
120	55	
175	55	
245	55	
275	55	
325	55	
375	55	
430	55	
475	55	
500	55	

What information will Donna need to determine the historical percentage of guests who eat breakfast?

8. Lakota Vela operates Hall's House, a mid-priced restaurant with a $30.00 check average. Her clientele consists of business people and tourists visiting her city. Based on the historical sales records she keeps, Lakota believes her

business will achieve a food sales increase next year of 4 percent per month for each of the first 6 months of the year. She feels this increase will be the result of increases in guest counts (not check averages).

At mid-year (July 1), Lakota intends to increase her menu prices (and thus her check average) by 2 percent. She feels that although these price increases could result in a slight, short-term reduction in her guest counts, the restaurant's guest counts will still increase 3 percent for the last 6 months of the year.

Taking into account her guest count growth estimates and mid-year price increases, Lakota would like to estimate her predicted year-end food revenues. Prepare the revenue estimates for Hall's House. (Spreadsheet hint: Use the ROUND function to zero decimal places in the Guest Count Forecast columns, cells D5 through D10 (D5:D10) and D15 through D20 (D15:D20). Use the SUM function for the totals, cells D11 and D21. Otherwise, your answers will not be correct.)

Months January Through June					
Month	**Guest Count Last Year**	**Guest Count % Increase Estimate**	**Guest Count Forecast**	**Original Check Average**	**Revenue Forecast**
January	6,270				
February	6,798				
March	6,336				
April	6,400				
May	6,930				
June	6,864				
6-Month Total					

Months July Through December					
Month	**Guest Count Last Year**	**Guest Count % Increase Estimate**	**Guest Count Forecast**	**New Check Average**	**Revenue Forecast**
July	6,845				
August	6,430				
September	6,283				
October	6,402				
November	6,938				
December	7,128				
6-Month Total					

9. Raktida is the manager of a popular Italian Restaurant on Mott Street. She wants to predict her guest counts for the first week of November so that she can estimate an accurate number of servers to schedule. Business is very good, but her sales history from last month indicates that fewer guests are served during the first few days of the week compared to last year, whereas more guests per day are served in the later part of the week. Raktida has entered the guest counts from last year and the estimated percentage change in guest counts for this year in the chart below. Because good service is so important to her, she wants to ensure that enough servers are scheduled to work each day. One server can provide excellent service to 50 guests. Help

Raktida calculate how many servers to schedule each day by completing the following chart. Note: Raktida always rounds the number of servers required up to the next whole number to ensure the best service possible for her guests! (Spreadsheet hint: Use ROUNDUP to 0 decimal places in the Estimated Guest Count This Year column and in the Number of Servers Needed This Year columns.)

How would Raktida's server scheduling efforts this year be affected if last year she had recorded only her weekly (not daily) guest counts?

Raktida's Guest Forecast and Server Scheduling Worksheet for the First Week of November

	Guest Count Last Year	Estimated Change This Year	Estimate Guest Count This Year	Number of Servers This Year
Sunday	625	–5%		
Monday	750	–5%		
Tuesday	825	0		
Wednesday	850	5%		
Thursday	775	5%		
Friday	1,250	10%		
Saturday	1,400	10%		

10. Arianna Grand operates the Grand Lotus Garden, a single unit Thai foods restaurant operated within a strip shopping center. Arianna's operation consists of 120 seats and a pickup counter. Ariana carefully monitors her sales. Currently she is preparing her sales forecast for February of the coming year.

    Based on her sales history, for the coming February she is forecasting a 3% increase in the number of on-premise diners and a 5% increase in the number of guests who choose carryout. Due to price increases she has instituted, her selling prices average 4% more than last year. Based on last year's sales and her price increases, help Arianna calculate her estimated customer counts, revenue, and check averages for February of the coming year, then answer the questions that follow

	Last Year			This Year		
February	Customer Count	Check Average	Revenue	Customer Count Forecast	Check Average Forecast	Revenue Forecast
On-premise Sales	895	$26.53	$23,744			
Carry-out Sales	476	$31.98				
Total Sales	1,371					

What will be Arianna's total forecasted customer count for February of the coming year? What will be her total revenue forecast for February of the coming year? What will be her estimated overall unit check average?

11. Marcia Curtis is the Vice President of Development for Tamales to Go, a rapidly expanding chain of restaurants featuring takeout tamales and other Mexican food specialties. Currently, the chain has 150 units and average unit volume is $850,000 per year. Annual per-unit revenue growth for opened units is 5 percent per year.

Tonya Gamez, the chain's president, has promised company stockholders that the chain will experience 12 percent overall revenue growth in the next year. Assuming that per-unit growth on existing units continues to average 5 percent, calculate how many total operating units Marcia will need to have open next year to meet Tonya's goal of a 12 percent overall increase in the chain's revenue.

	This Year	Next Year
Total Revenue	$127,500,000	
Per Unit Revenue	$850,000	
Total Number of Units	150	

12. Janice Bray owns the Little Sister's Bakeshop and Coffee House. She would like to make a daily forecast of her July 1 through July 7 sales based on her 4-week rolling average of prior daily sales. To be conservative, Janice wants to forecast each day's sales in the first week of July at 98 percent of each day's previous 4-week rolling sales average.

Using her June 3 through June 30 daily sales, help Janice complete the 4-Week Rolling Average/Forecast worksheet she has begun to prepare, rounding each day in July's forecasted amount to the whole dollar, then answer the questions that follow.

June 3–June 30 Sales: This Year					
Day	**June Date**	**Sales**	**Day**	**June Date**	**Sales**
Monday	3	$1,417	Monday	17	$1,571
Tuesday	4	$1,405	Tuesday	18	$1,522
Wednesday	5	$1,385	Wednesday	19	$1,329
Thursday	6	$1,330	Thursday	20	$1,510
Friday	7	$1,785	Friday	21	$1,885
Saturday	8	$1,810	Saturday	22	$1,890
Sunday	9	$1,770	Sunday	23	$1,900
Monday	10	$1,627	Monday	24	$1,482
Tuesday	11	$1,587	Tuesday	25	$1,491
Wednesday	12	$1,455	Wednesday	26	$1,610
Thursday	13	$1,490	Thursday	27	$1,555
Friday	14	$1,820	Friday	28	$2,305
Saturday	15	$1,775	Saturday	29	$2,015
Sunday	16	$1,810	Sunday	30	$1,885

Janice's 4-Week Average/Forecast Worksheet						
Sales	**Week 1**	**Week 2**	**Week 3**	**Week 4**	**4-Week Average**	**July 1–7 Forecast**
Monday	$1,417	$1,627	$1,571	$1,482		
Tuesday						
Wednesday						
Thursday						
Friday						
Saturday						
Sunday						
Total						
Daily Average						

In the past 4 weeks, on which day of the week did Janice average her lowest sales? In the past 4 weeks, on which day of the week did Janice average her highest sales? What will Janice forecast to be her average sales per day for the week of July 1 through 7? Are Janice's sales increasing or decreasing? Defend your answer.

CHAPTER 3

Purchasing and Receiving

OVERVIEW

In this chapter, you will learn the professional techniques and methods managers use to accurately estimate the future sales of their individual menu items. Doing so is important because those estimates, when used in conjunction with standardized recipes, give managers the information they must have to purchase the quality and quantity of ingredients needed to properly service their guests.

When ordered items are actually delivered, great care must be taken to ensure the delivered items exactly match those that management intended to buy. In this chapter, you will learn the procedures managers must implement to ensure all items delivered to them are identical in quantity, quality, and price to those items they originally ordered, and what to do if they are not.

Chapter Outline

- Forecasting Food Sales
- Forecasting Beverage Sales
- Importance of Standardized Recipes
- Purchasing Food
- Purchasing Beverages
- Purchase Orders
- Receiving Food and Beverage Products
- Technology Tools
- Apply What You Have Learned
- For Your Consideration
- Key Terms and Concepts
- Test Your Skills

LEARNING OUTCOMES

At the conclusion of this chapter, you will be able to:

- Use sales histories and standardized recipes to determine the amount of food and beverage products to buy in anticipation of forecasted sales.
- Purchase food and beverage products in a cost-effective manner.
- Implement proper procedures for receiving food and beverage products.

FORECASTING FOOD SALES

The continued post-COVID-19 pandemic shift in consumer behavior toward eating out resulted in the highest share of food spending dedicated to food away from home since the data series began in the 1930s. In 2023, nominal spending at food away from home establishments accounted for 58.5 percent of total food spending.[1] That percentage continues to grow each year.

Continued economic growth, gains in consumers' real disposable income, and changes in the lifestyles of today's busy American families are all spurring a rise in the number of meals served away from home. This is good news for your career as a hospitality manager.

This increasing growth activity, and consumer demand, however, will also create some guest service-related challenges for you. Consider the situation you would face if you used sales histories (see Chapter 2) to project that 300 guests will visit your restaurant for lunch today. Your restaurant serves only three entrée items: roast chicken, roast pork, and roast beef. The question you must consider is: "How many servings of each item should we produce so that we do not run out of any one item nor have too many of each item remaining after the lunch period is over?"

If you were to run out of one of your three menu items, guests who wanted that item would undoubtedly become upset. They might even leave your restaurant (and perhaps take with them dining companions who would have ordered the other items you had plenty of!). Producing too much of any one item would, on the other hand, result in unsold items that could cause your costs to rise to unacceptable levels unless these remaining items could be sold for their full price at a later time.

In this situation, it would be unwise to produce 300 portions of each item. Although you would never run out of any one item (i.e., each of your 300 estimated guests could order the same item and you would still have produced enough), you would also have 600 portions (900 portions produced − 300 portions sold = 600 portions) left over at the end of your lunch period.

What you would really like to do, of course, is instruct your staff to make the "right" amount of each menu item. The right amount would be the number of servings that minimizes your chances of running out of an item before lunch is over as well as minimizing your chances of having excessive numbers of portions unsold when the meal period is completed.

Just as estimating the number of guests who will arrive at your restaurant depends upon an accurate forecast of guest counts, the answer to the question of how many servings of roast chicken, pork, and beef you should prepare lies in accurate menu item sales forecasting.

Let us return to the three-item menu example cited previously. This time, however, assume that you were wise enough to have recorded last week's menu item sales on a form similar to the one presented in Figure 3.1.

A review of the data in Figure 3.1 indicates an estimate of 300 guests for next Monday makes good sense because the weekly sales total last week of 1,500 guests served averages 300 guests per day (1,500 guests served/5 days = 300/guests served per day).

[1] https://www.ers.usda.gov/amber-waves/2024/october/u-s-consumers-increased-spending-on-food-away-from-home-in-2023-driving-overall-food-spending-growth. Retrieved 6/20/2025.

FIGURE 3.1 Menu Item Sales History

Date: 1/2–1/6				Menu Items Sold			
Menu Item	**Mon**	**Tues**	**Wed**	**Thurs**	**Fri**	**Week's Total**	**Average Sold**
Roast chicken	70	72	61	85	77	365	73
Roast pork	110	108	144	109	102	573	115
Roast beef	100	140	95	121	106	562	112
Total	280	320	300	315	285	1,500	300

You also know that on an average day, you sold 73 roast chicken (365 sold/5 days = 73 per day), 115 roast pork (573 sold/5 days = 115 per day), and 112 roast beef (562 sold/5 days = 112 per day).

Once you know the average number of people selecting a given menu item, and you know the total number of guests who made their selections, you can compute each menu item's **popularity index**, or the percentage of total guests choosing a given menu item from a list of alternative or competing menu items.

In this example, you can improve your estimate about the quantity of each item to prepare if you use the sales history to help guide your production planning. If you assume that future guests will select menu items in a manner similar to that of past guests, and that each guest will buy one entrée, that information can be used to improve your predictions with the following formula:

$$\text{Popularity index} = \frac{\text{Total number of a specific menu item sold}}{\text{Total number of all menu items sold}}$$

In this example, the popularity index calculated for roast chicken last week would be 24.3 percent (365 roast chicken sold/1,500 total guests = 0.243, or 24.3%).

Similarly, 38.2 percent (573 roast pork sold/1,500 total guests = 38.2%) preferred roast pork, whereas 37.5 percent (562 roast beef sold/1,500 total guests = 37.5%) selected roast beef.

In most cases, if you know the number of menu items your future guests will select, then you are better prepared to make good decisions about the quantity of each item to produce. In this example, Figure 3.2 illustrates your best estimate of what your 300 guests are likely to order when they arrive.

The basic formula used for individual menu item forecasting, based on an item's individual sales history, is

$$\text{Number of guests expected} \times \text{Item popularity index} = \text{Predicted number to be sold}$$

The **predicted number to be sold** is simply the quantity of a specific food or beverage item likely to be sold given a good estimate of the total number of guests expected to be served.

Once you know what items your guests are likely to select, you can then determine how much of each menu item your production staff should be instructed to prepare or have ready to serve.

FIGURE 3.2 Forecasting Item Sales

Menu Item	Guest Forecast	Popularity Index	Predicted Number to Be Sold
Roast chicken	300	0.243	73
Roast pork	300	0.382	115
Roast beef	300	0.375	112
Total			300

It is important to recognize that, as a foodservice manager, you may face a great deal of uncertainty when attempting to estimate the number of guests who will arrive at your operation on any given day because a variety of factors can influence that number. Among these are:

1. Competition
2. Weather
3. Special events in your area
4. Holidays
5. Facility occupancy levels (in hospitals, dormitories, hotels, and the like) and/or the availability of parking (in selected situations such as shopping malls and strip shopping centers)
6. Your own advertising and promotional offers
7. Your competitor's advertising and promotional offerings
8. Quality of service
9. Changes in operating hours
10. Operational consistency

These factors, as well as others that can affect sales volume, can make accurate guest count prediction very challenging.

In addition, remember that sales histories track only the general trends of an operation. They are not typically able to estimate precisely the number of guests who may arrive on any given day. That is the job of the manager. Because they document what has happened in the past, however, sales histories can be a very helpful guide to what can be expected to happen in the future.

In Chapter 2, we examined the concept of sales forecasting. You learned that forecasting involves estimating the number of guests you expect and the dollar amount of sales you expect. Menu item forecasting helps predict the specific menu items guests will purchase when they arrive at your operation. Menu item forecasting is crucial if you are to effectively purchase and prepare the number of specific menu items you are likely to sell.

FORECASTING BEVERAGE SALES

Alcoholic beverages are those products that are meant for consumption as a beverage and that contain a significant amount of alcohol. These products are generally classified as follows:

1. **Beer:** Fermented beverages made from grain and flavored with hops.
2. **Wine:** Fermented beverages made from grapes, other berries, and fruits.
3. **Spirits:** Fermented beverages that are distilled to increase the alcohol content of the product; these are also referred to as **liquors**.

In most cases, estimating the predicted number to be sold for menu items is easier than for beverage items. This is so because the number of possible items available in the average bar or lounge is staggering.

Human imagination has few limits, thus the number of different alcoholic beverage mixtures a skilled bartender can produce makes forecasting popularity indexes for individual beverages a difficult process. Doing so effectively means utilizing the popularity index and predicted number to be sold formulas, but these formulas must be utilized somewhat differently when they are applied to alcoholic beverage sales.

FORECASTING BEER SALES

Given a choice of beverage products, some of your guests will likely choose beer. However, the questions you must answer to effectively manage your beer costs are "What percentage of my guests will choose beer?" and "Which kind of beer?" and "In what packaging format will they choose that beer?"

To illustrate, assume that you are the owner of LeRae, a small bar in a trendy section of a large West Coast city. Your clientele generally consists of upscale office and managerial professionals. For the time period January 2 through January 9, you served an alcoholic beverage to a total of 1,600 guests: 400 (25 percent) selected a beer product, 160 (10 percent) selected a wine product, and 1,040 (65 percent) selected some type of spirit-based drink.

You have monitored your historical sales data and found that, in the past, one out of four guests (25 percent) coming to LeRae's will order beer. But, given the several brands of beer that you offer, you need to determine the specific brand of beer your guests will buy. In addition, it is very likely that at least some of the beer you serve will come packaged in more than one form. That is, a specific beer brand might be sold in individual cans, bottles, or kegs. **Keg beer** is also known as **draft beer**, or beer in a form of packaging in which the beer is shipped to you in multi-gallon units for bulk sale.

By recording individual beer sales, you will know what your guests' beer preferences are. Figure 3.3 shows that, for the period January 2 to January 9, LeRae's served 400 beers to guests, and it details which specific beers those guests ordered as well as each product's percentage of the total number of sales.

By reviewing data compiled in your point of sales (POS) system, you will know exactly which beers, by brand and by packaging form, you have sold in the bar on a daily basis. This information can be used to estimate the number of specific beer products that will be ordered by your guests in the future. For example, from the data shown in Figure 3.3, if you estimated that 800 guests will purchase beer in a future time period, 90 of those guests would likely order Budweiser in a bottle (800 estimated guests × 0.1125 popularity index (decimal form) = 90 orders predicted to be sold).

FORECASTING WINE SALES

In many operations, the ability to accurately forecast wine sales is important. In most cases, the forecasting of wine sales will be divided into two parts:

1. Forecasting bottled-wine sales
2. Forecasting wine-by-the-glass sales

FIGURE 3.3 LeRae's Bar Detailed Beer Sales

Beverage Sales		Date: 1/2–1/9
Product: Beer	**Number Sold**	**Popularity Index (%)**
Budweiser bottles	45	11.25
Coors bottles	18	4.50
Miller Lite cans	61	15.25
Coors cans	68	17.00
Budweiser draft	115	28.75
Harp's draft	93	23.25
Total	400	100.00

FORECASTING BOTTLED-WINE SALES

When forecasting wines sold by the bottle, you treat an individual type of bottled wine exactly as you would treat a menu item. A wine listing or wine menu detailing the selection of wines available is presented to the guest, who then makes a choice.

Popularity indexes for bottled wines are computed exactly as they would be when analyzing food item sales. Unlike some food items, however, you may decide it is possible and quite desirable to offer wines with rather low sales levels because unopened wines in a bottle are not highly perishable. Thus, many operators develop extensive wine lists consisting of a large number of wines, many of which sell only infrequently.

FORECASTING WINE-BY-THE-GLASS SALES

Generally, forecasting the sale of **house wines** refers to estimating the sales of modestly priced wines offered by the glass (or carafe). Once you have estimated the number of guests who will select wine served by the glass, the type of wine they will select can then be forecasted.

The popularity of different wine types changes frequently, as newer styles and varieties gain in popularity and newer wine producing regions (such as Argentina, Australia, and South Africa) improve the quality and marketing of their locally produced wines.

Figure 3.4 details the by-the-glass sales of wine at LeRae's for the period January 2 to January 9. LeRae's manager can use that information to forecast future sales. For example, if LeRae's manager forecasts that 200 guests who purchase wine by the glass will be served in the future, 20 of those guests will likely select the house Merlot (200 estimated guests × 10.0% popularity index = 20 orders predicted to be sold).

If guests remain fairly consistent in their buying habits, managers can have a good idea of the total demand for their wine-by-the-glass products and will be better able to determine the amount of each wine type that must be on hand and available for sale on any given day.

FORECASTING SPIRIT SALES

Like beer and wine, the number of guests who order a **mixed drink** (a drink made with a spirit) can be recorded and analyzed. Unlike beer and wine, however, using that information to estimate the exact drink guests will request can be difficult. To illustrate, assume that two guests order bourbon and soda; one guest specifies Jack Daniel's brand of bourbon, whereas the other one prefers the lower-cost Heaven Hill brand. From the guests' point of view, they both ordered a bourbon and soda mixed drink. From the operator's point of view, two distinct items were selected.

FIGURE 3.4 LeRae's Bar Wine-by-the-Glass Sales

Beverage Sales		Date: 1/2–1/9
Product: Wine by the Glass	Number Sold	Popularity Index (%)
House Chardonnay	30	18.75
House Merlot	16	10.00
House Zinfandel	62	38.75
House Cabernet	52	32.50
Total	160	100.00

To make matters even more complicated, assume a third guest arrives and orders bourbon and soda without preference as to the type of bourbon used. This can result in a third type of bourbon and soda drink being served. Obviously, a different method of sales forecasting is necessary in a situation such as this, and, in fact, several are available.

One method requires that the bar operator view guest selection not in terms of the drink requested, such as bourbon and soda, gin and tonic, Sea Breeze, and the like, but rather in terms of the particular spirit product that forms the base of the drink. To illustrate, consider a table of four guests who order the following:

1. Kahlúa on the rocks
2. Kahlúa and coffee
3. Kahlúa and cream
4. Kahlúa and Coke

Each guest could be considered as having ordered the same item: a drink in which Kahlúa, a coffee-flavored liqueur from Mexico, forms the base of the drink. The purpose of this method is, of course, to simplify the process of recording guests' preferences and to aid in reordering needed spirit products.

An alternative method of tracking drink sales is to consider spirit types such as the following:

- Bourbon
- Scotch
- Vodka
- Gin
- Rum
- Tequila
- Coffee liqueur

When utilizing this approach, managers are recognizing that there are literally hundreds of different spirit drinks that can be made. These managers are more interested, for example, in knowing the total number of coffee liqueur drinks served than in knowing the number of coffee liqueur drinks served using Kahlúa. Figure 3.5 illustrates spirit sales and popularity index results at LeRae's utilizing such a system.

As the amount of time and effort required to track specific drink sales increases, so does accuracy. Each operator must determine the level of control appropriate for his or her own operation. This is because there must be a reasonable relationship between the time, money, and effort required to maintain a control system and its cost effectiveness.

FIGURE 3.5 LeRae's Bar Spirit Sales

Beverage Sales		Date: 1/2–1/8
Product: Spirits	**Number Sold**	**Popularity Index (%)**
Scotch	210	13.1
Bourbon	175	10.9
Vodka	580	36.3
Gin	375	23.4
Coffee liqueur	175	10.9
Tequila	85	5.3
Total	1,600	100.0

FUN ON THE WEB!

Increasingly, managers use **artificial intelligence (AI)** software and the mathematical manipulation of large sets of data (big data) to improve their forecasting ability. Essentially, AI is the ability of computers to understand the natural order of things (such as historical customer buying behavior) and then use that information to accomplish tasks (such as sales forecasting) that had previously been the job of managers.

To monitor advances in this increasingly important management tool, enter the words "artificial intelligence" and "POS sales forecasting" in your favorite search engine and view the results.

IMPORTANCE OF STANDARDIZED RECIPES

Although it is the menu that determines what is to be sold and at what price, the **standardized recipe** controls both the quantity and the quality of what a kitchen or bar will produce. A standardized recipe details the procedures to be used in preparing and serving each of your food or beverage items. A standardized recipe ensures that each time guests order a menu item they receive exactly what you intended them to receive.

A high-quality standardized recipe contains the following information:

1. Menu item name
2. Total recipe yield (number of portions produced)
3. Portion size
4. Ingredient list
5. Preparation/method section
6. Cooking time and temperature
7. Special instructions, if necessary
8. Recipe cost (optional)[1]

Figure 3.6 shows a sample standardized recipe for roast chicken. If this standardized recipe represents the quality and quantity management wishes its guests to have and if it is followed carefully each time, then guests will always receive the quality and value management intended.

Despite their tremendous advantages, some managers resist taking the time to develop and implement standardized recipes. The excuses they use are many, but the following list contains common arguments *against* using standardized recipes:

1. They take too long to use.
2. My employees don't need recipes; they know how we do things here.
3. My chef refuses to reveal his or her secrets.
4. They take too long to write up.
5. We tried them but lost some, so we stopped using them.
6. They are too hard to read, or many of my employees cannot read English.

Of the preceding arguments, only the last one, an inability to read English, has any validity. Even in that case, the effective operator should have recipes printed in the language of their production employees.

[1]This information is optional. If the recipe cost is not included in the standardized recipe, a standardized cost sheet must be developed for each recipe item (see Chapter 5).

FIGURE 3.6 Sample Standardized Recipe

Roast Chicken

Special Instructions: Serve with Crabapple Garnish (see Crabapple Garnish Standardized Recipe)

Recipe I.D: Entree-44

Serve on 10-in. plate.

Recipe Yield: 48 portions

Portion Size: ¼ chicken

Portion Cost: See cost sheet

Ingredients	Amount	Method
Chicken quarters (twelve 3–3½-lb. chickens)	48 ea.	Step 1. Wash chicken; check for pinfeathers; tray on 24 in. × 20 in. baking sheet.
Butter (melted)	1 lb. 4 oz.	Step 2. Clarify butter; brush liberally on chicken quarters; combine all seasonings; mix well; sprinkle all over chicken quarters.
Salt	¼ C	
Pepper	2 T	
Paprika	3 T	
Poultry seasoning	2 t	
Ginger	1½ t	
Garlic powder	1 T	Step 3. Roast in oven at 325°F for 1½ to 2 hours, or to an internal temperature of at least 165°F (75°C).

Standardized recipes have far more advantages than disadvantages. The best managers create and then enforce the use of standardized recipes because of the following reasons:

1. Accurate purchasing is impossible without the existence and use of standardized recipes.
2. Dietary concerns require some foodservice operators to know the exact ingredients and the correct amount of nutrients in each serving of a menu item.
3. Accuracy in menu laws requires that foodservice operators be able to tell guests about the type and amount of ingredients in their recipes.
4. The responsible service of alcohol is impossible without adhering to appropriate portion sizes in standardized drink recipes.
5. Accurate recipe costing and menu pricing is impossible without standardized recipes.
6. Matching food and beverages used to unit sales is impossible to do without standardized recipes.
7. New employees can be trained faster and better with standardized recipes.
8. The computerization of a foodservice operation is impossible unless the elements of standardized recipes are in place; thus, the advantages of advanced technological tools available to the operation are restricted or even eliminated if standardized recipes are not used.

Standardized recipes are essential to any serious effort to produce consistent, high-quality food and beverage products at a known cost.

Any recipe can be standardized. The process can sometimes be complicated, however, especially in the areas of baking and sauce production. To create a standardized recipe, it is always best to begin with a recipe of proven quality. For example, you may have a recipe designed to produce 10 servings, but you want to expand it to serve 100 people. In cases like this, it may not be possible to simply multiply the amount of each ingredient by 10. A great deal has been written regarding various techniques used to expand recipes. Computer software designed for that purpose is also readily available to managers. As a general rule, however, any menu item that can be produced in quantity can be standardized in recipe form.

FUN ON THE WEB!

A variety of good companies offer software programs that ease the process of standardized recipe development and measurement conversion. To examine such products, type "standardized recipe development software" in your favorite search engine and review the results.

ADJUSTING STANDARDIZED RECIPES

When adjusting standardized recipes to produce greater, or fewer, numbers of portions, it is important that recipe modifications be made properly. For example, weighing ingredients on a pound or an ounce scale is the most accurate method of measuring many ingredients. It is also important to note that the food item to be measured must be **recipe-ready**. That is, the item must be cleaned, trimmed, cooked, and generally made completely ready to be added to the recipe.

For liquid items, measurement of volume (e.g., ounce, cup, quart, or gallon) is usually preferred. Some operators, and many bakers, however, prefer to weigh all ingredients, even liquids, for improved accuracy.

When adjusting recipes for proper quantity (number of servings to be produced), two general methods are typically used:

1. Recipe conversion factor
2. Baker's math

RECIPE CONVERSION FACTOR

When using the factor method to adjust a recipe's yield, you first use the following formula to arrive at the proper **recipe conversion factor (RCF)**:

$$\frac{\text{Desired yield}}{\text{Current yield}} = \text{RCF}$$

If, for example, one of our current standardized recipes makes 50 portions and the number of portions we wish to make is 125, the RCF formula would be as follows:

$$\frac{\text{125 Desired portions}}{\text{50 Current portions}} = 2.5 \text{ RCF}$$

Thus, 2.5 would be the RCF. To produce 125 portions, we would multiply the amount of each ingredient in the original (current) recipe by 2.5 to arrive at the required amount of that ingredient.

Figure 3.7 illustrates the use of this method for a simple three-ingredient recipe.

FIGURE 3.7 Recipe Conversion Factor Method

Ingredient	Original Amount	Recipe Conversion Factor	New Amount
A	4 lb.	2.5	10 lb.
B	1 qt.	2.5	2½ qt.
C	1½ T	2.5	3¾ T

HERE'S HOW IT'S DONE 3.1

In some cases, a professionally produced standardized recipe will already include the ingredient amounts needed to make a number of different quantities. For example, a standardized recipe may indicate the ingredients needed to produce 25, 50, 75, and 100 portions of the recipe. In other cases, however, a standardized recipe must be adjusted from its current yield to produce the exact number of portions you need. Because that is true, you must know how to properly increase and decrease the amount produced by a standardized recipe.

It's easy to do.

Increasing Yields in Standardized Recipes

To illustrate the process used to *increase* the yield of standardized recipes, assume a recipe yields 80 portions. But you need to make 120 portions. Increasing the recipe yield from 80 portions to 120 portions is a two-step process.

Step 1: Calculate the RCF using the formula:

$$\frac{\text{Desired yield}}{\text{Current yield}} = \text{RCF}$$

In this example, it is:

$$\frac{120}{80} = 1.5$$

Step 2: Multiply each ingredient amount times the RCF. For example, if the original recipe required 12 eggs and the RCF is 1.5, the number of eggs needed would be 18 eggs.

In this example, the calculation is:

$$12 \text{ eggs} \times 1.5 \text{ RCF} = 18 \text{ eggs}$$

Decreasing Yields in Standardized Recipes

Decreasing the yield in a standardized recipe is also easy. To illustrate, assume a recipe yields 80 portions. But you need to make 40 portions. To decrease the recipe yield from 80 portions to 40 portions, you use the same two-step process.

Step 1: Calculate the RCF using the following formula:

$$\frac{\text{Desired yield}}{\text{Current yield}} = \text{RCF}$$

In this example, it is:

$$\frac{40}{80} = 0.5$$

Step 2: Multiply each ingredient amount times the RCF. For example, if the original recipe required 12 eggs and the RCF is .5, the number of eggs needed would be 6 eggs.

In this example, the calculation is

$$12 \text{ eggs} \times 0.5 \text{ RCF} = 6 \text{ eggs}$$

Here's a tip to help you remember how recipes are converted for yield.

If a recipe's yield is being *increased*, the RCF will always be *greater than 1.0*. If a recipe's yield is being *decreased* the RCF will always be *less than 1.0*.

BAKER'S MATH

Professional foodservice operators calculate recipe conversion factors (RCFs) for use in increasing and decreasing the size of their standardized recipes. Professional bakers accomplish the same task when they use specialized "baker's math" (sometimes referred to as "baker's percentage") ratios to adjust the size of their standardized recipes (often called "baker's formulas").

When using baker's math, bakers apply the following formula:

$$\frac{\text{Ingredient weight}}{\text{Flour weight}} \times 100\% = \text{Ingredient percentage}$$

Essentially, when using baker's math, a baker calculates the total amount of flour called for in a baking formula. Each formula ingredient's weight is then divided by the weight of the flour to determine that ingredient's percentage of the flour's weight.

For example, if a bread formula makes 8 loaves, and it calls for 5 pounds of flour and 1 pound of sugar, the sugar's ingredient percentage would be calculated as

$$\frac{\text{1 pound sugar}}{\text{5 pounds flour}} \times 100\% = 20\% \text{ sugar}$$

When the formula's sugar's percentage is known, it becomes easy for the baker to adjust the formula's yield for the desired number of loaves.

For example, if the baker wishes to make 12 loaves, the amount of flour required would be 7.5 pounds (5 pounds flour × (12 loaves desired yield/8 loaves current yield) = 5 × 12/8 = 7.5 pounds flour) and the amount of sugar required would be calculated as:

$$\text{7.5 pounds flour} \times .20 \text{ sugar} = \text{1.5 pounds of sugar required}$$

It is important to note that, because the weights of the various ingredients in a baker's formula are calculated as a percentage of the flour's weight, the sum of the formula's percentages will always exceed 100%.

The proper conversion of weights and measurements is important in recipe expansion or reduction. The judgment of the recipe writer is also critical because factors such as cooking time, cooking temperature, and even cooking utensil selection may vary as recipe sizes are increased or decreased. In addition, some recipe ingredients, such as spices or flavorings, may not respond well to mathematical conversions. In the final analysis, it is your assessment of product taste that should ultimately determine proper ingredient ratios in standardized recipes. But in all cases, every recipe in use in your operation should be standardized.

Cost Control Around the World

Some managers find it difficult to adjust recipes when they include measurements such as teaspoons, tablespoons, pints, and the like. While the general US population has been somewhat slow to adopt the international metric system for everyday use, those foodservice managers working internationally and increasingly those foodservice managers working in the United States often encounter standardized recipes developed using metric measurements rather than the Imperial (US) measurement system. When metric measurements are used standardized recipe adjustment is easier, even though metric amounts are often rounded for ease of use.

Because that is true, it is a good idea for all managers to memorize (or keep on file!) a table, such as the one below, that helps them convert from metric volume and weight measurements to those volume and weight measurements commonly used to develop standardized recipes in the United States.

Volume Measures		
United States	**Metric**	**Often Rounded To**
1.05 quarts	1 liter	
Gallon	3.79 liters	3.8 liters
Quart	.95 liters	
Pint	473 milliliters	500 milliliters
Cup (C)	237 milliliters	250 milliliters
Tablespoon (T)	14.8 milliliters	15 milliliters
Teaspoon (t)	4.9 milliliters	5 milliliters
Fluid ounce	28.35 milliliters	30 milliliters

Weight Measures		
United States	**Metric**	**Often Rounded To**
2.2 pounds	1 kilogram	
1 pound	453.6 grams	450 or 500 grams
¾ pound	340.2 grams	350 or 375 grams
½ pound	226.8 grams	250 grams
¼ pound	113.4 grams	120 or 125 grams
1 ounce	28.4 grams	30 grams

Note: 1 liter (L) = 1000 milliliters (ml)
1 kilogram (K) = 1000 grams (g)
1 pound (lb.) = 16 ounces (oz.)

It is important to recognize that managers only rarely must convert recipes from US measurements to their metric equivalents. This is so because standardized recipes will typically be written in only one (or both) of the systems. Thus, the need to do actual measurement conversions from one system to the other is quite rare.

Temperature Conversions

To convert Celsius to Fahrenheit temperatures, the formula is

$$(\text{Celsius temperature} \times 1.8) + 32 = \text{Fahrenheit temperature}$$

Example: Convert 100 degrees Celsius to Fahrenheit.

$$(100 \text{ degrees celsius} \times 1.8) + 32 = 212 \text{ degrees Fahrenheit}$$

FUN ON THE WEB!

If the need does arise, a number of websites provide precise metric to US measurement conversion tools. To visit such sites, enter "metric conversion calculator" in your favorite search engine and view the results.

PURCHASING FOOD

Menu item forecasts and the use of standardized recipes allow managers to know exactly what must be purchased to ensure their operations will have the available menu items their guests will want to buy. Entire books have been written about foodservice purchasing systems; however, proper purchasing is essentially a matter of determining three key pieces of information:

1. What should be purchased?
2. What is the best price to pay?
3. How can a steady supply be ensured?

WHAT SHOULD BE PURCHASED?

The question of *what* should be purchased is answered by the use of a product specification or "spec" for each needed item. A **product specification (spec)** is simply a detailed description of a recipe ingredient or menu item. A spec is a way for you to communicate with a vendor or supplier, in a very precise way, so that your operation receives the exact item you have requested, every time you request it.

It is very important to note that a product specification determines neither the best product to buy nor the product that costs the least. Rather, it is the product that you have determined to be the *most appropriate* product for its intended use in terms of both quality and cost.

A product specification that is not detailed enough can be a problem because the needed level of item quality may not be delivered. However, if your product specifications are too tight (that is, overly and unnecessarily specific), too few vendors may be able to supply the product you need, resulting in excessively high costs you will have to pay to receive that item.

Figure 3.8 is an example of a professionally prepared product specification for bacon used in a foodservice operation that produces bacon, lettuce, and tomato (BLT) sandwiches.

A properly prepared product specification will include the following information:

1. Product name or specification number
2. Pricing unit
3. Standard or grade
4. Weight range/size
5. Processing and/or packaging
6. Container size
7. Intended use
8. Other information such as product yield

PRODUCT NAME

This may seem self-explanatory, but, in reality, it is not. For example, a "mango" is a tropical fruit to those in the Southwestern United States, but the same term may mean a "bell pepper" to some in the Midwest part of the country. Further,

FIGURE 3.8 Sample Product Specification

Product Name:	Bacon, Sliced	Spec #: 117
Pricing Unit:	lb.	
Standard/Grade:	Select No. 1	
	Moderately thick slice	
	Oscar Mayer item 2040 or equal	
Weight Range:	14–16 slices per lb.	
Packaging:	2/10 lb. Cryovac packed	
Container Size:	Not to exceed 20 lb.	
Intended Use:	Bacon, lettuce, and tomato sandwiches	
Other Information:	Flat packed on oven-proofed paper, never frozen	
Product Yield:	60% yield	

bell peppers do not just come in green, their most common color, but can also be purchased in yellow, orange, and red. Thus, the product name used on a specification must be detailed enough to clearly and precisely identify the item you wish to buy.

When developing product specifications, you may find it helpful to assign a number, as well as a name, to the item you wish to buy. This can be useful when, for example, many forms of the same ingredient or menu item may be purchased.

To illustrate, a deli restaurant may use 10 to 20 different types of bread, depending on the intended use of the bread. In this example, each type of bread may be given a specific name and number. The same may be true with a number of other items, such as cheese, which may come in a variety of forms (e.g., block, sliced, shredded) and a variety of types (Colby, cheddar, Swiss, and others). Note that, in Figure 3.8, bacon is given a name and a specification number, indicating that more than one form of bacon is purchased by this operation.

PRICING UNIT

A product's pricing unit may be pounds, quarts, gallons, cases, or any other commonly used measurement unit in which the item is sold. Many pricing units, such as pounds and quarts, are easily understood. Some other pricing units are less well known. Parsley, for example, is typically sold in the United States by the bunch and grapes are sold by the lug.

Unless managers are very familiar with the pricing unit terminology used in their own geographic areas, they may not be able to buy products in an effective way. Figure 3.9 lists some of the more common pricing units for produce in the United States. You should insist that all your vendors supply you with definitions of each pricing unit upon which they base their selling prices.

STANDARD OR GRADE

A large number of food items are sold with varying degrees of quality. Because that is true, the US Department of Agriculture (USDA), the Bureau of Fisheries, and the Food and Drug Administration have developed standards, or grades, for many food items. In addition, grading programs are in place for many commonly used foodservice items. Trade groups such as the National Association of Meat Purveyors publish item descriptions for many of these products. Consumers are also aware of many of these distinctions. One example is the USDA's grades for beef used in most foodservice operations. In descending order of quality, they are listed as follows:

- Prime
- Choice
- Select
- Standard
- Commercial
- Utility

In some cases, managers may prefer to develop a specification identifying a particular product's brand name (e.g., Minor's beef base or A-1 Steak Sauce) or a specific point of origin (e.g., Maine lobster or Idaho potatoes) as an alternative or supplement to a product standard or grade.

WEIGHT RANGE/SIZE

Weight range or size is important when purchasing some meats, as well as some fish, poultry, fruits, and vegetables. In our standardized recipe example of roast chicken, the chicken quarters were to have come from chickens weighing in the

FIGURE 3.9 Selected Produce Container Net Weights

Items Purchased	Container	Approximate Net Weight (lb.)
Apples	Cartons, tray pack	40–45
Asparagus	Pyramid crates, loose pack	32
Beets, bunched	½ crate, 2 dozen bunches	36–40
Cabbage, green	Flat crates (1¾ bushels)	50–60
Cantaloupe	½ wirebound crate	38–41
Corn, sweet	Cartons, packed 5 dozen ears	50
Cucumbers, field grown	Bushel cartons	47–55
Grapefruit, FL	⅘-bushel cartons and wirebound crates	42½
Grapes, table	Lugs and cartons, plain pack	23–24
Lettuce, loose leaf	⅘-bushel crates	8–10
Limes	Pony cartons	10
Onions, green	⅘-bushel crates (36 bunches)	11
Oranges, FL	⅘-bushel cartons	45
Parsley	Cartons, wax treated, 5 dozen bunches	21
Peaches	2-layer cartons and lugs, tray pack	22
Shallots	Bags	5
Squash	1-layer flats, place pack	16
Strawberries, CA	12 one-pint trays	11–14
Tangerines	⅘-bushel cartons	47½
Tomatoes, pink and ripe	3-layer lugs and cartons, place pack	24–33

3- to 3½-pound range. This will make them very different in cost and cooking time than if they came from chickens in the 4- to 4½-pound weight range.

In the case of products requiring specific trim or maximum fat covering, that should be designated also, for example a 10-ounce strip steak, with maximum tail of 1 inch, and fat covering of ½ inch. Four-ounce hamburger patties, 16-ounce T-bone steaks, and ¼-pound hot dogs are examples of the type of items that require an exact size, rather than a weight range.

Count, in the foodservice industry, is a term that is often used to designate product size. For example, 16- to 20-count shrimp refers to the fact that, for this size shrimp, 16 to 20 of the individual shrimp would be required to equal 1 pound. In a like manner, 30- to 40-count shrimp means that it takes 30 to 40 of this size shrimp to equal 1 pound. In addition to sea foods such as shrimp and scallops, many fruits and vegetables are sold by count. In general, the larger the count, the smaller the size of the individual food items.

PROCESSING AND/OR PACKAGING

Processing and packaging refer to a product's state when you buy it. Apples, for example, may be purchased fresh, dried, canned, or frozen. Each form will carry a price appropriate for its processed or packaged state.

Food can come packed in a large number of forms and styles, including slab packed, layered cell packed, fiberboard divided, shrink-wrap packed, individually wrapped, and bulk packed. Although it is beyond the scope of this text to detail all of the many varieties of food processing and packing styles available to foodservice managers, it is important for you to know about them. Your vendors will be pleased to explain to you all the types of item processing and packaging alternatives offered for the products they sell.

Green and Growing!

"Farm to Fork" is a term of importance to foodservice managers. It refers to the path food follows from those who grow or raise it to those who will prepare and serve it. Ideally, this path is short, to maximize freshness, minimize health risks, and be environmentally friendly. For that reason, many foodservice operators prefer to seek out and buy locally grown foods whenever possible.

In addition to their freshness, locally grown foods are most often good for the environment because less storing, shipping, and packaging results in less energy to transport the items and less solid waste from excessive packing materials. Reduced transportation and packaging costs can also translate into lower prices charged to foodservice operators.

Locally grown and organic products also help support the same local economies from which many foodservice operations draw their customers. As a result, buying locally is an option that all foodservice managers should thoroughly examine for its benefits to their communities and the environment, as well as the health of their customers and their own bottom lines!

CONTAINER SIZE

Container size refers to the can size, number of cans per case, or weight of the container in which the product is delivered. Most operators know that a 50-pound bag of flour should contain 50 pounds of flour. Some may not be sure, however, what the appropriate weight for a "lug" of tomatoes would be (see Figure 3.9).

INTENDED USE

Different types of the same item are often used in the same foodservice operation but in a variety of ways. Consider, for example, the operator who uses strawberries. Perfectly colored and shaped large berries are best for chocolate-dipped strawberries served on a buffet table. Less-than-perfect berries, however, may cost less and be a perfectly acceptable form for sliced strawberries used to top strawberry shortcake. Frozen berries may make a good choice for a baked strawberry pie and may be of even lower cost.

Breads, dairy, apples, and other fruits are additional examples of foods that come in a variety of forms. It is important to recognize that the best form of a food product is not necessarily the most expensive form. Rather, it is the form that is best for a product's intended use.

OTHER INFORMATION SUCH AS PRODUCT YIELD

Additional information may be included in a specification if it helps the vendor understand exactly what you have in mind when your order is placed. An example is product yield. **Product yield** is the amount of product that you will have remaining after subtracting losses resulting from cleaning, trimming, cooking, and portioning. Knowing an item's product yield will help you and your vendor determine how much product you must purchase initially to have the desired product quantity you need after accounting for any product lost in an item's preparation. You will learn how to accurately calculate product yields in Chapter 5.

WHAT IS THE BEST PRICE TO PAY?

After purchase specifications have been developed and quantities to be purchased are known, your next step is to determine how to buy these items at the best price. Some would say that determining the best price should be a simple matter of finding the vendor with the lowest-cost product and placing an order with that vendor. In fact, that is almost always a sure sign of a manager who lacks understanding of the way business, and vendors, operate.

The best price, in fact, is more accurately stated as the lowest price that meets the product specifications of the operation and achieves the long-term goals of both the foodservice operation and its vendors.

When you have a choice of vendors, each supplying the same product (your operation's product specifications), it is possible to engage in comparison shopping. The management tool used to do this is called the **bid sheet,** or pricing sheet (see Figure 3.10).

The bid sheet includes vendor information, buyer information, item description, purchase unit, offered price, and date. It also includes the dates that the bid prices will be fixed, which means the dates that the supplier agrees to keep the bid price in effect.

Using the example shown in Figure 3.10, the bid sheet could be filled out using information found in your online account established by each of your meat vendors'

FIGURE 3.10 Bid Sheet

Vendor: ____________________ **Buyer:** ____________________

Vendor's Address: ____________________ **Buyer's Address:** ____________________

Vendor's Telephone #: ____________________ **Buyer's Telephone #:** ____________________

Vendor's E-mail: ____________________ **Buyer's E-mail:** ____________________

Item Description	Pricing Unit	Bid Price
109 Rib, 19–22 lb., Choice	Pound	
110 Rib, 16–19 lb., Choice	Pound	
112A Rib Eye lip on 9–11 lb., Choice	Pound	
120 Brisket, 10–12 lb. deckle off, Choice	Pound	
164 Steamship Round, 60 lb., square bottom, Choice	Pound	
168 Inside Round, 17–20 lb., Choice	Pound	
184 Top Butt, 9–11 lb., Choice	Pound	
189A Tender, 5 lb. avg.	Pound	
193 Flank Steak, 2 lb. avg., Choice	Pound	
1184b Top Butt Steak, 8 oz., Choice	Pound	
1190a Tenderloin Steak, 8 oz., Choice	Pound	
109 Rib, 19–22 lb., Certified Angus Beef, Choice	Pound	
123 Short Ribs, 10 lb.	Pound	
180 Strip, 10–12 lb., Choice	Pound	

Bid Prices Fixed From: ____________________ **To:** ____________________

Salesperson Signature: ____________________ **Date:** ____________________

websites. After you have identified the vendors' selling prices, you can compare those bids on a **price comparison sheet**, such as the one shown in Figure 3.11.

A price comparison sheet typically has a place to list the category being bid, namely, produce, dairy, and meat products. It also identifies the name of the vendors authorized to bid, the bid date, item descriptions, units of purchase, best bid price, best company quote, and last price paid. This information may then be used to select a vendor based on the prices charged for needed items. When this process is completed, a purchase order or PO (addressed later in this chapter) can be submitted for each vendor selected.

Bid sheets and price comparison sheets may be used to determine the specific vendor who can supply the *lowest* price, but they do not give enough information to determine the *best* price. This makes sense when you realize that your own guests do not necessarily go to the lowest-priced restaurants for all of their meals. If they did, there would be no hope of success for the operator who tried to provide a better food product, in a better environment, and with better service.

In most cases, any product can be sold a little cheaper if quality is allowed to vary. Even with the use of product specifications, vendor dependability, ease of order placement, ability to access important account information via the Internet, the quality of a vendor's service, accuracy in delivery, and payment terms can all be determining factors when attempting to determine the best price available from the best supplier.

In some cases, specific item vendors to be used by a foodservice operator have been determined in advance. This is often true in a large corporate organization or in a franchise situation. Contracts to provide goods may be established by the central purchasing department of these organizations. When that happens, the designated vendor may have a national or a long-term contract to supply, at a predetermined price, items that meet the operation's specifications.

FIGURE 3.11 Price Comparison Sheet

Vendors

A. Bill's Produce

B. Davis Foods

C. Ready Boy

Category: Produce

Date Bid: 1/2/20xx

Price Comparison Sheet							
Item Description	Unit	A	B	C	Best Bid ($)	Best Company	Last Price Paid
Lettuce, Iceberg, 24 ct.	Case	$19.70	$20.20	$19.95	$19.70	A	$19.50
Lettuce, Red Leaf, 24 ct.	Case	$20.50	$20.25	$20.75	$20.25	B	$20.10
Mushroom, Button, 10 lb.	Bag	$25.50	$25.75	$26.10	$25.50	A	$25.75
Mushroom, Portobello, 5 lb.	Bag	$29.00	$29.80	$28.90	$28.90	C	$29.00
Onion, White, Medium, 50 lb.	Bag	$30.25	$30.00	$30.50	$30.00	B	$29.80
Pepper, Green Bell, 85 ct.	Case	$38.90	$39.50	$40.10	$38.90	A	$37.00
Potato, Russet Idaho, 60 ct.	Case	$22.75	$22.50	$22.20	$32.20	C	$22.85

Reviewed By: B. Allesandro

HOW CAN A STEADY SUPPLY BE ENSURED?

Unfortunately, too little has been written in the field of foodservice about managing costs through *cooperation* with vendors. Your food and beverages salespersons can be some of your most important allies in controlling costs. Operators who determine their supplier only on the basis of lowest prices offered will most often find that they receive only the product they have purchased, whereas their competitors may be buying more than just food! Or, as one food salesperson said when asked why he should be selected as the primary food vendor for a business, "With my products, you also get my knowledge!"

Just as good foodservice operators know that their guests appreciate high-quality personal service levels, the best vendors also know that operators appreciate high-quality customer service. In many foodservice operations, suppliers can be of immense value in ensuring a steady supply of quality products at a fair price if managers remember the following points:

SUPPLIERS HAVE MANY PRICES, NOT JUST ONE

Unlike the restaurant business, which generally charges the same price to all who come in the door, suppliers have a variety of prices based on the specific characteristics of those customers to whom they are selling. The price you will pay will most often be based upon a variety of factors, including the total amount of food and other products you buy from the vendor and the promptness with which you pay your bills.

SUPPLIERS REWARD VOLUME CUSTOMERS

It is simply in the best interest of a supplier to give a better price to a high-volume customer. The cost to deliver a $2,000 food order to a restaurant is not that much different from the cost to deliver a $200 order. Each of these orders would require one truck and one driver. Those operators who decide to concentrate their business in the hands of fewer suppliers will, as a general rule, pay lower prices.

CHERRY PICKERS ARE SERVICED LAST

Cherry pickers is the term used by vendors and suppliers to describe the customers who look at prices from multiple vendors and then buy only the items that each vendor has "on sale" or for the lowest price. If an operator buys only a vendor's "on sale" items, the vendor will usually respond by providing limited service. It is a natural reaction to the foodservice operator's failure to take into account varying service levels, long-term relationships, dependability, or any other vendor characteristic except cheapest price.

SLOW PAY MEANS HIGH PAY

In most cases, operators who are slow to pay their bills will find that vendors will charge them more for products than they charge other operators who pay their bills in a timelier manner.

ONE VENDOR VERSUS MANY VENDORS

Many operators are faced with the decision of whether to buy from one vendor or many vendors. In general, the more vendors used, the more time managers must spend in ordering, receiving, and paying vendor invoices. Many operators, however, fear that if they give all their business to only one supplier, their costs may rise because of a lack of competition. In reality, the likelihood of this occurring is extremely small. Just as foodservice operators are unlikely to take advantage of their best guests (and, in fact, would tend to offer additional services not available to the occasional guest) so, too, does the vendor make extra effort on behalf of the operator who does most of his or her buying from that vendor. In fact, it makes good business sense for the vendor to do so.

Using one or perhaps two primary vendors tends to bring the average delivery size up as well, and that should result in lower per-item prices. On the other hand, giving one vendor all of the operation's business can be dangerous and costly if the items to be purchased from that vendor vary widely in quality and price. Staples such as flour, sugar, spices, and other nonperishables are most often best purchased in bulk from one vendor. Orders for meats, produce, and some bakery products are most often best split among several vendors, perhaps with a primary and a secondary vendor in each category so that you have a second alternative should the need arise. If you are comparing vendor prices, assessing at least three different vendors is advisable so that you have an adequate choice of prices and services.

The Move to Off-Premise Dining

In most cases, foodservice operators are free to purchase any food and beverage items they wish to sell. But that freedom does not always extend to the manner in which takeaway foods are packaged for delivery to off-premise diners.

One of the most notable long-term consequences of the Covid-19 pandemic has been the increased reliance on takeout foods to maintain and increase an operator's sales levels, but it is increasingly recognized that the use of Styrofoam, plastics, cardboard, and other types of takeaway food containers have a significant negative impact on the environment, and on wildlife.

Because it is rarely reused, significant amounts of food packaging is buried in landfills or carried along by water and wind into the environment. Also, some types of packaging are not biodegradable, or do not degrade quickly. The chemicals from these types of takeaway boxes and containers, including any ink from the packaging labels, can latch onto soil and water.

As a result of these packing-related issues, as of 2025, 11 US states and two territories have passed statewide legislation to explicitly ban the use of polystyrene foam (Styrofoam) takeout containers despite their popularity with foodservice operators. Twelve states ban single-use plastic bags and eight states ban the use of plastic straws.

Fortunately, using eco-friendly takeaway food containers can help solve many waste-related problems and increasingly foodservice operators can choose from takeout containers made using eco-friendly plant-based materials, including soy or water-based inks and petroleum-free boxes. These types of sustainable food containers are fully compostable, making them more environmentally friendly and thus their usage will likely continue to be required in increasing numbers of cities and states.

PURCHASING BEVERAGES

Most foodservice operators select only one quality level for food products. Once the proper quality level has been determined and a product specification written, then only that quality of meat, lettuce, milk, bread, and so on is purchased. In the area of alcoholic beverage products, however, several levels of quality and packaging format are generally chosen. This ensures that a beverage product is available for those guests who wish to purchase the very best and a product is also offered for those guests who prefer to spend less. Thus, the beverage buyer is faced with deciding not only whether, for example, to carry wine on the menu but also how many different kinds of wine and their quality levels. The same process is necessary for spirits and, to a lesser degree, for beers.

WINE PURCHASING

Determining which wines to buy is a matter of selecting both the right product and the right packaging form for your operation's needs. To do so, you must first determine if you will sell wine by the following:

- Glass
- Carafe
- Split or half-bottle
- Bottle

Wines sold by the glass or carafe can be served from bottles opened specifically for that purpose or they can be drawn from specially boxed wine containers. Wines in a box typically are house wines sold to the beverage manager in multi-liter-sized boxes to ease handling, speed, and service and to reduce packaging costs. Bottled wines of many sizes and varieties can also be purchased. The most commonly purchased wine bottle sizes for use in foodservice are shown in Figure 3.12.

As a good manager, you will build a **wine list**, the term used to describe the menu of wine offerings selected for your own particular operation and guest expectations. In wine list development, several points must be kept in mind. First, you must seek to provide alternatives for the guest who wants the best, as well as the guest who prefers to spend less. Second, wines that either complement the food or, in the case of a bar, are popular with guests must be available. Consumers' tastes in wines change, and you must keep up with these changes. Subscribing to at least one wine publication in the hospitality field is a good way to do so. These publications regularly list the most popular wines, the year(s) of best product production (the **vintage**), and the most successful wine producers (the **vintners**).

Operators must avoid the temptation to offer too many wines on a wine list. Excess inventory and use of valuable storage space make this a poor idea. In addition, when selling wine by the glass, those items that sell poorly can lose quality and flavor rapidly.

To optimize wine sales, you should strive to make the purchase of wine by the bottle a pleasant, nonthreatening experience. Wait staff, who help in selling wine, should be knowledgeable but not pretentious. You can enlist the aid of your own wine supplier to help you train your staff in the effective marketing of wine, whether it is sold by the glass, carafe, or bottle.

It is also important to remember that, if you must purchase specific wines for cooking, these products will be secured from your beverage wholesaler rather than your grocery wholesaler.

FUN ON THE WEB!

There are many websites operators can visit to gain information about wines, wine production, and the making of wine lists. To identify such sites, enter "restaurant wines" as well as "restaurant wine lists" in your favorite search engine. Pay special attention to any online publications your search identifies.

SPIRITS PURCHASING

Spirits have an extremely long **shelf life**, the period of time they can be stored with no significant reduction in quality. Thus, you can make a mistake and

FIGURE 3.12 Wine Bottle Sizes

Bottle Size (Capacity)	Common Name	Description
0.100 L	Miniature (mini)	A single-serving bottle
0.187 L	Split	¼ a standard bottle
0.375 L	Half-bottle	½ a standard bottle
0.750 L	Bottle	A standard wine bottle
1.5 L	Magnum	Two bottles in one
3.0 L	Double Magnum	Four bottles in one

Note: 1 US quart = 0.946 L
1 US gallon = 3.785 L

purchase the wrong spirit product without disastrous results, but only if that product can be sold over a reasonable period of time. Guest preference will dictate the types of liquors that are appropriate for any given operation, but it is your responsibility to determine appropriate product quality levels in the beverage area.

Consumer preferences concerning alcoholic beverages can change rapidly. A wise beverage buyer keeps abreast of changing consumption trends by reading professional journals and staying active in his or her professional associations. These associations can be major providers of information related to changes in consumer buying behavior. Nowhere is this more important than in the area of spirits.

Spirits used in foodservice are purchased by the bottle. In the United States, as in most parts of the world, bottle sizes for spirits are standardized. The bottle sizes shown in Figure 3.13 represent those most commonly offered for sale to buyers in the hospitality industry.

In the hospitality industry, the 750 milliliters and the 1 liter are the two most commonly purchased spirit bottle sizes. In general, restaurant operators will buy spirits in two major categories: well liquors and call liquors.

WELL LIQUORS

Well liquors are those spirits that are served when the guest does not specify a particular brand name when ordering. The name stems from the concept of the "well" or the bottle holding area in the bar. Beverage buyers must choose their well liquors very carefully. Guests who order well liquors may be price conscious, but that does not mean they are not quality conscious also.

Managers who shop for well liquors considering only the lowest possible price as the main criterion for selection will find that guest reaction can be extremely negative. Alternatively, if exceptionally high-quality products are chosen as well items, liquor costs may be excessive unless an adequate selling price structure is maintained.

CALL LIQUORS

Call liquors are those spirits that guests request (call for) by name, such as Jack Daniel's, Kahlúa, and Chivas Regal. Extremely expensive call liquors are sometimes referred to as **premium liquors**.

You will generally charge a higher price for those drinks prepared with call or premium liquors. Guests understand this and, in fact, by specifying call liquors indicate their preference to pay the higher prices these high-quality products command.

FIGURE 3.13 Spirit Bottle Sizes and Capacities

Common Bottle Name	Metric Capacity	Fluid Ounce Capacity
Miniature	50 ml	1.7
Half-pint	200 ml	6.8
Pint	500 ml	16.9
Fifth	750 ml	25.4
Quart	1.0 L	33.8
Half-gallon	1.75 L	59.2

Note: 1 liter (L) equals 1000 milliliters (ml)

FUN ON THE WEB!

The Internet is a great source of information about cocktails and bartending. If you are 21 or over, you can also explore numerous informational sites developed by liquor companies. To learn more about spirits and spirit production, go to your favorite search engine and enter the words "liquor and spirits." From there, you will readily be launched into a world of "spirited" information!

BEER PURCHASING

Beer is the most highly perishable of alcoholic beverage products. The **pull date**, or expiration date, on these products can be as short as a few months or even a few weeks. Because of this, it is important that the beverage operator only stock those items that will sell relatively quickly. This generally means selecting both appropriate brands and popular packaging forms.

Beverage operators typically carry between 3 and 10 types of beer. Some operations, however, stock as many as 50 or more! Generally speaking, geographic location, clientele, ambience, and menu help determine the beer product that will be selected. Obviously, we would not expect to see the same beer products at Hunan Gardens Chinese Restaurant that we would at Three Pesos Mexican Restaurant.

Most foodservice operators find that one or two brands of light beer, two or three national domestic brands, one or two craft beers, and one or two quality import beers meet the great majority of their guests' demand. You must be very careful in this area not to stock excessive amounts of products that sell poorly.

Beer is typically sold to foodservice operators in cans, bottles, or kegs. Although each of these containers has its advantages and disadvantages, many foodservice operators with active beverage operations will select a variety of these packaging methods. It makes little sense to carry the same beer product in both bottles and cans; however, many operators choose to serve some brands of their bottled or canned beer in draft form as well.

Many beer drinkers prefer draft beer (beer from kegs) over bottled and canned beer, and the per-glass cost to you is lower with beer sold from a keg. Special equipment and serving techniques are, however, required if quality draft beer is to be sold to guests. Also, the shelf life of keg beer is the shortest of all packaging types, ranging from 30 to 45 days for an untapped keg, that is, one that has not yet been opened by the bartender, and even fewer days for a tapped (opened) keg. Also, full kegs can be difficult to handle because of their weight, and it can be hard to keep an exact count of the product served without special metering equipment. Despite these drawbacks, many operators serve draft beer. Draft beer is sold in a variety of keg and barrel sizes, as listed in Figure 3.14.

FIGURE 3.14 Draft Beer Containers

Size (US Gal)	Size (L)	No. of 12-oz. Drinks	No. of 16-oz. Drinks	No. of 20-oz. Drinks	Weight	Common Name
1.32	5	14	10	8	—	Mini-keg (single use/ recyclable)
5	18.9	53	40	32		Home keg
5.23	19.8	55	41	33		Sixth barrel
6.6	25	70	52	42		Half barrel (Europe)
7.75	29.3	82	62	49	90 lb.	Quarter barrel/pony keg
13.2	50	140	105	84		Import keg (standard European barrel)
15.5	58.7	165	124	99	140–170 lb.	Half barrel/full keg

FUN ON THE WEB!

Beers, and especially locally crafted beers, are increasingly popular with foodservice guests. To learn more about the beers brewed in your own area, enter the words "craft beer" and your state name into your favorite search engine. Also search the Web to find regularly produced online publications that are devoted to beers and information about how they are produced.

Green and Growing!

Eco-oriented drink programs present challenges and opportunities for restaurant and bar owners. Increasingly, customers ask for (or even demand!) organic beverages. Available organic spirits products include vodka, gin, tequila, scotch, and rum. Although more expensive than nonorganic versions of the same product, these items are increasingly available for purchase from mainline beverage distributors. Organic wines and beers are also becoming increasingly common.

Concerns about costs can plague green beverage initiatives because organic product costs are historically higher, and in the past, the number of interested customers has been smaller. Today, however, using seasonal and local produce and juices for drink production can help moderate every operation's costs while increasing quality. In addition, consumers are increasingly willing to pay the extra costs incurred by operations that choose to go green. The result? Delighted guests and increased profit margins.

PURCHASING ETHICS

Ethics have been defined as the choices of proper conduct made by individuals in their relationships with others. Ethics come into play in purchasing foodservice products because of the tendency for some suppliers to seek an unfair advantage over their competition by providing "personal" favors to buyers. These favors can range from small holiday gifts given in appreciation for another year's business to outright offers of cash bribes or kickbacks to the buyer in exchange for volume purchases.

Foodservice buyers often face ethical dilemmas related to gifts or favors from vendors. It is the job of foodservice managers to establish and communicate to their employees and to their suppliers the ethical standards of purchasing for their own operations. Some large foodservice organizations have formal standards, or codes, of ethical conduct for their buyers. When these are in place, they should, of course, be carefully reviewed and followed by all managers. If your organization does not have a formal set of ethical guidelines for buying, the following self-test can be helpful in determining whether a considered course of action with your suppliers is indeed ethical.

Ethical Guidelines

1. Is it legal?

   Any course of action that violates written law or company policy and procedure is wrong.
2. Does it hurt anyone?

   Are benefits that rightfully belong to the owner of the business accruing to the buyer? Discounts, rebates, and free products are the property of the business, not the buyer.
3. Am I being honest?

   Is the activity one that you can comfortably say reflects well on your integrity as a professional, or will the activity actually diminish your reputation with other suppliers?

4. Would I care if it happened to me?

   If you owned the business, would you be in favor of your buyer behaving in the manner you are considering? If you owned multiple units, would it be good for the business if all of your buyers followed the considered course of action?
5. Does it compromise my independence as a buyer?

   If your action negatively influences the way you perform as a buyer, then you shouldn't do it. For example, you accept a gift from a supplier and then the supplier provides you with poor service. Will you feel comfortable reprimanding the supplier even though he has given you a gift? If you would feel uncomfortable, then maybe you shouldn't accept the gift.
6. Would I publicize my action?

   A quick way to review the ethical merit of a situation is to consider whom you would tell about it. If you are comfortable telling your boss and your other suppliers about the considered course of action, it is likely ethical. If you would prefer that your actions go undetected, you are probably on shaky ethical ground. If you wouldn't want your action to be read aloud in a court of law (even if your action is legal), you probably shouldn't do it.

PURCHASE ORDERS

Regardless of the food, beverage, and takeaway packaging items to be purchased, managers should use a purchase order to request delivery of needed items. A **purchase order (PO)** is a detailed listing of products requested by a buyer. A PO may include a variety of product information but must always include the quantity ordered and the price quoted by the supplier. Figure 3.15 is an example of a simple yet effective PO.

When ordering products, you may be able to choose from a variety of ways to tell suppliers what you want to buy. These may include ordering via an Internet account, by telephone, or by placing orders in person. Regardless of your communication method, however, it is critical that it includes a written PO because the PO serves as the record of what you have decided to buy.

In most cases, the PO should be available in triplicate (three viewable copies). One copy is used by the receiving clerk in the receiving area. One copy is used by management for bookkeeping purposes. The original is submitted to the vendor.

When products are ordered online, the PO will be retained in the operation's online vendor account. In all cases, however, it is important to place all orders using a formal PO. If this is not done, the receiving clerk will have no record of what should be coming in on the delivery.

POs can be simple or complex, but should always contain space for the following information:

Purchase Order Information

1. Vendor information
2. Purchase order number
3. Date ordered
4. Delivery date
5. Name of person who placed order
6. Name of person who received order
7. Name of ordered item
8. Item specification #, if appropriate
9. Quantity ordered

FIGURE 3.15 Sample Purchase Order

Vendor: ______________________ **Purchase Order #:** ______________

Vendor's Address: ______________ **Delivery Date:** ______________

Vendor's Telephone #: ______________

Vendor's E-mail: ______________

Item Purchased	Spec #	Quantity Ordered	Quoted Price	Extended Price
1.				
2.				
3.				
4.				
5.				
6.				
7.				
8.				
9.				
10.				
11.				
12.				
13.				
14.				
15.				
16.				
Total				

Order Date: ______________ **Buyer's Account #** ______________

Ordered By: ______________ ______________

Received By: ______________ ______________

Delivery Instructions: ______________ ______________

10. Quoted price (per unit)
11. Extension price
12. Total price of order
13. Delivery instructions
14. Buyer's account #

The advantages of a written PO are many and include the written verification of the following:

1. Quoted price
2. Quantity ordered
3. Receipt of all goods ordered
4. Special instructions to the receiving clerk, as needed
5. Conformance to product specifications
6. Authorization to prepare vendor invoice for payment

Figure 3.16 shows a PO developed after a thorough inspection of a produce walk-in at a mid-sized hotel. The chef has used a sales forecast to determine the quantity of products needed for next Thursday's delivery. A check of the produce walk-in lets the chef know what is on hand and the quantity, if any, of each product that must be purchased. This information allows for the accurate preparation of the PO. In this case, the order is then placed using the vendor's website. The receiving clerk at the hotel is now prepared with the information necessary to effectively receive the products that have been ordered.

FIGURE 3.16 Purchase Order

Vendor: Scooter's Produce

Purchase Order #: 56

Vendor's Address: 123 Anywhere Street, City, State

Delivery Date: 1/18

Vendor's Telephone #: 1-800-999-0000

Vendor's E-mail: scootersproduce@isp.org

Item Purchased	Spec #	Quantity Ordered	Quoted Price	Extended Price
1. Bananas	81	30 lb.	$0.44 lb.	$13.20
2. Parsley	107	4 bunches	$1.80/bunch	$7.20
3. Oranges	101	3 cases	$41.50/case	$124.50
4. Lemons	35	6 cases	$39.20/case	$235.20
5. Cabbages	85	2 bags	$23.80/bag	$47.60
6.				
7.				
8.				
9.				
10.				
11.				
12.				
13.				
14.				
15.				
16.				
Total				$427.70

Order Date: 1/15

Buyer's Account # 5625

Ordered By: Joshua Jenn

Submitted on 1/15

Received By: ______

Transmitted by Joshua Jenn

Delivery Instructions: After 1:00 P.M.

FUN ON THE WEB!

Foodservice operators seeking to buy high-quality products have an important ally in the US Department of Agriculture (USDA), which provides the quality grade marks usually seen on beef, lamb, chicken, turkey, butter, and eggs. For many other products, such as fresh and processed fruits and vegetables, the grade mark is not always visible on the retail product. However, quality grades are widely used—even if they are not prominently displayed—as a "language" among food buyers and sellers.

The USDA's Agricultural Marketing Service (AMS) currently grades fresh fruits and vegetables, processed fruits and vegetables, poultry, eggs, livestock and meat, and dairy products. To learn about the specific characteristics of food items that result in their various quality grades, enter "grades" in the search bar at www.usda.gov and review the results.

RECEIVING FOOD AND BEVERAGE PRODUCTS

After a PO has been submitted to a supplier, a foodservice operator must ensure the operation is ready to accept (receive) the order. In a large foodservice operation receiving tasks may be performed by a designated receiving clerk. However, in most smaller operations the responsibilities of receiving fall upon the operation's owner, manager, or other designated individual.

In all cases, however, it is a good idea for an operation's manager to establish the purchasing and receiving functions so one individual places orders and a different individual is responsible for verifying delivery and acceptance of the orders. If this is not done, the potential for purchasing fraud or theft can be substantial.

In fact, foodservice **auditors** (those individuals or entities responsible for reviewing and evaluating proper operating procedures in a business) who discover purchasing and receiving-related fraud frequently find that a buyer ordered a product, signed for its acceptance, and authorized payment for it when, in fact, no product was ever delivered! In this case, the buyer could be receiving cash payments from the supplier without the business owner's knowledge. In other cases, a buyer may purchase items, accept delivery of the items, but then remove the items from inventory for their own personal use.

When purchasing duties are split among two or more individuals in the purchasing chain, these types of fraud are much less likely to happen. If it is not possible to have more than one person involved in the buying process, the work of buyers and receiving clerks must be carefully monitored by business owners to prevent fraud.

Even when no fraud is occurring, proper receiving methods must be used to ensure that only delivered products are paid for and that all delivered products have met pre-established specifications (specs). Finally, proper receiving records must be in place to ensure that the operation will pay only the prices that were agreed upon when POs were initially submitted. If significant errors occur in any of these procedures the negative effects on product costs can be substantial.

Proper receiving is important, but there is probably no area of the foodservice establishment more ignored than the physical area in which receiving takes place. This is unfortunate because this is the area where you ensure the quality and quantity of the products you have ordered were indeed delivered. The requirements for effective receiving include:

1. Proper location
2. Proper tools and equipment
3. Proper delivery schedules
4. Proper staff training

PROPER LOCATION

An operation's back door, which is most often used for receiving, is frequently no more than that, just an entrance to the kitchen. In fact, the receiving area must be adequate to handle the job of receiving, or product loss and inconsistency will result.

First, the receiving area must be large enough to allow for properly checking products delivered against both the **delivery invoice**, which is the seller's record of what is being delivered, and the PO, the buyer's record of what was originally ordered. In addition to the space required to count and weigh incoming items, accessibility to the equipment required to move products to their proper storage area and to dispose of excess packaging is important. A location near refrigerated storage areas is desirable for maintaining deliveries of refrigerated and frozen products at their optimal temperatures.

You should make sure the receiving area stays free of trash and clutter, which can make it too easy to hide delivered food items that may be taken home at the end of a dishonest employee's shift. It is important to remember that the delivery person can also be a potential thief. Although most suppliers are extremely careful to screen their delivery personnel for honesty, it is a fact that a delivery person has access to products and has a truck available to remove as well as deliver goods. For this reason, it is important that the receiving clerk work in an area that has a clear view of delivery personnel and their vehicles.

The receiving area should be kept extremely clean; you do not want to contaminate incoming food or provide a carrying vehicle for pests. Sometimes suppliers deliver goods that can harbor roach eggs or other insects. A clean receiving area makes it easier to both prevent and detect this type of problem. The area should be well lit and properly ventilated. Too little light may cause product defects to go unnoticed and excessive heat in the receiving area can quickly damage delivered goods, especially if they are refrigerated or frozen products.

Flooring in receiving areas should be light in color and of a type that is easily cleaned. In colder climates, it is important that the receiving area be warm enough to allow the receiving clerk to carefully inspect products. For example, if the outside temperature is below freezing, the outside loading dock is no place for an employee to conduct a thorough inspection of incoming products!

PROPER TOOLS AND EQUIPMENT

Although the tools and equipment needed for effective receiving vary by type and size of operation, some items are standard in any receiving operation. These include the following.

- *Scales:* Scales should be of two types: those accurate to the fraction of a pound (for large items) and those accurate to the fraction of an ounce (for smaller items and preportioned meats). Scales should be calibrated (adjusted) regularly to ensure their accuracy.
- *Wheeled equipment:* These items, whether hand trucks or carts, should be available so that goods can be moved quickly and efficiently to their proper storage areas.
- *Box cutters:* These tools, properly maintained and used, allow the receiving clerk to quickly remove excess packaging and thus accurately verify the quality of delivered products. Of course, care must be taken when using this tool, so proper training is essential to minimize any safety hazards.
- *Thermometers:* Foods must be delivered at their proper storage temperatures. You must establish the range of temperatures you deem acceptable for product delivery. For most operators, the temperatures are:

Acceptable Temperature Range for Delivered Products		
Item	°F	°C
Frozen foods	10°F or less	−12°C or less
Refrigerated foods	30°F to 45°F	−1°C to 7°C

- *Calculator:* Vendor calculations should always be checked, especially if the invoice has been prepared by hand. It is most helpful if the calculator used has a physical tape (as on an adding machine) that can be used by the receiving clerk when needed. The calculator should also be available in case the original invoice is either increased or decreased in the amount due because of incorrect vendor pricing or because items listed on the invoice were not delivered. In addition, invoice totals will change when all or a portion of the delivery is rejected because the items were of substandard quality.

PROPER DELIVERY SCHEDULES

In an ideal world, you would accept delivery of products only during the hours you choose. These times would likely be during your slow meal periods, when there would be plenty of time for a thorough checking of the products delivered. In fact, some operators are able to demand that deliveries be made only at certain times, say between 9:00 A.M. and 10:30 P.M. These are called **acceptance hours**. The operation may refuse to accept delivery of products at any other time. Some large operations prefer to establish times in which they will *not* accept deliveries, say between 11:00 A.M. and 1:00 P.M. These are called **refusal hours**. A busy lunchtime may make it inconvenient to accept deliveries at this time, and some operators will simply not take deliveries then. In both cases, however, the assumption is that the operator is either a large enough or a good enough customer to make demands such as these. If your operation does establish either acceptance hours or refusal hours, these should be enforced equally with all vendors.

In many operations, it is the supplier who will determine when goods are to be delivered. Although this may seem inconvenient (and often is), remember that all foodservice units would like to have their deliveries made during the slow periods—between their peak meal times. In many cases, it is simply not possible for the supplier to stop his or her trucks for several hours to wait for a good delivery time. In fact, in remote locations, some foodservice operators will be told only the day a delivery will be made, not a specific time of day!

The key to establishing a successful delivery schedule with suppliers is, quite simply, to communicate with them. Although every relationship between operator and supplier is somewhat different, both sides, working together, can generally come to an acceptable agreement regarding the timing of deliveries.

PROPER STAFF TRAINING

Receiving personnel should be properly trained. When they are well trained, they can readily verify the following key characteristics of delivered items:

1. Weight
2. Quantity
3. Quality
4. Price

WEIGHT

One of the most important items to verify when receiving food products is their weight. Doing so is not always easy. For example, a 14-pound package of ground beef will look exactly like a 15-pound package. There is no easy way to tell the difference without putting the product on the scale. Because that is true, receiving clerks should be required to weigh all meat, fish, and poultry delivered. The only exception to this rule would be unopened Cryovac (sealed) packages containing items such as hot dogs, bacon, and the like. In this situation, the entire case should be weighed to detect shortages in content.

Often, meat deliveries consist of several items, all of which are packaged together. When the receiving clerk or supplier is very busy, the temptation exists to weigh all of the products together. The following example shows why it is important to weigh each item separately, rather than the entire group of items as a whole.

Assume that you ordered 40 pounds of product from Bruno's Meats. The portion of the PO detailing the items you ordered is displayed in Figure 3.17.

When the Bruno's delivery person arrived, all three items were in one box, and the deliverer was in a hurry. He therefore suggested that your receiving clerk simply weigh the entire box. Your receiving clerk did just that and found that the contents weighed 40.5 pounds. Since the box itself weighed about ½ pound, she signed for the delivery. When she began to put the meat away, however, she weighed each item individually and found the information in Figure 3.18.

If you called the supplier to complain about the overcharge ($608.50 total order price per vendor price list – $532.75 actual value of delivered items = $75.75 overcharge), you would likely be told that 40 pounds of meat were ordered and 40 pounds were delivered, thus the misdelivery was simply a mistake caused by human error. It may well have been, but the lesson here is to always instruct your receiving personnel to weigh delivered items individually, even if your staff member is in a hurry.

When an item is ordered by weight, its delivery should be verified by weight. It is up to the operator to train receiving clerks to always verify that the operation is charged only for the product weight delivered. Excess packaging, ice, or water in the case of produce can all serve to increase the delivered weight. The effective receiving clerk must be aware of and be on guard against deceptive delivery practices.

QUANTITY

The proper counting of products is as important as proper weighing. Suppliers typically make more mistakes in not delivering products than they do in excessive

FIGURE 3.17 Ordered Items

Item Ordered	Unit Price	Total Ordered	Total Price
Ground beef	$4.25/lb.	10 lb.	$42.50
New York strip steak	$19.00/lb.	20 lb.	$380.00
Corned beef	$18.60/lb.	10 lb.	$186.00
Total		40 lb.	$608.50

FIGURE 3.18 Delivered Items

Item Ordered	Unit Price	Total Delivered	Actual Value
Ground beef	$4.25/lb.	15 lb.	$63.75
New York strip steak	$19.00/lb.	10 lb.	$190.00
Corned beef	$18.60/lb.	15 lb.	$279.00
Total		40 lb.	$532.75

delivery. Products delivered but not charged for cost the supplier money. Products not delivered but charged for cost you money.

If you order five cases of Scotch, then, of course, you want to receive and pay for five cases. This is important for two reasons. First, you only want to pay for products that have been delivered. Second, and just as important, if you have prepared your PO correctly, you truly need five cases of Scotch. If only three cases are delivered, you may not be able to prepare enough of the Scotch-based drinks your guests seek to order. If this means you will run out of an item or have to make a substitute, you may be forced to deal with unhappy guests.

Shorting is the term used in the industry to indicate that an ordered item has not been delivered as requested. When a vendor shorts the delivery of an item you ordered, that item may or may not appear on the delivery invoice. If it does not appear, it must be noted so that management knows that the item is missing, and appropriate reorder action can be taken.

If the undelivered item is, in fact, listed on the delivery invoice, the delivery driver should sign a **credit memo**. A credit memo indicates an adjustment to a delivery invoice must be made. Credit memos should also be prepared in triplicate (three copies). One copy is retained in the receiving area to be filed. One copy is retained by management for proper bill payment. The original is sent to the vendor.

Figure 3.19 is an example of a credit memo. Note that the credit memo has a place for the signature of a representative from your operation, as well as the vendor. It must be signed by both.

FIGURE 3.19 Credit Memo

CREDIT MEMO

Unit Name: ____________________

Vendor: ____________________ Delivery Date: ____________________

Invoice #: ____________________ Credit Memo #: ____________________

Correction					
Item	Quantity	Short	Refused	Price	Amount
Total					

Original Invoice Total: ____________________

Less: Credit Memo Total: ____________________

Adjusted Invoice Total: ____________________

Additional Information: ____________________

____

____

Vendor Representative: ____________________

Vendor Representative Telephone # or e-mail: ____________________

Operation Representative: ____________________

Operation Representative Telephone # or e-mail: ____________________

The credit memo is a formal way of notifying a vendor that an item listed on the original delivery invoice is missing, and, thus, the value of that item should be reduced from the invoice's total. If a supplier consistently shorts your operation, that supplier is suspect in terms of both honesty and lack of concern for your operation's long-term success.

The counting of boxes, cases, bags, and the like must be routine behavior for the receiving clerk. Counting items such as the number of lemons or oranges in a box should be done on a periodic basis, but the value of counting items such as these on a regular basis is questionable. Although an unscrupulous supplier might be able to remove one or two lemons from each box delivered, the time it would take for an employee to detect such behavior is often not worth the effort required to spot it. It is preferable to do a thorough item count of such items only periodically and to work with reputable vendors.

The direct delivery of products to a foodservice operator's storeroom or holding area is another area for concern. The delivery person may deliver some items, such as bread, milk, and carbonated beverage mixes, directly to the storage areas, thus bypassing the receiving clerk. This should not be allowed because, after such an activity, it may be impossible to verify the accurate quantity of items delivered.

If this process must be used, product dates on each item can help ensure that all products listed on the delivery invoice were received.

QUALITY

No area of your operation should be of greater concern to you than that of the appropriate quality of products delivered. If you take the time to develop product specifications but then accept delivery of products that do not match these specifications, you are simply wasting time and effort. Without product specifications, verification of quality is difficult because the receiving staff, and management itself, will be unsure of the quality level that is desired. Suppliers know their products. They also know their customers. Some customers will accept only those items they have specified. Others will accept virtually anything because they do not inspect or verify deliveries. If you were a supplier and you had a sack of onions that was getting a bit old, which customer would you deliver it to?

Unscrupulous suppliers can cost your operation guests because of both overcharging and shortchanging. Consider, for example, the restaurant manager who specifies that no more than a ¼-inch fat cover on all New York strip steaks ordered is acceptable. Instead, the meat company delivers steaks with a ½-inch fat cover. The operation will, of course, pay too much for the product because steaks with a ¼-inch fat covering are sold by vendors at a higher price per pound than those with a ½-inch covering.

Checking for quality means checking the entire shipment for conformance to established product specifications. If only the top row of tomatoes in the box conforms to the operation's specification, it is up to the receiving clerk or manager to point that out to the vendor. If the remaining tomatoes in the box do not meet the specification, the box should be refused. The credit memo can then be used to reduce the total on the invoice to the proper amount.

Sometimes, quality deficiencies are not discovered until after a delivery driver has left your establishment. When that is the case, you should notify the vendor that a thorough inspection has uncovered substandard product. The vendor is then instructed to pick up the nonconforming items. Many managers use the Additional Information section of the credit memo form shown in Figure 3.19 to record this requested pickup. When the product is picked up, the pickup information is recorded. Alternatively, a separate memo to the vendor requesting a product pickup could be produced. It is best, however, to keep the number of cost control forms to a minimum whenever possible, especially when minor modifications of one form will allow that form to serve two purposes.

FUN ON THE WEB!

Foodservice operators purchasing fresh seafood face some unique challenges as they inspect and receive requested products. Simply put, much of the seafood sold in the United States (and other parts of the world) is mislabeled.

In fact, in a study summary released by the UK-based *The Guardian* newspaper in 2021*, it was found that 36 percent of over 9,000 (seafood) products were mislabeled. And the issue isn't just afflicting other parts of the world: Of the countries included, the United States was the third worst, with 38 percent of seafood mislabeled—behind only the United Kingdom and Canada (both of which had a 55 percent mislabeling rate). The problem goes beyond ill-informed grocery shoppers, as well. One study published in 2018 collected samples from 180 restaurants across 23 European countries and found that a third of those restaurants had sold mislabeled seafood—with the odds of some menu items being authentic equal to a coin toss.

While some amount of mislabeling of seafood is likely caused by innocent supplier error, foodservice operators must be extremely careful when purchasing fresh seafood. To find out more about this continuing industry problem enter "mislabeled seafood in the restaurant industry" in your favorite search engine and view the results.

*https://www.theguardian.com/environment/2021/mar/15/revealed-seafood-happening-on-a-vast-global-scale. Retrieved 1/23/2026.

Experienced managers know that the skill required to receive beverage products is somewhat less than what is needed for receiving food. The reason is that alcoholic beverage products do not typically vary in quality in the same manner as food products do.

As with food, the beverage receiving clerk needs a proper location, tools, and equipment. In addition, proper delivery schedules must be maintained. The training required in beverage receiving, however, is reduced due to the consistent nature of the product received. A case of freshly produced Coors beer, for example, will be consistent in quality regardless of the vendor. And if the product clearly displays a pull date, then very little inspection is required to ensure that the product is exactly what was ordered.

In fact, when matching a beverage PO to the vendor's delivery invoice, only container size, quantity ordered, and price must be verified, unlike food deliveries that require the verification of weight, quantity, quality, and price.

When receiving beverage products, the following items are of special concern and should be verified:

Key Beverage Receiving Checkpoints

1. Correct brand
2. Correct bottle size
3. No broken bottles or bottle seals
4. Pull dates (beer)
5. Correct vintage/year produced (wine)
6. Correct unit price
7. Correct price extension
8. Correct invoice total

Whether the item purchased is a food item or an alcoholic beverage, vendors sometimes run out of a product, just as you may sometimes run out of a menu item. In such cases, the receiving clerk must know whether it is management's preference to accept a product of higher quality, lower quality, or no product at all as a substitute. If this information is not known, one can expect that suppliers will be able to downgrade quality simply by saying that they were out of the requested product and did not want the operator to be "shorted" on the delivery.

Training to assess and evaluate quality products is a continuous process. The effective receiving clerk should develop a keen eye for quality. This is done to ensure that both the operation and the guests it serves receive the quality products intended by management.

Consider the Costs

"Come on Gloria, give me a break. I'm already behind because my truck broke down this morning, and I've got people all over town calling my boss to scream about their deliveries. It wasn't my fault I'm late. I'm just the driver," said Monte.

Monte, who makes deliveries for Raider Produce, was talking to Gloria, the receiving clerk for the High-Five Restaurant. The delivery he was making was a big one, and it was 2 hours late.

"You know, Gloria," Monte continued, "you all take longer to accept a delivery than any other restaurant on my route. Nobody else inspects and weighs like you do. And you hardly ever find any problems. Look, I know it's a big delivery, but just this once can't you just sign the invoice and let me get going? I want to see my son's ball game, and I won't make it if you take forever to inspect this load."

"I don't know Monte," replied Gloria. "We've got procedures to follow here, and I'm supposed to use them every time."

"Look, just sign the ticket. If you find a problem later, I'll take care of it. I promise you," said Monte, who seemed to be increasingly flustered.

1. What do you think Gloria will most likely do about Monte's request that she simply sign the invoice without inspecting and weighing the delivery?
2. Assume you were this restaurant's owner. If you were personally accepting the delivery, what would you say to Monte?

PRICE

When training staff to verify proper pricing, two major concerns must be addressed:

1. Matching PO unit price to delivery invoice unit price
2. Verifying price extensions and total

Matching PO Unit Price to Delivery Invoice Unit Price

When the person responsible for purchasing food for the operation places a PO, the confirmed quoted price should be carefully recorded because it is never safe to assume that the delivered price will match the price listed on the PO. As an ethical manager, you should not be happy with either an overcharge or an undercharge for a purchased product. Just as you would hope that a guest would inform you if a waiter forgot to add the correct price of a bottle of wine to the dinner check, a good receiving clerk works with the supplier to ensure that the operation is neither under- nor overcharged for all items delivered. Honesty and fair play must govern the actions of both the operation and the vendor.

If the receiving clerk has access to the original PO, it is a simple matter to verify the quoted price and the delivered price. If these numbers do not match, management should be notified immediately. If management notification is not possible, both the driver and the receiving clerk should initial the Comments section of the PO, showing the difference in the two prices, and a credit memo should be prepared.

Obviously, if the receiving clerk has no record of the quoted price, from either a PO or an equivalent source, price verification of this type is not possible. An inability to verify the quoted price and the delivered price at the time of delivery is a sure indication of a poorly designed food cost control system.

Verifying Price Extensions and Total

Price extension is just as important for you to monitor as is the ordered/delivered price. Price extension is the process by which you compute an extended price. **Extended price** is simply the unit price multiplied by the number of units delivered. For example, if the price for a case of lettuce is $18.00 and the number of cases delivered is six, the extended price is calculated as follows:

Unit price × Number of units delivered = Extended price
$18.00 per case × 6 cases = $108.00

Extended price verification is extremely important. It is critical that the receiving clerk verify the following:

- Unit prices
- Number of units delivered
- Extended price computations (unit price × number of units delivered)
- Invoice totals

There seem to be two major reasons why operators do not always insist that the receiving clerk verify the extended prices and invoice totals. The most common reason is the belief that there is not enough time to do so. The driver may be in a hurry, and the operation may be very busy. If that is the case, the process of verifying the extended price can be moved to a slower time. Why? Because there is a written record provided by the vendor of both the unit price and the number of units delivered. Thus, extension errors are vendor errors, recognized by the vendor in the vendor's own handwriting! Or, more accurately, today they are in the vendor's own computer system.

The second reason operators sometimes ignore price extensions is related to these same computers. Some operators believe that if an invoice is machine-generated, the mathematics of price extension must be correct. Nothing could be further from the truth. Anyone familiar with the process of using computers knows that there are many possible entry errors that can result in extension errors (even formulas entered into Excel-type spreadsheets can be entered in error). Once all extension prices have been verified as correct, receiving clerks should check the invoice total against the sum of the individual price extensions. Managers must ensure that delivery clerks verify both extension prices and invoice totals. If this cannot be done at the time of delivery, it must be done as soon thereafter as possible.

RECEIVING RECORD OR DAILY RECEIVING SHEET

Some large operations use a receiving record when receiving food and beverage products. This method, although taking extra administrative time to both prepare and monitor, has some advantages.

A receiving record generally contains the following information:

1. Name of supplier
2. Invoice number
3. Item description
4. Unit price
5. Number of units delivered
6. Total cost
7. Storage area (unit distribution)
8. Date of activity

Figure 3.20 is an example of a receiving record, specifically designed for the receiving area of a large hotel. Note that some items are placed directly into

FIGURE 3.20 Hotel Pennycuff

Receiving Report **Date: 1/2**

						Distribution				
Supplier	**Invoice #**	**Item**	**Unit Price**	**No. of Units**	**Total Cost**	**A**	**B**	**C**	**D**	**E**
Dairy O	T-16841	½ pt. milk	$0.44	800	$352.00	75	600	125	—	
Tom's Rice	12785	Rice (bags)	$22.00	3	$66.00	—	1	—	2	—
Barry's Bread	J-165	Rye	$3.62	25	$90.50	—	25	—	—	—
		Wheat	$3.51	40	$140.40	20	—	20	—	—
		White	$2.48	90	$223.20	40	10	—	40	—
Total Units				958		135	636	145	42	0
Total Cost					$872.10					

Distribution Key:

A = Coffee Shop
B = Banquet Kitchen
C = Direct Use
D = Storeroom
E = Sundry (nonfood items)

Comments: ______________________

production areas (direct use), whereas others may be used in specific units or sent to the storeroom. Sundry items, such as paper products, in-room guest amenities, and cleaning supplies, may be stored in specific non-food areas. Note also that subtotals for storage areas can be determined in terms of either units or dollars, as the operator prefers. In all cases, the sums for each distribution area should equal the total for all items received during the day.

Receiving reports can be useful to management if it is important to record where items are to be delivered or where they have been delivered. Although some foodservice operators will find the receiving report useful, some will not because much of the information it contains is also included in the receiving clerk's copy of the PO.

Figure 3.21 summarizes the steps to be taken to ensure proper receiving in a foodservice operation, and management control of each step is essential to the operation's cost control-related efforts when receiving products.

FIGURE 3.21 Steps in Proper Receiving

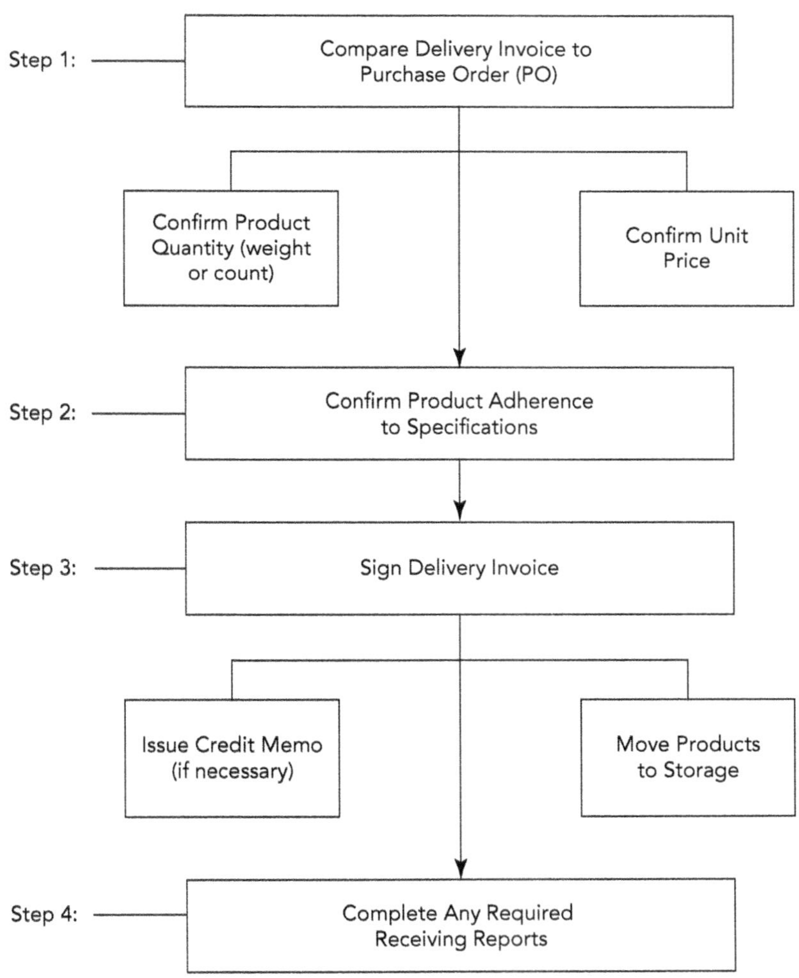

Technology Tools

This chapter focused on managing food- and beverage-related costs by controlling the areas of forecasting future item sales, using standardized recipes, effective purchasing, and proper receiving. A variety of location-based and cloud-based software applications are available to foodservice operators in these areas. They include the following:

Sales Forecasting Software

1. Estimate total unit revenue by partial hour, hour, shift, day, week, and month based on historical sales levels.
2. Adjust sales forecasts based on holidays and local events information supplied by management.
3. Estimate ideal inventory levels based on sales forecasts and vendor delivery schedules.
4. Create suggested POs based on established par levels and on-hand inventory levels.
5. Estimate ideal production schedules based on sales forecasts.
6. Estimate ideal staffing patterns based on hourly sales forecasts.

Recipe Software

1. Maintain standardized recipe files.
2. Provide real-time standardized recipe cost based on product vendors' current price lists.
3. Allow for computer-assisted recipe quantity conversion.
4. Maintain and supply dietary information by recipe and by portion.
5. Create ingredient lists for purchase or issue from storeroom.
6. Suggest alternative cooking methods.
7. Supply color pictures of finished products.
8. Create POs based on production forecasts.
9. Calculate per-portion food cost percentage based on current menu prices.
10. Determine suggested selling prices based on current product costs.

Purchasing Software

1. Establish minimum and maximum inventory par values by ingredient.
2. Create vendor POs based on forecasted ingredient needs.
3. Maintain vendor price histories.
4. Compare bids and make purchase recommendations based on best cost/best value.
5. Create POs and auto-orders based on established par levels with automatic electronic order confirmation.
6. Interface one unit with other units in a system to create aggregate POs (in multiunit settings).
7. Create purchase recaps (purchase amounts and price) by item and/or vendor.
8. Maintain a "red-flag" system for product purchasing variances that exceed forecasted usage.
9. Utilize POs to compare actual product usage to planned-for usage to determine product loss rates and their impact on unit profitability.
10. Purchase directly from participating vendors using vendor-supplied purchasing software.

Receiving Software

1. Prepare a daily receiving report.
2. Compare products received (receiving records) with original POs in the following areas:
   a. Weight
   b. Quantity

c. Quality/grade
d. Price
3. Maintain outage and shortage records by vendor.
4. Verify vendor price extensions.
5. Maintain receiving histories by date and vendor.
6. Capture pictures of vendor invoices (via smart devices) for placement in an operation's online file cabinet.

As you can see, managing the purchasing of food and beverage products is a process in which a great number of advanced technology tools are available. Remember, however, that the amount of information that can be generated in this area is vast, and you should elect to gather and maintain only information that is of real value to your own operation.

Apply What You Have Learned

Tonya Johnson is the Regional Manager for Old Town Buffets. Each of the five units she supervises is in a different town. Produce for each unit is purchased locally by each buffet manager. One day, Tonya gets a call from Danny Trevino, one of the buffet managers reporting to her. Danny states that, as a thank you for his business, one of the local produce suppliers he uses has offered Danny the use of two season tickets to the local university's home football games. Danny likes football and wants to accept the tickets.

1. Would you allow Danny to accept the tickets? Why or why not?
2. Would you allow your managers to accept a gift of any kind (including holiday gifts) from a vendor?
3. Draft a "gifts" policy that you would implement in your region. Would you be subject to the same policy?

For Your Consideration

1. Experienced managers know that the use of standardized recipes is critical to their cost control efforts, but standardized recipes also play an important role in quality control. How does the use of standardized recipes directly affect a foodservice manager's quality control efforts?
2. Strictly enforcing the use of product specifications is important to all food and beverage managers, but the enforcement of the use of product specifications is especially essential for those professionals responsible for the operation of multiunit chain restaurants. Why do you think that would be so?

Key Terms and Concepts

The following are terms and concepts addressed in the chapter that are important for you as a manager. To help you review, define the terms below.

Popularity index	Alcoholic beverages	Spirits
Predicted number to be sold	Beer	Liquors
	Wine	Keg beer

Draft beer
House wines
Mixed drink
Artificial intelligence (AI)
Standardized recipe
Recipe-ready
Recipe conversion factor (RCF)
Baker's math
Product specification (spec)
Count
Product yield
Bid sheet
Price comparison sheet
Cherry pickers
Wine list
Vintage
Vintner
Shelf life
Well liquors
Call liquors
Premium liquors
Pull date
Ethics
Purchase order
Auditors
Delivery invoice
Acceptance hours
Refusal hours
Shorting
Credit memo
Extended price

Test Your Skills

You may download the Excel spreadsheets for the Test Your Skills exercises from the student companion website at http://www.wiley.com/go/dopson/foodandbeveragecostcontrol8e. Complete the exercises by placing your answers in the shaded boxes and answering the questions as indicated.

1. Nikki Decker is the Food and Beverage manager at Harrison Residence Hall, the largest student housing facility on State University campus. Once each month Nikki features an Italian Night, a themed dinner, for her students. She serves the four menu items listed below. Nikki has maintained sales histories showing the number of individual menu items selected for each of her last three Italian Nights. Calculate the total number of each menu items served over the past 3 months and then use that information to calculate each menu item's popularity index.

	Month 1	Month 2	Month 3	3-Month Total	Popularity Index
Menu Item					
Lasagna	215	130	164		
Pasta Carbonara	65	52	74		
Seafood Shells	35	62	64		
Pork Milanese	98	145	116		
Total Served					

Over the entire 3-month period, which Italian themed menu item was the most popular with Nikki's students? Which menu item was the least popular?

2. Dan Flint is the head baker at the Raised Mitten bakery. Dan uses 10 pounds of cake flour to make his famous Lemon Chiffon Cake. The formula calls for 55 percent sugar and it produces 16 cakes. Dan wants to make 48 cakes. Complete the form below to tell Dan how much cake flour and sugar he will need to order, or have in inventory, to make the 48 cakes.

Formula Yield (Cakes)	**16**	**48**
Ingredient	**Ingredient % (decimal form)**	
Cake flour	1.00	
Sugar	0.55	
Ingredient	**Ingredient weight (pounds)**	
Cake flour	10	
Sugar		

3. Jean Alexander is the F&B Director at the City Plaza Hotel. This year Jean's hotel was selected as the host site for the Delmar Electronics company's annual holiday party. Jean asked her bartenders to keep careful records of the alcoholic and nonalcoholic beverages they served the guests during the event. This year 1,350 drinks were served. Calculate the popularity indexes for each of the individual drink options offered to the party attendees as well as the total popularity indexes for the beer, wine, cocktails, and nonalcoholic drink categories.

Delmar Electronics Holiday Party Drink Consumption		
Drink Type	**Number Served**	**Popularity Index**
Beer		
Bottled	145	
Draft	210	
Beer Total		
Wine		
Red		
White	150	
Wine Total	360	
Cocktails		
Whiskey	85	
Vodka	248	
Gin	45	
Rum	65	
Cocktail Total		
Nonalcoholic		
Coffee	62	
Soda	130	
Nonalcoholic Total		
Total Drinks	1,350	100%

Why would this information be important to Jean if the City Plaza Hotel is chosen to host the same event next year?

4. Alli Katz buys Honeycrisp apples to make Old-Fashioned Apple pie. The baker's standardized recipe calls for 16 pounds of apples to make 8 apple pies. The baker wants to make 35 apple pies. Calculate the RCF the baker will use to adjust the recipe and then determine the number of pounds of apples required to make the baker's 35 desired Old-Fashioned Apple pies.

Old Fashioned Apple Pie Recipe	**Current Yield (number of pies)**	**Current Yield Pounds of Apples Required**	**Desired Yield (number of pies)**	**RCF**	**Desired Yield Pounds of Apples Required**
	8	16	35		

5. Rob Stewart is the Director of Food Service at the 600-bed Memorial Hospital. Rob carefully maintains cumulative records of individual entrées served in the past when these menu items were offered, and these are listed in the "Number Served" column below. Today Rob is preparing kitchen

production orders for tomorrow's dinner meal where he will offer the same four menu items listed. He estimates that 500 dinners will be served. Help Rob forecast the number of each entrée his patients will likely request when they make their menu selections for tomorrow's dinner.

Entrée	Number Served	Popularity Index	Number to Produce
Roast Turkey	3,848		
Pasta Primavera	1,332		
Broiled Tilapia	3,700		
Sliced Beef with Mushroom Sauce	5,920		
Total Served		100%	

6. Joshua is the Chef at the Mayberry House restaurant. At a private dinner tomorrow, Joshua will be serving Medallion of Veal Sauté to 24 guests. The standardized recipe Joshua has for Medallion of Veal Sauté only makes 12 portions. Calculate the RCF Joshua will utilize to convert the recipe to 24 servings and then calculate the amounts of ingredients he will need to have on hand to prepare this entrée for tomorrow's guests.

Standardized Recipe Worksheet: Medallion of Veal Sauté					
Medallion of Veal Sauté	**For 12 Servings**			**For 24 Servings**	
Ingredients	**Amount**	**Unit**	**RCF**	**Amount**	**Unit**
Veal loin, boneless, 6 oz. each	12	Each			Each
Flour	2	Cups			Cups
Butter, clarified	2/3	Cups			Cups
Mushrooms, fresh, sliced	6	Cups			Cups
Shallots, finely chopped	3	T			T
White wine, dry	1	Cups			Cups
Heavy cream	1½	Cups			Cups
Veal velouté	2	Cups			Cups
Salt/pepper	To taste			To taste	

Method

1. Season veal medallions with salt and pepper. Dip in flour and sauté in clarified butter over medium heat until golden brown. Remove to tray.
2. Sauté sliced mushrooms and shallots until just wilted.
3. Add white wine, cream, and velouté sauce and then simmer to reduce sauce by half.
4. Season sauce with salt and pepper and spoon onto individual plated veal portions.
5. Garnish with fluted mushroom cap and sprig of fresh rosemary.

7. Cynthia is the F&B Director at the Shady Grove assisted living facility. This Saturday Cynthia is serving Chutney Raisin Chicken Salad to the residents of the facility. Her standardized recipe makes 8 pounds of salad, but she only needs to make 6 pounds. Calculate the RCF Cynthia will utilize to convert the recipe from 8 pounds to 6 pounds and then calculate the amounts of ingredients she will need to have on hand to prepare this item for her residents.

Standardized Recipe Worksheet: Chutney Raisin Chicken Salad					
Chutney Raisin Chicken Salad	For 8 Pounds			For 6 Pounds	
Ingredients	Amount	Unit	RCF	Amount	Unit
Cooked chicken	5	lb.			lb.
Celery, diced	8	oz.			oz.
Green onions, chopped	4	oz.			oz.
Mango chutney	12	oz.			oz.
Raisins	1	Cup			Cup
Mayonnaise	1	lb.			lb.
Seedless grapes	12	oz.			oz.

Method

1. Remove bones, fat, and skin from cooked chicken and dice into quarter-inch pieces.
2. Combine the chicken, celery, green onions, mango chutney, raisins, and mayonnaise in a bowl. Mix well.
3. Slice grapes in half. Add grapes to the chicken mixture and combine gently.
4. Refrigerate for 2 hours before serving.

8. Karis Noel manages Burger! Burger!, an upscale sandwich shop that features 4 ounce (Regular) and 7 ounce (Jumbo) burgers. Karis buys fresh ground beef, her meat spec # 48, in 10-kilogram cartons. Complete the chart below to show how many 4-ounce or how many 7-ounce burgers she can make if she buys 1 carton of ground beef and if she buys 5 cartons. Round your answers down to the nearest whole number of burgers that can be produced.

Burger! Burger!

Ground Beef spec # 48	Cartons Purchased	Kilograms per Carton	Kilograms Purchased	Pounds per Kilogram	Pounds Purchased	Ounces Purchased	4-ounce Burgers Produced	7-ounce Burgers Produced
	1	10	10	2.2				
	5	10		2.2				

9. Renée Radloff is the owner of an intimate Italian bistro located in Northern California. While traveling in Tuscany on vacation, Renée Radloff was served a garden salad with garlic croutons. Renée was so impressed with the croutons; she asked the restaurant's manager for the recipe, and he was delighted to give it to her. Renée wants to use the recipe to prepare the croutons in her own restaurant but must first convert the recipe ingredient amounts from metric to US measurements. Help Renée convert the recipe into a measurement form that can be easily used by her own production staff. (Spreadsheet hints: Use the ROUND function to 0 decimals for ounces, tablespoons, and teaspoons. Use the ROUND function to 1 decimal for pounds.)

Standardized Recipe: Tuscan Garlic Croutons				
Tuscan Garlic Croutons	**Yield: 900 g**		**US Conversion**	
Ingredients	**Metric Amount**	**Metric Unit**	**US Amount**	**US Unit**
Butter, unsalted	170	g		oz.
Garlic, chopped	15	ml		T
Sourdough bread cubes	680	g		lb.
Parmesan Reggiano, grated	30	g		oz.
Fresh basil	30	ml		T
Fresh oregano	5	ml		t

Method

1. Melt the butter in a small sauté pan and add the garlic. Sauté the garlic in butter over low heat for 5 minutes.
2. Place the bread cubes in a mixing bowl. Add the Parmesan Reggiano and fresh herbs.
3. Pour the garlic butter over the bread cubes and toss gently to combine.
4. Spread bread cubes in a single layer on a sheet pan and bake, stirring occasionally, at 175°C until dry and lightly brown (approximately 15 minutes).

At approximately what temperature (in Fahrenheit) should Renée's staff bake the croutons?

10. Billie Mendoza is the purchasing manager for a medium-sized all-suite hotel with a restaurant and a banquet hall. She needs to create the food POs for tomorrow, and she wants to make sure that she is getting the best price for the best quality and service. At the beginning of this week, Billie received bids on produce items from Village Produce, City Produce, and Country Produce. She has listed these prices in the following price comparison sheet.

    Identify the best bid price and best company for each of Billie's produce items. (Spreadsheet hint: Use the MIN function in the Best Bid $ column.)

    If Billie were going to create a PO just for Village Produce, what items would she likely put on the PO based solely on bid prices?

Price Comparison Sheet								
Item	**Description**	**Unit**	**Village Produce**	**City Produce**	**Country Produce**	**Best Bid ($)**	**Best Company Quote**	**Last Price Paid**
Avocados	48 ct.	Case	$61.80	$60.30	$59.46			$57.94
Cauliflower	12 ct.	Case	$12.80	$12.90	$13.27			$13.26
Cucumbers	Medium	Case	$11.10	$11.52	$10.91			$11.34
Grapes	Red seedless	Lug	$19.32	$19.50	$19.14			$18.72
Lettuce	Green Leaf, 24 ct.	Case	$19.53	$19.84	$20.27			$20.02
Lettuce	Romaine, 24 ct.	Case	$17.75	$17.82	$18.22			$18.10
Pears	D'Anjou	Case	$20.82	$20.58	$20.64			$20.62
Peppers	Green Bell, med.	Case	$28.30	$28.38	$29.28			$29.02
Pineapples	7 ct.	Case	$20.50	$20.38	$20.68			$20.08
Potatoes	B Reds	50-lb. bag	$25.06	$24.82	$24.88			$24.98
Potatoes	Peeled, large	25-lb. bag	$17.52	$17.22	$17.28			$17.18
Squash	Yellow #2	30-lb. case	$28.55	$28.71	$28.98			$29.10
Strawberries	Driscoll	Flat	$18.29	$18.06	$17.10			$18.30

11. Loralei operates the foodservice for a large elementary school. She buys produce from Shady Tree Produce Company. Wayne, Shady Tree's owner, makes many errors when he prepares invoices for his customers, but he is the only vendor who has the ability to deliver daily. Loralei needs daily delivery. Because she knows Wayne is careless with the invoices, Loralei checks each one carefully. For Monday, Loralei needs the items listed below.

Monday Order		
Item	Unit Price	Number of Cases Needed
Tomatoes	$18.50	3
Potatoes	$12.90	6
Carrots	$18.29	4

Using this information, complete her PO for Monday below.

Purchase Order					
Supplier				Purchase Order	
Company Name:	Shady Tree			PO Number	123
Street Address:	123 Somewhere			Order Date:	1/15
City, State, Zip:	Village, TX 12345			Delivery Date:	1/16
Phone Number:	555–5555			Phone Number:	555–5557
Fax Number:	555–5556			Fax Number:	555–5558
E-Mail:	Shady@isp.org			E-Mail:	Lora@isp.org
Contact:	Wayne			Buyer:	Loralei
Item	**Description**	**Unit**	**Quantity**	**Unit Price $**	**Extension**
Tomatoes	4 x 5, layered	Case			
Potatoes	Peeled, large	Case			
Carrots	Julienne, 5-lb. bag	Case			
Total					

Upon delivery, Loralei checks her invoice from Wayne to see if he has, once again, made any errors.

Review the following invoice extended prices and total.

Shady Tree Produce Company Invoice for Monday Order			
Item	**Unit Price**	**Number Delivered**	**Extended Price**
Tomatoes	$18.50	3	$55.55
Potatoes	$19.20	6	$115.20
Carrots	$18.92	4	$75.68
Total Amount Due			$256.43

Does the invoice contain errors? If so, what are they? What is the total amount of error on the invoice, if any? How can Loralei detect such errors in the future?

12. Both Fraheen and Winthrop operate franchised casual dining restaurants and offer identical menus. Last month Fraheen's unit sold 3,800 servings of the operation's most popular menu item and she incurred total product costs for that item of $18,430. Winthrop's operation sold 4,100 servings of the same menu item and incurred $21,115 in total product cost for the item.

    Fraheen's operation consistently utilizes standardized recipes to aid in serving the menu item's correct portion size, but Winthrop's operation does not. Calculate the product cost per portion, per portion product cost variance, and per portion % cost variance for the menu item served in these two operations.

	Total Product Cost	Portions Served	Product Cost per Portion	Per Portion Product Cost Variance	Per Portion % Cost Variance
Fraheen: With standardized recipes	$18,430				
Winthrop: Without standardized recipes	$21,115				

How much money did Winthrop's operation lose last month because it does not utilize standardized recipes for this menu item?

CHAPTER 4

Managing Inventory and Production

OVERVIEW

This chapter begins with an overview of the procedures used to properly store food and beverage products until they are ready for use. The careful control of products while in storage is important to help minimize product loss due to waste, spoilage, and theft. The chapter then addresses the steps managers must take to ensure product inventories are kept secure. When stored products are needed by production staff, these products must be properly issued and their issue amounts carefully recorded to assist in future cost calculation and purchasing efforts. Finally, managers must ensure the procedures used for producing food and beverage items for their guests ensure the desired product quality and proper cost control. This chapter directly addresses these important procedures.

Chapter Outline

- Product Storage
- Storing Food Products
- Storing Beverage Products
- Inventory Control
- Product Issuing and Restocking
- Managing Food Production
- Managing Beverage Production
- Technology Tools
- Apply What You Have Learned
- For Your Consideration
- Key Terms and Concepts
- Test Your Skills

LEARNING OUTCOMES

At the conclusion of this chapter, you will be able to:

- Implement an effective product storage and inventory system.
- Control the issuing of products from storage.
- Manage the food and beverage production process.

PRODUCT STORAGE

In Chapter 3, you learned the procedures used to properly purchase and receive foodservice products. In most cases, after purchased items have been received, they must be immediately placed into storage. Food and beverage product quality rarely improves with increased storage time. In fact, the quality of most products you buy will be at its peak when the product you ordered is delivered to you. From then on, these products will often decline in freshness, quality, and nutritional value.

It is important to understand that all of the stored food and beverage products that make up your **inventory** are the same as money to you. In fact, you should think of items placed in storage in exactly that same way. The Cosmic Crisp apple in a produce walk-in is not just an apple. It represents, at a selling price of $3.00, that amount in revenue to the airport foodservice director, who hopes to sell a fresh Cosmic Crisp apple to a weary traveler. If the apple disappears, revenue of $3.00 will disappear also. When you think of inventory items in terms of their sales value, it becomes clear why all food and beverage products must be stored and maintained carefully.

STORING FOOD PRODUCTS

The ideal situation for you and your operation would be for you to store only the food you will use between the time of a vendor's delivery and the time of that vendor's next delivery. This is true because storage costs money, in terms of providing both the storage space and energy costs for holding inventory items and the money that is tied up in inventory items and thus unavailable for use elsewhere. It is always best, whenever possible, to order only the products that are absolutely needed by the operation. In that way, the vendor's storeroom actually becomes your storeroom because the vendor then absorbs the costs of storing your needed products. In all cases, however, you must have an adequate supply of products on hand to service your guests. They are your main concern. If you are doing your job well, you will have many guests and will need many items in storage!

After receiving staff have properly accepted the food products you have purchased, the next step in the control of food costs is that of properly storing those items. In most cases, the storage process consists of three main tasks:

1. Placing products in storage
2. Maintaining food and beverage storage areas
3. Maintaining product security

PLACING PRODUCTS IN STORAGE

Food products are highly perishable items. As such, they must be moved quickly from your receiving area to the area selected for storage. This is especially true for refrigerated and frozen items. An item such as ice cream, for example, can deteriorate substantially in quality if it remains at room temperature for only a few minutes. Most often, in foodservice, this high perishability means that the same individual responsible for receiving the items is the one responsible for placing them in storage.

Consider the situation of Kathryn, the receiving clerk at Fairview Estates, an extended-care facility with 200 residents. She has just taken delivery of seven loaves of whole-wheat bread. They were delivered in accordance with the purchase order her manager prepared. The bread must now be put in storage. When Kathryn stores these items, she must know whether management requires her to use the **LIFO (last in, first out)** or **FIFO (first in, first out)** method of product rotation.

LIFO SYSTEM

When using the LIFO storage system, the storeroom manager intends to use the most recently delivered product (last in) before he or she uses any part of that same product previously on hand. If Kathryn decides, for example, to use the just-delivered bread *prior* to using other bread already in storage, she would be utilizing the LIFO system.

In all cases, you must strive to maintain a consistent product standard. In the case of some bread, pastry, fruit, and vegetable items, the storeroom clerk could be instructed to utilize the LIFO system. With LIFO, you will need to take great care to order only the quantity of product needed between deliveries. If too much product is ordered, loss rates will be very high. For most of the items you will buy, however, the best storage system to use is the FIFO system.

FIFO SYSTEM

Using a FIFO (first in, first out) storage system means that you intend to rotate your stock in such a way that products already on hand are sold prior to the sale of more recently delivered products. With FIFO, older products (first in) are used before newer products. When the FIFO system is used, the storeroom clerk must take great care to place new stock behind, underneath, or at the bottom of old stock. It is often the tendency of employees not to do this. Consider, for example, the storeroom clerk who must put away six cases of tomato sauce. The cases weigh about 40 pounds each. The FIFO method dictates that these six cases be placed *under* the five cases already stacked in the storeroom. Will the clerk take the time and effort to place the six newly delivered cases underneath the five older cases? In many instances, the answer is no. Unless management strictly enforces the FIFO rule, employees may be tempted to use the faster and easier, but improper, way of storing the newly delivered products on top of those already in storage. Figure 4.1 demonstrates the difference between LIFO and FIFO when dealing with storing cases of food products.

FIGURE 4.1 FIFO and LIFO Storage Systems

Oldest	Newest
Oldest	Newest
Oldest	Newest
Newest	Oldest
Newest	Oldest
Newest	Oldest
FIFO	LIFO

FIFO is the preferred storage technique for most perishable and nonperishable items. Failure to implement a FIFO system of storage management can result in excessive product loss due to spoilage, shrinkage, or deterioration of product quality.

Decisions about storing food items according to the LIFO or FIFO method are management decisions. Once these decisions have been made, they should be communicated to storeroom staff and monitored on a regular basis to ensure compliance.

To assist them in their efforts, some foodservice managers require the storeroom clerk to mark or tag each delivered item with the date of delivery. These markings provide a visual aid in determining which products should be used first. This is especially critical in the area of highly perishable and greater cost items such as fresh meats and seafood. Special meat and seafood date tags should be available from your vendor. These tags contain a space for writing in the item's name, quantity, and delivery date. The use of these physical tags or an alternative date tracking system is strongly recommended. If the supplier has computerized his or her delivery, the box or case may already bear a barcode strip or **QR code**. A QR code is a type of two-dimensional barcode, meaning it encodes information in both horizontal and vertical directions, allowing it to store significantly more data than a traditional barcode. QR codes are scanned using a mobile device camera.

Figure 4.2 shows products in a dry-storage area when management requires that the storeroom clerk marks each stored item with its date of delivery.

FIGURE 4.2 FIFO Stocking System

Tomato sauce	2/10	Beets	2/10	Egg noodles	2/18
Tomato sauce	2/18	Beets	2/10	Egg noodles	2/18
Tomato sauce	2/18	Beets	2/18	Egg noodles	2/22
Tomato sauce	2/22	Beets	2/22	Egg noodles	2/22

Some operators prefer to go even further when labeling products in storage. These operators date the item and then also indicate the day (or hour) on which the product should be pulled from storage, thawed, or even discarded (disc.).

Figure 4.3 is an example of a typical storage label utilized for this process.

FIGURE 4.3 Product Storage Label

	DATE	HOUR
PULL		
THAW		
DISC.		

MONDAY

FUN ON THE WEB!

Daydots is a line of labeling products from Ecolab developed especially for the restaurant industry. To learn more about how these storage labels can help your operation ensure product quality and save money, type "day dot food service labels" in your favorite search engine.

MAINTAINING FOOD STORAGE AREAS

To maintain the quality and security of delivered food items, they should be immediately placed into one of three storage areas:

1. Dry storage
2. Refrigerated storage
3. Frozen storage

Dry Storage

Dry-storage areas should be maintained at a temperature ranging between 65°F and 75°F (18°C and 24°C). Temperatures below that range can be harmful to food products. More often, however, dry-storage temperatures get very high, exceeding by far the upper limit of temperature acceptability. This is because storage areas are frequently located in poorly ventilated and closed-in areas of an operation. Excessively high or low temperatures will damage dry-storage products.

Shelving in dry-storage areas must be easily cleaned and sturdy enough to hold the weight of dry products. Slotted shelving is preferred over solid shelving when storing food because slotted shelving allows for better air circulation around stored products.

All shelving should be placed at least 6 inches above the ground to allow for proper cleaning beneath the shelving and to ensure proper ventilation.

When placed into storage all can and case labels should face out for easy identification. Bulk items such as flour or sugar should be stored in wheeled bins whenever possible so that heavy lifting and the potential for employee injuries related to lifting can be avoided.

Most importantly, dry-storage space must be sufficient in size to handle your operation's needs. Cramped and cluttered dry-storage areas tend to increase costs because inventory cannot be easily rotated, maintained, located, or counted. Hallways leading to storage areas should always be kept cleared and free of excess storage materials and empty boxes. This helps both in accessing the storage area and in reducing the number of potential hiding places for insects and other pests.

The Move to Off-Premise Dining

One of the many changes resulting from the move to off-premise dining is the need for the additional dry storage space for off-premise packaging materials. The boxes and cases containing takeout-related items such as hot and cold cups, lids, pre-wrapped utensils, and food containers can require a significant amount of storage space.

In many cases, traditional restaurants were built with sufficient refrigeration, freezer, and dry food storage spaces, but they lack dry storage areas close enough to production and delivery areas, and this can make the storage of needed takeout packaging items a real challenge.

As they meet these challenges, foodservice operators must remember that, like all dry-storage items, takeaway packaging should never be stored directly on the floor and it should not be stored:

- In locker rooms
- In toilet rooms
- In dressing rooms
- In garbage rooms
- In mechanical rooms
- Under sewer lines that are not shielded to intercept potential drips
- Under leaking water lines, including leaking automatic fire sprinkler heads, or under lines on which water has condensed
- Under open stairwells

Refrigerated Storage

Time/temperature control for safety (TCS) foods (formerly referred to in the foodservice industry as potentially hazardous food [PHF]) are those that require time/temperature control for safety to limit harmful pathogenic microorganism growth or toxin formation. Most common TCS foods include meat products, eggs, fish and shellfish, dairy products, cooked vegetables, potato dishes, protein-rich plants, raw sprouts, cut leafy greens, sliced melons, and tomatoes. TCS foods must be kept out of the **temperature danger zone** (41°F–135°F, 5°C–57°C).

Refrigerator temperatures used to store TCS foods should be maintained at 41°F (5°C) or less. In most cases, the lower areas tend to be coldest in refrigerators because warm air rises and cold air falls. In fact, the refrigerator itself may vary as much as 4° (F) between its coldest spot (near the bottom) and its warmest spot (near the top). In most cases, it is advisable to set refrigerator thermostats at 40°F (4.5°C) to ensure proper cold food holding temperatures are maintained in all parts of refrigerated storage areas.

Refrigerators should be opened and closed quickly when used, both to lower operational costs and to ensure that the items in the refrigerator stay at their peak of quality. Refrigerators should be properly cleaned on a regular basis. Condensation drainage systems in refrigerators should be checked at least weekly to ensure they are kept clean and are functioning properly.

Freezer Storage

Freezer temperatures should be maintained at 0°F (−18°C) or less. Newly delivered products should be carefully checked with a thermometer when received to ensure that they are solidly frozen and have been delivered at the proper temperature. In addition, these items should be carefully inspected to ensure that they have not been thawed and then refrozen. This is because refrigerators and frozen-food holding units remove significant amounts of stored product moisture, causing shrinkage and **freezer burn** in meats and produce. Freezer burn refers to deterioration in product quality resulting from poorly wrapped or stored items kept at freezing temperatures for extended periods of time.

Unless they are built in, frozen-food holding units as well as refrigerators should be high enough off the ground to allow for easy cleaning around and under them to prevent cockroaches and other insect pests from living beneath them. Standalone units should be placed 6 to 10 inches away from walls to allow for the free circulation of air around, and efficient operation of, the units.

Frozen-food holding units must be regularly maintained, a process that includes cleaning them inside and out, and have constant temperature monitoring to detect possible improper operation. A thermometer permanently placed in the unit or one easily read from outside the unit is best. It is also a good idea to periodically check that gaskets on freezers, as well as on refrigerators, tightly seal the food cabinet. This will not only reduce operating costs but also maintain peak food quality for a longer period of time.

FUN ON THE WEB!

The United States Food and Drug Administration (FDA) produces and regularly (every four years) updates a food code containing information about the proper handling of foods stored and prepared for sale in foodservice operations. The 2022 edition of the food code made some significant changes to the 2017 edition (which for the first time included a requirement that the person in charge (PIC) of a foodservice operation be a Certified Food Protection Manager (CFPM)).

To review food safety recommendations included in the 2022 food code produced by the FDA, go to: www.fda.gov/Food/GuidanceRegulation/RetailFoodProtection/FoodCode.

STORING BEVERAGE PRODUCTS

Although the storage life for most beverage products is relatively long, alcoholic beverages, and especially wine, must be stored carefully. Beverage storage rooms should be easily secured from unauthorized entrance because beverages are often expensive items and are a favorite target for both employee and guest theft.

Many beverage managers use **recodable electronic locks** for beverage storage areas. These locks include features that allow management to issue multiple keys or entry codes and to precisely identify the time an issued code was used to access the lock and to whom that code was issued. For beverage storage areas, the use of recodable electronic locks is highly recommended.

FUN ON THE WEB!

The recodable electronic lock was introduced by VingCard in 1979. Today, their locks are popular in hotels, where guest room security is extremely important, as well as in many other situations (such as liquor and wine storage areas), where traceable access and high levels of security are critical. To view their current product offerings, visit the ASSA ABLOY company website.

BEER STORAGE

Beer in kegs should be stored at refrigeration temperatures of 36°F to 38°F (2°C to 3°C). Keg beer is unpasteurized and thus must be handled carefully to avoid spoilage. Pasteurized beer in either cans or bottles should be stored in a cool, dark room at 50°F to 70°F (10° to 21°C) and does not require refrigeration.

Beer storage areas should be kept clean and dust free. Canned beer, especially, should be covered when stored to eliminate the chance of dust or dirt settling on the rims of the cans. Product rotation is critical if beer is to be served at its maximum freshness, and it is important that you and your storage team devise a system to ensure that this happens. The best method is to date each case or package as it is delivered. In this manner, you can, at a glance, determine whether proper FIFO product rotation is occurring.

WINE STORAGE

Wine storage is the most complex and time-consuming activity required of beverage storage personnel. Depending on the type and sales volume of the restaurant, extremely large quantities of wine may be stored. Regardless of the amount of wine to be stored, the principles of proper wine storage must be followed if the quality of the wine is to be maintained and product losses are to be kept at a minimum. Despite the mystery too often associated with wine storage, in most cases you will find that proper wine storage can be achieved if you monitor the following areas:

1. Temperature
2. Light
3. Cork condition

TEMPERATURE

A great deal of debate has centered on the proper temperature at which to store wine. But, generally speaking, most experts would agree that wines should be stored at a **cellar temperature** or approximately 50°F to 57°F (12°C to 14°C). White wines, however, are often stored at refrigerator temperatures. If you find that your wines must be stored at higher temperatures than these, the wine storage area should be as cool as can reasonably be achieved, and it is important to remember that, although many wines may improve with age, it improves only if it is properly stored. Excessive heat is always an enemy of proper wine storage.

LIGHT

Just as wine must be protected from excess heat, it must be protected from direct sunlight. In olden times, this was achieved by storing wines in underground cellars or caves. In your own foodservice establishment, this means using a storage area where sunlight cannot penetrate and where the wine will not be subject to excessive fluorescent or incandescent lighting. The general rule for wine storage is that wine bottles should be exposed only to the minimum amount of light needed.

CORK CONDITION

It is the wine's cork that protects it from oxygen, its greatest enemy, and from the effects of oxidation. **Oxidation** occurs when oxygen comes in direct contact with bottled wine. You and your guests can detect a wine that has been overly oxidized because it smells somewhat like vinegar. Oxidation deteriorates the quality of bottled wines. Keeping oxygen out of the wine is a prime consideration of the vintner, and it should be important to you as well.

While many wine makers now seal their bottles with screw-type caps, cork continues to be the bottle sealer of choice for many wine producers. Quality wines demand quality corks, and traditionally the best wines were fitted with corks that will last many years if the cork is not allowed to dry out. This is the reason wine should be stored so that the cork remains in contact with the wine and, thus, it stays moist. In an effort to accomplish this, most foodservice managers store wines on their sides.

Corks should be inspected at the time wine is received and periodically thereafter to ensure that there are no leaks resulting in oxidation and, thus, damaged products. If a leak is discovered at delivery, the wine should be refused. If the leak occurs during storage, the wine should be examined for quality and then either consumed or discarded, as appropriate.

In general, you can effectively manage the storage of wines if you think about how you should treat the cork that is protecting the wine. Always keep the cork (and thus the wine):

1. Cool
2. Moist
3. In the dark

LIQUOR STORAGE

Spirits should be stored between 70°F and 80°F (21°C to 27°C) and in a locked, or highly secure, and dry-storage area. Since these products do not generally require refrigeration, they may be stored along with food products, if necessary. An organized, well-maintained area for storing spirits will also ensure that purchasing decisions will be simplified because no product is likely to be overlooked or lost in inventory.

MAINTAINING PRODUCT SECURITY

Inventoried items are the same as money to you. In fact, you should think of inventory items in exactly that way. As already noted, the apple in a produce walk-in is not just an apple, it represents potential revenue. When you think of inventory items in terms of their sales value, it becomes clear why product security is of the utmost importance.

All foodservice establishments will experience some amount of theft, in its strictest sense. The reason is very simple. Some employee theft is impossible to detect. Even the most sophisticated, computerized control system is not able to determine if an employee or vendor's employee walked into the produce walk-in and ate one green grape.

Similarly, an employee who takes home two small sugar packets per night will likely go undetected. In neither of these cases, however, is the amount of loss significant, and certainly not enough to install security cameras in the walk-in or to search employees as they leave for home at the end of their shifts. What you do want to do, however, is to make it difficult to remove significant amounts of food or beverage items from storage without authorization, so that you can know when those items have been removed. Good cost control systems must be in place if you are to achieve this goal.

It is amazing how large the impact of theft can be on profitability. Consider the following example. Jesse is the receiving clerk at the Irish Voice sports bar. On a daily basis, Jesse takes home $2.00 worth of food products. How much, then, does Jesse cost the bar in one year? The answer is a surprising $14,600 in sales revenue! If Jesse or others pilfer $2.00 per day for 365 days, the total theft amount is $730 (365 days × $2.00 per day = $730). If the bar makes an after-tax profit of 5 percent on each dollar of food products sold, to recover the lost $730, the operation must make additional sales of $1.00/0.05 × $730 = $14,600! In the case of a smaller operation, $14,600 may well represent several days' sales revenue. Clearly, small thefts can add up to large dollar losses!

INVENTORY CONTROL

You should attempt to carefully control access to the location of your stored products. In some areas, this may be done by a process as simple as keeping the dry-storage area locked and requiring employees to get the keys or code from a manager or supervisor when products are needed. In other situations, cameras may be mounted in both storage areas and employee exit areas. Sometimes the physical layout of the foodservice operation may prevent management from being able to effectively lock and secure all storage areas, but too easy access is sure to cause theft problems. This is not because employees are basically dishonest. Most are not. Theft problems develop because of the few employees who feel either that management will not miss a few of whatever is being stolen or that they falsely feel they "deserve" to take a few things because they work so hard.

It is your responsibility to see to it that your product inventories remain secure. As a general rule, if storerooms are to be locked, only one individual should have access to them during any shift. In reality, however, it may be difficult to keep all inventory items under lock and key. Some items must be received and immediately sent to the kitchen for processing or use. As a result, most managers find that it is impossible to operate under a system where all food products are locked away from all employees. Storage areas should, however, not be accessible to guests or vendor employees. If proper control procedures are in place, employees will know that you can determine if theft has occurred. Without such control, employees may feel that theft will go undetected. This must, and can, be avoided.

Proper inventory management seeks to provide appropriate **working stock,** which is the amount of an ingredient you anticipate using before purchasing that item again, and a minimal **safety stock,** the extra amount of that ingredient you decide to keep on hand to meet higher-than-anticipated demand. Demand for a given menu item can fluctuate greatly between supplier deliveries, even when the delivery occurs daily. With too little inventory on hand, you may run out of products and, therefore, reduce guest satisfaction and increase employee frustration. When too much inventory is on hand, however, waste, theft, and spoilage can become excessive. The ability to effectively manage the inventory process is one of the best skills a foodservice manager can acquire.

DETERMINING PROPER INVENTORY LEVELS

The actual amount of the products you should keep in your inventory will vary based on a number of factors such as the following:

1. Storage capacity
2. Item perishability
3. Vendor delivery schedule
4. Potential savings from increased purchase size
5. Operating calendar
6. Relative importance of stock outages
7. Value of inventory dollars

STORAGE CAPACITY

Inventory items must be purchased in quantities that can be adequately stored and secured. Many times, kitchens may lack adequate storage facilities. This may mean more frequent deliveries and holding less of each product on hand than would otherwise be desired. When storage space is too great, however, the tendency by some managers is to fill the space. It is important that this not be done because increased inventory of items generally leads to greater spoilage and loss due to theft. Moreover, large quantities of goods on the shelf tend to send a message to employees that there is plenty of everything. This may result in careless use of valuable and expensive products. It is also unwise to overload refrigerators or freezers because these units will then operate less efficiently, and stored items may be difficult to locate and thus suffer loss of quality.

ITEM PERISHABILITY

If all food products retained the same level of freshness, flavor, and quality while in storage, you would have little difficulty in maintaining proper product inventory levels. The shelf life (see Chapter 3) of food products, however, varies greatly.

Figure 4.4 lists the difference in shelf life of some common foodservice items when properly stored in a dry storeroom or a refrigerator. Figure 4.5 lists the typical shelf life of some common foodservice items when properly stored in a refrigerator or freezer.

Because food items have varying shelf lives, you must balance the need for a particular product with the optimal shelf life of that product. Serving items that are too old is a sure way to produce guest complaints. In fact, one of the quickest ways to determine the overall effectiveness of a foodservice manager is to "walk the boxes." This means to take a tour of a facility's storage areas. If many products, particularly in the refrigerated areas, are moldy, soft, overripe, or rotten, it is a good indication of a foodservice operation that does not have an understanding of proper inventory levels based on the shelf lives of the items kept in inventory. It is also a sign that menu item sales forecasting methods are either not in place or are not working well.

FIGURE 4.4 Shelf Life of Selected Products

Item	Storage Area	Shelf Life
Canned vegetables	Dry storeroom	12 months
Flour	Dry storeroom	3 months
Potatoes	Dry storeroom	14–21 days
Sugar	Dry storeroom	3 months
Bacon	Refrigerator	30 days
Butter	Refrigerator	14 days
Ground beef	Refrigerator	2–3 days
Lettuce	Refrigerator	3–5 days
Milk	Refrigerator	5–7 days
Steaks (fresh)	Refrigerator	14 days
Tomatoes	Refrigerator	5–7 days

VENDOR DELIVERY SCHEDULE

It is the fortunate foodservice operator who lives in a large city with many vendors, some of whom may offer the same services and all of whom would like to have the operator's business. In many cases, however, you will not have the luxury of daily delivery. Your operation may be too small to warrant such frequent stops by a vendor, or your operation may be in such a remote location that daily delivery is not possible.

Consider, for a moment, the difficulty you would face if you were the manager of a foodservice facility located on an oil rig operating off the coast of Louisiana. Clearly, in a case like that, a vendor willing to provide daily doughnut delivery is going to be hard to find! In all cases, it is important to remember that the cost to the vendor for frequent deliveries will be reflected in the prices the vendor charges you.

POTENTIAL SAVINGS FROM INCREASED PURCHASE SIZE

Sometimes you will find that you can realize substantial savings by purchasing needed items in large quantities. This certainly makes sense if the total savings actually outweigh the added costs of receiving and storing the larger quantity. Remember, however, that there are costs associated with extraordinarily large purchases. These may include storage costs, spoilage, deterioration, insect or rodent infestation, or theft.

As a general rule, you should determine your ideal product inventory levels and then maintain your stock within those ranges. Only when the advantages of placing an extraordinarily large order are very clear should such a large purchase be undertaken.

OPERATING CALENDAR

When an operation is involved in serving meals seven days a week to a relatively stable number of guests, the operating calendar makes little difference to inventory level decision making. If, however, the operation opens on Monday and closes on Friday for two days, as is the case in many school foodservice accounts, the operating calendar plays a large part in determining desired inventory levels. In general, it can be said that an operator who is closing down either for a weekend or for a season (as in the operation of a youth summer camp or a seasonal hotel) should attempt to greatly reduce overall perishable inventory levels as the closing period approaches.

FIGURE 4.5 Recommended Refrigeration and Freezer Storage Period Maximums

Cold Storage Chart		
These short, but safe, time limits will help keep potentially hazardous foods from spoiling or becoming dangerous to eat. Because proper freezing keeps potentially hazardous foods safe indefinitely, recommended storage times are for quality only.		
Product	**Refrigerator: 41°F (5°C) or less**	**Freezer 0°F: (–18°C) or less**
Eggs		
Fresh, in shell	3 to 5 weeks	Do not freeze
Raw yolks and whites	2 to 4 days	1 year
Hard cooked	1 week	Does not freeze well
Liquid pasteurized eggs, egg substitutes		
Opened	3 days	Does not freeze well
Unopened	10 days	1 year
Mayonnaise		
Commercial, refrigerate after opening	2 months	Do not freeze
Frozen dinners & entrées		
Keep frozen until ready to heat	—	3 to 4 months
Deli and vacuum-packed products		
Store-prepared (or homemade) egg, chicken, ham, tuna, and macaroni salads	3 to 5 days	Does not freeze well
Hot dogs and luncheon meats		
Hot dogs		
Opened package	1 week	1 to 2 months
Unopened package	2 weeks	1 to 2 months
Luncheon meat		
Opened package	3 to 5 days	1 to 2 months
Unopened package	2 weeks	1 to 2 months
Bacon and Sausage		
Bacon	7 days	1 month
Sausage, raw—from chicken, turkey, pork, beef	1 to 2 days	1 to 2 months
Smoked breakfast links, patties	7 days	1 to 2 months
Hard sausage—pepperoni, jerky sticks	2 to 3 weeks	1 to 2 months
Summer sausage labeled "Keep Refrigerated"		
Opened	3 weeks	1 to 2 months
Unopened	3 months	1 to 2 months
Ham, corned beef		
Corned beef, in pouch with pickling juices	5 to 7 days	Drained, 1 month
Ham, canned labeled "Keep Refrigerated"		
Opened	3 to 5 days	1 to 2 months
Unopened	6 to 9 months	Do not freeze
Ham, fully cooked, vacuum sealed at plant, undated, unopened	2 weeks	1 to 2 months
Ham, fully cooked vacuum sealed at plant, dated, unopened	"Use-by" date on package	1 to 2 months
Ham, fully cooked		
Whole	7 days	1 to 2 months
Half	3 to 5 days	1 to 2 months
Slices	3 to 4 days	1 to 2 months

Product	Refrigerator: 41°F (5°C) or less	Freezer 0°F: (–18°C) or less
Hamburger, ground, and stew meat		
Hamburger and stew meat	1 to 2 days	3 to 4 months
Ground turkey, veal, pork, lamb, and mixtures of them	1 to 2 days	3 to 4 months
Fresh beef, veal, lamb, pork		
Steaks	3 to 5 days	6 to 12 months
Chops	3 to 5 days	4 to 6 months
Roasts	3 to 5 days	4 to 12 months
Variety meats—tongue, liver, heart, kidneys, chitterlings	1 to 2 days	3 to 4 months
Prestuffed, uncooked pork chops, lamb chops, or chicken breasts stuffed with dressing	1 day	Does not freeze well
Soups and Stews		
Vegetable or meat added	3 to 4 days	2 to 3 months
Cooked meat leftovers		
Cooked meat and meat casseroles	3 to 4 days	2 to 3 months
Gravy and meat broth	1 to 2 days	2 to 3 months
Fresh poultry		
Chicken or turkey, whole	1 to 2 days	1 year
Chicken or turkey, pieces	1 to 2 days	9 months
Giblets	1 to 2 days	3 to 4 months
Cooked poultry leftovers		
Fried chicken	3 to 4 days	4 months
Cooked poultry casseroles	3 to 4 days	4 to 6 months
Pieces, plain	3 to 4 days	4 months
Pieces covered with broth, gravy	1 to 2 days	6 months
Chicken nuggets, patties	1 to 2 days	1 to 3 months
Pizza, cooked	3 to 4 days	1 to 2 months
Stuffing, cooked	3 to 4 days	1 month

RELATIVE IMPORTANCE OF STOCK OUTAGES

In many foodservice operations, not having enough of a single food ingredient or menu item is simply not that important. In other operations, the shortage of even one menu item might spell disaster. For example, it may be fine for the local French restaurant to run out of one of their specials on Saturday night, but it is not difficult to imagine the problem of the McDonald's restaurant manager who runs out of French fried potatoes on that same Saturday night!

For small operators, a mistake in the inventory level of a minor ingredient that results in an outage can often be corrected by a quick run to the local grocery store. For larger facilities, such an outage may well represent a substantial loss of sales or guest goodwill. In the restaurant industry, when an item is no longer available on the menu, you "86" the item, a reference to restaurant slang originating in the early 1920s (86 rhymed with "nix," a Cockney term meaning "to eliminate"). If you, as a manager, "86" too many items on any given night, the reputation of your restaurant may suffer.

A strong awareness and knowledge of how critical outage factors are will help you determine appropriate product inventory levels. A final word of caution is, however, necessary. The foodservice operator who is determined never to run out of anything must be careful not to set inventory levels so high as to actually end up costing the operation more than if more realistic levels were maintained.

VALUE OF INVENTORY DOLLARS

In some cases, operators elect to remove dollars from their bank accounts and convert them into product inventory. When this is done, the operator is making the decision to value products more than dollars. If the dollars used to purchase inventory must be borrowed from the bank, rather than being available from operating revenue, a greater cost to hold the inventory is incurred because interest must be paid on the borrowed funds. In addition, a foodservice company of many units that invests too much of its money in inventory may find that funds for acquisition, renovation, or marketing the company are reduced. In contrast, a state institution that is given its entire annual budget at the start of its **fiscal year** (a year that is a full 12 months long but does not begin on January 1) may find it advantageous to use its purchasing power to acquire large amounts of nonperishable inventory at the beginning of the year and at very low prices.

DETERMINING AMOUNTS IN FOOD INVENTORY

It is important for you to continually monitor the amount of products you have in inventory. Inventory levels may be determined by counting the item, as in the case of cans, or by weighing items, as in the case of meats. Volume, that is, gallons, quarts, and the like, is another method of establishing product amounts. If an item is purchased by the pound, it is generally weighed to determine the amount on hand. If it is purchased by the piece or case, the appropriate unit to determine the item amount may be either pieces or cases. If, for example, canned pears are purchased by the case, with six cans per case, management might decide to count the item in terms of either cases or cans. That is, three cases of the canned pears might be considered as three items (when counted by the case) or as 18 items (when counted by the can). Properly used, either method will correctly establish the amount of product you have on hand and is acceptable.

FUN ON THE WEB!

Counting food and beverage inventories has historically been a very time-consuming task. However, advances in inventory-counting systems are occurring regularly. To see examples of various products designed to help restaurant managers control their inventory costs and speed up their inventory taking, enter "restaurant inventory management systems" in your favorite search engine and review the results.

DETERMINING AMOUNTS IN BEVERAGE INVENTORY

Determining amounts in inventory for liquor products is most often more difficult than determining these same levels for food. Food items, as well as beers and wines, can most often be counted for inventory purposes. Opened containers of spirits, however, must be valued also. It is the process of determining the amounts of product in these opened liquor containers that will present a challenge to you. Two inventory methods are commonly used to accomplish this goal:

1. Weight
2. Visual estimate

LIQUOR INVENTORY BY WEIGHT

The weight method of determining spirits inventories uses a scale or smart-device app to weigh open bottles and determine their remaining contents. The product weight, in ounces or milliliters, is then multiplied by the product cost per ounce or milliliter to arrive at the value of the opened container's contents. This system is effective only if you remember to subtract the weight of the empty bottle itself from the total product's weight or if that feature is built into the smart-device app being used.

VISUAL ESTIMATE

Counting full bottles of spirits by sight is easy, but determining the quantities and values of partial bottles visually is more difficult. You can do it rather quickly and fairly accurately, however, if you estimate the amount remaining in open bottles by using the **tenths system**. This system requires that the inventory taker assign a value of 10/10 to a full bottle, 5/10 to a half bottle, and so on. Then, when inventory is taken, the partial bottle is visually examined and the appropriate tenth is assigned, based on the amount left in the bottles. Some operators use rulers to assist in this effort since bottle shapes and sizes vary.

Although this system results only in an approximation of the actual amount in a bottle, many managers feel the tenths system is accurate enough for their purposes. It does have the advantage of being a rather quick method of determining inventory levels of open bottles held in bar areas.

In general, it is important to take liquor inventories at a time when the operation is closed so that product quantities on hand do not change when the inventory is being taken. It is also important that product contained in the lines of mechanical drink-dispensing systems be counted if the quantity of product in these lines is deemed to be significant.

FUN ON THE WEB!

Today's beverage managers benefit from the development of advanced technology systems that assist them in determining and managing their liquor inventory levels.

Wisk is one good example of a company that provides software that helps operators take accurate bar inventories, interfaces with an operation's POS to provide accurate usage data, and provides sales forecasts to estimate future product usage.

Enter "Wisk bar and restaurant inventory management software" in your favorite browser to learn more about the ways advanced technology such as theirs is being utilized to assist bars and restaurants in the completion of their liquor inventory management tasks.

DETERMINING INVENTORY VALUES

Proper inventory control requires that you monitor both the amount and value of items in your inventory. Valuing or establishing a dollar value for an inventory item is performed using the following item inventory value formula:

Item amount × Item value = Item inventory value

An item's actual value can often be more complicated to determine than its amount. This is because the price you pay for an item may vary slightly each time you buy it. For example, assume that you bought Belgian endive for $10.00 per pound on Monday, but the same item cost you $12.50 per pound on Wednesday. On Friday, you see that you have 1 pound of the curly endive in your refrigerated walk-in. Is the value of the item $10.00 or $12.50?

Item value must be determined by using either the LIFO or the FIFO method. When the LIFO method is in use, the item's value is said to be the price paid for the least recent (oldest) addition to the item amount. If the FIFO method is in use, the item value is said to be the price paid for the most recent (newest) product on hand. In the hospitality industry, most operators value inventory at its most recently known value; thus, FIFO is the more commonly used inventory valuation method.

Inventory value is determined using a form similar to the **inventory valuation sheet** shown in Figure 4.6. This form has a place for entering all inventory items, their quantity on hand, and the unit value of each. There is also a place for the date the inventory was taken, a spot for the name of the person who counted the product, and another space for the person who extends (calculates) the monetary value of the inventory. It is recommended that these two be different individuals to reduce the risk of inventory valuation fraud.

The inventory valuation sheet should be completed each time the inventory is taken (counted). It can be manually prepared but is now most often produced as part of an inventory valuation software program. Regardless of the system used, each item's total inventory value is determined using the inventory value formula.

FIGURE 4.6 Inventory Valuation Sheet

Unit Name: ______________________ **Inventory Date:** ______________________

Counted By: ______________________ **Extended By:** ______________________

Item	Unit	Item Amount	Item Value	Inventory Value
			Page Total	

Page __________ **Of** __________

Thus, if we have five cases of fresh beets in our inventory and each case has a value of $35.00, the inventory value of our beets is:

Item amount × Item value = Item inventory Value

or

5 cases × $35.00 per case = $175

The process of determining inventory value requires that you or a member of your staff count all food products on hand and multiply the value of the item by the number of units on hand. The process becomes more difficult when one realizes that the average foodservice operation has hundreds of items in inventory. Thus, taking the inventory can be a very time-consuming task.

A **physical inventory**, one in which the food and beverage products are actually counted, must, however, be taken to determine your actual product usage. Some operators take this inventory monthly, others weekly or even daily to determine their inventory amounts on hand and as an aid to calculating their actual food and beverage usage and their product cost percentages (see Chapter 5).

HERE'S HOW IT'S DONE 4.1

Determining the total value of your inventory items is actually an easy process. It simply requires that you multiply the number of units you have in inventory by the value of each unit. Then you sum each of the individual product values. In the foodservice industry, there are a variety of different units you will buy. Some examples include gallons, cases, pounds, bags, boxes, and crates. Regardless of the unit in which an item is purchased, however, managers always use the same formula to establish the individual item's inventory value:

Item amount × Item value = Item inventory value

When using this formula, the item amount is equal to the amount held in inventory. The item value is equal to the item's unit cost. Thus, the amount in inventory multiplied by the item's unit cost equals its inventory value. To illustrate, assume an inventory item is purchased in gallons. If 1.5 gallons are currently in inventory and if each gallon costs your operation $18.00, the calculation is

1.5 gallons in inventory × $18.00 cost per gallon = $27.00 inventory value

As you can see from the table below, regardless of the unit size used to count the amount in inventory, the process used to calculate item values is exactly the same.

After the value of each individual inventory item is calculated, it is easy to use addition to determine the value of your operation's total inventory.

As you will learn in Chapter 5, knowing the value of your inventory at the beginning and at the end of an accounting period is essential when you want to accurately calculate your operation's overall food cost for that accounting period.

Inventory Item	Amount on Hand	Item Value	Total Item Value
2-oz. coffee packets	1.5 cases	$28.00/case	$42.00
Pure maple syrup	1.5 gal.	$48.00/gal.	$72.00
Roquefort cheese	1.5 lb.	$38.00/lb.	$57.00
Sugar packets	1.5 boxes	$18.00/box	$27.00
Balsamic vinegar	1.5 bottles	$18.00/bottle	$27.00
Baking powder	1.5 cans	$18.00/can	$27.00
Asparagus (fresh)	1.5 crates	$38.00/crate	$57.00
		Total Inventory Value	**$309.00**

PRODUCT ISSUING AND RESTOCKING

Issuing is the formal process of removing needed beverage, food, and supply products from inventory. In smaller properties, issuing may be as simple as entering a locked or unlocked storeroom, selecting the product, and closing the door behind you. In a more complex operation, especially one that serves alcoholic beverages, this method may be inadequate to achieve appropriate control.

The issuing of products should occur only after the product has been requisitioned. A **requisition** is a formal request to have products issued from storage. The act of requisitioning products from the storage area need not be unduly complex. Sometimes foodservice managers create difficulties for their workers by developing a requisition system that is far too time-consuming and complicated. The difficulty in such an approach usually arises because management hopes to equate products issued with products sold without taking a physical inventory. In reality, this process is difficult, if not impossible, to carry out.

Consider, for example, that you manage the bar area of Scotto's Supper Club. Scotto's is a high-volume steakhouse with an upscale clientele. If, on a given night, you attempt to match liquor issued to liquor sold, you would need to assume that all liquor issued on a given day is sold on the same day and that no liquor issued on a prior day was sold on the given day. This, of course, will not be the case. Similarly, in the kitchen, some items issued today, for example, ice cream, may be sold over several days; thus, in the same way as for liquor, food products issued will not relate exactly to products sold. It is simply good management to view an issuing system as one providing basic product security and to view a total inventory control/cost system as a separate entity entirely. Given that approach, let us examine how an issuing system can be designed to help protect the security of food and beverage products.

Employees should always requisition food and beverage items based on management-approved estimates of future sales and product production schedules. Although special care must be taken to ensure that employees use the products for their intended purpose, maintaining product security can be achieved with relative ease if a few key principles are observed:

1. Food, beverages, and supplies should be requisitioned only as needed based on approved production schedules.
2. Needed items should be issued only with management approval.
3. If a written record of issues is to be kept, each person removing food, beverages, or supplies from the storage area must sign, acknowledging receipt of the products.
4. Products that do not ultimately get used should be returned to the storage area and their return should be recorded.

Some larger foodservice operators employ a full-time storeroom person to operate their requisition program. This process can sometimes be helpful because requisition schedules for tomorrow's food, for instance, can be submitted today, thus allowing storeroom personnel the time to gather the items from inventory before delivering them to the kitchen. Occasionally, products are even weighed and measured for kitchen personnel, according to the standardized recipes to be prepared. When this system is in place, the storeroom is often called an **ingredient room**.

Figure 4.7 illustrates the kind of food and supplies requisition form you might use at Scotto's Supper Club.

Note that the requested amounts and issued amounts may vary somewhat. In the case of rice, it is relatively easy to issue exactly the amount requisitioned because the rice can be weighed. In the cases of broccoli and strip steak, however, the nature of the product itself may make it impossible to exactly match the amount

FIGURE 4.7 Storeroom Requisition

Unit Name: Scotto's Supper Club
Requisition #: 0221
Date: 1/15

Item	Storage Unit	Requested Amount	Issued Amount	Unit Cost	Total Cost
Rice	1 lb.	5 lb.	5.0 lb.	$0.55/lb.	$2.75
Broccoli	1 lb.	30 lb.	28.5 lb.	$0.90/lb.	$25.65
Rib roast	1 lb.	100 lb.	103.5 lb.	$18.40/lb.	$1904.40
Total					$1932.80

To: Kitchen X
Bar ______

Requisition Approved By: S.A.R.
Requisition Filled By: T.A.P.

issued with the amount requisitioned. In other cases, the storeroom may be out of a requested item completely.

Note also that the total cost of issues is arrived at by computing the value of the issued amount, not the requisitioned amount. This is so because it is the value of the issued amount that has actually been removed from inventory and, thus, will ultimately be considered when determining amounts to purchase in the future and the cost of the products you have sold.

It is important that a physical or electronic copy of the storeroom requisition form be sent to the person responsible for purchasing after it has been prepared so that this individual will know about the movement of products in and out of the storage areas and thus the amount of product that remains in inventory.

SPECIAL CONCERNS FOR ISSUING BEVERAGES

The basic principles of product issuing that apply to food and supplies also apply to beverages. There are, however, special concerns that must be addressed when issuing beverage products from liquor storerooms and wine cellars:

1. Liquor storeroom issues
2. Wine cellar issues

LIQUOR STOREROOM ISSUES

Although various systems could be used for issuing liquor, many managers favor the **empty for full system** of liquor replacement. In this system, each bartender is required to hold empty liquor bottles in the bar or a closely adjacent area. At the conclusion of the shift, or at the start of the next shift, each empty liquor bottle is replaced with a full one. The empty bottles are then either broken or disposed of as local beverage law requires.

It is important to note that all liquor bottles issued from the liquor storage area should be visibly marked in a manner that is not easily duplicated. This allows management to ensure, at a glance, that all liquor to be sold is the property of the foodservice operation, and not that of a bartender who has produced his or her own bottle for the purpose of going into an illicit partnership with the operation. In a partnership of this type, the operation supplies the guest while the bartender provides the liquor and then pockets the product sales and profit! Although bottle marking will not completely prevent dishonest bartenders from bringing in their own liquor, it will force them to pour their product into the operation's marked bottles and to dispose of their own empties, which makes revenue-related theft of this type much more difficult.

WINE CELLAR ISSUES

The issuing of wine from a wine cellar is a special case of product issuing because these sales cannot be predicted as accurately as sales of other alcoholic beverage products. That is, you may know that a given percentage of your guests are likely to select wine, but you are not likely to know the specific wines they will select. This is especially true in an operation where a large number of wines are contained on a wine list. If the wine storage area contains products valuable enough to remain locked, it is reasonable to assume that each bottled wine issued should be noted. You can use a form such as the one shown in Figure 4.8 to record your operation's wine-issuing activity.

A form of this type may be used to identify wine sold to guests, wine transferred to a bar area to fill guests' wine by the glass orders, or for use in the kitchen. If the wine is to be sent to a guest as complimentary or a (no charge) "**comp**," that can be noted as well, along with the initials of the management personnel authorizing the comp. In the case of the wine cellar issue, the paper or electronic form itself should remain in the wine cellar for use by the wine-purchasing agent and a copy should, on a daily basis, be available to management for review.

RESTOCKING INVENTORY

Regardless of the methods used by employees to requisition food and beverage products or the systems management uses to issue these products, inventory levels in an operation will be reduced as sales are made. It will be your responsibility to monitor reductions in product inventories and purchase additional products as needed. Restocking your inventory is critical if product shortages are to be avoided and if products needed for menu item preparation are to be available.

Nothing is quite as traumatic for the foodservice manager as being in the middle of a busy meal period and finding that the operation is out of a necessary ingredient or frequently requested menu item. For most managers, it would be very time consuming to monitor the amount of each ingredient, food product, and all supplies in inventory on a daily basis. The average foodservice operation stocks hundreds of ingredients and items, each of which may or may not be used every day. The task could be overwhelming.

Imagine, for example, the difficulty associated with monitoring, on a daily basis, the use of each sugar packet in a high-volume restaurant. Taking a daily inventory of the use of such a product would be akin to spending $10 to protect a penny! The effective foodservice manager knows that proper control involves spending time and effort where it is most needed and can do the most good. It is for this reason that many operators practice the ABC method of inventory control. To understand

FIGURE 4.8 Wine Cellar Issues

Product	Vintage	Number of Bottles	Check #	Removed By
1. Bolla Soave	2016	2	60,485 L	T.A.
2. Glen Ellen Cabernet Sauvignon	2015	1	60,486 L	S.J.
3. Barton & Guestier Medoc	2005	1	Manager "comp"	S.A.R.
4. Copperridge Cabernet	Current stock	1	Kitchen	S.A.R.
5. Lindeman's Chardonnay	2017	1	60,500 M	S.J.
6. Copperridge Cabernet	Current stock	1	Bar stock	S.A.R.
7.				
8.				
9.				
10.				
11.				
12.				

Remarks: #4 Requested by Chef 1/15

#6 House Wine Sent to Bar Area 1/15

the principles utilized by the ABC inventory control system, however, you first must first understand the concepts of physical inventory and perpetual inventory.

PHYSICAL AND PERPETUAL INVENTORIES

A physical inventory system is one in which an actual physical count and valuation of all inventory on hand is taken at the close of an accounting period. A **perpetual inventory** system is one in which the entire inventory is counted and recorded, and then additions to and deletions from total inventory are recorded as they occur.

Both physical and perpetual inventories have advantages and disadvantages for the foodservice operator. The physical inventory, properly taken, is the most accurate of all because each item is actually counted and then valued. It is the physical inventory, taken at the end of the accounting period (the ending inventory), that is used in conjunction with the beginning inventory (the ending inventory value from the prior accounting period) to compute an operation's cost of food or beverages sold (see Chapter 5).

Despite its accuracy, however, the physical inventory suffers from a significant disadvantage: It can be extremely time-consuming. Even with the use of software programs that can extend inventory (multiply the number of units on hand by each unit's value) or handheld scanners that assist in the process, counting each food and beverage item in storage can be a cumbersome and time-consuming task. This is so even for the well-trained individuals who will carefully count and weigh the inventoried items.

Perpetual inventory seeks to eliminate the need for frequent counting by adding to the inventory when appropriate (via receiving slips) and subtracting from inventory when appropriate (via requisitions or issues). Perpetual inventory is especially popular in monitoring liquor and wine inventories, where each product may have its own inventory sheet or, in some cases, a bin card.

A **bin card** is simply a series of entries on an electronic spreadsheet (or on a physical card) that details additions to and deletions from a given product's inventory level. When using electronic or hard copy bin cards to monitor inventory levels, managers need only review the cards, rather than take a physical inventory, to determine how much product is on hand at any point in time. Figure 4.9 illustrates the use of such a card. The accurate use of a bin card, or any other perpetual inventory system, requires that every addition to, and subtraction from, the products in inventory be carefully recorded.

As bar-code-reading hardware and software programs become more popular, additions and deletions to and from inventory are increasingly recorded electronically. In the ideal situation, perpetual inventory systems, regardless of the form they take, must be verified, on a regular basis, by an actual physical inventory.

Some managers prefer to use the same form for recording both the quantity and the price of an inventory item. This can be done by using a perpetual inventory card. **Perpetual inventory cards** are simply modified bin cards, similar to the one shown in Figure 4.9, which also includes the product's price. A new perpetual inventory card or spreadsheet line is created each time the product's purchase price changes, with the quantity of product on hand entered on the new card. This system allows for continual tracking of quantity of items on hand as well as their value.

Today, managers increasingly use computer spreadsheets and specialized inventory software to maintain perpetual inventories. In addition, many modern point of sale (POS) systems include built-in inventory management components.

The accurate use of a perpetual inventory system requires that each change in product quantity be noted. However, employees, when in a hurry, may simply forget to update the perpetual system as they add or remove inventory items. Mistakes such as these will reduce the accuracy of the perpetual inventory. For this reason, it is not wise to depend solely on a perpetual inventory system for accurate cost calculations. There are, however, several advantages to the perpetual inventory system, among them the ability of the purchasing agent to quickly note the quantity of product on hand without resorting to a daily physical inventory count.

The question of which of the two systems is best arises when making the decision about whether to use a physical or perpetual inventory system. Experienced managers know that neither is best in all cases, so they select the best features of each system as needed by their own operations.

FIGURE 4.9 Bin Card

Product Name: Canadian Club			Bottle Size: 750 ml
Balance Brought Forward: 24			Date: 1/31
Date	In	Out	Total on Hand
2/1	4		28
2/2		6	22
2/3		5	17
2/4	12		29

ABC INVENTORY CONTROL

Utilizing the best features of physical and perpetual inventory systems is exactly what the ABC inventory system was designed to do. It separates inventory items into three main categories:

> **Category A** items are those that require tight control and the most accurate record keeping. These are typically high-value items, and though they are few in number, they can make up 70 to 80 percent of the total inventory value.
>
> **Category B** items are those that make up 10 to 15 percent of the inventory value and require only routine control and record keeping.
>
> **Category C** items make up only 5 to 10 percent of the inventory value. These items require only the simplest of inventory control systems.

Returning to the hypothetical example of Scotto's Supper Club, assume that the following 10 items are routinely held in your inventory:

1. Precut New York strip steak
2. Prepared horseradish
3. Eight-ounce chicken breasts (fresh)
4. Garlic salt
5. Onion rings
6. Crushed red pepper
7. Dried parsley
8. Lime juice
9. Fresh tomatoes
10. Rosemary sprigs

As can be seen, even with this short list, you have a variety of items in inventory. Some of these items, like the New York strip steak, are very valuable, highly perishable, and critical for the execution of your menu. Others, like the crushed red pepper, are much less costly, not highly perishable, and may less dramatically affect the operation if you ran out between deliveries. Clearly, these two items should not be treated the same for inventory purposes. The simple fact is that they are not equally critical to the operation's success. The ABC system helps you distinguish between those items that deserve special, perhaps daily, attention, and those items that you can safely spend less time managing.

To develop the A, B, and C categories, follow these steps:

1. Calculate monthly usage in units (e.g., pounds, gallons, cases, and the like) for each inventory item.
2. Multiply total item usage by its purchase price (unit value) to arrive at the total monthly dollar amount of product usage.
3. Rank items from highest dollar usage to lowest.

In a typical ABC analysis, approximately 20 percent of the items held in inventory will account for about 70 to 80 percent of the total monthly product cost. These represent the A product category. It is not necessary that the line separating A, B, and C products be drawn the same for every operation. Many operators use the following guide, but it can be adapted as you see fit:

Category A: Top 20 percent of items
Category B: Next 30 percent of items
Category C: Next 50 percent of items

It is important to note that, although the percentage of items in category A is small, the percentage of total monthly product cost these items represent is large. Alternatively, the number of items in category C is large, but the total dollar value of product cost these items account for is rather small. It is important to note that the ABC inventory system is concerned with the monetary value of products, not the number of items. Returning to the Scotto's example may help make this distinction of the ABC system clear. One item on your menu is New York strip steak. The preparation of this item is simple. Your cook sprinkles the steak with seasoning salt and cooks it to the guest's specification. The steak is then garnished with one large, deep-fried onion ring (which you buy frozen). In this example, these inventory items would likely be grouped as follows:

Scotto's Strip Steak	
Ingredient	**Inventory Category**
New York strip steak	A
Onion ring	B
Garlic salt	C

Figure 4.10 shows the complete result of performing an ABC analysis on the 10 ingredient items listed previously and then ranking those items in terms of their inventory value.

The ABC inventory system specifically directs your attention to the areas where it is most needed. Controlling product costs, especially for category A items, is extremely important. Figure 4.11 details the differences in how you would manage items in the A, B, and C inventory categories.

The ABC system focuses management's attention on the essential few items in inventory, while focusing less attention on the many low-cost, slower-moving items. Again, it is important to note that management's time is best spent on the items of most importance. In the case of inventory management, these are the category A and, to a lesser degree, category B items.

The ABC system can also be used to arrange storerooms or to determine which items should be stored in the most secure areas. Regardless of the inventory management system used, however, whether it is the physical, perpetual, or ABC

FIGURE 4.10 ABC Inventory Analysis on Selected Inventory Items

Item	Monthly Usage	Purchase Price	Monthly Usage	Category
Precut New York strip steak	300 lb.	$18.40/lb.	$5,520.00	A
8-oz. chicken breasts (fresh)	450 lb.	$3.10/lb.	$1,395.00	A
Fresh tomatoes	115 lb.	$1.95/lb.	$224.25	B
Onion rings	30 lb.	$2.20/lb.	$66.00	B
Rosemary sprigs	10 lb.	$4.50/lb.	$45.00	B
Prepared horseradish	4 lb.	$3.85/lb.	$15.40	C
Lime juice	2 qt.	$4.10/qt.	$8.20	C
Garlic salt	2 lb.	$4.95/lb.	$9.90	C
Crushed red pepper	1 oz.	$16.00/lb.	$1.00	C
Dried parsley	4 oz.	$4.00/lb.	$1.00	C

FIGURE 4.11 Guide to Managing ABC Inventory Items

Category	Inventory Management Techniques
A	1. Order only on an as-needed basis. 2. Conduct perpetual inventory on a daily or, at least, weekly basis. 3. Have a clear idea of purchase point and estimated delivery time. 4. Conduct monthly physical inventory.
B	1. Maintain normal control systems; order predetermined inventory (par) levels. 2. Monitor more closely if sale of this item is tied to sale of an item in category A. 3. Review status quarterly for movement to category A or C. 4. Conduct monthly physical inventory.
C	1. Order in large quantity to take advantage of discounts if item is not perishable. 2. Stock consistent levels of product. 3. Conduct monthly physical inventory.

inventory, management must be strict in monitoring both withdrawals from inventory and the process by which inventory is restocked.

PURCHASE POINTS

When your inventories are properly controlled and issues from inventories are carefully recorded, you will be in an excellent position to determine when your inventories should be restocked. You can do so by creating a purchase point for each item you hold in inventory. A **purchase point** is the inventory level at which an item should be reordered. An item's purchase point can be identified by one of two methods:

1. As needed (just-in-time)
2. Par level

AS NEEDED

When you choose to use the **as-needed (just-in-time)** method of determining an item's purchase point, you are basically purchasing food items based on your prediction of future sales and the amount of the ingredients (from standardized recipes) necessary to produce those sales. Then, no more than the absolute minimum of needed inventory level is purchased from your vendors. When this system is used, a buyer compiles a list of needed ingredients and submits it to management for approval to purchase. For example, in a hotel foodservice operation, the demand for 500 servings of a raspberries and cream torte dessert, to be served to a group in the hotel next week, would cause the responsible person to check the standardized recipe for this item (see Chapter 3) and determine the amount of raspberries and other key item ingredients that should be ordered. Then that amount, and no more, would be ordered from the appropriate vendors.

PAR LEVEL

Foodservice operators also set predetermined purchase points, called **par levels**, for some inventory items. A par level is the amount of an item that should be held in inventory at all times. In the case of the raspberries and cream torte dessert referred to previously, it is likely that making the torte will require a few tablespoons of vanilla extract. It does not make sense, however, to order vanilla extract by the needed tablespoon. In fact, you are likely to find that you are restricted in the quantity of it you can buy due to the vendor's delivery minimum, namely, one bottle or one case. In instances such as this, or when demand for a product is relatively constant, you may decide that the use of an as-needed ordering system will not work as well as identifying par levels for some ingredients or items you sell.

When determining par levels, you actually establish both minimum and maximum amounts to be held in inventory. Many foodservice managers establish a minimum par level by computing working stock and then adding 25 to 50 percent more for safety stock. Then, an appropriate purchase point, or point at which additional stock is purchased, is determined. If, for example, you have decided that the inventory level for coffee should be based on a par system, the decision may be made that the minimum amount (based on your past usage) of coffee that should be on hand at all times is four cases. This would be the minimum par level. To ensure you never ran out of coffee, however, assume that you set the maximum par level of ten cases. Although the inventory level in this situation would vary from a low of four cases to a high of ten cases, you would be assured that you would never have too little or too much of this particular menu item.

When cases of coffee were to be ordered under this system, you would always attempt to keep the number of cases on hand between the minimum par level (four cases) and the maximum par level (ten cases). The purchase point in this example might be six cases; that is, when your operation had six cases of coffee on hand, an order would be placed with the coffee vendor. The intention would be to get the total stock of coffee up to ten cases before your supply got below four cases. Delivery might take one or two days, so six cases might be an appropriate purchase point.

Whether you use an as-needed, a par-level, or a combination of both systems, each ingredient or menu item should have a preset and desired management-designated inventory level or amount. As a rule, highly perishable items should be ordered on an as-needed basis, whereas items with a longer shelf life can often have their inventory levels set using a par-level system. The answer to the question "How much of each ingredient should I have on hand at any point in time?" must come from you and from your estimate of future sales.

Restocking inventory requires the use of the purchase order (PO) described in Chapter 3. Before you can place an order, however, you will need to find out what you need! The process is much like going through your own home and making a shopping list before you go to the grocery store. Some managers say that they intuitively know what they need without looking at their storage areas, inventory valuation sheets, or records of requisitions and issues. If you have ever gone grocery shopping without a list, however, you know that you can easily miss a few items. For that reason, making a detailed assessment of your present inventory prior to placing orders is always a good idea.

A good way to make sure that you have checked all your in-storage items is to use a **daily inventory sheet** (Figure 4.12). A daily inventory sheet will have the items listed in your storage areas, their unit of purchase, and their par values preprinted on the sheet or in a inventory management software program. In addition, the form will have the following columns:

- On hand
- Special order
- Order amount

Having a preprinted list of all your items in storage and units of purchase is important so that you will not have to write down the items every time you take the inventory. The list should be arranged in the same order that you store the items so that you can quickly and easily locate your products. The par value should be listed so that you know how much inventory you should have in storage at any given time.

To use the daily inventory sheet, you would physically walk through your storage areas to determine which items (and amounts) you should order. Under the

FIGURE 4.12 Daily Inventory Sheet

Item	Description	Unit	Par Value	On Hand	Special Order	Order Amount
Hot wings	15 lb. IQF	Case	6	2	3	7
Baby back ribs	2–2 ½ lb.	Case	3	1.5	1	3
Sausage links	96, 1 oz.	Case	5			
Drummies	2–5 lb.	Case	4			
Bologna	10 lb. avg.	Each	3			
Beef pastrami	10 lb. avg.	Each	4			
Slice pepperoni	10 lb. avg.	Each	6			
All-beef franks	4/1, 10 lb.	Case	7			

On Hand column, you would record how many of each item you have on hand, that is, sitting on the shelf. You also need to list any "Special order" amounts needed above the par level. An example of this would be extra cases of perishables, such as raspberries, ordered for a banquet. Then, you must calculate the order amount as follows:

$$\text{Par value} - \text{On hand} + \text{Special order} = \text{Order amount}$$

In Figure 4.12, the order amount for hot wings would be calculated as follows:

Order Amount for Hot Wings

$$6 \text{ cases par value} - 2 \text{ cases on hand} + 3 \text{cases special order} = 7 \text{ cases to be ordered}$$

For example, if you have less than a whole purchase unit on hand, such as a $\frac{1}{2}$ case of baby back ribs as in Figure 4.12, then you may need to round the order amount up to a full case as follows:

Order Amount for Baby back Ribs

$$3 \text{ cases par value} - 1.5 \text{cases on hand} + 1 \text{ case special order} = 2.5 \text{ cases to be ordered}: \text{(rounded up to 3 cases)}$$

Although the form in Figure 4.12 is called the daily inventory sheet, this does not mean that you always have to check your inventory or prepare purchase orders on a daily basis. Rather, you should always make your grocery lists on, or as close as reasonably possible to, your normal ordering days.

MANAGING FOOD PRODUCTION

After you have ordered and received the food and beverage products you forecast your guests will buy, your concern will turn to the important function of controlling the food and beverage production process. If any single activity stands at the heart of foodservice management and cost control, this is it. To study this process, assume again that you are the manager of Scotto's Supper Club. As you prepare for another week of business, you review your sales history, sales forecasts, purchase orders, and menu specials. You do these things to begin taking the first step in controlling your food production costs.

CONTROLLING FOOD PRODUCTION COSTS

Often, those individuals who manage restaurants do so because they enjoy managing **back of house (BOH)**, or kitchen production area, activities. Managing the BOH to produce tasty, attractive, nutritious, and cost-effective menu items is one of the most challenging and enjoyable aspects of foodservice management. Properly managing the food production process means controlling five key areas:

1. Waste
2. Overcooking
3. Overportioning
4. Improper carryover utilization
5. Inappropriate make or buy decisions

WASTE

Food losses through simple product waste will cause excessive food costs. This waste may be easy to observe, such as when an employee does not use a rubber spatula to get all of the salad dressing out of a 1-gallon jar, or as difficult to detect as the shoddy work of the salad preparation person who trims the lettuce just a bit more than management would prefer, resulting in a reduced amount of usable lettuce available to make salads and in higher per-portion salad costs.

Management must demonstrate its concern for the value of products on a daily basis. Each employee should be made to realize that wasting food directly affects the profitability of the operation and, thus, their own economic well-being. In general, food waste is most often the result of poor training or management's inattentiveness to detail. Unfortunately, some managers and employees feel that small amounts of food waste are unimportant. Your primary goal in reducing waste in the food production area should be to maximize product utilization and minimize the dangerous "it's only a few pennies worth so it doesn't really matter" syndrome.

OVERCOOKING

Cooking is simply the process of exposing food to heat. Excessive cooking, however, most often results in reduced product volume, whether the item cooked is roast beef or vegetable soup. This is because many foods have high moisture contents and heating usually results in moisture loss. To minimize this loss, cooking times and methods listed on standardized recipes must be carefully calculated and meticulously followed.

It is important to recognize that, in many ways, excess heat is the enemy of both well-prepared foods and a manager's cost control efforts. Too much time on the steam table line or in the holding oven or cabinets extracts moisture from products, reduces product quality, and results in fewer portions that are available for sale and service. The result is increased food cost. To illustrate, note that Figure 4.13

FIGURE 4.13 Effect of Overcooking on Portion Cost

Portion Cost of 50 lb. (800 oz.) of Roast Prime Rib of Beef			
			50 lb. cost = $900
Preparation State	Ending Weight (oz.)	Number of 8-oz. Portions	Portion Cost
Properly prepared	800	100	$9.00
Overcooked 15 min.	775	97	$9.28
Overcooked 30 min.	750	94	$9.57
Overcooked 45 min.	735	92	$9.78
Overcooked 60 min.	720	90	$10.00
Overcooked 75 min.	700	88	$10.23
Overcooked 90 min.	680	85	$10.59

details the change in **portion cost**, which is the cost of producing one serving of a menu item, when the ending weight of a roast prime rib of beef is reduced due to overcooking.

If we assume that a properly cooked pan of roast prime ribs of beef yields 50 servable pounds at a cost of $18.00 per pound, the total product cost would equal $900.00 (50 lb. × $18.00/lb. = $900.00).

When properly cooked the roast beef would yield 100 eight-ounce portions, for a cost of $9.00 per portion ($900 product cost/100 portions = $9.00 portion cost). As you can see in Figure 4.13, increased cooking time or an excessive final product temperature will result in moisture loss, cause product shrinkage, and increase the item's portion cost.

Although the difference between a portion cost of $9.00 and $10.59 may seem small, it is the small numbers that can really add up and increase costs. Thus, it is the control of this type of production issue that often separates the good foodservice manager from the outstanding one.

To control product loss due to overcooking, you must strictly enforce standardized recipe cooking times. This is especially true for high moisture content items such as meats, soups, stews, baked goods, and the like. Moreover, extended cooking times can result in total product loss if properly prepared items are placed in an oven, fryer, steam equipment, or broiler and then "forgotten" until it is too late! It is, therefore, advisable to supply kitchen production personnel with small, easily cleanable timers and thermometers for which they are responsible. These can help substantially in reducing product losses due to overcooking.

OVERPORTIONING

Probably no other area of food and beverage cost control has been analyzed and described as fully through articles, speeches, and even books as the control of portion size. There are two reasons for this. First, overportioning on the part of service personnel has the effect of increasing operational costs and may cause the operation to mismatch its production schedule with anticipated demand.

For example, assume that 100 guests are expected and thus 100 products to be served to them are produced, yet overportioning causes you to be out of the product after only 80 guests have been served. The remaining 20 guests will be left clamoring for "their" portions, which, of course, no longer exist due to overportioning; these portions have already been served to other guests.

Also, overportioning must be avoided because guests always want to feel that they have received fair value for their money. If portions are large one day and small the next, guests may feel that they have been cheated on the second day. Consistency is a key to operational success in foodservice. Guests want to know exactly what they will get for their money, and they want to get that amount every time they buy.

HERE'S HOW IT'S DONE 4.2

Portion cost is the amount it costs an operation to produce one serving of a menu item. Portion cost is one important factor in establishing the selling prices of menu items. For that reason and for many others, the ability to calculate a portion cost correctly is an important management skill.

In some cases, the calculation of the portion cost is very easy. For example, assume a manager sells fresh apples. The manager buys apples in 10-pound bags that contain 30 apples. Each bag costs $21.60. The manager would use division to calculate the portion cost for one apple:

$$\frac{\$21.60 \text{ cost per bag}}{30 \text{ apple portions}} = \$.72 \text{ portion cost}$$

In other situations, managers utilize the cost required to produce a standardized recipe when they calculate their portion costs. For example, assume a manager sells custard pie. The manager's standardized recipe for custard pie produces seven pies. Each pie will be cut into eight pieces. Thus, the standardized recipe produces 7×8 or 56 pie portions.

If the cost of producing the standardized recipe, including the pastry for the crust, eggs, sugar, salt, vanilla, milk, and nutmeg, is $25.20, the manager would again use division and the same portion cost formula to calculate the cost of producing one slice of custard pie:

$$\frac{\$25.20 \text{ recipe cost}}{56 \text{ pie portions}} = \$0.45 \text{ portion cost}$$

Because the vast majority of items served by restaurateurs should always be produced from standardized recipes (see Chapter 3), it is easy to see why adherence to proper recipe production and proper portioning is so critical to accurately calculating and controlling an operation's portion costs.

It is not possible to state what a "correct" portion size for menu items should be because it is not advantageous to establish the same standard portion sizes for all foodservice operations. For example, the proper portion size of a specific entrée in a college residence hall feeding only male athletes should clearly be different from that same item served in an extended-care facility whose typical resident might be over 75 years old.

It is important for you to consider your clientele, facility ambiance, pricing structure, and desired quality standards prior to establishing appropriate portion sizes for your own operation. Once portion size for an item has been established, however, it is up to you to strictly enforce it.

Often, employees resist management's efforts to control portion size. When this is the case, it is a clear indication that management has failed in its obligation to provide employees with a basic understanding that underlines the foodservice industry. Employees must be made to see that strict adherence to predetermined portion size is a benefit both to the guest and to the operation. Management must be sensitive, also, to the fact that it is the server who often must deal with the guest who complains about the inadequacy of portion size. Therefore, servers must be made to feel comfortable about predetermined portion sizes so that management, along with the employees, will want to maintain them.

In many cases, tools are available that will help employees serve the proper portion size. Whether these tools consist of scales, food scoops, ladles, dishes, or spoons, employees must have an adequate number of easily accessible portion control devices if they are to use them. One commonly used tool that assists in portion control is the food scoop. **Food scoops** are sized based on the number of servings per quart of product served. Thus, a #12 scoop will yield 12 servings per quart, a #20 scoop will yield 20 servings per quart, and so on.

It is important to understand that many portion sizes are directly tied to the purchasing function. To serve a ¼-pound hot dog, for instance, one must purchase ¼-pound hot dogs. In a similar vein, if one full banana is sliced for addition to breakfast cereals, the purchasing agent must have been diligent in ordering and accepting only the banana size for which management has developed a specification.

The constant checking of portion sizes served is an essential task of management. When incorrect portion sizes are noticed, they must be promptly corrected. If not, considerable cost increases can be incurred. Returning again to our example of Scotto's Supper Club, consider that you purchase, on occasion, and in accordance with your menu plan, 3-pound boxes of frozen yellow corn to be served as your vegetable of the day. Each box costs \$8.80. With a total of 48 ounces (3 lb. × 16 oz. = 48 oz.) available and an established portion size of 3 ounces, you know that you should average 48 oz./3 oz. = 16 servings per box of corn. Figure 4.14 demonstrates the effect on total portion cost if one, two, or three servings are lost due to overportioning.

As Figure 4.14 shows, a small amount of overportioning on an item as inexpensive as corn costs the operation only a few cents per serving. Those few cents per serving, however, multiplied time after time, can mean the difference between a profitable operation and one that is only marginally successful. If your portion cost for corn should have been 55.0 cents (\$8.80/16 portions = 55.0 cents), but, due to overportioning, it rises to 67.7 cents (\$8.80/13 portions = 67.7 cents), then your food costs are 12.7 cents higher than they should be on this item. If Scotto's is open seven days a week and serves an average of 200 portions of corn per day, your total loss for a year would be 365 days × 200 portions served per day × \$0.127 lost per portion = \$9,271 total loss, an amount that would be more than enough to buy your staff proper portioning tools. It is also an amount worthy of your attention and correction.

IMPROPER CARRYOVER UTILIZATION

As addressed earlier in this text, predicting future guest counts is often an inexact process. Because this is true, and because most foodservice operators want to offer the same broad menu to the evening's last diner as was offered to its first, it is inevitable that some prepared food will remain unsold at the end of the operational day. These items are called **carryovers** or, in some operations, leftovers.

In some segments of the hospitality industry, carryovers are a potential problem area; in others, it is less of a concern. Consider, for example, the operation of a gelato shop. At the end of the day, any unsold gelato is simply held in the freezer until the next day with no measurable loss of either product quantity or quality. Contrast that situation, however, with a full-service cafeteria. If closing time is 8:00 P.M., management wishes to have a full product line, or at least some of each menu item, available to the guest who walks through the door at 7:55 P.M. Obviously, in five more minutes, a large number of displayed items will become carryovers. This cannot be avoided. A manager's ability to effectively integrate carryover items on subsequent days can make the difference between profits and losses in some operations.

You have learned that in almost every case, food products are at their peak of quality when they are delivered to your restaurant. From then on, storage, preparation, and holding activities often diminish product quality. These forces are especially at work in the area of carryovers. It is for this reason that production schedules must note carryover items on a daily basis. If this is not done, these items tend to get stored and then may get lost in walk-in refrigerators or freezers.

You should have a clear use in mind for each menu item that may have to be carried over. For example, broiled or sautéed fish may later be used to prepare seafood chowder or bisque. Today's prime rib roast may be the key ingredient in tomorrow's beef Stroganoff, and so on. Menu specials, substitutions, and employee meals can be sources of utilization for products such as these. This utilization process can be a creative one and, if you involve your staff, quite effective.

It is important to understand that carryover foods seldom can be sold for their original value. Today's beef stew made from yesterday's prime rib will not likely be sold at prime rib prices. Carryovers generally mean reduced income relative to product value and less profits; thus, it is critical that you strive to minimize carryovers.

Carryover items that are to be reused should be properly labeled, wrapped, and rotated so that these items can be found and reused easily, resulting in both greater employee efficiency in locating carryover items and reduced energy costs because refrigerator doors will be opened for shorter periods of time. Managers will find that requiring carryover foods to be properly labeled and stored in clear glass or plastic containers helps greatly in this regard.

INAPPROPRIATE MAKE OR BUY DECISIONS

Many foodservice operators choose to buy some food products that are preprepared in some fashion. These items, called **convenience foods**, or **ready foods**, can be of great value to your operation. Often, they can save dollars spent on labor, equipment, and hard-to-secure food products. They can also add menu variety beyond the skill level of the average kitchen crew. A disadvantage, however, is that these items tend to cost more on a per-portion basis. This is to be expected because these items most often will include a built in charge for the labor needed to make the item, enhanced packaging and, of course, the food item itself.

Convenience item usage is not, of course, an all-or-nothing operational decision. Nearly all foodservice operations today use canned products, sliced bread, precut produce, and the like—items that would have been considered convenience items years ago. Therefore, the question is not whether to use convenience items but, rather, it is which convenience items are best to use in your own operation.

In general, the following guidelines are of value when determining whether to adopt the use of a specific convenience product:

1. Is the product's quality acceptable? This question must be answered from the point of view of the guest, not management alone.
2. Will the product save labor costs? Identifiable labor savings must be discovered if management is to believe that the convenience item will reduce overall costs.
3. Would it matter if the guest knew? If an operation has built its image on featuring made-on-premise items, guests may react negatively if they know that an item has actually been prepared from a boil-in-a-bag package or simply heated in a microwave oven.

FIGURE 4.14 Portion Cost Chart for Frozen Corn

Corn Portion Cost @ $8.80 per 3-lb. Box		
Number of Portions per 3-lb. Box	Portion Size (oz.)	Portion Cost (cents)
16	3.0	55.0
15	3.2	58.7
14	3.4	62.9
13	3.7	67.7

4. Does the product come in an acceptable package size? If convenience items are not sold in a size that complements the operation, excessive waste can result.
5. Is appropriate storage space available? Many convenience items must be stored in a refrigerated or frozen form. Your facility must have the storage capacity needed for these items.

It is important to recognize that a kitchen's physical design will also greatly impact an operator's ability to produce food items cost-effectively. A properly designed kitchen facility will:

1. Reflect the operation's unique concept
2. Support the operation's specific menu offerings
3. Match the operation's estimated volume
4. Match the operation's guest-service style (e.g., cafeteria, counter-service, table-service, drive-thru, or carryout)
5. Provide ample space for needed production equipment
6. Include adequate holding, refrigerator, freezer, and dry goods storage space
7. Provide ample employee work space
8. Include adequate space for bulk hot and cold food preparation
9. Provide sufficient electrical outlets, utility hook-ups, and ventilation
10. Be easily cleanable

Green and Growing!

A foodservice operation's cost of food is affected by a variety of factors. One such factor relates to the source reduction decisions made by the operation's suppliers. Where recycling occurs within the foodservice operation and seeks to reuse materials, suppliers utilize source reduction to minimize the amount of resources initially required to package, store, and ship the items they sell. The result of effective source reduction is a lessened impact on the environment and lower product costs.

Consider, for example, the difference in the producer's cost between packaging, storing, and shipping frozen orange juice concentrate (which foodservice operators will reconstitute on-site) and ready-to-serve juice purchased in plastic or glass containers. Not only will packaging costs be higher in the second case (because of the greater product volume), but the item's greater weight will also mean higher transportation costs. Storage costs, too, will be greater with the ready-to-serve item, as will the cost of disposing of used product containers.

Packaging and energy costs add up, and these costs will inevitably be passed on from the manufacturer and/or vendor to the foodservice operation. The result is higher food costs and potentially lowered profits. Creative source reduction efforts, however, can result in fewer wasted natural resources and lower food costs. Foodservice operators should encourage these efforts by purchasing quality products from those suppliers who take their source reduction responsibilities seriously and act upon them.

FOOD PRODUCTION SCHEDULES

Foodservice managers are in charge of kitchen production. How much of each item to prepare each day may be a joint decision between you and your chef or kitchen production manager, but it is you who must ultimately take the responsibility for proper production decisions. A complete food production schedule process will require you to do the following:

1. Maintain sales histories.
2. Forecast future sales levels.
3. Purchase and store needed food and beverage supplies.
4. Plan daily production schedules.
5. Issue needed products to production areas.
6. Manage the food and beverage production process.

Item #4 above (planning daily production schedules) is important because you will want to have both the products and the staff needed to properly service your guests. If, for example, you forecast that 50 chocolate cakes will be needed on a given day for the college residence hall you manage, then you must have both the products and the staff necessary to produce that number of cakes. In a similar manner, if you know that today 50 pounds of ground beef patties must be prepared for guests that will be coming to your burger restaurant, then the ground beef and the staff to cook it must be available.

Ideally, the process of determining how much of each menu item to prepare on a given day will be based on your sales forecast and would look as follows:

Prior-day carryover + Today's production = Today's sales forecast ± Margin of error

The margin of error amount should be small; however, projecting sales and guest counts is an imprecise science at best, thus most foodservice managers will find that they should produce a small amount more than they anticipate selling each day. This minimizes the chances of running out of an important menu item. Of course, with some menu items, preparation does not begin until the sale is made. For example, in most cases a New York strip steak will not actually be cooked until the guest orders it. An order for coconut cream pie, however, cannot be filled in the same manner because the pies must be made ahead. It is because of items like coconut cream pie that production sheets are necessary. Figure 4.15 illustrates a sample production sheet in use at Scotto's Supper Club.

From Figure 4.15, you can see that 15 servings of prime rib left were over from the prior day's operation. You would know that by looking at the carryover column of the *prior* day's production sheet or by taking a physical inventory. If you anticipate sales of 85 servings of prime rib it might seem that only 70 servings should be prepared (70 new servings plus 15 carryovers from the prior day equals 85). In fact, you would most likely prepare a number of new servings that is slightly higher than anticipated demand. The reason is simple: If you have more guests come in than anticipated or if more of the guests than you forecast actually select prime rib, you do not want to run out of the item. There is no industry-wide standard percentage that should be overproduced for a given item. The amount you plan to make will depend on a variety of factors, including your own knowledge of your guests and the importance of running out of a given item. Common overages tend to be 5 to 10 percent above normal sales forecasts. For purposes of this example, assume that five extra servings of prime rib is the amount of overproduction you feel is appropriate for this item. With estimated sales of 85 servings, you actually want to have 90 servings available.

In the case of broccoli, you make the decision not to carry over any broccoli that was not sold on the prior day. If any such product exists, it could be used to make soup or, if there is no appropriate use for it, discarded because the quality of previously cooked broccoli is not likely to be at the same high level as that of freshly cooked broccoli.

FIGURE 4.15 Production Schedule

Unit Name: Scotto's Supper Club

Date: 1/15

Menu Item	Sales Forecast	Planned Overage	Prior-Day Carryover	New Production	Total Available	Number Sold	Carryover
1. Prime rib	85	5	15	75	90		
2. Broccoli	160	10	0	170	170		
3. Coconut cream pie	41	0	70	0	70		
4.							
5.							
6.							
7.							
8.							
9.							
10.							
11.							
12.							
13.							
14.							
15.							
16.							
17.							
18.							

Special Instructions: Thaw turkeys for Sunday preparation

Production Manager: S. Anthony

Regardless of the type of operation you manage, you will likely find that some of your menu items simply do not retain their quality well enough to be carried over for sale in their original form. These items must be discarded or utilized as an ingredient in a different dish. Again, in the case of broccoli, proposed production exceeds anticipated demand by a small margin (10 servings). In the case of the coconut cream pie, you make the decision to produce none on this particular day. This is because this item is made in large quantities, but not each day. With 70 servings available and an anticipated demand of 41, you have more than enough to sustain your operation through this day and, perhaps, the next as well, if the quality of the carried-over pies remains high.

At the end of the evening service period, you would enter the number sold in the appropriate column and make a determination about how much, if any, of each product you will carry over to the next day. Some foodservice managers preprint their production sheets listing all menu items and thus ensure that production levels for each major menu item are assessed on a daily basis. Others prefer to use the production sheet on an as-needed basis. When this is the case, it is used daily but only for the items to be prepared that day. Either method is acceptable, but production schedules are always critical to operational efficiency.

Consider the Costs

"How many dozen should I put in the proofer?" asked Elizabeth, the new baker at the Sands Cafeteria.

Rami El-Hussieny was the day-shift operations manager, and, unfortunately, he did not know how to answer Elizabeth's question. What she wanted to know was simple enough: How many dozen rolls should be placed in the proofer in anticipation of the night's dinner business?

The problem was that the frozen dinner roll dough used at the Sands Cafeteria needed to proof for at least 2 hours prior to being baked for 15 minutes. If too many rolls were proofed, they would never be needed, but they would still have to be baked and made into bread dressing or even tossed out. If too few dozen were proofed and the night was busier than anticipated, they would run out of "Fresh Baked Rolls" (one of the restaurant's signature items) and Rami knew that the night manager would be really upset. It was a daily guess, and sometimes Rami missed the guess!

He wondered if purchasing prebaked rolls, with a shelf life of 3 or 4 days, would, despite not having been baked on site, be the best long-term solution to this problem.

1. What do you think is the main cause of Rami's difficulty?
2. Most foodservice managers would agree that fresh baked goods are very high quality and greatly enjoyed by their guests, yet many of these bake few, if any, products on-site. What are two specific reasons why higher-quality, made-on-site baked items are often not produced?
3. What is the relationship between product quality and product availability in this case study? Which product characteristic do you think is more important from the guest's point of view?

MANAGING BEVERAGE PRODUCTION

In its simplest but also its least desired form, beverage production can consist simply of a bartender that will **free-pour** drinks. Free pouring is making a drink by pouring liquor from a bottle, but without carefully measuring the poured amount. In a situation such as this, it is very difficult to control product costs. At the other end of the control spectrum are automated beverage-dispensing systems that are extremely sophisticated control devices. The following are alternative beverage production control systems that you can use based on the amount of control you feel is appropriate for your operation.

FREE-POUR

The lack of control resulting from free-pouring alcohol is significant. It should never be allowed in the preparation of the majority of drinks your bartenders will serve. It is appropriate in some settings, however, for example in wine-by-the-glass sales. In this situation, the wine glass itself serves as a type of product control device. Large operations, however, may even elect to utilize an automated dispensing system for their wines by the glass. Also, it is most often necessary for a bartender to free-pour when he or she must add extremely small amounts of a product as a single ingredient in a multi-ingredient drink recipe. An example would be a bartender who must add a very small amount of dry Vermouth when making a Dry Martini (typically 5 parts gin to 1 part dry vermouth).

FUN ON THE WEB!

Bartenders faced with a guest order for a drink with which they are unfamiliar can quickly refer to the drink's recipe by using one of the many popular applications found on smart devices that can access a virtually unlimited number of drink recipes. A variety of companies offer such apps. To review them, enter "mixology apps" in your favorite search engine and view the results.

JIGGER POUR

A **jigger** is a device (like a small cup) used to measure alcoholic beverages, typically in ounces and fraction of ounce quantities. Because jiggers are inexpensive, this control approach is also inexpensive. It is also quite portable. It is a good system to use in remote serving locations, such as a pool area, beach, guest suite, or banquet room. The disadvantage, of course, is that there is still room for employee overpouring error as well as the potential for bartender fraud (see Chapter 5).

METERED POUR SPOUT

In some situations, you may determine that a **metered pour spout**, designed to dispense a predetermined amount of liquor each time the bottle is inverted, makes good sense. Metered pour spouts are inserted into bottles and are available to dispense a variety of different quantities. When using a metered pour spout, the predetermined portion of product is dispensed whenever the bartender is called upon to serve that product.

FUN ON THE WEB!

Many managers find that metered pour spouts are very effective cost control tools. Metered spouts that dispense a predetermined amount of liquor ranging from ⅞ ounce to 2 ounces are commonly available. In electronically monitored systems, every liquor shot served is preportioned and then recorded. At the end of each shift, managers can produce a pouring report that indicates how much money bartenders should have entered into the POS system based on the amount of liquor served. To review such systems, enter "alcoholic beverage control systems" in your favorite search engine, and then review the results to see how electronically based metered pour spout systems can help managers control their liquor costs.

BEVERAGE GUN

In some large operations, beverage guns are connected directly to liquor bottles or beverage containers of various sizes. The gun may be activated by pushing a mechanical or electronic button built into the gun or the POS. In either case, the bartender may find, for example, that pushing a gin and tonic button on a gun device will result in the dispensing of a predetermined amount of both gin and tonic. Although the control features built into gun systems are many, their cost, lack of portability, and possible negative guest reaction can be limiting factors in their selection.

TOTAL BAR SYSTEM

The most expensive, but also the most complete, beverage production control solution is a **total bar system**. This system combines sales information with automated product dispensing information to create a complete revenue and production management system. Depending on the level of sophistication and cost, the total bar system can perform one or all of the following tasks:

1. Record beverage sale by product brand.
2. Record the individual who made the sale.

3. Record sales dollars and/or post the sale to a guest room folio (bill) in a hotel.
4. Measure and dispense liquor for drinks.
5. Add predetermined amounts of mixes to drinks.
6. Reduce liquor values from beverage inventory value totals as drink sales are made.
7. Prepare liquor requisitions.
8. Compute liquor cost by brand sold.
9. Monitor bartender labor costs.
10. Identify payment method (e.g., cash, digital wallet [e-wallet], credit or debit card).
11. Record guest sales by table or check number.
12. Record date and time of product sales.

RESPONSIBLE ALCOHOLIC BEVERAGE SERVICE

Alcoholic beverage sales are important in many foodservice operations. People have long been fond of alcoholic beverages because they add greatly to the enjoyment of food and friends. In moderate doses, **ethyl alcohol,** the type of alcohol found in drinkable alcoholic beverages, acts as a mild tranquilizer (depressant). In excessive doses, however, it can become toxic, causing impaired judgment and, in some cases, even death. Clearly, a foodservice manager whose establishment serves alcoholic beverages must take great care in the serving of alcoholic beverages and the monitoring of their guests' alcohol intake.

As well, most states have now enacted third-party liability legislation, which, under certain conditions, holds your business and, in some cases, you personally responsible for the actions of your guests who consume excessive amounts of alcoholic beverages in your operation. These legislative acts are commonly called **dram shop laws**. *Dram shop* is derived from the word "dram," which refers to a small drink, and the "shop" where such a drink was sold.

Dram shop laws shift some of the liability for acts committed by an individual under the influence of alcohol from that individual to the server or beverage operation that supplied the intoxicating beverage.

The Move to Off-Premise Dining

The sale of individual servings of alcohol (i.e., beer, wine, and cocktails) "to go," once banned in most states, became seen as a necessity during the COVID 19 pandemic. It has since become a common practice for restaurants and bars in many cities and towns in the United States. According to the Distilled Spirits Council's tracking, as of 2025, 29 states allow the sale of cocktails-to-go on a permanent basis. Additionally, 16 states allow restaurants and bars to use third-party delivery services for cocktails-to-go.*

Each state has its own laws regarding customer pickup and delivery, the use of third-party delivery services, how many drinks can be sold, and if food must be purchased along with the alcohol.

While it certainly helps build operational revenue, alcohol-to-go sales pose special challenges. To address some of these challenges, the National Restaurant Association (NRA) recommends that foodservice operators should:

1. **Check the alcohol-to-go laws in their own your locale.** Laws vary not only by state, but may vary by county or city as well.
2. **Find out if they are allowed to deliver alcohol, with or without food.** A number of states limit alcohol to go to pickup only, requiring customers to order and pick up drinks directly from the restaurant or bar.

In states or localities that allow delivery, some allow third-party delivery, while others require that restaurant employees deliver orders.

3. **Serve to-go alcohol in appropriate containers.** Laws vary by state and municipality.
4. **Use delivery drivers who are at least 21.****

It is also important to recognize that, though there are a few exceptions, delivery drivers (even third-party drivers), face many of the same liability issues as restaurants and bars, and thus should only be used if they have been properly trained in the responsible service of alcohol.

*https://pluribusnews.com/news-and-events/raising-the-bar-states-serve-up-permanent-to-go-alcohol-laws/#:~:text=According%20to%20the%20Distilled%20Spirits. Retrieved 10/25/2025.

**https://restaurant.org/education-and-resources/resource-library/safe-serving-tips-for-alcohol-to-go/. Retrieved 10/25/2025.

The responsible service of alcohol requires careful control of the amount of alcohol served to each guest. The proper management of a beverage operation requires strict adherence to control procedures for several reasons:

1. Beverage operations are subject to tax audits to verify sales revenue. In some states, these audits are unannounced.
2. Beverage operations can, in some cases, be closed down on the spot for the violation of a liquor law.
3. Employees in a bar may deceptively seek to become operational "partners" by bringing in their own beverage products to sell and then keeping sales revenue.
4. Detecting the disappearance of small amounts of beverage products is extremely difficult, as, for example, the loss of 8 ounces of beer from a multigallon keg.

For these reasons and others, serving standard-sized drink portions is an absolute must for the beverage operation. Although direct reference by a bartender to a standardized recipe may be necessary for only a few types of drinks, standardized recipes that detail the quantity of beverage product management has predetermined as appropriate must be strictly adhered to. That is, if management has determined that bourbon and water should be a 2-ounce portion of bourbon and a 2-ounce portion of water, then both items should be measured by a jigger or another measuring device used to dispense these predetermined amounts.

Each drink sold in a bar must have a standardized recipe or, if the preparation of the alcoholic beverage is simple, such as a glass of wine or draft beer, a standardized portion size. These standards directly affect responsible guest service, guest satisfaction, and the ultimate profitability of the operation.

Cost Control Around the World

The responsible service and consumption of alcohol is of concern to hospitality managers worldwide. As result, it is an important issue to members of hospitality associations worldwide. HOTREC is the umbrella group for trade associations representing hotels, restaurants, cafés, and similar establishments in Europe. HOTREC consists of 40 national trade associations representing the hospitality sector in 29 countries, from Portugal to Estonia and from Ireland to Cyprus. One important initiative HOTREC actively supports is the Wine in Moderation Programme.

The Wine in Moderation Programme is a pan-European initiative designed to promote responsibility and moderation in wine consumption. It seeks to educate the public about the social and health risks of excessive alcohol consumption and misuse. HOTREC members recognize that education is a powerful tool that can be used in efforts to encourage responsible alcohol service and consumption. You can learn more about the Wine in Moderation Programme and how managers working in every country can support its principles and its educational efforts to promote responsible drinking by entering "wine in moderation programme" into your favorite search engine and then reviewing the information on the program's website.

Technology Tools

In this chapter, you learned about cost control concepts related to product storage, inventory management, product issuing, and production. There are a variety of advanced technology programs available to assist managers in these key areas. These include those that:

1. Maintain current product inventory values by food category (e.g., produce, meat, dairy, and the like).
2. Create daily shopping lists resulting from the comparison of current inventory levels to planned production schedules.
3. Report daily par stock levels, storeroom issues, and total product usage.
4. Maintain electronic perpetual inventory values.
5. Compute current inventory values based on LIFO or FIFO valuation systems.
6. Maintain custom inventory product values based on management-preferred category, such as vendor, storeroom location, product type, or alpha order.
7. Identify and daily report any below-par product inventory levels.
8. Identify and report daily monetary values of cost of goods issued or sold.
9. Interface with handheld bar or QR code readers for accurate inventory count and price extension.
10. Compute inventory loss rates.
11. Develop production schedules based on management's hourly, weekly, daily, or monthly sales forecasts.
12. Create product requisition (issues) lists based on forecasted sales.

In some cases, these software programs are designed to be self-standing. In other cases, the programs are designed to operate in conjunction with, or are a built-in part of, advanced POS and back-of-house accounting systems currently on the market.

Apply What You Have Learned

Assume you are the manager of a restaurant (serving casual Italian food) that is part of a national chain. Beverage sales account for 35 % of your total sales with one-half of those sales coming from diners and one-half from guests drinking in the bar area.

An e-mail message from your supervisor arrives asking your opinion about the company converting from the manual bartending system currently in use to one that is fully automated. The system essentially controls and accounts for the quantity of alcohol poured when making drinks or serving beer or wine. Your written response to the following questions is requested.

1. How would such a system likely affect the control procedures in place at your restaurant?
2. How would guests sitting in the dining room likely perceive the system?
3. How would guests sitting at the bar likely perceive the system?
4. How would the system likely be perceived by your bartenders?

For Your Consideration

1. Confirmed cases of food-borne illness can be devastating to the reputation of a foodservice operation. How can a restaurant manager's inventory control and production systems assist in helping to reduce the chances of a food-borne illness outbreak?
2. Bartender theft is significant in many beverage operations. What are some ways that effective beverage inventory, requisition, and issuing procedures can help reduce bartender theft?

Key Terms and Concepts

The following are terms and concepts addressed in the chapter that are important for you as a manager. To help you review, define the terms below.

Inventory
LIFO (last in, first out)
FIFO (first in, first out)
QR code
Time/temperature control for safety (TCS) foods
Temperature danger zone
Freezer burn
Recodable electronic lock
Cellar temperature
Oxidation
Working stock
Safety stock
Fiscal year
Tenths system
Inventory valuation sheet
Physical inventory
Issuing
Requisition
Ingredient room
Empty for full system
Comp
Perpetual inventory
Bin card
Perpetual inventory card
Purchase point
As needed (just-in-time)
Par level
Daily inventory sheet
Back of house (BOH)
Portion cost
Food scoop
Carryover
Convenience foods (ready foods)
Free-pour
Jigger
Metered pour spout
Total bar system
Ethyl alcohol
Dram shop laws

Test Your Skills

You may download the Excel spreadsheets for the Test Your Skills exercises from the student companion website at www.wiley.com/go/dopson/foodandbeveragecostcontrol8e.

Complete the exercises by placing your answers in the shaded boxes and answering the questions as indicated.

1. Olan Haynes is responsible for managing the inventory at the Dobie Road Retirement Center. Olan purchases a variety of items by the case, but he prefers to inventory all of these items in other ways.

   For example, he purchases beets by the case. Each case of beets contains six cans. But Olan wants to calculate his inventory value based on the number of cans of beets he has in inventory, not the number of cases. Using the chart below, calculate the preferred inventory unit value $ (e.g., $ per can, jar, bottle, and the like) Olan should use for each item listed, based on the purchase unit cost of each selected item Olan holds in his inventory.

Inventory Item	Purchase Unit	Purchase Unit Cost	Preferred Inventory Unit	Inventory Units per Case	Preferred Inventory Unit Value $
Beets	1 case	$51.95	#10 can	6	
Chicken base	1 case	$106.12	Jar	12	
Catsup	1 case	$98.20	Bottle	30	
Soy sauce	1 case	$59.95	Gallon	4	
Cocoa mix	1 case	$91.55	Bag	8	

2. Karis Bashar is the kitchen manager at the Big Texan Steakhouse. Her operation offers its guests five different steak choices. Because she uses the ABC method of inventory, and steaks are a Category A item, each morning she inventories and utilizes a spreadsheet to calculate the total value of all the steaks on hand.

   This morning, Karis has already weighed the steaks in each of her five steak categories. Using the spreadsheet below, help Karis calculate this morning's total value of steaks in inventory.

   Big Texan Steakhouse: Steak Inventory; Total Value This Morning:

Inventory Item	Purchase Unit	Purchase Unit Cost	Amount in Inventory (pounds)	Total Value in Inventory
Filets	Pound	$22.11	48.5	
New York Strips	Pound	$16.52	32	
Ribeyes (10 oz.)	Pound	$18.45	65	
Ribeyes (14 oz.)	Pound	$18.45	45.5	
Top Sirloin	Pound	$11.40	36	
Steaks: Total Value				

3. Jack Stills operates the Spartan Restaurant. He is taking the end-of-month inventory in the restaurant's bar. When calculating the value of his spirits, he uses the tenths method of estimating product value. He has accounted for all of the spirits product held in his liquor storage area and is now calculating the value of his behind-the-bar inventory. He has completed this process for all of the products he serves except gin. Help him arrive at the ending inventory value for gin by completing the following portion of his spirits inventory value sheet.

Product	Cost Per 750 ml Bottle	Amount on Hand (750 ml)	Product Value
Seagram	$13.99	3.3	
Well Gin	$9.50	4.6	
Bombay	$24.50	2.2	
Bombay Sapphire	$28.95	2.9	
Tanqueray	$19.50	2.0	
Tanqueray Ten	$31.40	1.7	
Tanqueray Rangpur	$23.00	2.3	

Product	Cost Per 750 ml Bottle	Amount on Hand (750 ml)	Product Value
Gordon's	$14.75	1.5	
Hendricks	$31.95	0.8	
Total Value			

4. Kyle Minoge manages the four-person ingredient room staff at the 800-bed Parkview Hospital. One tool Kyle has developed to help monitor his inventory levels on a daily basis is a requisitions and issues record that records daily all of the inventory items requested by the kitchen's production staff and issued by Kyle's storeroom staff. The record Kyle maintains also indicates the monetary value of the issues made. Help Kyle complete part of yesterday's requisitions and issues record below.

Parkview Hospital Requisitions and Issues Record						
Item	**Inventory Unit**	**Unit Cost**	**Requested Amount**	**Issued Amount**	**Issued Unit**	**Total Cost**
Flour	25-lb. bag	$25.00	25 lb.	1	Bags	
Beef base	1-lb. jar	$13.15	3 lb.	3	Jars	
Roast beef (top round)	lb.	$12.24	50 lb.	54	Pounds	
Potatoes (redskin)	40-lb. box	$65.50	78 lb.	2	Boxes	
Broccoli (frozen)	3-lb. box	$5.55	61 lb.	21	Boxes	

Why do you think the amount of roast beef, potatoes, and broccoli requested by the kitchen staff yesterday was different than the amount issued by Kyle's ingredient room staff?

5. Robin Christopher uses bin cards to manage the wine inventory at the Michelin starred Riverside Restaurant. Each of the wines held in inventory has its own bin card. The bin card for Hundred Acre Cabernet, listed on the menu for $450 per bottle, is presented below, along with its purchase and sales activity for the month of October.

   Make the appropriate entries on the card for the month of October to determine how much of this wine Robin should have in inventory on each of the days listed, including the end of the month.

Bin Number: 237 **Hundred Acre Cabernet**		**Bottle Size: 750 ml**	
Date	**In**	**Out**	**Amount on Hand**
Oct. 1			23
Oct. 3		2	
Oct. 8		1	
Oct. 10		5	
Oct. 12	12		
Oct. 14		2	
Oct. 15		2	

(continued)

Bin Number: 237 Hundred Acre Cabernet		Bottle Size: 750 ml	
Date	In	Out	Amount on Hand
Oct. 17		4	
Oct. 19		5	
Oct. 21	12		
Oct. 23		2	
Oct. 25		3	
Oct. 30		3	
Oct. 31		1	

What would it mean if Robin took a physical inventory at the end of October and there was actually one bottle less in inventory than indicated on the wine's bin card? What would it mean if there was actually one bottle more than indicated on the wine's bin card?

6. Kesha Avril operates Rio Bravo. Rio Bravo is a small fast casual Mexican-style restaurant. Keisha would like to set up an ABC inventory system for monitoring some of her inventoried products. Keisha has listed below last month's usage for 20 key items she holds in inventory.

   a. Calculate Monthly Usage $, then see b. and c. below.

Item	Monthly Usage	Purchase Unit	Purchase Price per Unit	Monthly Usage $
Chicken breast	1,023	Pound	$1.99	
Avocado	85	Case	$39.50	
Corn tortillas	230	Pound	$0.78	
Skirt steak	527	Pound	$14.15	
Green peppers	7	Case	$38.50	
Onions	2	Bag	$64.00	
Parsley	4	Crate	$28.00	
Tortilla chips	265	Pound	$1.49	
Chili powder	12	Pound	$9.60	
White cheddar cheese	28	Pound	$8.99	
Sofrito	42	Jar	$6.10	
Rice	6	Bag	$45.00	
Cumin	10	Pound	$12.35	
Iceberg lettuce	5	Case	$36.25	
Cilantro	6	Crate	$24.10	
Fresh tomatoes	16	Lug	$19.50	
Sour cream	21	5-lb. carton	$9.28	
Pork butt	483	Pound	$5.78	
Flour tortillas	200	Pound	$1.88	
Tilapia fillets	61	Pound	$8.95	

b. After calculating Monthly Usage $ above, copy the table above and paste it below. After pasting, sort the table by Monthly Usage $, from largest to smallest.

c. Assign each inventory item to an A, B, or C inventory category where the top cost items (75 percent of the cost) are assigned to category A, the next highest cost items (15 percent of the cost) are assigned to category B, and the remaining items (10 percent of the cost) are assigned to category C.

Item	Monthly Usage	Purchase Unit	Purchase Price per Unit	Monthly Usage $	Inventory Category (ABC)
Don't forget to calculate the total after you sort ——→			Total		

7. Lesly is the purchasing manager of an organic food restaurant with a banquet room. She needs to create the food purchase orders for tomorrow. Her par value, on-hand, and special-order requirements are as follows. Help Lesly determine the amount of each item she needs to order. (Spreadsheet hint: Use the ROUNDUP function with zero decimals in the Order Amount column.)

Daily Inventory Sheet						
Item	**Description**	**Unit**	**Par Value**	**On Hand**	**Special Order**	**Order Amount**
Avocados	48 ct.	Case	6	2	3	
Cauliflower	12 ct.	Case	5	1		
Cucumbers	Medium	Case	3	0		
Grapes	Red seedless	Lug	4	1		
Lettuce	Green leaf, 24 ct.	Case	5	2	4	
Lettuce	Romaine, 24 ct.	Case	3	0.5		
Pears	D'Anjou	Case	3	1		

(continued)

Daily Inventory Sheet						
Item	**Description**	**Unit**	**Par Value**	**On Hand**	**Special Order**	**Order Amount**
Peppers	Green bell, med.	Case	7	2	5	
Pineapples	7 ct.	Case	4	0		
Potatoes	B reds	50-lb. bag	3	0.5		
Potatoes	Peeled, large	25-lb. bag	4	1	4	
Squash	Yellow #2	30-lb. case	5	2	2	
Strawberries	Driscoll	flat	3	1.5		

8. Maxine Wilson is the Executive Chef at the Plainview Air Force Base. Fajita Spiced Whitefish is one of Maxine's most popular items. Using the standardized recipe information below, calculate the total recipe cost and the per portion cost for Maxine's Fajita Spiced Whitefish.

Fajita Spiced Whitefish Yield in portions: 50 Portion size: 6 oz.				
Ingredient	**Amount**	**Recipe Cost**	**Procedure**	
Boneless whitefish fillet: 6 oz.	50 fillets	$113.20	Pat whitefish fillets dry.	
Potato flakes; dry	1 lb. 8 oz.	$1.58	Mix potato flakes, cornstarch, flour, and seasoning to make a breading. Coat each fillet with the breading mixture.	
Cornstarch	10 oz.	$0.96		
Flour, all-purpose	4 oz.	$0.25		
Fajita seasoning	10 oz.	$4.18		
Salt	4 oz.	$0.10		
Vegetable oil	4 cups	$2.73	Heat oil in fry pan. Place breaded fillet into hot oil, presentation side down. Cook until browned, then turn to brown other side. Remove from frying pan and drain.	
	Total Recipe Cost		**Portion Cost**	

9. Sue Trimble is the head bartender at Lamont's Bar and Grill. Jack Nesbitt is Lamont's general manager. Each week Jack asks Sue to take a full bar inventory. Part of that inventory involves counting the well liquors the bar offers for sale. Sue counts the inventory including cases, bottles, and tenths of a bottle held in the well; and Jack extends the inventory to calculate the inventory's total value. Complete the inventory valuation form below that Sue has now submitted to Jack for the week ending January 15th.

Lamont's Bar and Grill: Beverage Inventory									
Weekly Beverage Inventory: Well Liquors							Week Ending: 1-15		
		Storage Area							
Well Item	Purchase Unit (Bottle)	Bar Area (Bottles)	Liquor Storeroom (Cases)	Liquor Storeroom (Bottles)	Total Inventory (Bottles)	Cost per Case	Purchase Units per Case	Cost per Purchase Unit	Inventory Value
Bourbon	750 ml	4.5	2	6		$81.50	12		
Gin	750 ml	3.1	3	3		$78.25	12		
Rum	750 ml	6.2	1	8		$68.70	12		
Scotch	750 ml	5.5	2	10		$83.55	12		
Vodka	750 ml	5.1	4	2		$65.65	12		
								Total Value	
Inventory Taken:	*Sue T.*					Inventory Valued:		*Jack N.*	

What was the total inventory value of all the well liquors counted by Sue when she took this inventory?

10. Helen Lang is the manager at the Prime Rib House restaurant. Her best-selling menu item is the 10-ounce regular cut prime rib. Helen pays $14.97 a pound for the standing rib roasts she uses. When properly trimmed, cooked, and sliced, each 22-pound rib roast she buys will produce 25 regular cut slices weighing 10 ounces each. Tonight, Helen's production forecast calls for the cooking of two rib roasts. Calculate the portion cost of this item that Helen's operation would achieve if the following number of 10-ounce portions are actually served from the two cooked roasts.

Number of roasts	
Rib roast weight in lb.	
Total number of lb.	
Cost per lb.	
Total cost of ribs	

Rib Portions Served	Portion Cost
50	
48	
46	
44	
42	
40	

What two key areas of controlling production costs would Helen want to monitor to ensure the two rib roasts cooked and served by her production staff have the lowest portion cost?

11. Mario Armando Perez is the kitchen manager at the Asahi Sushi House. Mario's restaurant offers five popular types of sushi roll. Mario keeps careful records of the number of each roll type sold, from which he computes each item's popularity index. For March 1, Mario estimates 150 guests will be served.

Based on his experience, and to ensure he does not run out of any item, Mario would like to have extra servings (planned overage) of selected menu items available for sale. Using planned overage, the popularity index of his menu items, and his prior day's carryover information, help Mario determine the amount of new production needed and the total available for the day.

At the end of the day, Mario also records his actual number sold in order to calculate his carryover amount for the next day. Using all of the information given, help Mario complete his kitchen production schedule for March 1.

Mario's Kitchen Production Schedule: March 1

Menu Item	Popularity Index	Sales Forecast	Planned Overage	Prior Day Carryover	New Production	Total Available	Number Sold	Carryover
Spicy Tuna Roll	34%		5	6			52	
Salmon Special Roll	10%		5	2			12	
Spicy Shrimp Roll	30%		5	5			50	
California Roll	20%		5	2			30	
BBQ Eel Roll	6%		0	0			5	
Total to Serve	100%	150	20					

12. Warren Stills is the manager of Salads and More. Warren's operation does a significant amount of takeout business. At the beginning and end of each month Warren takes inventory of his packaging items. Rather than count each item in a case, he uses an inventory valuation system assigning a value of 1.0 to a full case, .75 for a ¾-full case, .5 to a ½-full case, and .25 to ¼-full case.

He also records the number of cases of each packaging item he has purchased in the month. He is interested in knowing his total monthly packaging costs.

Help Warren calculate his total monthly packaging costs by completing the inventory spreadsheet he has started.

Salads and More Packaging Items

Item	Item Case Cost	Cases in Inventory First Day of Month	Cases Purchased in Month	Total Cases Available for Use	Cases in Inventory Last Day of Month	Monthly Usage	Monthly Item Cost
12 oz. hot cups	$48.49	2.5	7	9.5	2	7.5	
12 oz. cold cups	$42.99	1	5		1.5		
Lids	$36.25	1.25	3		2		
Straws	$19.95	1.25	1		0.5		
Clam shells (small)	$105.00	2.25	6		3.5		
Clam shells (large)	$29.75	3.5	9		4		
Total Monthly Packaging Costs							

CHAPTER 5

Monitoring Food and Beverage Product Costs

OVERVIEW

This chapter teaches you how to calculate the product-related costs of making the menu items you sell. In addition, you will learn how to compare the product cost results you actually achieve with those you planned to achieve. The chapter concludes by identifying steps you can take to help ensure that you meet your operation's profit goals by monitoring your product costs and then reducing those costs if they are too high.

Chapter Outline

- Cost of Sales
- Computing Cost of Food Sold
- Computing Cost of Beverage Sold
- Computing Costs with Transfers
- Utilizing the Cost of Sales Formula
- Reducing the Cost of Sales Percentage
- Technology Tools
- Apply What You Have Learned
- For Your Consideration
- Key Terms and Concepts
- Test Your Skills

LEARNING OUTCOMES

At the conclusion of this chapter, you will be able to:

- Accurately calculate food and beverage costs and their cost percentages of total sales.
- Compare product costs achieved in an operation against the product costs the operation planned to achieve.
- Apply strategies to reduce an operation's cost of sales and cost of sales percentage to meet the operation's financial goals.

COST OF SALES

In Chapters 3 and 4, you learned how to properly purchase, receive, inventory, and issue food and beverage products. You also learned in Chapter 1 that, when it follows the Uniform System of Accounts for Restaurants (USAR), an operation's food and beverage costs are reported two ways: (1) as a total dollar amount and (2) as a percentage of the operation's total sales.

The USAR recommends reporting food costs separately from alcoholic beverage costs. When the amounts of an operation's individual food and beverage costs are combined, they are referred to as the operation's total **cost of goods sold** (COGS) or **cost of sales**, as both terms refer to the direct costs associated with producing and selling goods or services.

The ability to accurately calculate and report an operation's cost of sales is an important management skill. In nearly all foodservice operations, a manager's ability to control cost of sales will be used as a primary measure of that manager's competence and ability.

All restaurant managers must be able to calculate their actual food costs. In those operations that sell alcoholic beverages, the manager must also know how to calculate the operation's actual beverage costs. As well, managers should know how to estimate, or forecast, their future product costs. When they can do that accurately, they are able to compare the cost of sales their operations actually achieve to their operations' cost of sales forecasts. If significant differences exist, they can take the steps needed to improve their operations product usage and, as a result, help to meet their profit goals.

COMPUTING COST OF FOOD SOLD

It is important to recognize that the cost of sales incurred by an operation in an accounting period is most often *not* equal to the amount of food purchases in that same accounting period. Because that is true, managers must use a very specific process to accurately calculate their cost of food sales and cost of food sales percentages. The cost of food sold formula they use to do so is shown in Figure 5.1.

It is important to recognize that an operation's cost of food sold is actually the dollar amount of all food sold *plus* the costs of any food that was thrown away, spoiled, wasted, or stolen. To use the cost of food sold formula properly, managers must fully understand each of its individual parts.

FIGURE 5.1 **Formula for Cost of Food Sold**

Beginning inventory
PLUS
Purchases = Food available for sale
LESS
Ending inventory = Cost of food consumed
LESS
Employee meals = Cost of food sold

BEGINNING INVENTORY

When calculating the cost of food sold, the **beginning inventory** is the dollar value of all food on hand at the beginning of an accounting period. It is determined by completing a physical inventory, which is an actual count and valuation of all foods in storage and in production areas (see Chapter 4).

If, when taking a physical inventory, products are undercounted, your food costs will ultimately appear higher than they actually are. If, on the other hand, you erroneously overstate the value of products in inventory, a process called **padding inventory**, your costs will appear artificially low.

It is important that managers take accurate physical inventories and, for the purpose of calculating food or beverage costs, recognize that the beginning inventory for an accounting period is always the ending inventory amount from the prior accounting period. Thus, for example, an operation's ending inventory for January 31st will become the operation's beginning inventory for February 1st.

PURCHASES

Purchases are the sum cost of all food bought during the accounting period. The purchases amount is determined by properly summing the amounts on all delivery invoices and any other bills for products purchased in the accounting period.

FOOD AVAILABLE FOR SALE

Food available for sale is the sum of the beginning inventory and the purchases made during a specific accounting period. Some managers refer to food available for sale as **goods available for sale** because this term was commonly used prior to the publication of the most recent edition of the USAR. Regardless of the term used, each represents the value of all food that was available for sale during that accounting period.

ENDING INVENTORY

Ending inventory refers to the dollar value of all food on hand at the end of the accounting period. It also must be determined by completing an accurate physical inventory.

COST OF FOOD CONSUMED

The **cost of food consumed** is the actual dollar value of all food used, or consumed, by the operation. Again, it is important to note that this is not merely the value of all food sold but rather the value of all food no longer in the establishment due to sale, spoilage, waste, or theft. Cost of food consumed also includes the cost of any complementary meals served to guests as well as the value of any food (meals) eaten by employees.

EMPLOYEE MEALS

Employee meal cost is actually a labor-related, not food-related, cost (see Chapter 7). Free or reduced-cost employee meals are a benefit offered by many employers, much in the same manner as they may provide medical insurance or paid vacations. Therefore, the value of this benefit, if provided, should not be recorded as a cost of food. Instead, the dollar value of all food eaten by employees is *subtracted* from the cost of food consumed and *added* to the cost of labor to more accurately reflect an operation's true cost of food sold. Food products do not have to be consumed as a full meal in order to be valued as a labor cost. Soft drinks, snacks, and other food items consumed by employees at work are all considered employee meals for the purpose of computing cost of food sold. If records are kept on the number of employees consuming food each day, monthly employee meal costs are easily determined.

Some operators prefer to assign a fixed dollar value to each employee meal served in an accounting period rather than record the amount of food eaten by each employee. Thus, for example, if an operator assigns a constant value of $6.00 to each employee meal served and 1,000 meals are served in a month, the value of employee meals for that month would be $6,000 ($6.00 per meal × 1,000 meals served = $6,000).

COST OF FOOD SOLD

Cost of food sold, or cost of goods sold, is the actual amount of all food expenses incurred by the operation *minus* the cost of employee meals. It is not possible to accurately determine this number unless a beginning physical inventory has been taken at the start of an accounting period, followed by another physical inventory taken at the end of the accounting period.

Calculating actual cost of food sold on a regular basis is important because it is not possible to improve your food cost control efforts unless you first know what your food costs actually are. In nearly all operations, cost of food sold is calculated on at least a monthly basis because it is reported on the operation's income statement (P&L). In many operations, these same costs may be calculated on a weekly or even on a daily basis!

The Move to Off-Premise Dining

Foodservice operators doing a significant amount of off-premise takeout or delivery sales must make a decision about how to account for the cost of the packaging materials used for these off-premise items.

Historically, the USAR has recommended that the cost of paper and packaging used for off-premise food and beverage sales be listed as a direct operating expense (see Chapter 8), rather than a cost of food or beverage sales.

Increasingly, however, some foodservice operators are electing to include required packaging items as part of their COGS expense. The view of these operators is that packaging that is directly related to the cost of producing a menu item, such as boxes or wrappers, should be listed separately, but included in COGS. Thus, for example, for these operators the cost of the box used when selling a carryout pizza would be considered part of their foodservice operations' COGS.

Items like paper or plastic bags, foam containers, napkins, utensils, plastic cups, lids, straws, and sauce packets can often cost as much as 3 to 5 percent of an operation's total sales. Thus, whether the decision is made to record these expenses as a COGS or as a direct operating expense, it is important to understand how the control of these costs affects an operation's profitability.

FOOD COST PERCENTAGE

To properly analyze an operation's cost of sales for a specific accounting period, managers must first determine the amount of food used in that period and the amount of sales achieved in the same period. When they have done so, they can calculate their **food cost percentage**.

HERE'S HOW IT'S DONE 5.1

Managers calculate their food cost percentages by first accurately determining their cost of food sold during an accounting period and then dividing that amount by their food sales:

$$\frac{\text{Cost of food sold}}{\text{Food sales}} = \text{Food cost \%}$$

Some foodservice managers, however, are more interested in how much it costs to serve each of their guests rather than their operations' food cost percentages. Operations in which this average cost of meals served value is important include military bases, hospitals, senior living facilities, school and college foodservice operations, and business organizations that provide free meals to their employees.

Whether the guests served are soldiers, patients, residents, students, or workers, calculating an operation's cost per meal is easy because it uses a variation of the basic food cost percentage formula. The formula used to calculate the average cost of meals served in an operation is as follows:

$$\frac{\text{Cost of food sold}}{\text{Total meals served}} = \text{Cost per meal}$$

For example, assume that a manager's operation incurred $65,000 in cost of food sold during an accounting period. In that same accounting period, the manager served 10,000 meals. To calculate this operation's cost per meal, the manager applies the cost per meal formula. In this example:

$$\frac{\$65{,}000}{\$10{,}000} = \$6.50 \text{ per meal}$$

It is essential to understand that an operation's food cost percentage represents that portion of food sales that was spent on all food expenses.

To illustrate, assume that you managed an operation that achieved $50,000 in food sales in the current accounting period. In the period you spent $15,000 for food. Your food cost percentage for the period would be calculated as follows:

$$\frac{\$15{,}000}{\$50{,}000} = 30\%$$

Those operations that utilize a modern point of sale (POS) system (see Chapter 2) will have the option of purchasing POS components that interface their inventory systems with their sales records. Doing so allows these managers to automatically produce a variety of accurate COGS metrics, including their food cost percentages.

COMPUTING COST OF BEVERAGE SOLD

The proper computation of a beverage cost percentage is identical to that of a food cost percentage with one important difference. Typically, there is no equivalent for employee meals because the consumption of alcoholic beverage products by employees who are working should be strictly prohibited. Thus, "employee drinks" would never be considered as a reduction from overall beverage cost. However, it is important to recognize that an operation's **cost of beverage sold** is actually the dollar amount of all beverage products sold as well as the costs of all beverages that were given away, wasted, or stolen. The cost of beverage sold formula is shown in Figure 5.2.

FIGURE 5.2 Formula for Cost of Beverage Sold

Beginning inventory
PLUS
Purchases = Beverages available for sale
LESS
Ending inventory = Cost of beverage sold

BEVERAGE COST PERCENTAGE

To properly analyze an operation's **beverage cost percentage,** managers must first determine the amount of beverages used in an accounting period and the amount of beverage sales achieved in the same period.

The formula used to compute an operation's beverage cost percentage is as follows:

$$\frac{\text{Cost of beverage sold}}{\text{Beverage sales}} = \text{Beverage cost \%}$$

To illustrate, assume that you managed an operation that achieved $100,000 in beverage sales in the current accounting period. In the period you spent $18,000 on beverages. Your beverage cost percentage for the period would be calculated as follows:

$$\frac{\$18,000}{\$100,000} = 18\%$$

COMPUTING COSTS WITH TRANSFERS

In some operations, other reductions from, and additions to, food and beverage expenses must be considered when accurately calculating the operation's COGS. To illustrate, consider the situation you would face as a manager of Rio Lobo's, a popular Tex-Mex style restaurant that has a high volume of alcoholic beverage sales. To prepare your drinks, your bartenders use a large quantity of limes, lemons, and fruit juices issued from the kitchen. You would, of course, like your beverage cost percentage to reflect all of the product costs associated with serving your beverage items, including the food ingredients used in your drink preparations. As a result, you must transfer the cost of any food ingredients used in drink production away from your cost of food sold and add them to your costs of beverage sold.

Assume further that, in this operation, your kitchen produces several standardized recipes that require the use of wines that are taken from your wine inventory as needed, as well as beer used in recipes that will also be obtained from the bar. In this case, you must transfer the cost of these wines and beers away from your cost of beverage sold and add them to your cost of food sold.

The monetary value of transfers to and from one operating department to another should be recorded on a **product transfer form.** The form includes a space for the amount of product transferred, the product's value, and the individuals authorizing the transfers. Figure 5.3 shows the way in which a product transfer form would be used at Rio Lobo's to record product transfers from January 1st through January 7th.

FIGURE 5.3 Rio Lobo Weekly Transfer Record

Location: Rio Lobo

Date: 1-1 to 1-7

Product Value

Date	Item	Quantity	Cost To Bar	Cost From Bar	Issued By	Received By
1/1	Lemons	12	$6.85		T. S.	B. H.
	Limes	8 (large)	$5.45		T. S.	B. H.
	Cream	2 qt.	$14.22		T. S.	B. H.
1/2	Chablis	1 gal.		$23.56	B. H.	T. S.
1/3	Coffee	2 lb.	$31.20		T. S.	B. H.
1/4	Cherries	½ gal.	$9.85		T. S.	B. H.
1/4	Lemons	4	$2.28		T. S.	B. H.
	Limes (small)	12	$7.65		T. S.	B. H.
	Ice cream (vanilla)	1 gal.	$14.58		T. S.	B. H.
1/5	Pineapple juice	1 can	$1.45		T. S.	B. H.
1/6	Tomato juice	6-can pack	$11.25		T. S.	B. H.
1/7	Sherry	750 ml		$15.85	B. H.	T. S.
1/7	Celery	1 bunch	$2.45		B. H.	T. S.
Total Product Value			$107.23	$39.41		

After all appropriate transfers have been made, accurate cost of beverage sold, and cost of food sold calculations can be made.

Figure 5.4 illustrates the formulas you would use to calculate cost of food sold and cost of beverage sold at Rio Lobo's including transfers for the *entire month* of January.

UTILIZING THE COST OF SALES FORMULA

When managers truly understand the cost of sales formulas for food and beverages, they are in a good position to perform a variety of tasks that can improve the management of their operations. These include the following:

1. Calculating cost of sales for individual product categories
2. Estimating daily cost of sales
3. Comparing actual costs to attainable costs

CALCULATING COST OF SALES FOR INDIVIDUAL PRODUCT CATEGORIES

In some cases, managers are interested in knowing the cost of sales for their individual inventory items or groups of items in addition to their overall cost of

FIGURE 5.4 Cost of Food Sold and Cost of Beverage Sold with Transfers

Accounting Period: 1/1–1/31	
Unit Name: Rio Lobo's	
Cost of Food Sold	
Beginning inventory	$25,000
PLUS	
Purchases	$40,000
= Food available for sale	$65,000
LESS	
Ending inventory	$27,000
= Cost of food consumed	$38,000
LESS	
Food transfers from kitchen	$5,000
PLUS	
Beverage transfers to kitchen	$3,000
LESS	
Employee meals	$1,000
= Cost of food sold	**$35,000**
Cost of Beverage Sold	
Beginning beverage inventory	$12,000
PLUS	
Purchases	$18,000
= Beverages available for sale	$30,000
LESS	
Ending inventory	$14,000
= Cost of beverage consumed	$16,000
PLUS	
Food transfers from kitchen	$5,000
LESS	
Beverage transfers to kitchen	$3,000
= Cost of beverage sold	**$18,000**

sales. To illustrate, consider the manager of Benjamin's steakhouse. In the month of January, Benjamin's achieved $190,000 in food sales. Figure 5.5 details Benjamin's food usage in five major inventory categories for the month.

In this case, the manager has decided to analyze their product usage in terms of five categories of food items. This is because, although they are interested in their overall food cost and food cost percentage, they are also interested in their meat cost percentage, as well as their product usage in other specific areas.

Calculating individual costs for the various product categories simply requires that they first identify the desired product subcategories. They would then take

FIGURE 5.5 Inventory by Product Category

Unit Name: Benjamin's						Date: January 31
	Meat	Seafood	Dairy	Produce	Other	Total
Beginning inventory	$26,500	$4,600	$7,300	$2,250	$23,000	$63,650
Purchases	$33,800	$17,700	$4,400	$15,550	$1,800	$73,250
Goods available	$60,300	$22,300	$11,700	$17,800	$24,800	$136,900
Ending inventory	$28,000	$10,900	$6,000	$4,500	$21,000	$70,400
Cost of food consumed	$32,300	$11,400	$5,700	$13,300	$3,800	$66,500
Employee meals	$900	$200	$100	$250	$50	$1,500
Cost of food sold	$31,400	$11,200	$5,600	$13,050	$3,750	$65,000

FIGURE 5.6 Food Cost Category Percentage and Proportion of Total

Benjamin's Sales: $190,000			
Category	Cost of Food Sold	Food Cost Percentage	Proportion of Total Product Cost
Meat	$31,400	16.5%	48.3%
Seafood	$11,200	5.9%	17.2%
Dairy	$5,600	2.9%	8.6%
Produce	$13,050	6.9%	20.1%
Other	$3,750	2.0%	5.8%
Total	$65,000	34.2%	100.0%

beginning and ending inventories for each product category and then apply the cost of sales formula to determine the cost of the goods (products) sold in that category.

Because this manager understands the cost of sales formula, they can calculate food cost percentages by product category as well as calculate each category's proportion of their total product cost as shown in Figure 5.6.

Figure 5.6 shows that Benjamin's food cost percentage for January was 34.2 percent ($65,000 ÷ $190,000 = 34.2%). To calculate the food cost percentages in each category, the manager at Benjamin's uses $190,000 as the denominator in their food cost percentage calculations and the sum total of the cost of food sold in each product category as the numerator.

In some cases, managers want to know the proportion of total food cost accounted for by each product category. To calculate the proportion of total product cost percentage, the manager at Benjamin's would use the following formula:

$$\frac{\text{Cost in product category}}{\text{Total cost in all Categories}} = \text{Proportion of total product cost}$$

Thus:

$$\frac{\text{Cost in meat category}}{\text{Total cost in all categories}} = \text{Meat cost proportion of total product cost}$$

In this example that would be:

$$\frac{\$31,400}{\$65,000} = 48.3\%$$

By using the categories listed in Figure 5.6, this manager may be better able to determine when their costs are above those they would expect. Using the category food cost proportion approach, they would know, for example, that meats accounted for 48.3 percent of their total food usage in the month of January. They could compare this figure to the meat expense of prior months to determine whether their meat cost proportion is rising, declining, or staying constant. If their category cost or their overall percentages are higher than they anticipated, they must find out why and then take corrective action, as necessary.

HERE'S HOW IT'S DONE 5.2

All managers should be interested in their overall product costs and cost percentages, but sometimes managers are also interested in their COGS percentages for individual food or beverage product categories. To see why, consider a manager who calculated their beverage cost percentage for last week and found it to be 22 percent. Utilizing a minor variation of the beverage cost percentage formula, the manager can calculate individual beverage cost percentages for beer, wine, and spirits sales as shown below.

Product	Sales	Cost of Beverage Sold	Beverage Cost (%)
Beer	$8,000	$1,600	20%
Wine	$12,000	$3,600	30%
Spirits	$10,000	$1,400	14%
Total	$30,000	$6,600	22%

Using beer as an example, the formula the manager would use to calculate their beverage cost percentage for beer is:

$$\frac{\text{Cost of beverage sold for beer}}{\text{Beer sales}} = \text{Beverage cost \% for beer}$$

In this illustration, the manager may well gain more critical information about their operation by knowing the beverage cost percentages of their individual beverage categories than they would by knowing only their overall beverage cost percentage.

Today's modern POS systems routinely report an operation's product sales by category. Because that is so, managers utilizing the same product categories when taking their beginning and ending inventories can calculate individual product category cost percentages as well as the overall cost percentages achieved in their operations.

In many cases, restaurant managers calculate separate cost of sales percentages or proportions only for certain items, such as meats and sea foods. Similarly, beverage managers often calculate separate cost of subcategory sales percentages for their beer, wine, and liquor.

ESTIMATING DAILY COST OF SALES

The best managers want to know their cost of sales on a daily or weekly basis, rather than merely on a monthly basis. When they do, they can better address cost of sales-related problems as they occur. But you now know that the accurate calculation of an operation's actual cost of sales requires its managers to conduct a physical inventory.

FIGURE 5.7 Six-Column Form

Date: ____________

Weekday	Today	To Date	Today	To Date	Today	To Date

You could, of course, take a physical inventory every day. If you did, you could compute your operation's daily cost of sales. But taking an accurate physical inventory in most operations is a very time-consuming task. It would be convenient if you could have a close estimate of your food usage on a daily or weekly basis without the extra effort of a daily inventory count. Fortunately, such an approximation method exists. Figure 5.7 illustrates a six-column form that you can use for a variety of purposes. One of them is to estimate your food cost percentages on a daily or weekly basis. Note that Figure 5.7 has one column for the day or date and six columns for Today and To Date data entries.

To illustrate how the form is used, assume that you own an Italian restaurant that serves no liquor and caters to a family-oriented clientele. You would like to monitor your food cost percentage on a more regular basis than once per month, which is your regular accounting and physical inventory period. You have decided to use a six-column form to estimate your food cost percentage. Since you keep track of both daily sales and daily purchases, you can easily do so. In the space above the first two data entry columns, you put the word "Sales." Above the middle two columns, you write "Purchases," and above the last two columns, you enter "Cost %."

You then proceed each day to enter your daily sales revenue in the column labeled "Sales Today". Your delivery invoices for food purchases are totaled daily and entered into the column titled "Purchases Today". Dividing the "Purchases Today" column by the "Sales Today" column yields the figure that is placed in the "Cost % Today" column.

Purchases to date (the cumulative purchases amount) are divided by sales to date (the cumulative sales amount) to yield the cost percentage to date figure. A quick summary of the form is below:

Six-column food cost % estimate

1. $\frac{\text{Purchases today}}{\text{Sales today}} = \text{Cost \% today}$

2. $\frac{\text{Purchases to date}}{\text{Sales to date}} = \text{Cost \% to date}$

Figure 5.8 shows this information for your operation for the time period January 15 (Monday) to January 21 (Sunday).

As you can see, you buy most of your food at the beginning of the week, but sales are highest in the later part of the week. This is a common occurrence at many foodservice establishments. As can also be seen, your daily cost percentage ranges from a high of 130 percent (Monday) to a low of 0 percent (Sunday), when no deliveries are made. In the "Cost % To Date" column, however, the range is only from a high of 130 percent (Monday) to a low of 39.20 percent (Sunday).

In order for the six-column food cost to be an accurate estimate, you must make one important assumption: *for any time period you are evaluating, the beginning inventory and ending inventory amounts are the same.* In other words, over any given time period, you will have approximately the same amount of food on hand at all times. If this assumption is correct, the six-column food cost estimate is, in fact, a good indicator of your food usage.

FIGURE 5.8 Six-Column Food Cost Estimate

Date: 1/15-1/21

	Sales		Purchases		Cost %	
Weekday	**Today**	**To Date**	**Today**	**To Date**	**Today**	**To Date**
Monday	$850.40	$850.40	$1,106.20	$1,106.20	130.0%	130.0%
Tuesday	$920.63	$1,771.03	$841.40	$1,947.60	91.4%	110.0%
Wednesday	$1,185.00	$2,956.03	$519.60	$2,467.20	43.8%	83.5%
Thursday	$971.20	$3,927.23	$488.50	$2,955.70	50.3%	75.3%
Friday	$1,947.58	$5,874.81	$792.31	$3,748.01	40.7%	63.8%
Saturday	$2,006.41	$7,881.22	$286.20	$4,034.21	14.3%	51.2%
Sunday	$2,404.20	$10,285.42	0	$4,034.21	0%	39.2%
Total	$10,285.42		$4,034.21		39.2%	

If you assume that your inventory is basically constant (e.g., that beginning inventory equals ending inventory), the cost of sales formula indicates that your cost of food sold for the 1-week period in this example would be a little less than $4,034.21 (39.20 percent of sales). Why a little less? Because a proper cost of food sold formula requires you to subtract the value of employee meals, if any are provided, since they are an employee benefit cost, not a food expense.

How accurate is the six-column form for estimating product usage? For most operators, it is quite accurate and has the following advantages:

1. It is very simple to compute, requiring 5 minutes or less of data entry per day for most operations (the cells in a six-column Excel spreadsheet can be formulated to do the math computations for you).
2. It records both sales history and purchasing patterns.
3. It identifies potential problems before the end of the monthly accounting period.
4. By the ninth or tenth day, the degree of accuracy in the "To Date" columns is very high.
5. It is a daily reminder to both management and employees that there is a very definite relationship between sales and expenses.

Those managers whose operations use an issuing system (see Chapter 4), rather than daily purchases, for recording their daily food usage can simply substitute "Issues" for "Purchases" in the six-column form and achieved the same cost of sales estimate results.

The use of a six-column food cost estimator is highly recommended for the operator who elects to conduct a physical inventory less often than every 2 weeks. Because it keeps costs uppermost in the minds of managers and employees alike, it is recommended that this cost estimate be posted where food production staff can see it. It communicates daily to employees both sales and the costs required to generate those sales. It also provides management with a visual reminder of the importance of controlling product usage.

COMPARING ACTUAL COSTS TO ATTAINABLE COSTS

When managers can calculate their overall cost of sales amounts and percentages, and can even estimate these costs on a daily or weekly basis, they are in a good position to answer several important operational questions:

1. What are our actual product costs?
2. What should be our attainable product costs?
3. How close are we to our product cost goals?

PRODUCT COSTS

In Chapter 3, you learned that every menu item sold in a restaurant should be produced from a standardized recipe to ensure product quality and consistency, and to aid in product purchasing.

In addition to having a standardized recipe for each item, managers should maintain standardized recipe costs for each item. A **standardized recipe cost sheet** is simply a record of the ingredient costs required to produce a standardized recipe. In an electronic or paper format, managers use standardized recipe cost sheets to calculate their total recipe costs and their individual portion costs.

A standardized cost sheet can be created using any basic spreadsheet software. The spreadsheet simply multiplies the cost of each ingredient by the amount of the ingredient used in the standardized recipe. For example, if the cost of an ingredient in a recipe is $2.00 per cup and the standardized recipe calls for three cups, then the cost for this recipe ingredient is calculated as follows:

$2.00 cost per cup × 3 cups required = $6.00 ingredient cost

Spreadsheet programs are an excellent means of creating these records, doing the mathematical calculations and keeping the recipe costs current. Properly maintained, recipe cost sheets provide you with up-to-date information that can help with pricing decisions in addition to comparing your actual food and beverage costs with those you should incur.

Figure 5.9 shows the format you might use for a standardized recipe cost sheet if you operate Steamer's, a small soup and sandwich carryout kiosk. The recipe in this example is for beef stew, and it yields a total recipe cost of $84.55 and a cost per serving (portion cost) of $2.20. Note that all ingredients are listed in their edible portion forms, a concept addressed in detail later in this chapter.

A standardized recipe cost sheet can be produced in seconds today using computer-based spreadsheets and smart device apps. This formerly tedious task has become so simplified that there is just no reason for management not to have accurate, up-to-date costing data on all its recipes. As a result, it is easy for managers

FIGURE 5.9 Standardized Recipe Cost Sheet

Unit Name: Steamer's

Menu Item: Beef stew | Recipe Number: 146

Special Notes: All ingredients weighed as edible portion | Recipe Yield: 40

Portion Size: 8 oz.

Portion Cost: $2.20

Ingredients		Ingredient Cost	
Item	Amount	Unit Cost	Total Cost
Corn, frozen	3 lb.	$1.60 lb.	$4.80
Tomatoes	3 lb.	1.40 lb.	$4.20
Potatoes	5 lb.	0.90 lb.	$4.50
Beef cubes	5 lb.	13.76 lb.	$68.80
Carrots	2 lb.	0.56 lb.	$1.12
Water	2 gal.	N/A	—
Salt	2T	0.30 lb.	$.02
Pepper	2t	12.00 lb.	$.12
Garlic	3 cloves	.80 of a clove	$2.40
Tomato juice	1 qt.	9.00 gal.	$2.25
Onions	4 lb.	1.00 lb.	$4.00
Total Cost			$84.55

Total Recipe Cost: $88.07 | Recipe Type: Soups/Stews

Portion Cost: $2.20 | Date Costed: 4/15

Previous Portion Cost: $2.01 | Previous Dated Costed: 1/15

to know the precise portion costs that should be attainable when their standardized recipes are followed carefully.

> **FUN ON THE WEB!**
>
> A number of companies offer recipe costing programs and apps. These programs can greatly assist managers with portion cost calculation and much more.
>
> To review programs of this type, enter "recipe costing software" or "recipe costing apps" in your favorite search engine. Pay special attention to those programs that interface (connect electronically) with an operation's POS and inventory management systems.
>
> One especially popular product offering of this type is Food Trak, developed and marketed by Systems Concept Inc. (SCI). SCI is one of the world's first companies to specialize in the field of food and beverage cost management automation. Watch for their products in your search results.

Accurately calculating portion costs based on standardized recipe costs is important. But some managers have difficulty computing total recipe costs because most recipes contain ingredient amounts that are used in a different quantity than they are purchased. For example, you may purchase soy sauce by the gallon, but your recipes may call for it to be added by the cup or tablespoon.

In other cases, ingredients may be purchased in pounds or gallons but added to recipes in grams and liters. When situations such as this arise, an ingredient conversion table or chart similar to the one presented in Figure 5.10 must be used. This is because recipe ingredient weights, measures, and sizes must be precisely computed if your recipe costs are to be accurate.

Calculating recipe costs and portion costs accurately is important. For you to correctly calculate recipe costs, however, you must first understand the concept of product yield and how it affects costs.

PRODUCT YIELD

Most foodservice products are delivered in their **as-purchased (AP)** state. This refers to the weight or count of a product as it is received by the foodservice operator. For example, if a case of iceberg lettuce containing 24 heads is delivered to an operation, the lettuce will be delivered in its AP state.

Edible portion (EP) refers to the weight of a product after it has been cleaned, trimmed, cooked, and portioned. For example, after the 24 heads of lettuce delivered in AP state have been trimmed, washed, and chopped or otherwise prepared, the heads will now be in the EP state. Many foodservice professionals refer to the EP state as the "recipe-ready" state.

AP refers to food products as the operator receives them and EP typically refers to food products in their recipe-ready state, or as the guest receives them. Foodservice buyers purchase ingredients at AP unit costs, but these items are served in their EP forms. To determine actual recipe costs, it is often necessary to conduct a **yield test** to determine an ingredient's EP, rather than its AP, cost. A yield test is a procedure used for computing your actual EP costs on any AP product that will experience significant weight or volume loss during processing or preparation.

FIGURE 5.10 Ingredient Conversion Table

Weight and Measure Equivalents		
Item	**Equivalent**	
60 drops	1 teaspoon	
3 teaspoons	1 tablespoon	
2 tablespoons	1 liquid ounce	
4 tablespoons	¼ cup	
16 tablespoons	1 cup	
2 cups	1 pint	
2 pints	1 quart	
4 quarts	1 gallon	
4 pecks	1 bushel	
16 ounces	1 pound	
Select Spices*		
Pepper: 4.20 tablespoons	1 ounce	
Salt: 1.55 tablespoons	1 ounce	
Common Can Sizes		
Can Size	**Quantity**	
No. 303	1¼ cups	
No. 2	2½ cups	
No. 2½	3½ cups	
No. 5	7⅓ cups	
No. 10	13 cups	
Conversion Formulas		
To Convert	**Multiply By**	
Ounces to grams	Ounces	28.35
Grams to ounces	Grams	0.035
Liters to quarts	Liters	0.950
Quarts to liters	Quarts	1.057
Pounds to kilos	Pounds	0.45
Kilos to pounds	Kilos	2.20
Inches to centimeters	Inches	2.54
Centimeters to inches	Centimeters	0.39

*Spices have different conversions based on their individual weights.

To illustrate how a yield test results in the determination of actual product costs, assume that you purchased 10 pounds of fresh carrots (AP) to be used for stew. You know that you will have product losses due to the peeling and trimming of the carrots. As a result, the original 10 pounds of fresh carrots for your stew will yield less than 10 pounds when the carrots are peeled and sliced into their EP (recipe-ready) state.

The formula used to calculate product yields is:

Product as purchased (AP) − Losses due to preparation = Product edible portion (EP)

HERE'S HOW IT'S DONE 5.3

Foodservice operators utilizing standardized recipes know that ingredient amounts called for in their recipes are typically measured by weight or volume, and that the two most common measurement systems in use in recipes are the metric system and the Imperial (US) system. But these same managers also know that recipe ingredients are often purchased in one (larger) unit of measurement and utilized in their recipes in other, smaller recipe-ready units.

For example, salt will likely be purchased by the pound, but a recipe may call for a number of ounces, or tablespoons, of salt to be used. To properly calculate their recipe costs, managers must be able to convert the cost of larger purchase units into the cost of smaller, recipe-ready units.

Making these conversions from larger purchase unit to smaller unit when using the metric system is very easy:

Because managers converting units of metric weight know that 1 kilogram (kg) equals 1,000 grams (g), the cost per gram for an item purchased by the kilogram is always calculated as follows:

$$\frac{\text{Cost per kg}}{1{,}000} = \text{Cost per g}$$

Similarly, when managers converting units of metric volume know that 1 liter (L) equals 1,000 milliliters (ml), the cost per ml for an item purchased by the liter is always calculated as follows:

$$\frac{\text{Cost per L}}{1{,}000} = \text{Cost per ml}$$

Converting weight and volume measurements into recipe-ready measurements when using the US system is more challenging. This is because, when using the US system, a larger purchase unit can most often be viewed as being equivalent to a number of different smaller recipe-ready units. For just one example, consider that 1 gallon (gal) of milk equals:

4 quarts (qt.)
8 pints (pt.)
16 cups (c.)
128 fluid ounces (fl. oz.)
256 tablespoons (T)
768 teaspoons (t)

Despite its increased complexity, it is important to recognize that when the purchase unit and the recipe-ready unit are known, measurement conversions of both weight and volume when using the US system are calculated in the same manner. The conversion formula is as follows:

$$\frac{\text{Cost per purchase unit}}{\text{Number of recipe-ready units per purchase unit}} = \text{Cost per recipe-ready unit}$$

For example, if an ingredient is purchased by the pound, but is utilized in a recipe by the ounce, the cost per ounce for the item is calculated as follows:

$$\frac{\text{Item cost per pound}}{\text{16 oz. per pound}} = \text{Item cost per ounce}$$

Thoughtful readers will immediately recognize that metric conversions require knowledge of only two measurement equivalents (kilograms/grams and liters/milliliters) and that both of these equivalents utilize ratios of 1000 to 1 for conversion.

The US system, however, requires knowledge of literally dozens of weight and volume equivalents (and even more when fluid ounces vs. ounces of weight are considered!). It is no wonder that thoughtful chefs and bakers (as well as those who must calculate recipe costs) increasingly appreciate the elegance of the metric system of measurement for cooking, baking, and recipe costing.

FUN ON THE WEB!

A number of measurement conversion websites make it easy for foodservice managers to convert metric and Imperial (US) purchase unit weights and volumes to their equivalent recipe-ready amounts. To find one such website, simply enter "measurement conversion calculator" in your favorite search engine, then take a moment to familiarize yourself with how it works.

Returning to the example of 10 pounds of fresh carrots, if one half pound of carrots is lost in their preparation, the product's EP yield would be calculated as follows:

$$\text{10 lb. AP} - \text{0.5 lb. loss due to preparation} = \text{9.5 lb. EP}$$

Many fresh produce items as well as many meats and seafoods purchased in their AP state will yield significantly reduced amounts when converted to their EP states. For meats and seafoods, managers conduct a **butcher's yield test** that considers loss due to trimming inedible fat and bones, as well as losses due to cooking or slicing when they calculate their EP costs. A sample butcher's yield test showing the EP yield for the beef chuck used in the beef stew recipe previously presented is shown in Figure 5.11.

When performing yield tests, managers are interested in a product's waste percentage and yield percentage. **Waste percentage (waste %)** is the percentage of product lost due to cooking, trimming, portioning, or cleaning. For example, the yield test shown in Figure 5.11 indicates that 8 pounds, or 128 ounces (8 lb. × 16 oz. = 128 oz.) of beef chuck will lose 3 pounds, or 48 ounces (3 lb. × 16 oz. = 48 oz.), during the preparation process. You can compute this item's waste % using the following formula:

$$\frac{\text{Product loss}}{\text{AP weight}} = \text{Waste \%}$$

In this example:

$$\frac{48 \text{ oz.}}{128 \text{ oz.}} = 0.375, \text{ or } 37.5\%$$

Once its waste % has been determined, it is possible to compute an ingredient's yield percentage. **Yield percentage (yield %)** is the percentage of product you will have remaining after cooking, trimming, portioning, or cleaning.

Waste % + Yield % = 100%, so yield % can be calculated as shown in the following formula:

$$1.00 - \text{Waste \%} = \text{Yield \%}$$

In the beef cubes example, the yield % is computed as follows:

$$1.00 - 0.375 = 0.625, \text{ or } 62.5\ \%$$

FIGURE 5.11 Butcher's Yield Test Results

Unit Name: Steamer's **Item:** Beef Cubes **Date Tested:** 1/15

Specification: # 842 **Item Description:** Beef Chuck

AP Amount Tested: 8 lb.

Price per Pound AP: $8.60

Loss Detail	Weight	% of Weight
AP	8 lb. 0 oz.	100.0%
Fat loss	1 lb. 2 oz.	14.1%
Bone loss	1 lb. 14 oz.	23.4%
Cooking loss	0	0%
Carving loss	0	0%
Total product loss (waste)	**3 lb. 0 oz.**	5%

Product yield: 62.5% **EP cost per pound:** $13.76

Yield test performed by: L. D.

Yield percentage is important because, when you know a product's yield %, you can compute the AP weight you must buy to obtain the EP amount needed in your recipes. You can determine the amount of an AP ingredient you must have on hand by using the following formula:

$$\frac{\text{EP required}}{\text{Product yield \%}} = \text{AP required}$$

In the example recipe shown in Figure 5.9, with an EP required of 5 pounds of beef cubes, or 80 ounces (5 lb. × 16 oz. = 80 oz.), and a product yield % of 62.5 percent, or 0.625, the computation to determine the appropriate AP amount of beef chuck to buy to produce the beef stew recipe is as follows:

$$\frac{80\text{ oz.}}{0.625} = 128\text{ oz.}$$

This can then be converted to the AP amount required when the weight is divided by the number of ounces (16) in 1 pound:

$$\frac{128\text{ oz.}}{16\text{ oz./lb.}} = 8\text{ lb. AP required}$$

To check the preceding figures to verify that you should use a yield % of 0.625 when purchasing this item, you can proceed as follows:

$$\text{EP required} = \text{AP required} \times \text{Yield \%}$$

In this example,

$$\text{EP required} = 8\text{ lb.} \times 0.625 = 5\text{ lb.}$$

Another way to determine product yield % is to compute it directly using the following formula:

$$\frac{\text{EP weight}}{\text{AP weight}} = \text{Product yield \%}$$

In the beef cubes example, the EP weight needed in the recipe is equal to the product's AP weight of 8 lb. less the product's preparation loss of 3 lb.

Thus, EP weight equals 5 lb. and product yield % is computed as follows:

$$\frac{5\text{ lb. EP weight}}{8\text{ lb. AP weight}} = 62.5\%$$

Managers can only calculate recipe costs accurately if they know their EP costs. The **EP cost** is the portion cost of an ingredient or item after its cooking, trimming, portioning, or cleaning. The EP cost must be known because it represents the true cost of an ingredient or item based on its product yield.

It is important to note that in some cases the same product may have different yields when purchased from different vendors. As a result, managers should always use EP cost rather than AP prices to compare product prices offered by various suppliers. In general, you want to choose the supplier that offers the lowest EP cost for the same product, assuming that the same specification or quality is being purchased.

To compute a product's EP cost per purchase unit, you simply divide the AP price per pound or purchase unit by the product's yield %.

In our beef cubes example, with an AP price per pound for beef chuck of \$8.60 and a product yield of 0.625, the EP cost per pound would be \$13.76. This EP cost is computed using the EP cost formula for items purchased by the pound:

$$\frac{\text{AP price per lb.}}{\text{Product yield \%}} = \text{EP cost (per lb.)}$$

In this example:

$$\frac{\$8.60 \text{ AP cost per lb.}}{0.625 \text{ Product yield}} = \$13.76 \text{ EP cost per lb.}$$

You know now that your actual EP cost per pound when buying beef chuck with this particular specification, and from this supplier, is \$13.76.

You should conduct additional butcher yield tests if you are considering changing suppliers, modifying your beef chuck purchase specifications, or varying the quality level of the beef used in your recipe. In addition, it is usually a good idea to conduct yield tests on all of the meat and seafood items that require on-site trimming at least twice per year.

HERE'S HOW IT'S DONE 5.4

The formula managers use to calculate a product's yield percentage is:

$$\frac{\text{EP weight}}{\text{AP weight}} = \text{Product yield \%}$$

Managers must understand the concept of a product yield percentage and how this percentage is calculated because it helps them answer two very important questions:

1. How much of a product should I buy?
2. How much will the product cost?

How Much to Buy?

To calculate how much of a product to buy, managers simply divide the EP amount that is needed in a recipe by the product's yield percentage. For example, if 15 pounds of a product is needed in its recipe-ready EP state and the product's yield is 75 percent, the AP amount the manager must buy is calculated as:

$$\frac{\text{15 lb. EP needed}}{\text{0.75 yield}} = \text{20 lb. AP}$$

When the manager buys 20 pounds of the product, and achieves a 75 percent yield, 15 pounds of product in its EP state will be available.

What Will it Cost?

To find their actual costs, managers must always price the ingredients used in their recipes in each ingredient's EP state, even when the product is purchased in its AP state.

To calculate what a product will actually cost, managers divide the product's AP price by the product's yield percentage. For example, if a product is purchased by the pound in its AP state, the ingredient costs \$6.00 per pound, and the product's yield is 75 percent, then the product's actual recipe-ready EP cost per pound will be:

$$\frac{\$6.00 \text{ per lb. AP cost}}{0.75 \text{ yield}} = \$8.00 \text{ per lb. EP cost}$$

When managers know how to calculate product yield percentages, they are better prepared to do a good job purchasing and accurately calculating the true portion costs of their standardized recipes.

Waste percentage and yield percentage can be known if good records are kept on meat and seafood cookery, the cleaning and processing of vegetables and fruits, and the unavoidable losses that can occur in some products during portioning. Most recipes assume consistency in these areas, and good foodservice managers take the losses into account when making purchasing decisions. For example, in the beef stew recipe example, knowing that you will experience a yield percentage of 0.625 will help you determine exactly the right amount of meat to purchase.

Good vendors are often an excellent source for providing information related to trim and loss rates for standard products they sell. With this information, some savvy operators even go so far as to add a minimum or required yield percentage as a part of their product specifications (see Chapter 3).

DETERMINING OPERATIONAL EFFICIENCY

It is good for managers to be able to calculate their actual product cost. To manage your operation exceptionally well, however, you must be able to compare how well you are doing with how well you *should* be doing. This process of assessing how well you are doing begins with determining your attainable (or ideal) product cost.

Attainable product cost is defined as the COGS figure that should be achievable in your operation if avoidable losses were completely eliminated. Some managers refer to attainable product cost as standard or ideal costs because these would be the costs they achieved if all of their units' operational standards were met.

When you compare attainable product cost to your actual product cost, you get a ratio measure of your operation's efficiency. The formula for this **operational efficiency ratio** is:

$$\frac{\text{Actual product cost}}{\text{Attainable product cost}} = \text{Operational efficiency ratio}$$

To illustrate, assume again that you own Steamer's, the small soup and sandwich carryout kiosk referred to in the previous example. By calculating your recipe and portion costs, you determine your attainable product cost for a day to be \$1,850. That is, if your operation is very efficient in its food usage, your cost of sales for the day should be \$1,850. Assume further that you actually achieve a product cost of \$1,850 for that day. Applying the operational efficiency ratio, your results would be:

$$\frac{\$1850 \text{ actual product cost}}{\$1850 \text{ attainable product cost}} = 100\%$$

These results represent perfection in the relationship between your attainable and actual operational results. In this example, your operation incurred no avoidable product loss.

Assume, however, that actual product costs for the day were \$1,950. In that case, the operational efficiency ratio formula would be computed as:

$$\frac{\$1{,}950 \text{ actual product cost}}{\$1{,}850 \text{ attainable product cost}} = 105.4\%$$

In this scenario, you would know that your actual product usage, and thus your cost, was 5.4 percent (105.4 actual product cost − 100.0 attainable product cost = 5.4) *higher* than your attainable product cost goal.

A 100 percent operational efficiency ratio is rarely achieved for a number of reasons. One key reason is that some product loss in a foodservice operation is virtually unavoidable. Consider, for example, the cost of a product such as a pot of freshly brewed coffee. Although you may be able to compute the portion cost of producing an 8-ounce cup of coffee, it is much more difficult to estimate the amount of product that must be discarded on a regular basis to ensure that guests are getting only freshly brewed coffee. Further, at the end of the day, there will likely be some brewed coffee that must be discarded.

An additional reason why 100 percent operational efficiency ratios are rarely achieved is that every foodservice operation employs people, and people make errors at work. Sometimes these errors can result in product loss. Despite these challenges, however, managers should strive to ensure that their actual product costs are as close as possible to their attainable product costs.

In general, operational efficiency ratings in the range of 100 to 110 percent are attainable. Variance beyond that, however, can indicate serious product control problems. Ratios that are too high, that is, ratios above 110 percent, could be a good indication of excessive waste, ingredient theft, spoilage, overportioning, or inaccurate recipe cost sheet computation.

Operating efficiency ratings that are too low, that is, ratings in the 80 to 90 percent range must also be avoided. Efficiency ratios such as these could be the result of miscalculation of the number of items sold, inaccurate ingredient costing, underportioning, incorrect standardized cost sheets, or errors in valuing inventory.

Many managers find that it is helpful to share operational efficiency ratio results with food production employees because an operation's staff should also be very interested in the answer to the fundamental question of "How well are we doing?"

ACCEPTABLE AND UNACCEPTABLE VARIANCE

In Chapter 2, you learned how managers calculate their sales variance and their sales variance percentages. Using a similar approach, managers can calculate their product cost variances and product cost variance percentages. When comparing their actual product costs with their attainable product costs, they use the following formula:

$$\frac{\text{Actual product cost} - \text{Attainable product cost}}{\text{Attainable product cost}} = \text{Cost percentage varience}$$

For example, assume that a manager calculated their attainable product cost to be $8,000 for a week. The manager's actual product cost was $8,500 for the week, resulting in a variance of $500 ($8,500 actual product cost − $8,000 attainable product cost = $500).

In this example, the manager's cost percentage variance would be calculated as follows:

$$\frac{\$8{,}500 - \$8{,}000}{\$8{,}000} \quad \text{or} \quad \frac{\$500}{\$8{,}000} = 6.25\%$$

Managers calculate and analyze their cost percentage variances because they realize they should not treat all variances from expected results in the same way. For example, a variance of one or two dollars from an expected result of several thousand dollars is small enough that it constitutes a very acceptable level of variation. A variance of one or two dollars from an expected result of nine dollars, however, might well be cause for concern and, thus, is likely very unacceptable.

Figure 5.12 displays the operational efficiency ratios, dollar variances, and percentage variances that might result if you were to compute individual efficiency ratios on a variety of individual products in a steakhouse. The important concept here is a simple one: As a manager, your attention should be directed toward those areas where the need is greatest. In most cases, this means addressing large dollar variances and/or large percentage variances from your expected results.

FIGURE 5.12 Acceptable and Unacceptable Variance

Item	Actual Cost	Attainable Cost	Efficiency Ratio	Dollar Variance	Percentage Variance
Steaks	$1,010	$1,000	1.01	$10.00	1%
Coffee	$20	$10	2.00	$10.00	100%
Bolla Soave	$550	$500	1.10	$50.00	10%
Horseradish	$22	$20	1.10	$2.00	10%
Parsley	$45	$50	0.90	–$5.00	–10%

Results such as those shown in Figure 5.12 call upon you to make decisions about the acceptability of the variances you may encounter. Efficiency ratios in this example range from a low of 0.90 to a high of 2.00.

The $10.00 variance in actual steak cost is likely within an acceptable range because the expected cost was $1,000 and the efficiency ratio was 1.01. The $10.00 variance in coffee costs, however, is likely unacceptable because it represents actual costs that were twice as high as the expected coffee costs and that efficiency ratio is 2.00.

In a similar manner, the 10 percent variance in the cost of Bolla Soave wine represents $50 and thus may well be worth investigating, whereas the 10 percent parsley cost variation of $5.00 may be too small to merit your immediate attention.

Although it is not possible to determine one range of variance acceptability that is appropriate for all food facilities, it is important for you to establish variances from attainable costs that are acceptable for your own operation. When you do so, you will be able to easily identify those variations that must be investigated and then corrected.

HERE'S HOW IT'S DONE 5.5

In most cases, when managers calculate a cost percentage variance they will generate a positive number. That is true because, in many cases, an operation's actual product costs will exceed its attainable or ideal product costs. In some cases, however, the cost variance percentage will be negative.

For example, assume that a manager estimated they would incur $500 in salad dressing cost after serving 1,000 salads. After serving 1,000 salads, the manager's actual salad dressing cost was $450. To calculate his cost percentage variance, the manager applies the cost percentage variance formula:

$$\frac{\text{Actual product cost} - \text{Attainable product cost}}{\text{Attainable product cost}} = \text{Cost percentage variance}$$

In this example, the manager's cost percentage variance for salad dressing is:

$$\frac{\$440 - \$500}{\$500} = -12.0\%$$

In this example, the manager's salad dressing costs were 12 percent *less* than they had anticipated. There can be several reasons why an operation could achieve a negative cost percentage variance. One reason could be that the price paid for salad dressing by the operation was less than originally estimated. In that case, a negative cost percentage variance actually indicates good news for the operation.

It could be instead, however, that the operations' wait staff was serving less than the standardized amount of dressing on each guest's salad. While this would cause a short-term decrease in the operation's per-salad dressing costs (because the amount served was smaller than the amount desired by management), it is a practice that may cause a longer-term decrease in guest satisfaction. It is for that reason managers must investigate negative cost percentage variances indicating reduced costs just as thoroughly as they examine those positive cost percentage variances that indicate excessive costs.

Green and Growing!

Increasingly, foodservice operators recognize that most human activity, including their own businesses, leaves a carbon footprint on the planet. A carbon footprint is a measure of the impact human activities have on the environment in terms of the amount of greenhouse gases (carbon dioxide) produced. This term is useful for individuals and organizations to conceptualize their personal (or organizational) impact in contributing to global warming.

Examples of activities in which foodservice operations can engage to help reduce their carbon footprint include the following:

- Buying food products locally
- Monitoring efficient energy usage regularly
- Avoiding the sale of bottled waters where the packaging and shipping of these items result in the unnecessary production of carbon dioxide
- Using cold water for washing and cleaning when practical and sanitary
- Reducing, reusing, and recycling!

Reduction of a business's carbon footprint is good for the environment and for business. Keep watching the hospitality trade press for an ever-increasing number of new strategies you can consider for reducing your own operation's carbon footprint!

REDUCING THE COST OF SALES PERCENTAGE

In some cases, managers who evaluate their product usage will find that their operations are extremely effective at utilizing their food and beverage products. In other cases, managers may find that their product-related costs are simply too high to allow them to meet their profit goals. When that occurs, corrective action must be taken.

To illustrate, assume that a manager has a food cost percentage goal, or target, of 35 percent. In a specific accounting period, however, the manager's operation actually achieved a 38 percent food cost. In this example, the manager must take steps to reduce the excessive cost of sales percentage. Recall, however, that sales to customers mean costs are incurred. Increased sales mean increased costs. Therefore, despite missing their target, this manager's goal would not simply be to reduce overall product cost. Rather, the goal would be to maintain current (and increasing!) sales levels while achieving product costs that are at their proper proportion of sales. To do just that, restaurant managers must address three important tasks:

1. Minimizing product loss in the kitchen
2. Minimizing product loss in the bar
3. Optimizing overall cost of sales percentage

MINIMIZING PRODUCT LOSS IN THE KITCHEN

Excessive costs can result when control systems fail to prevent losses incurred while products are in storage, losses due to overcooking or overportioning, and losses due to employee theft. It's especially important to have programs in place that will minimize the chances of product theft by employees. Although all kitchens can expect to experience small amounts of employee-related product loss, such as an employee eating an apple in secret or nibbling a carrot where the supervisor cannot see it, excessive product loss must be prevented.

Most kitchen-related theft involves the removal of food products, rather than money, from the premises. This is because, unlike cashiers and bartenders, few kitchen production workers handle cash. Kitchen workers can, however, work with service personnel to defraud the operation (see Chapter 11). In addition, many kitchen workers will have direct access to valuable food and beverage products.

The following product security tips are helpful when designing control systems to ensure the safety and security of food (and beverage) products:

Product Security Tips

1. Keep all storage areas locked and secure.
2. Issue food only with proper authorization and management approval.
3. Monitor the use of all carryovers.
4. Do not allow food to be prepared unless a POS entry, guest check, or other written request precedes the preparation.
5. Maintain an active inventory management system.
6. Ensure that all food received is signed for by the appropriate receiving clerk.
7. Do not pay suppliers for food products without an appropriate and signed delivery invoice.
8. Do not use petty cash to pay for food items unless a receipt and the product can be produced.
9. Conduct systematic physical inventories of all A, B, and C level products.
10. Do not allow employees to remove food from the premises without management's specific approval.

MINIMIZING PRODUCT LOSS IN THE BAR

Employee product theft can occur in either the bar or the kitchen production areas, but it is often more prevalent in the bar areas. Experienced food and beverage managers seem to have an endless supply of stories related to theft in bar operations. Indeed, bar theft is one of the most frequent types of thefts in the foodservice industry. Although it may well be impossible to halt all kinds of bar theft, the following are activities you should watch for to help ensure that the opportunities for product theft in your bar areas are minimized.

ORDER FILLED BUT NOT ENTERED

In the "order filled but not entered" scenario, a bartender delivers a drink as requested by a guest or server, but the drink sale is never recorded in the POS system, and the bartender simply pockets the amount of the sale. As a result, the sales as recorded in the POS will be equal to the amount of money in the cash drawer, despite the fact that beverage inventory has been reduced by the amount of product served in the drink.

All drink sales should be entered into the POS system prior to drink preparation to prevent this type of theft. Management's vigilance is critical to ensure that no drink is prepared until *after* the order is properly recorded by entering it into the POS.

OVERPOURING AND UNDERPOURING

Overpouring occurs when more alcohol than management desires is served in a drink. **Underpouring** occurs when less alcohol than management desires is served in a drink. When bartenders overpour, they are stealing from the operation. When they underpour, they are shortchanging the guest. Remember that your bartenders will pour the appropriate amount if you always insist that they do so. When they overpour, bar staff may be doing so for their friends or for the extra tips this activity may yield. When they underpour, they may be making up for drinks they have given away or sold but have not rung up. In either case, management must prevent such behavior.

Proper portion size when serving spirits area is ensured through the enforced use of jiggers, metered devices, or other mechanical or electronic equipment (see Chapter 4). In the case of draft beer, **head size**, that is, the amount of foam on top of the glass directly affects portion size and portion cost, and, thus, it too must be controlled.

FUN ON THE WEB!

Product loss related to draft beer can be due to either employee theft or improper dispensing issues (e.g., wrong service temperature, too much carbonation, improperly cleaned beer lines, and the like). As a result, draft beer dispensing systems and their maintenance are a critical area of concern for bar managers. There is a variety of quality dispensing systems on the market. To review them, enter "draft beer product control" in your favorite search engine, and then review the results.

PRODUCT DILUTION

Often called "watering down the drinks," product dilution theft involves adding water to a liquor to make up for the amount of the spirit that has either been stolen or given away. It is especially easy to water down products such as gin, vodka, rum, or tequila because these clear spirits will not change color with the addition of water. Detection of this type of theft is often difficult.

Periodic sampling of a known-proof alcohol against bar stock by a knowledgeable food and beverage professional is one of the defenses against such bartender fraud. Since each alcohol product has a particular specific gravity or weight associated with it, you may also check for product dilution through the use of a **hydrometer**, which identifies specific gravity. If water has been added to a bottle of liquor, the content's specific gravity will change from the value originally associated with that liquor.

PRODUCT SUBSTITUTION

When a specific call brand liquor has been ordered and paid for by a guest, it should, of course, be served to that guest. If, however, a bartender substitutes a less expensive well liquor for the call brand, while charging the guest for the higher priced call liquor, the bartender may intend to keep the difference in the prices paid for the two items. This has the effect of shortchanging guests, who paid higher prices for premium liquor drinks they did not receive. Conversely, if guests have ordered well drinks, but the bartender makes their drinks from premium or super premium call liquors, guests have received more value than they will pay for, and the operation is shortchanged.

Although it is impossible to list all types of product-related bar thefts, it is important to note that they can and do occur. Conscientious managers should hire honest bartenders, train them well, and demand that they follow all operational policies.

To control drink production, it is also important to designate portion size and glassware to be used. This helps ensure that the portion size of the drink is appropriate and consistent with the guest's visual perception of a full glass. The ice to be used in drinks is also important for controlling drink production.

What does ice have to do with standardized beverage production? A lot! Large ice cubes will leave space between them when scooped into a glass. This will permit a larger amount of mixers to be added, which may dilute the drink more than intended. By contrast, smaller cubes or (especially) shaved ice will pack a glass and permit less of the mixer to be added. This may, in turn, give the impression of a stronger drink. A second difference created by ice size is that large cubes have less exposed surface area and will melt (and dilute the drink) at a slower pace than smaller cubes. As a result, in the bar area, you should select an ice machine that makes ice in a form that you feel best fits your view of proper spacing in the glass and that possesses the most desirable melting characteristics.

FUN ON THE WEB!

A number of companies sell high-quality commercial ice machines. Scotsman is one of those. To learn more about machines that produce ice cubes, nuggets, and flakes, enter "Scotsman ice machines" in your favorite search engine, go to their site, and then explore their site to review some ice production options available to foodservice managers.

DIRECT THEFT

Alcohol is a highly desirable product for many employees; therefore, its direct theft is always a possibility. This is especially true in a beverage service area that is secluded or in which bartenders have direct access to product inventory and ease of exit. Proper controls as well as strict rules limiting the access of employees to beer, wine, and liquor storage areas should help deter and detect this type of theft.

It is also important to remember that not all alcohol sales take place in a traditional restaurant or bar area. Because of this, additional problem areas, some of which occur only in hotels, can include the following:

1. In-room minibars
2. Bottle sales
3. Open bars
4. Banquet operations
5. Room service operations

IN-ROOM MINIBARS

Minibars, typically containing bottled and canned beer, 100-milliliter wine bottles, and 50-milliliter liquor bottles, are popular in hotels that cater to upscale travelers. The control issue here is one of matching requests by the hotel's housekeeping department for bottle replacement with actual guest usage. Some large hotels deal with this issue by having a single individual or even a department charged with the responsibility of refilling the minibars. Today most modern minibars record liquor sales electronically as products are removed by guests, but such records still must be carefully monitored to ensure that items issued from storage are indeed used to restock the minibars and are not lost to employee theft.

BOTTLE SALES

When liquor sales are made by the bottle, either through room service in a hotel or at a reception area, the control issue is one of verifying bottle count. The guest and the operation must both be treated fairly in such a transaction. In the case of full-bottle sales to a guest room, the guest should be required to sign a receipt confirming acceptance of the product. This is the only way to avoid potential misunderstandings about cost.

In the case of receptions or banquets, guests should be charged only for empty bottles, or, in the case of the purchase of a specified number of bottles, should be shown both full and empty bottles equal to the number used and charged for the event. In an effort to protect both the establishment and the guest from employee theft, the thoughtful beverage manager will mark the bottles for receptions or banquets in a way that is not easily duplicated. Marking bottles helps prevent employees from bringing in their own empty bottles and then removing full ones at the event host's expense.

OPEN BARS

Open bars (hosted bars) are those in which no charge is made to guests for the individual drinks they consume but, when the bar is closed, one total amount is charged to the host or sponsor of the open bar. Open bars are common in hotels and some restaurants and especially at events such as weddings, special-occasion parties, and cocktail receptions. Unfortunately, because the individual drinker is not paying for each drink, the open-bar situation can sometimes create an "all you care to drink" mindset among guests and bartenders. Because that can be the case, consumption by individual drinkers at open bars must be carefully monitored.

The production control issues associated with open bars fall into one of two main categories: portion size and accountability. At open bars, guests can sometimes cajole bartenders into serving drinks with larger than normal amounts of alcohol. This must, of course, be prevented. Bartenders, as well as guests, must understand that, although it may be an open bar, someone will be paying the bill at the end of the event. The hosts of the open bar have the right to expect reasonable portion control if they are paying on a per-drink or per-bottle-used basis. If the foodservice operation has established a per-person charge for the open bar, overportioning costs will have to be absorbed by the operation. This means strict control of portion size and total liquor consumption per guest must be carefully monitored.

As great an issue as overportioning is, accountability looms larger and larger on the horizon as an area of legitimate cost control concern for the effective beverage manager. Recall from Chapter 4 that dram shop legislation increasingly holds liquor sellers responsible for the actions of those they serve. As a result, bartenders who work open bars should be specially trained to spot signs of guest intoxication. As difficult as it may sometimes be, guests should be made aware that it is illegal, in all states, to serve an intoxicated guest. To do so puts the entire food and beverage operation at risk.

Some managers have virtually eliminated the open-bar concept, preferring to implement a coupon system where each coupon issued is good for one drink, and the number of coupons issued, rather than the number of drinks served, can be controlled. Although the possibility exists that coupons can be shared and, thus, improperly given to an intoxicated guest, the coupon system does demonstrate an attempt by management to exercise reasonable care, an effort that may prove vital in their defense in the event of dram shop-related litigation.

FUN ON THE WEB!

Dram shop laws are often referred to as third-party liability laws. In third-party liability legislation, the first party is the individual or business that served the alcohol, the second party is the individual who consumed the alcohol and later caused harm or injury. The third party is the individual who was harmed or injured. To avoid liability, foodservice operators must train their beverage service staff well.

Many large and small beverage operations experience employee turnover that results in a constant need for responsible alcoholic beverage service training. Increasingly, state and local dram shop laws mandate or recommend such training. Fortunately for beverage managers, many companies now offer training in the service, sale, and consumption of alcoholic beverages online. To review online training programs of this type, enter "responsible alcohol service training" in your favorite search engine and review the results.

BANQUET OPERATIONS

The sale of alcoholic beverages during a seated banquet usually takes the form of bottled-wine sales. Guests may be provided with a set number of bottles on the table, to be shared by those seated at the table. Alternatively, as they consume their wine, they can be served by the banquet wait staff. It is the latter method that presents the greatest product control problems because the host of the event will be charged by either the number of bottles served or the number of guests served. If the payment is based on the number of bottles served, the bottles should be marked and the empties made available for inspection by either the guest or the banquet captain. If the sale is based on the number of glasses poured, then both the host of the event and the beverage operation must be in agreement as to the desired portion size and the total number of portions allowed to be served to each guest, table, or the entire group.

ROOM SERVICE OPERATIONS

Hotels that offer alcoholic beverages as part of their room service menu face special control issues. Room service initiated sales of individual drinks and/or full bottles of alcoholic beverages should, of course, be recorded prior to their delivery to a guest's room. The collection of any unopened bottles (by law opened bottles must be disposed of) remaining after guest departure requires coordination between a hotel's housekeeping department and its food and beverage department. Such control is needed to ensure that unopened bottles are returned to the food and beverage area after the departed guest's room has been cleaned and that the return of the items is properly recorded so their quantity and value can be added back to the food and beverage department's beverage inventory.

Perhaps a manager's best beverage control tactic of all is to simply be vigilant. It is a good idea to watch bar areas carefully, or enlist the aid of a **spotter**, a professional who, for a fee, will observe the bar operation. Commonly known as **mystery shoppers**, spotters pose as anonymous customers but, during their unannounced visits, they observe workers carefully and later report to management any unusual or inappropriate behavior by bartenders or beverage service staff.

OPTIMIZING OVERALL COST OF SALES PERCENTAGE

After you have determined what your operation's costs actually are and have compared them to what your operation's costs should be, it may often be found that Walt Kelly's Pogo comic character was correct when he said, "We have met the enemy, and he is us!"

Foodservice managers (and their bosses) often seem to be on a never-ending quest to reduce food and beverage production costs. Although you must remember to guard against inappropriate cost-cutting efforts, you may, in some cases, find yourself managing an operation where food and beverage production costs, when expressed as a total or as a percentage of sales, are simply too high and thus they must be reduced. As you design strategies to reduce your product costs, it is helpful to review potential solutions based on the cost of sales equation.

You have learned that the product cost percentage equation is deceptively easy to state. In its simplest form, it can be represented as:

$$\frac{A}{B} = C$$

where

A = Cost of products sold

B = Sales

C = Cost percentage

This formula can, however, become extremely complex. Its analysis occupies many a food organization staff meeting and can give the foodservice operator many sleepless nights! Six basic cost reduction strategies are used to influence this rather simple formula. A quick algebra lesson, however, prior to our discussion of these strategies can be useful.

The rules of algebra reveal the following characteristics about any A/B = C formula:

- If A is unchanged and B increases, C decreases.
- If A is unchanged and B decreases, C increases.
- If A increases at the same proportional rate B increases, C remains unchanged.
- If A decreases and B is unchanged, C decreases.
- If A increases and B is unchanged, C increases.

Put into foodservice management terms, these same algebraic statements can be translated as follows:

- If product costs can be kept constant while sales increase, the product cost percentage goes down.
- If product costs remain constant but sales decline, the cost percentage increases.
- If product costs go up at the same rate sales go up, the cost percentage will remain unchanged.
- If product costs can be reduced while sales remain constant, the cost percentage goes down.
- If product costs increase with no increase in sales, the cost percentage will go up.

In general, foodservice managers work to control the variables that impact product cost percentage and seek to reduce the overall value of C in the cost percentage equation. Six basic cost reduction approaches used to optimize overall product cost percentage, along with a summary of each approach, are presented here to help you devise your own cost reduction strategies:

How to Reduce Overall Product Cost Percentage

1. Ensure that all products purchased are sold.
2. Decrease portion size relative to price.
3. Vary recipe composition.
4. Alter product quality.
5. Achieve a more favorable sales mix.
6. Increase price relative to portion size.

To reduce your product costs, you will ultimately select an appropriate strategy from this relatively small number of alternatives. These strategies can be applied to excessive food costs or to excessive beverage costs. It is the careful selection and mixing of these approaches to cost control that differentiates the successful operator from the unsuccessful one.

To illustrate, assume that you own and operate a family-style restaurant. You compute your actual food cost percentage for last month and have determined that it was 4 percentage points higher than you had planned. If you have approximately six cost-reducing options available to you, by the mathematics law of permutations, this means you have (6×5×4×3×2×1) or 720 possible combinations of these six different cost reduction methods. No wonder, then, there is so much information written about reducing product costs!

It is not the authors' contention that all product cost reduction methods are exhausted by these six approaches, but rather the approaches are presented here as a means of systematically analyzing the various cost percentage reduction alternatives that are available to most foodservice managers.

ENSURE THAT ALL PRODUCT PURCHASED IS SOLD

These seven words have tremendous implications. They include all phases of professional purchasing, receiving, storage, inventory, issuing, production, service, and cash control. Perhaps the hospitality industry's greatest challenge in the area of cost control is ensuring that all products, once purchased, do indeed generate sales that are ultimately deposited into the operation's bank account!

DECREASE PORTION SIZE RELATIVE TO PRICE

Product cost percentages are directly affected by portion size. Too often, foodservice managers and bar operators assume that their standard portion sizes must conform to some unwritten rule of uniformity. This is simply not the case. Most guests would prefer a smaller portion size of higher-quality ingredients than the reverse. In fact, one problem some restaurants have is that their portion sizes are simply too large. The result is excessive food loss because uneaten products left on plates must be thrown away. It is important to remember that portion sizes are determined by the foodservice manager and, as a result, they are variable.

To see the impact of portion size on (liquor-only) product cost percentage, review the data in Figure 5.13. This figure presents the significant effect on liquor cost percentage of varying the standard drink size served in an operation using $21.00 per liter as the standard cost of liquor and assuming 0.8-ounce evaporation per 33.8-ounce (1-liter or 1,000-milliliter) bottle and a standard $15.00 selling price per drink.

Note that, in this example, the manager's (rounded) product cost percentage ranges from 4.2 percent when a 1-ounce portion is served, to a 8.5 percent product cost when a 2-ounce portion is served.

Portion sizes of both food and drink items directly affect product cost percentage as well as guest perceptions of value delivered. As a result, when establishing your portion sizes, you should take all of the variables affecting your operation into account. These may include your location, service levels, competition, and the clientele you seek to attract.

FIGURE 5.13 Impact of Drink Size on Liquor Cost Percentage at Constant Selling Price of $15.00 per Drink

Drink Size	Cost per Liter	Drinks Per Liter	Cost Per Drink	Sales Per Liter	Liquor Cost % Per Liter
2 oz.	$21.00	16.5	$1.27	$247.50	8.5%
1¾ oz.	$21.00	18.9	$1.11	$283.50	7.4%
1½ oz.	$21.00	22.0	$0.95	$330.00	6.4%
1¼ oz.	$21.00	26.4	$0.80	$396.00	5.3%
1 oz.	$21.00	33.0	$0.64	$495.00	4.2%

Cost Control Around the World

Some foodservice managers in the United States operate on a "bigger is better" philosophy when it comes to determining ideal portion sizes. But the same is not true worldwide.

For example, at McDonalds, a small-size drink is 89% bigger in the United States than in the UK. The medium is 55% bigger. And the large is 89% bigger in the United States compared to the UK. In fact, the US medium size is actually 24% bigger than the UK large. Similarly, the US large fry is 46% bigger than in the UK.*

At Kentucky Fried Chicken (KFC) the largest bucket sold in the UK contains 14 pieces, However, in the United States, the largest bucket contains 16 pieces. In the United States, the smallest bucket that can be ordered from KFC contains 8 pieces, but in the UK a 6-piece bucket can be purchased.

In the UK, Domino's largest pizza measures 13.5 inches across, while in the United States Domino's largest pizza measures 16 inches across.

The important take away for foodservice managers, regardless of where they operate, is that there is no universally accepted portion size for nearly any food item sold. As a result, effective foodservice managers carefully evaluate portion sizes that make sense for their own customers and then price their menu items accordingly.

*Source: https://www.businessinsider.com/fast-food-portion-sizes-us-vs-uk-mcdonalds-kfc-dominos-starbucks-2023-1. Retrieved 8/23/2025.

VARY RECIPE COMPOSITION

Experienced managers know that even the simplest standardized recipes can often be varied somewhat. For example, what is the proper ratio of beef to carrot to be used when making 100 servings of high-quality beef stew? Because the cost of one pound of beef far exceeds the cost of one pound of carrots, 100 servings of beef stew made with increased amounts of beef will cost more than 100 servings made with increased amounts of carrot. The question of what constitutes an ideal recipe composition must be answered by management. The answer to that question, for each standardized recipe utilized in an operation, will directly impact the operations' overall food cost percentage.

Similarly, the proportion of alcohol to mixer has a profound effect on beverage cost percentages. In some cases, the amount of alcohol used in drinks can be reduced, yet overall drink sizes can actually be increased. This can be done, for example, by increasing the drink's proportion of lower cost standardized drink recipe ingredients such as milk, juices, and soda and reducing the proportion of higher cost spirit products. Utilization of this beverage cost reduction strategy often contributes to a feeling of satisfaction by the guest, while allowing the operator to reduce their beverage cost percentages and increase profitability.

ALTER PRODUCT QUALITY

In nearly all cases, higher-quality products cost more than lower-quality products. That is certainly true for managers purchasing food and beverages. As a result, one way to achieve cost savings is to reduce product quality. This area must be approached with great caution, however, because you do not want to offer inferior quality. Rather, you should always strive to purchase the quality of product appropriate for its intended use.

For example, managers may find that less expensive chopped and frozen asparagus can be excellent when used in a baked casserole dish, but the same frozen product is an unacceptable substitution for the freshly cooked asparagus accompanying a $45.00 prime rib-eye steak.

Similarly, a specific coffee liqueur such as Kahlua, when called for by a guest, must, of course, be used to prepare that guest's drink. It may be quite acceptable, however, to use a nonbrand name (well brand) and lower cost coffee liqueur for the many specialty drinks (e.g., Black Russian, Brave Bull, Black Magic, or Sombrero) that include coffee liqueur as a major or minor ingredient. In this example, a lower-cost generic-type coffee liqueur might be used for these specific drinks with totally satisfactory results.

When managers find that an appropriate ingredient, rather than the highest-cost ingredient, provides good quality and good value to guests, product costs may be able to be reduced using product substitution. With appropriate experience and care, you can determine the quality of ingredients that best serve your operation and your guests and then purchase only that specific quality. But recall that managers must be very careful in this area. Lower-quality products may cost an operator less, but customers may also perceive that menu items made from these lower-quality ingredients provide reduced levels of value to them and that negative reaction by guests must always be avoided.

Consider the Costs

"You're our food guru. They are offering the same size pizza as we do, but for $3.00 less. How can they do that?" asked Kevin Gustafson to Sara Leiboda. Sara was the Director of Operations for the seven-unit Brooklyn Pizza House restaurants. Kevin was the company President. Brooklyn Pizza House was known for its mid-priced, but very high-quality, thin crust pizza.

Located near the main campus at State University, the company's target market is students who attend the school. The advertising programs in use by the company were clever and effective. Business and profits were good. But now Kevin and Sara were discussing the new $10.99 pizza promotion that had just been announced by their major competitor.

"They changed their cheese topping formulation Kevin," replied Sara. "They increased their use of pizza cheese by another 25 percent. One of my friends works in their central production kitchen. That's where they process and prepackage the pizza ingredients for delivery to their stores. My friend says they changed their 75/25 mozzarella/pizza cheese ratio to a 50/50 ratio. As you know, we only use 100 percent mozzarella cheese on all of our pizzas."

"Why would they do that?" asked Kevin.

"Well," replied Sara, "It's simple. Pizza cheese is a processed cheese product that costs quite a bit less per pound than high-quality mozzarella. Pizza cheese can actually contain as little as 51 percent real cheese and still be sold as pizza cheese. It melts OK, but it can't compare to the taste of real mozzarella. By using it, they can reduce their per-pizza costs and, because their costs are lower, they can reduce their selling prices. That's what they did. I guess they think it will increase their sales, reduce their food cost percentage, and ultimately increase their profits."

1. Why do you think this competitor has decided to reduce the quality of cheese used in its pizzas?
2. Do you think students at the university will notice the change? Would you advise Kevin and Sara to make the same change to their pizza formulation? Why?

ACHIEVE A MORE FAVORABLE SALES MIX

Experienced managers know that their customers' item selection decisions will have a direct and significant impact on the product cost percentages that will be achieved in these managers' operations. This is because an operation's product cost percentages are determined in large part by the operation's sales mix.

Sales mix is defined as the series of guest purchasing decisions that result in a specific food or beverage cost percentage. Sales mix affects overall product cost percentage anytime guests have a choice among several menu selections, each selection having its own unique product cost percentage.

To see a simple example of how sales mix can directly affect an operation's overall product costs, assume that a quick-service restaurant offered only three menu items for sale. In this restaurant, the manager also offers a special "value meal" that includes one of each item, as shown in Figure 5.14.

From the data in Figure 5.14, it is easy to see that if, on a specific day, 100 percent of the restaurant's customers bought a hamburger and nothing else, the operation's product cost for that day would be 35.8 percent. If, on another day, 100 percent of the restaurant's customers purchased a soft drink and nothing else, the product cost percentage for that day would be 17.6 percent.

FIGURE 5.14 Four-Item Menu

Menu Item	Product Cost	Selling Price	Product Cost %
Hamburger	$2.50	$6.99	35.8%
French fries	$0.95	$2.99	31.8%
Soft drink	$0.35	$1.99	17.6%
Value meal	$3.80	$10.99	34.6%

Similarly, if every customer purchased only the "value meal" on a specific day, the operation would achieve a 34.6 percent product cost on that day. In this example, it is easy to see that the operation's actual product cost will be largely determined by the mix of the individual product costs resulting from the menu item choices made by the operation's customers.

To show a more realistic example of the effect of sales mix on product costs, assume that you are the food and beverage director at the Raider Resort, a 400-room beachfront hotel property on the Gulf Coast of Texas. In addition to your regular restaurant, you monitor your beverage sales in three serving areas:

1. Banquets: The beverages served in receptions prior to meal events. These sales are made in your grand ballroom, foyer, or outdoors. Banquet beverages are also served during banquet meal functions.
2. The Starlight Bar: An upscale bar with soft piano music that typically appeals to hotel guests who are 55 years and older.
3. Harry O's: A bar with indoor and poolside seating. Contemporary Top-40 music played in the evenings draws a younger crowd interested in mingling and dancing.

You compute a separate beverage cost percentage for each of these beverage outlets. Figure 5.15 details the separate operating results recently achieved in each location. It also shows the combined (total) beverage cost percentage achieved in all three locations when calculated using the beverage cost percentage formula introduced earlier in this chapter.

In this example, assume that you know each beverage location uses the same portion size for all standardized drink recipes. Well and call brand liquors, as well as wine-by-the-glass brands, are the same in all three locations. In this resort setting, you dislike the difficulty associated with serving draft beer, thus beer is sold in cans or bottles only. All three bars offer the same beer choices.

In addition, bartenders are typically rotated on a regular basis through every serving location. Given all that, should you be concerned that your beverage cost percentage varies so greatly by service location? The answer, in this case, is that you most likely have no cause for concern. In this situation, it is clear that your sales mix, and not poor control systems, has primarily determined your beverage cost percentage in each individual location.

A closer examination of the three outlets as shown in Figure 5.16 reveals how this can happen.

Although product cost percentages are the same in each location for the individual beverage categories, the overall beverage cost percentages in the three locations are not the same. The reason that each unit varies in total beverage cost percentage is due to sales mix; the guests' overall preference for the specific product choices offered for sale. In other words, the hotel's guests, and not management alone, have helped to determine the final beverage cost percentages in each of the locations.

Managers can directly help to influence guest selection and sales mix by such techniques as strategic pricing, effective menu design, and creative marketing, but to some degree it will always be the guest who will determine an operation's overall product cost percentage. This is because it is customers and their choices that determine an operations' sales mix.

FIGURE 5.17 Raider Resort

Beverage Sales Percentage Recap				
Unit	Beer	Wine	Spirits	Total Sales
Banquets	12.5%	50.0%	37.5%	100%
Starlight Bar	33.0%	11.0%	56.0%	100%
Harry O's	67.0%	10.5%	22.5%	100%

As indicated, each beverage outlet operates with a unique sales mix. Figure 5.17 shows that in the banquet area the sales mix is heavy in wines and spirits, the choice of many guests when they are at a reception or dining out.

The Starlight Bar clientele is older, and their preferred drink tends to be spirits. Harry O's, on the other hand, caters to a younger crowd that, in this example, prefers beer. It is important to remember that, despite controls that are in place, costs that are in line, and effective management policies, variations in product cost percentages can still occur due to sales mix, rather than other factors. Experienced managers recognize this fact, and as a result, they monitor sales mix carefully to determine its impact on the overall product cost percentages of both food and beverages.

Because you now understand sales mix, you can recognize that the effective marketing and promotion of good-cost (i.e., lower-cost) items can help you reduce your product cost percentage and increase your profitability, while allowing the portion size, recipe composition, and product quality of your menu items to remain constant.

INCREASE PRICE RELATIVE TO PORTION SIZE

In many cases, managers facing rising product cost percentages feel they simply must increase their selling prices. While increasing prices is often relatively easy to do, this area must be approached with the greatest caution of all. There is no bigger temptation in foodservice than to raise prices in an effort to counteract management's ineffectiveness at controlling product costs. This temptation must be resisted.

There are times, of course, when selling prices on selected menu items must be increased. This is especially true in inflationary times or in times of unique product shortages. Price increases should be considered, however, only when all other alternatives and needed steps to control costs have been considered and effectively implemented. Any price increases passed on to guests should reflect only increases in your costs, not your inefficiency.

While in most cases it is important to keep menu prices low, some operators may be afraid to charge 25 or 50 cents more than their competitors for similar menu items. In some instances, keeping prices in line with competitors' prices is a good strategy. More frequently, however, decor, quality of product, and superior service levels can allow you to charge higher prices than your competitors. Experienced managers know that, given the proper facility ambiance, most guests will not react negatively to small variances in prices because it is perceived value, not price alone, which drives a foodservice guest's purchase decision.

You have learned that managing the food and beverage production process is a complex task and that it must be accomplished with the utmost skill. In fact, the goal of all your purchasing, production, and cost control systems should be to deliver to guests only high-quality products sold at a fair price. It is important that your food and beverage products are prepared correctly and at an appropriate cost to you, but it is just as important to ensure that these products are sold to guests at prices that help ensure your operation's profitability. For that reason, the important task of properly pricing your menu items will be the topic of the next chapter.

The Move to Off-Premise Dining

The control of production and service costs is important to all foodservice managers, but especially so for those whose operations generate a significant amount of off-premise revenue. While on-premise production costs can be calculated rather easily, identifying *total* COGS sold for off-premise sales requires the careful monitoring of three distinct cost areas.

1. Products
2. Packaging
3. Condiments

Products: Drive-thru (and pickup) services have evolved into a complex system where efficiency and accuracy are paramount. Recent benchmark data from the 23rd Annual Drive-Thru Study underscores this, revealing an average order accuracy rate of 86%, meaning that nearly 15% of drive-thru orders are improperly filled.* While error rates are directly affected by inexperienced staff and increasingly complex menus, it is important to recognize that improperly filled orders often result in increased costs due to replacing items or offering full refunds. The result of these errors directly affects these products' total COGS.

Packaging: The cost of some packaging items, such as bags, boxes, and wrappers, can be easily calculated. Additional takeaway foods costs, however, can include items such as napkins, cups, straws, disposable cutlery, or other items required to complete a guest's order. As previously addressed in this chapter, the cost of these items can be significant, and they are increasingly considered as part of COGS.

Condiments: Nearly everyone who has ever asked for ketchup with their fries when going through a drive-thru has probably experienced looking in the bag and finding five or even 10 or more ketchup packets! The same can be true with higher-cost condiments such as mayonnaise, Ranch dressing, or dipping sauces. In many cases, foodservice operators have not directly addressed the issue of how many condiments should be given as "free" to those guests who request them. The result can often be that workers most often error on the side of giving too many rather than the "right" number. Also, in many cases the condiments are viewed by workers as having little monetary value, so giving away too many is not seen as a major cost issue. It is, however, and this important cost must be addressed if total COGS are to be properly controlled.

*https://foodchainmagazine.com/news/customer-frustration-at-drive-thru-75-error-rates. Retrieved 10/31/2025.

Technology Tools

In the past, foodservice operators were slow to install advanced technological tools in kitchen areas where production staff could easily use them. Increasingly, however, the use of computerized cost control-related programs and smart device apps are being undertaken. In a professional kitchen, cost control efforts are often shared between management and the production staff. Advanced technology programs available for kitchen production use include those that can help both you and your production staff members:

1. Calculate total recipe costs and per portion costs.
2. Automatically update ingredient prices to current levels
3. Compute actual product yield percentages.
4. Compute actual product waste (trim loss) percentages.
5. Compute actual versus ideal costs based on product issues.
6. Compare portions served to portions produced to monitor overportioning and waste.
7. Estimate and compute hourly, meal period, or daily food cost.
8. Estimate and compute per meal costs (i.e., cost per served meal).
9. Maintain product usage records by:
   a. Vendor
   b. Product type
   c. Food category
   d. Individual menu item
   e. Individual menu ingredient
10. Calculate operational efficiency ratios based on actual food and beverage usage and sales.

FUN ON THE WEB!

Many of the software programs and apps available for assisting in foodservice operations are geared toward commercial restaurants. Some of the very best cost control tools, however, have been developed for the noncommercial (institutional) foodservice market.

To view the product features offered by two of these companies, search the Internet to locate the website of Computrition, a company that was originally formed to assist healthcare facility foodservice operators. Next, locate the home page of the CBORD Group, a company that was initially formed to assist foodservice managers operating campus dining facilities.

Apply What You Have Learned

Jennifer Tye is the manager of a fine-dining restaurant that serves its guests a variety of traditional Southern dishes prepared with the very finest ingredients. It is most famous for its upscale version of chicken fried steak.

At Jennifer's restaurant, this dish (beef steak, seasoned, breaded, and then pan fried) is prepared using USDA Choice Ribeye steak. As the price of ribeye has increased, Jennifer's food cost percentage on this item has risen dramatically.

Currently, Jennifer is considering whether the selling price of this dish should be increased significantly or if she should take other steps to help reduce her overall product cost percentage.

1. What issues should Jennifer consider prior to implementing a price increase?
2. What alternatives to raising prices do you think are available to Jennifer?
3. What would you advise Jennifer to do?

For Your Consideration

1. In most cases, managers who buy fresh fruits, vegetables, and meat products that have been cleaned, trimmed, chopped, portioned, sliced, or in other ways preprocessed will pay more per purchase unit than managers who do their own on-site processing of these items. What key information do you think managers must possess to best determine whether it is better to buy preprocessed products or to do their own processing of such products on-site?
2. Why is a thorough understanding of the relationship between sales mix and overall product cost ratios so important to a manager's cost control efforts?

Key Terms and Concepts

The following are terms and concepts addressed in the chapter that are important for you as a manager. To help you review, define the terms below.

Cost of goods sold (COGS)
Cost of sales
Beginning inventory
Padding inventory
Purchases
Food available for sale
Goods available for sale
Ending inventory
Cost of food consumed
Employee meal cost
Cost of food sold
Food cost percentage
Cost of beverage sold
Beverage cost percentage
Product transfer form
Standardized recipe cost sheet
As-purchased (AP)
Edible portion (EP)
Yield test
Butcher's yield test
Waste %
Yield %
Edible portion (EP) cost
Attainable product cost
Operational efficiency ratio
Overpouring
Underpouring
Head size
Hydrometer
Open bar (hosted bar)
Spotter
Mystery shopper
Sales mix

Test Your Skills

You may download the Excel spreadsheets for the Test Your Skills exercises from the student companion website at http://www.wiley.com/go/dopson/foodandbeveragecostcontrol8e. Complete the exercises by placing your answers in the shaded boxes and answering the questions as indicated.

1. Belinda Gates is the manager at the Roasted Pepper restaurant. Belinda records her monthly food and beverage sales separately. She also follows the USAR recommendation and thus separately calculates her product costs for food, beverages, and her overall cost of sales. Help Belinda complete this portion of her monthly operating records by calculating her total cost of sales and all of her cost of sales percentages for the month.

Roasted Pepper: Sales and Cost of Sales		
	This Month	Percentage
SALES		
Food	$90,000	75%
Beverages	$30,000	25%
Total Sales	$120,000	100%
COST OF SALES		
Food	$23,200	
Beverages	$6,600	
Total Cost of Sales		

2. Mark Chaplin is a pastry chef at the Raised Mitten bakery. Mark makes his famous buttercream icing for use in a variety of items sold at the bakery. He wants to know the cost per ounce of his icing. Complete the recipe cost sheet so Mark can know his actual icing cost.

Buttercream Icing		Recipe Cost Sheet		# 105
Ingredient	Purchase Unit	Purchase Unit Cost	Recipe Amount	Ingredient Cost
Confectioners' sugar	25 lb. bag	$28.45	1.5 lb.	
Whipping cream	Quart	$11.20	1 pt.	
Butter	Pound	$4.85	7 oz.	
Vanilla	Quart	$37.15	3 T	
			Total cost	
Recipe #	105			
Recipe yield (oz.)	36		Portion cost	
Portion size (oz.)	1		Costed on	1-15-20xx

What will be Mark's portion cost for buttercream icing used to frost cupcakes that require 1.25 ounces of buttercream icing per cupcake?

3. Saint John's Hospital foodservice director, Herman Zindu, has a problem. He has the following information about his operation for the month of April but has forgotten how to compute cost of food sold for the month. Use Herman's figures to compute actual cost of food sold for his operation.

   Could Herman have computed this figure if he had not taken a physical inventory on April 30? Why or why not?

Inventory March 31			$22,184.50
April Purchases:			
Meats		$11,501.00	
Dairy		$6,300.00	
Fruits and vegetables		$9,641.00	
All other foods		$32,384.00	
Total Purchases			
Number of employees eating monthly	2,550		
Cost per employee meal	$1.25		
Inventory on April 30			$23,942.06

Beginning inventory	
Purchases	
Goods available for sale	
Ending inventory	
Cost of food consumed	
Employee meals (30 days)	
Cost of food sold	

4. Spike Dykes operates the student foodservice in a dormitory at Clairmont College. Spike is interested in calculating his food cost per student meal served. Data about his costs and meals served for the spring semester can be found in the following table. Help Spike complete the table.

   In which months did Spike achieve his lowest and highest cost per meal? Spike wants his average cost per meal for the semester to be below $6.00. Did he achieve this goal?

	Jan	Feb	March	April	May	Total
Beginning inventory	$22,500	$21,750	$26,500	$25,500	$16,000	
Purchases	$65,000	$64,750	$63,000	$64,500	$64,300	
Goods available for sale						
Ending inventory	$21,750	$26,500	$25,500	$16,000	$12,000	
Cost of food consumed						
Employee meals	$5,750	$5,500	$5,250	$5,000	$4,850	
Cost of food sold						
Meals served	10,750	10,100	11,500	11,250	9,000	
Cost per meal						

5. Jana Foster is the General Manager for a new restaurant in the Champos Restaurants chain. This new facility is located in a beachfront resort town, and sales there are excellent. The problem, according to Jana's Regional Manager, is that the new operation is consistently operating at a beverage cost percentage higher than the company average. Jana's Regional Manager has flown to Jana's town to see why her beverage cost percentage is too high. The prices set by the company for all restaurants are $6.00 for beer, $7.00 for wine, $10.00 for spirits (nonfrozen well drinks), and $14.00 for frozen specialty drinks.

   Help Jana compare her beverage cost percentages with the company averages in the chart that follows.

Company Averages				
Product	Product Mix	Cost of Beverages	Beverage Sales	Beverage Cost %
Beer	30%	$24,336	$121,680	
Wine	20%	$20,280	$81,120	
Spirits (nonfrozen specialty drinks)	30%	$30,420	$121,680	
Spirits (frozen specialty drinks)	20%	$24,336	$81,120	
Total				
Jana's Beachfront Restaurant				
Product	Product Mix	Cost of Beverages	Beverage Sales	Beverage Cost %
Beer	15%	$10,700	$53,500	
Wine	5%	$6,125	$24,500	
Spirits (nonfrozen specialty drinks)	25%	$30,500	$122,000	
Spirits (frozen specialty drinks)	55%	$75,500	$250,000	
Total				

Look at the sales mixes and the beverage cost percentages of both the company and the beachfront restaurant. Explain why Jana's total beverage cost percentages are consistently different from the company averages. What would you advise Jana to tell her regional manager?

6. "Fast Eddie" Green operates a restaurant located within a casino in Biloxi, Mississippi. He is checking over the work of his assistant manager, who has been newly hired. One of the jobs of the assistant manager is to complete the daily six-column food cost estimate. "Fast Eddie" finds that, although the purchase and sales data are there for the first 10 days of the accounting period, the form has not been completed. Complete the form for "Fast Eddie" so that he can go home.

	PURCHASES		SALES		COST %	
Weekday	Today	To Date	Today	To Date	Today	To Date
1/1	$1,645.80		$3,842.50			
1/2	$2,006.40		$2,970.05			
1/3	$1,107.20		$2,855.20			
1/4	$986.24		$3,001.45			
1/5	$1,245.60		$3,645.20			
1/6	$2,006.40		$4,850.22			
1/7	$0.00		$6,701.55			
1/8	$1,799.90		$3,609.20			
1/9	$851.95		$2,966.60			
1/10	$924.50		$3,105.25			
Total						

7. Elaine is the Director of Foodservice at a large retirement center, and she has asked Gerry, one of her managers, to investigate the costs involved in adding a carving station to the regular Sunday brunch menu. Gerry is trying to decide which carved meats could be served. He must first determine the EP costs and yields of the various kinds of meats. Help him calculate the EP cost and yield of an inside round of beef.

Butcher's Yield Test Results

Unit Name: Elaine's   Date Tested: May 20

Item: Inside Round

Specification: # 138

AP Amount Tested: 20 lb.

Price per Pound AP: $7.50

Loss Detail	Pounds	Ounces	Total Ounces	% of Original
AP weight	20	0		
Fat loss	3	6		
Bone loss	2	4		
Cooking loss	1	12		
Carving loss	0	8		
Total Production Loss				
EP Weight				

Net product yield %:

Yield test performed by: GW

EP cost per pound:

8. Lebron operates Ham from Heaven, a sandwich shop specializing in slow roasted spicy ham sandwiches. Currently, Lebron buys bone-in hams and then roasts them in-house. These bone-in hams cost $5.99 per pound and produce a 58 percent usable yield. Semi-boneless hams of the same quality cost $6.99 per pound with a 78 percent yield. Boneless hams of the same quality would cost $9.99 per pound and produce a 96 percent yield. Calculate the EP cost per pound of each ham alternative.

   Choose the alternative you would recommend to Lebron and explain your reason for choosing it. Specifically, what factors would influence your decision?

Ham from Heaven			
	AP Price per Pound	**Yield**	**EP Cost per Pound**
Bone-in ham			
Semi-boneless ham			
Boneless ham			

9. Wayne Gasslen is the chef at Gratzi, an upscale Northern Italian restaurant. Wayne is costing the recipe for his famous Braised Beef Ragu. He utilized an online measurement conversion site to obtain the information he needs to convert some of his purchase unit amounts to their recipe-ready equivalents. His purchase unit costs were taken from his most recent delivery invoices. Wayne rounds his recipe ingredient costs to the nearest tenth of a cent. Help Wayne complete his recipe costing spreadsheet for Braised Beef Ragu.

One	**Equals**	
Kilogram	2.2	Pounds
Liter	1000	Milliliters
Pound	16	Ounces
Pound	32	Tablespoons
#10 Can	13	Cups
Gallon	4	Quarts
Quart	32	Ounces

Recipe:	**Braised Beef Ragu**		**Portion Size 8 oz.**		**Portions 25**	
Ingredient	**Purchase Unit**	**Purchase Unit Cost**	**Recipe-Ready Unit**	**Recipe-Ready Unit Cost**	**Recipe-Ready Unit Quantity**	**Total Cost**
Beef chuck (boneless)	Kilogram	$17.95	lb.		7	
Canola oil	Liter	$9.99	ml		125	
Red wine	750 ml	$7.99	ml		275	
Onion (pre-diced)	Pound	$0.69	oz.		14	
Flour	Pound	$0.98	oz.		5	
Tomato puree	# 10 can	$21.05	cup		1.5	
Garlic (minced)	Pound	$16.00	tbl.		3	
Beef stock	Gallon	$14.98	qt.		3.5	
Dijon mustard	Quart	$23.15	oz.		6	
Gemelli pasta (dry)	Pound	$1.90	lb.		6.25	
				Total Recipe Cost		
				Portion Cost		

What is Wayne's current portion cost for Braised Beef Ragu? What would be Wayne's portion cost for Braised Beef Ragu if he changed the portion size from 8 ounces to 10 ounces?

10. Maxine Wilson is the Director of Dietary Services at State College. Each semester, Maxine carefully monitors the average cost of serving her students dinner in the residence halls. Maxine's attainable costs and actual food costs for the most recently completed semester are shown in the below table. Calculate the efficiency ratio, dollar variance, and percentage variance in each of Maxine's product categories and her total dinner cost per student.

    What was the overall percentage variance for Maxine's dinner cost per student? Is this cause for concern?

    Which product category showed the highest percentage variance? What do you think may have caused this variance?

Product Category	Actual Cost	Attainable Cost	Efficiency Ratio	Dollar Variance	Percentage Variance
Meat	$1.09	$1.06			
Seafood	$0.55	$0.41			
Fruits	$0.98	$1.02			
Vegetables	$0.84	$0.79			
Dairy products	$0.79	$0.77			
Desserts	$0.50	$0.55			
Cost per student (dinner)	$4.75	$4.60			

11. Nancy operates a takeout fresh-made cookie store in the local shopping mall. Business is good, and guests seem to enjoy the products. However, her employees (mostly teenagers), are a problem because they seem to like the products also (and they secretly eat a lot of them!)

    Nancy takes a physical inventory on a weekly basis. This week, her total cost of goods sold figure for cookies was $5,220.25. Nancy has determined that this week she will also compute her attainable food cost and her operational efficiency ratio. Help Nancy by completing the following information using the attainable product cost form she has created.

Attainable Product Cost

Unit Name: Nancy's
Date Prepared: 1/22 **Time Period:** 1/15 to 1/21
Prepared By: S.L.

Item	Number Sold in Dozens	Cost per Dozen	Total Cost	Menu Price per Dozen	Total Sales
Chocolate chip	85	$9.25		$36.00	
Macadamia	60	$11.20		$36.00	
Coconut chip	70	$8.88		$36.00	
Fudge	45	$10.22		$36.00	
M & M	68	$9.17		$36.00	
Oatmeal	63	$8.35		$32.00	
Sugar	81	$7.96		$32.00	
Total					
Attainable product cost					
Actual product cost	$5,220.25				
Operational efficiency ratio					

Does Nancy have a significant cost control problem related to her cookie production? Explain your answer.

12. Judy Kavanaugh is the banquet coordinator at the White House Hotel and Event Center. Judy had an initial meeting with a client couple who have requested carved-to-order roast beef tenderloin for their rehearsal dinner to be held at the hotel. Because this is not a standard banquet item, Judy must calculate the price per-person she will charge for that entrée.

The hotel's Food and Beverage Director informs Judy that the beef tenderloins that would be purchased to produce this item are sold in 10-kilogram boxes and that each box will cost the hotel $390.60. The director states that the waste resulting from trimming and cooking the tenderloin will be 35%. Based on historical records, the director also estimates that the average guest will be served 5 ounces of roasted beef tenderloin when they go through the buffet line.

Judy knows that she wants to price the hotel's meat entrees to produce a 25% food cost. Using the pricing form she has developed below, what is the *minimum* price per-person for roasted beef tenderloin that Judy should quote to these clients when she meets with them?

	Cost per 10-Kilogram Box	AP Cost per Pound	EP Cost per Pound	EP Cost per Portion	Selling Price @ 25% COGS
Beef tenderloin	$390.60				

CHAPTER 6

Managing Food and Beverage Pricing

OVERVIEW

This chapter begins by teaching you about the different types of menus you will encounter as a hospitality manager. The types of menus offered directly affect the prices managers can charge for their menu items. Knowledge of various menu formats can also help you reduce costs through effective utilization of your food and beverage products as well as better utilization of your staff. The chapter examines and analyzes, in detail, the factors that influence the prices you will charge for the menu items you will sell.

Finally, the chapter explains the procedures used to assign individual menu item prices based on cost and collected historical sales data. By understanding the pricing process, you can help to ensure that your menu items will generate the sales revenue you need to meet your profit goals.

Chapter Outline

LEARNING OUTCOMES

At the conclusion of this chapter, you will be able to:

- Choose the best menu format for a specific foodservice operation.
- Identify the variables to be considered when establishing menu prices.
- Assign menu prices to menu items based on the items' costs, popularity, and profitability.

MENU FORMATS

If you have determined that your purchasing, receiving, storing, issuing, and production controls are well in line, you have an excellent chance of reaching your profit goals. It is possible to find, however, that even when these areas are well managed, food and beverage costs are still too high to achieve your financial goals. When this is the case, the problem may well lie in the fundamental areas of menu format, product pricing, or both.

Menus are one of the most effective ways managers can communicate with their guests. In a foodservice operation, the term "menu" actually means two things. First, it refers to all of the items available for sale to guests. Secondly, it refers to the specific way items available for sale are presented or communicated to guests. Alternatives for menu presentation include websites, apps, and QR codes that direct users to a website or app, printed paper menus, on-premise menu display boards or blackboards, and oral recitation by servers.

Some facilities create separate menus for their food and alcoholic beverage products. Other operators find it more appropriate to have menus that combine both their food and beverage offerings. Regardless of the choices a business makes, the menu provides an excellent opportunity to build impulse sales or to communicate the special products and services a foodservice operation has to offer.

With the advent of inexpensive printing capabilities and advanced technology, many operators find that they can create their own menus for a very low cost. As a result, they can change their menus more frequently than they would have been able to in the past. In addition, they can create menu **tip-ons**, which are smaller menu segments clipped to more permanent menus. Tip-ons can prove very effective in influencing your customers' impulse buying.

Menus and their design can vary greatly so it makes good sense to examine alternative menu formats before addressing the topic of how to price menu items. Menu formats in foodservice establishments generally fall into one of the following three major categories:

1. Standard menu
2. Daily menu
3. Cycle menu

Each of these formats can help your efforts to control food costs if they are used in the proper setting.

STANDARD MENU

The **standard menu** is the same every day. It can be displayed on a screen, printed and handed to guests, displayed in the operation, or recited by service staff. It is the most common menu format in commercial restaurants. Examples of commercial restaurants that utilize the standard menu include Cracker Barrel, Red Lobster, IHOP, Applebee's, and Texas Roadhouse. You may periodically add or delete a menu item, but the standard menu generally remains unchanged on a day-to-day basis. There are many operational advantages to such a menu format. First, the standard menu simplifies your ordering process. The menu remains constant each day, so it is easy to know which products must be purchased to produce the menu items offered for sale.

Second, guests tend to have a good number of choices when selecting from a standard menu. This is true because virtually every item that management wants to offer can be produced by the kitchen and is available for selection by each guest.

A third significant advantage of the standard menu is that guest preference data are easily obtained because the total number of menu items that will be served stays constant and, thus, is generally smaller than in some alternative menu formats.

As you learned in Chapter 3, menu item sales histories can be used to accurately compute menu item popularity and that makes it easier to create production schedules and buy needed ingredients. In addition, standard menus often become marketing tools for your operation because guests soon become familiar with and return for their favorite menu items.

The standard menu tends to be most popular in those segments of the foodservice business where the guest selects the location of the dining experience, as contrasted with situations where a guest's choice is restricted. Examples of restricted-choice situations include cafeterias in college residence halls where students are required to dine in on-campus, a hospital where patients during their stay must choose their menu selections from that hospital only, and an elementary school's cafeteria.

Despite its many advantages, the standard menu does have drawbacks from a cost control standpoint. First, standard menus often do not allow for easy utilization of carryovers. In fact, in many cases, items that are produced for a standard menu and remain unsold must be discarded because the next day their quality will not be acceptable to guests. An example is a quick-service restaurant (QSR) that produces too many hamburgers for a busy dinner period and does not sell all of them. Indeed, for some QSRs, a burger that is made but not sold within 5 minutes (or less) is discarded. Contrast that cost control strategy with one that says that cooked burgers not sold within 5 minutes will be chopped and added to the house specialty chili, and it is easy to see how menu design and the specific items placed on the menu can directly affect food costs and food cost control.

A second disadvantage of the standard menu is its lack of ability to respond quickly to market changes and product cost changes. For example, a restaurant that does not list green beans on the menu cannot take advantage of the seasonal harvest of green beans, a time when they are at their peak in freshness and can be purchased at a very low price per pound.

Conversely, if management has decided that its two house vegetables sides will be broccoli and corn, even considerable price increases in these two items will have to be absorbed by the operation because these two menu items are listed on the permanent (standard) menu. This could cause serious problems if the ingredients needed to produce listed menu items increase significantly in price.

An extreme example of this kind of problem was found in a quick-service seafood restaurant chain that found it was paying almost three times what it had the previous year for a seafood item that constituted approximately 80 percent of its menu sales. A foreign government had restricted fishing for this product off its shores, and the price the operation paid for the fish it offered on its standard menu skyrocketed. This chain was nearly devastated by this turn of events. In response, management quickly moved to add chicken, lower cost seafood products, and even seafood pastas to the menu to counter the effect of this very large price increase.

Whenever possible, you should monitor food prices with an eye to making seasonal adjustments that emphasize foods when their costs are low and deemphasize those foods when their costs are high. The standard menu, however, makes this quite difficult. Some restaurant groups respond to this problem by changing their standard menu on a planned schedule. They develop a standard menu for the summer, for example, and another for the winter. In this way, they can take advantage of seasonal cost savings and add some variety to their menus (by introducing new items and removing less popular ones) but still maintain the core menu items for which they are known. By doing so, they continue to enjoy the cost-related benefits of offering a standard menu while minimizing the disadvantages of using such a menu format.

DAILY MENU

In some restaurants, you might elect to operate without a standard menu and, instead, implement a **daily menu**, that is, a menu that changes every day. This

FIGURE 5.15 Raider Resort

Monthly Beverage Percentage Report			
Location	Cost of Beverages	Beverage Sales	Beverage Cost %
Banquets	$20,500	$80,000	25.6%
Starlight Bar	$10,350	$45,500	22.7%
Harry O's	$16,350	$67,000	24.4%
Total	$47,200	$192,500	24.5%

FIGURE 5.16 Raider Beverage Outlets

Monthly Beverage Percentage Recap			
#1 Outlet Name: Banquets			Month: January
Product	Cost of Beverages	Beverage Sales	Beverage Cost %
Beer	$2,500	$10,000	25.0%
Wine	$12,000	$40,000	30.0%
Spirits	$6,000	$30,000	20.0%
Total	$20,500	$80,000	25.6%
#2 Outlet Name: Starlight Bar			
Product	Cost of Beverages	Beverage Sales	Beverage Cost %
Beer	$3,750	$15,000	25.0%
Wine	$1,500	$5,000	30.0%
Spirits	$5,100	$25,500	20.0%
Total	$10,350	$45,500	22.7%
#3 Outlet Name: Harry O's			
Product	Cost of Beverages	Beverage Sales	Beverage Cost %
Beer	$11,250	$45,000	25.0%
Wine	$2,100	$7,000	30.0%
Spirits	$3,000	$15,000	20.0%
Total	$16,350	$67,000	24.4%

In the example of the Raider Resort, it is easy to analyze the sales mix by examining Figure 5.17, a detailing of the beverage products selected by guests in each beverage outlet.

Each sales percentage in Figure 5.17 was computed using this formula:

$$\frac{\text{Item dollar sales}}{\text{Total beverage salees}} = \text{Item \% of total beverage sales}$$

Therefore, in the case of beer sales in the banquets area, and using the data from Figure 5.16:

$$\frac{\text{Banquet beer sales}}{\text{Total banquet beverage sales}} = \text{\% Banquet beer sales}$$

or

$$\frac{\$10,000}{\$80,000} = 12.5\%$$

format is especially popular in upscale restaurants where the chef's daily creations are viewed with great anticipation, and even some awe, by eager guests. The daily menu offers some advantages over the standard menu. For example, management can respond very quickly to changes in ingredient or item prices. In fact, that is one of the daily menu's great advantages. In addition, carryovers are less of a problem because any product unsold from the previous day has at least the potential of being incorporated, often as a new dish, into the next day's menu.

Every silver lining has its cloud, however. For all of its flexibility, the daily menu is recommended only for very special situations due to the tremendous control drawbacks associated with its implementation. First, item popularity data can be difficult to obtain because the items served on any given day may never before have been served in that particular combination. Thus, the preparation of specific items in certain quantities can be pure guesswork, and this is most often an inefficient way of determining efficient production schedules.

Second, it may be difficult to plan to have the necessary ingredients on hand to prepare the daily menu if the menu is not known well ahead of time. How does one decide on Monday whether one should order tuna steaks or sirloin steaks for the menu on Thursday? Obviously, this situation requires that even the daily menu be planned far enough in advance to allow an operation's purchasing agent to select and order the items necessary to produce the daily menu.

Third, the daily menu may sometimes serve as a marketing tool, but it can just as often serve as a disappointment to guests who had a wonderful menu item the last time they dined at a particular establishment and have now returned only to find that their favorite item is not being served that day. On a positive note, it is very unlikely that any guest will get bored with a routine at a daily menu restaurant because the routine is, in fact, no routine at all.

Both the standard menu and the daily menu have control advantages and disadvantages. A third menu format, the cycle menu, is an effort by management of selected types of operations to enjoy the best aspects of both of these approaches and to minimize their respective disadvantages.

Cost Control Around the World

In the United States, the **à la carte** menu is by far the style that is most commonly used by restaurants. When managers use an à la carte menu guests select individual menu items, and each menu item is priced separately. As a result, for example, guests ordering from an à la carte menu may, or may not, choose an appetizer to begin their meal. Similarly, guests may, or may not, choose a dessert to complete their meals. In all cases, with the à la carte menu, guests choose what they want and are charged only for the items they have selected.

In many other parts of the world, including most of Europe, the **prix fixe** (pronounced "prefix") **menu**, is the more common style. When an operation offers a prix fixe menu, guests choose from a predetermined list of items presented as a multicourse meal. The items included in the meal are then sold at one set price. Thus, for example, a guest selecting roast chicken for an entrée might also receive crab bisque as an appetizer, crème brûlée as a dessert, and a choice of coffee or tea as a beverage. In this example, the guest would pay one set (fixed) price for all of the items included in their prix fixe menu selection.

Each menu style has its advantages and disadvantages, and each is popular in different parts of the world. But managers must be very aware of the differences between these two menu styles because the differences directly affect the way managers using each format make appropriate pricing decisions for their operations.

CYCLE MENU

A **cycle menu** is a menu that is in effect for a specific time period. The length of the cycle refers to the length of time the menu is in effect. Thus, we refer to a 7-day cycle menu, a 21-day cycle menu, a 30-day cycle menu, or one of any other length of time.

FIGURE 6.1 Sample Cycle Menu Rotation

Days	Cycle
1–7	A
8–14	B
15–21	C
22–28	D
29–35	A
36–42	B
43–49	C
50–56	D
57–63	A

Cycle menus are utilized most often by institutions such as colleges and universities, hospitals, assisted-living facilities, correctional facilities, and other settings where the same guest, or the same type of guest, is served every day.

Typically, a cycle menu is repeated on a regular basis. For example, a particular cycle menu could consist of four 7-day periods. If each of the four periods were labeled as A, B, C, and D, the cycle periods could rotate as illustrated in Figure 6.1.

Within each cycle, the individual menu items offered to guests will vary on a daily basis. For example, cycle menu A might consist of the following seven dinner items:

Day 1	Monday	Cheese Enchiladas
Day 2	Tuesday	Turkey and Bread Dressing
Day 3	Wednesday	Corned Beef and Cabbage
Day 4	Thursday	Fried Chicken Strips
Day 5	Friday	Stir-Fried Lobster
Day 6	Saturday	Lasagna
Day 7	Sunday	Beef Pot Roast

These same menu items will be served again when cycle menu A repeats itself on days 29 through 35.

Day 29	Monday	Cheese Enchiladas
Day 30	Tuesday	Turkey and Bread Dressing
Day 31	Wednesday	Corned Beef and Cabbage
Day 32	Thursday	Fried Chicken Strips
Day 33	Friday	Stir-Fried Lobster
Day 34	Saturday	Lasagna
Day 35	Sunday	Beef Pot Roast

In the typical case, cycle menus B, C, and D would consist of many (or all) different menu items. In this manner, no menu item will be repeated more frequently than desired by management.

Cycle menus make the most sense when guests dine with you on a very regular basis, either through the choice of the individual, such as a college student or summer camper eating in a dining hall, or through the choice of an institution, such as a hospital or correctional facility. In cases like this, menu variety is very important. The cycle menu provides a systematic method for incorporating that variety into the menu. At a glance, the foodservice manager can determine how often, for example, fried chicken strips will be served per week, month, or year, and how frequently bread dressing rather than saffron rice is served with baked chicken. In this respect, the cycle menu can offer more choices to the guest than will the standard menu. With cycle menus, production personnel can be trained to produce a wider variety of foods than with the standard menu, thus expanding their skills, but requiring fewer skills than might be needed with a daily menu concept.

Cycle menus also have the advantage of allowing managers to systematically incorporate today's carryovers (see Chapter 4) into tomorrow's finished products. This is an extremely important cost-related advantage. Also, because of its cyclical nature, management should have a good idea of guest preferences and, thus, be able to schedule and plan production to a greater degree than with the daily menu. Purchasing, too, is simplified because the menu is known ahead of time, and menu ingredients that will appear on all the different cycles can be ordered with ample lead time.

Inventory levels are easier to maintain as well because, as is the case with the standard menu, product usage is well known. Even price reductions in item costs (e.g., incorporating the use of lower cost and locally grown produce during peak harvest periods) can be incorporated into cycle menu planning, thus lowering costs and maximizing product quality.

To illustrate the differences and the impact of operating under the three different menu systems, consider the case of Larry, Moe, and Curly Jo, three foodservice operators who wish to serve roast turkey and dressing for their dinner entrée on a Saturday night in April. Larry operates a restaurant with a standard menu. If he is to print a standard menu that allows him to serve turkey in April, and does not do a menu reprint, he will be required to offer the same item in January and June. If he is to utilize any carryover parts of the turkey, he must offer a second turkey item on his menu, which also must be made available every day. Larry is not sure all the trouble and cost is worth it!

In addition, consider the expense involved in a national chain restaurant with thousands of outlets. If this turkey dinner menu item needed to be changed in any way, reprints of menus reflecting the change would likely be very costly. If Larry is the **CEO** (**chief executive officer**), the highest ranking member of a national or international foodservice organization, any decision he makes to change the standard menu to incorporate new items would indeed be a major undertaking.

Moe operates a restaurant with a daily menu. For him, serving roast turkey and dressing on a Saturday in April is quite easy. His problem, however, is that he has no idea how much to produce because he has never before served this item at this time of the year and on this specific evening in his restaurant. Also, few of the guests he has served in the past year are likely to know in advance about his decision to serve the menu item. What if no one orders it?

Curly Jo operates on a cycle menu. She can indeed put roast turkey and dressing on the cycle. If it sells well, she will keep it on the cycle menu. If it does not, it will be removed from the next cycle. Curly Jo makes a note to herself that she should record how well it sells and leaves a space in the cycle for the utilization of any carryover product that might exist within the next few days. The advantages of the cycle menu, in this specific situation, are quite apparent.

MENU SPECIALS

Regardless of the menu format used, you can generally incorporate relatively minor menu changes on a regular basis. This can be accomplished through the offering of daily or weekly **menu specials**, that is, menu items that will appear on the menu as you desire and then are removed when the items are sold, or their sale is discontinued. These daily or weekly specials are most often an effort on the part of management to provide menu variety, take advantage of low-cost raw ingredients, utilize carryover products, or test-market the potential of new menu items.

The menu special is a powerful cost control tool. Properly utilized, it helps shape the future menu by testing guest acceptance of new menu items while providing opportunities for you to respond to the challenges of using carryovers or other food and beverage products you have in inventory.

FACTORS AFFECTING MENU PRICING

Regardless of the menu format utilized, managers must carefully determine their menu items' selling prices. A great deal of important information has been written in the area of menu pricing and menu pricing strategy. A great deal of nonsense has also been written. For the serious foodservice operator, menu pricing is a topic that deserves its own significant consideration and study. Menu pricing is directly related to cost control by virtue of the basic profit formula presented in Chapter 1:

$$\text{Revenue} - \text{Expenses} = \text{Profit}$$

FUN ON THE WEB!

Revenue management is the term hospitality professionals use to describe the process of monitoring demand and determining selling prices. Revenue management is an important process in hotels and in restaurants. Effective revenue management requires an understanding of consumer buying behavior, product costs, and customer-centric revenue optimization. John Wiley publishes a very popular book specifically on this topic. Two of the book's chapters focus exclusively on revenue management for foodservice operations. To review the entire book's contents, go to the Amazon website.

When you arrive, enter "Revenue Management for the Hospitality Industry; Hayes, Hayes, et al." in the search bar.

When foodservice operators find that profits are too low, they frequently question whether their prices (and thus their revenues) are too low. It is important to remember, however, that revenue and price are not synonymous terms.

Revenue means the amount spent by all guests. Thus, it is the amount of money an operation collects from all guests. **Selling price** (price) refers to the amount charged to one guest for their purchase. Thus, total revenue is generated by the following formula:

$$\text{Price} \times \text{Number sold} = \text{Total revenue}$$

From this formula, it can be seen that there are actually two components of total revenue. Price is one component, and the other is the number of items sold. It is most often true that as an item's selling price increases the number of that item

FIGURE 6.2 Alternative Results of Selling Price Increases

Old Price	New Price	Number Sold	Total Revenue	Revenue Result
$1.99		200	$398	
	$2.99	200	$598	Increase
	$2.99	175	$523	Increase
	$2.99	133	$398	No change
	$2.99	125	$374	Decrease

sold will decrease. For this reason, price increases must be evaluated based on their impact on the number of items that will be sold, and not on selling price alone.

To illustrate how price and number sold affect total revenue, assume that you own a QSR chain. You are considering raising the price of small fountain drinks from $1.99 to $2.99. Figure 6.2 illustrates the various possible effects of this price increase on total revenue in a single one of your units. Note especially that, with at least one alternative result, increasing your selling price will actually have the effect of *reducing* your total revenue.

Experienced foodservice managers know that increasing prices without giving added value to guests can result in higher prices but, frequently, lower revenue because of a reduction in the number of guest purchases. This is true because guests demand a good **price/value relationship** when making a purchase. The price/value relationship simply reflects guests' view of how much value they are receiving for the prices they are paying.

Perhaps no area of hospitality management is less understood than the area of pricing. This is not surprising when you consider the many factors that play a part in the pricing decision. For some foodservice operators, inefficiency in cost control is passed on to guests in the form of higher prices. In fact, sound pricing decisions should always be based mainly on establishing a positive price/value relationship in the mind of the guest.

It is important to recognize that most foodservice operators face fairly similar product costs when they purchase the ingredients needed to produce the items they sell. Whether the product on the menu is pancakes or prime rib, wholesale prices for needed raw ingredients may vary only slightly from one supplier to the next. In some cases, this variation is due to volume buying, whereas in others, it is the result of the relationship established with the vendor. Regardless of the source, the fact remains that variations in prices operators pay for ingredients are often small relative to potential variations in menu prices. This becomes easier to understand when you realize that a menu item's selling price is most often a function of much more than its raw product cost. In fact, menu prices are significantly affected by all of the following factors:

Factors Influencing Menu Price

1. Economic conditions
2. Local competition
3. Service levels
4. Guest type
5. Product quality
6. Portion size
7. Ambience
8. Meal period
9. Delivery method
10. Location
11. Sales mix

ECONOMIC CONDITIONS

The economic conditions that exist in a local area or even in an entire country can have a significant impact on the prices restaurant managers can charge for their menu items.

A variety of economic conditions can affect sales in a foodservice operation. An extreme example of this, of course, was the COVID-19 pandemic of 2020. The pandemic triggered a wave of closures that devastated the restaurant industry. Forced shutdowns, capacity constraints, and safety concerns left many restaurants unable to survive financially. By the end of 2020, more than 110,000 restaurants had closed permanently or long-term in the United States alone, according to the National Restaurant Association.[1] These closures meant that, regardless of prices to be charged, normal levels of sales were not possible.

When the local economy is robust and growing, managers generally have a greater ability to charge higher prices for the items they sell. When a local economy is in recession or is weakened by other events, a manager's ability to raise or even maintain prices in response to rising product costs may be more limited. In most cases, managers will not have the ability to directly influence the strength of their local economies. It is the manager's job, however, to monitor local economic conditions and to carefully consider these conditions when establishing menu prices.

LOCAL COMPETITION

While the prices charged by competitors may be important, this factor is often too closely monitored by the typical foodservice operator. It may seem to some managers that the average guest is vitally concerned with low price and nothing more. In reality, small variations in price generally make little difference in the buying behavior of the average guest.

For example, if a group of young professionals goes out for pizza and beer after work, the major determinant will not likely be whether the selling price for the beer is $7.00 in one establishment or $7.75 in another. This small variation in price is simply not likely to be the major factor in determining which operation the group of young professionals will choose to visit. Other factors will be more important.

Your competition's selling prices are somewhat important when establishing your own prices, but it is a well-known fact in foodservice that someone can always sell a lesser quality product for a lesser price. The prices competitors charge for their products can be useful information in helping you arrive at your own selling prices. They should not, however, be the only determining factor in your own pricing decisions.

Successful foodservice operators spend their time focusing on building guest value in their own operations and not in attempting to mimic the efforts of the competition. Despite the fact that many managers feel customers only want low prices, it is important to remember that, in the consumer's mind, higher prices are most often associated with higher-quality products and thus with a *better* price/value relationship.

SERVICE LEVELS

The service levels an operation provides its guests directly affect the prices the operation can charge. Most guests expect to pay more for the same product when service levels are higher. Thus, for example, the can of soda sold from a vending machine is generally less expensive than a similar sized soda served by a human being.

Different groups categorize restaurants differently. While there is no universal agreement on how restaurants should be categorized, it is interesting to note that

[1] https://thebestchefawards.com/area-talks/the-impact-of-the-pandemic-on-the-restaurant-industry/. Retrieved 1/31/2026.

the most popular way to do so is to view them based on the service levels they provide. The following are the most common service-based restaurant classifications:

Quick-service (restaurant) (QSR): Sometimes referred to by those outside the foodservice industry as "fast food," these operations most often offer lower-priced menu items, counter service, and drive-thru windows.

Fast Casual: Similar to those QSRs offering counter service, but typically with "build-your-own menu" item offerings such as sandwiches, pizza, wraps, or burritos that are prepared fresh to order. Some fast casuals may perform limited table service by delivering orders, but there is no serving staff.

Casual: Sometimes called "family dining," these operations provide table service to guests and employ service staff to take orders and deliver food.

Fine dining: Similar to casual family dining, but offering the finest foods, the highest level of service, and the most elegant décor, but at higher prices.

Because service levels impact pricing directly, as the personal level of service increases in an operation, selling prices may also be increased. This personal service may range from the speedy delivery of products served in a QSR's drive-thru to a full-service restaurant manager's decision to improve customer service by increasing the number of servers in a busy dining room and, thus, improving service quality by reducing the number of guests each worker must serve.

These examples should not imply that extra income from increased menu prices must go only to pay for extra labor required to increase service levels. Guests are willing to pay more for increased service levels, but this higher price should cover extra labor cost and provide for extra profit as well. In the hospitality industry, those companies that have been able to survive and thrive over the years have done so because of their uncompromising commitment to high levels of guest service, and they can charge menu prices that reflect those enhanced service levels.

GUEST TYPE

All guests want good value for their money. But some guest types are simply less price sensitive than others. The question of what represents good value varies by the type of clientele an operation serves. An example of this can clearly be seen in the pricing decisions of convenience stores across the United States. In these facilities, food products such as premade sandwiches, fruit, drinks, cookies, and the like are sold at relatively high prices. The guests these stores cater to, however, value speed and convenience above all else. For this speed and convenience, the customers are willing to pay premium prices.

In a similar manner, guests at an expensive steakhouse restaurant are less likely to respond negatively to small variations in drink prices than guests at a corner neighborhood tavern. A thorough understanding of who their guests are, and what these guests value most, is critical to the ongoing success of foodservice managers as they established their operations' menu prices.

PRODUCT QUALITY

In nearly every instance, a guest's quality perception of any specific menu item offered for sale in the foodservice business can range from very low to very high. These perceptions are the direct result of how the guest views a restaurant's menu offerings. These perceptions are directly affected by food quality, but they should never be shaped by the guest view of a menu item's wholesomeness or safety. All foods must be wholesome and safe to eat. In most cases, guests' perceptions of quality will be based on a variety of factors, only one of which is actual product quality. Visual presentation and stated or implied grade of ingredients, as well as portion size and service level are additional factors that impact a guest's view of overall product quality.

To illustrate, consider that, when most foodservice guests think of a hamburger they actually think not of one product, but of a range of products. A hamburger, in these guests' minds, may include a rather small burger patty on a regular bun, wrapped in waxed paper, and served in a paper sack. If so, expectations will be that its price will be low.

If, however, the guests' thoughts turn to an 8-ounce gourmet "Kobe steakburger" presented with avocado slices and alfalfa sprouts on a fresh-baked, toasted, whole-grain bun and served for lunch in a white-tablecloth restaurant, the purchase price expectations of the guests will be much higher and so will their expectations of service levels provided.

As an effective foodservice manager, you will often choose from a variety of quality levels and delivery methods when developing product specifications and, consequently, planning your menus and establishing your prices. For example, if you select the market's cheapest bourbon as your well brand, you will likely be able to charge less for drinks made from it than your competitor who selects a better, higher-priced brand. Your guests' perceptions of your drink quality level, however, may also be perceived to be lower than your competitors'. To be successful, managers should select the product quality levels that best represent their target markets' anticipated desires as well as their operations' own pricing and profit goals.

PORTION SIZE

Portion size plays a large role in determining menu pricing. Portion size is an often-misunderstood concept, yet it is probably the second most significant factor (next to sales mix) in overall pricing. The great chefs know that people "eat with their eyes first!" This relates to presenting food that is visually appealing. It also impacts product size and pricing. A pasta dish that fills an 8-inch plate may well be lost on an 11-inch plate. On the 8-inch plate, guests may perceive high levels of value. On the 11-inch plate, guest perceptions of value delivered for price paid may be significantly reduced. Portion size, then, is a function of both product quantity and presentation.

In a traditional table-service operation, management must control portion size because the larger the portion size, the higher the product costs. One very good way to determine whether portion sizes are too large is simply to watch the dishwashing area in an operation to see what comes back from the dining room as uneaten. Significant return amounts of uneaten food may indicate portion sizes are too large. In this regard, the dishroom operator can become an important player in the cost control team.

Many of today's health-conscious consumers prefer lighter food with more choices of fruits and vegetables. The portion sizes of these items can often be boosted at a fairly low increase in cost. At the same time, average beverage sizes are increasing, as are the size of many side items such as French fries. Again, some of these tend to be lower-cost items. This can be good news for the foodservice operator if prices can be increased to adequately cover the costs of providing the larger portion sizes.

Every menu item served should be analyzed with an eye toward determining if the quantity (portion size) being served is the "optimum" quantity. You would like to serve this amount, of course, but no more. The effect of portion size on menu price is significant, and it will be your job to establish and maintain control over your desired portion sizes.

AMBIENCE

If people ate only because they were hungry, few restaurants would be open today. People eat out for a variety of reasons, some of which have little to do with the food itself.

Fun, companionship, time limitations, adventure, and variety are just a few reasons diners cite for eating out rather than eating at home. For the foodservice operator who provides an attractive ambience, menu prices can reflect this. **Ambience** is the feeling or overall mood created in an operation. Ambience is affected by an operation's decor, staff uniforms, music, and other factors that directly affect its atmosphere. When ambience is a major customer draw, managers find they are selling much more than food and, thus, may justly charge higher menu prices. In most cases, however, foodservice operations that count too heavily on ambience alone to carry their business generally start out well but are not ultimately successful. Excellent product quality with outstanding service goes much further in the long run than do clever restaurant designs. Ambience may, however, draw guests to a service location for the first time. When this is true, prices may be somewhat higher if the quality of products provided to guests supports the higher pricing structure.

MEAL PERIOD

In some cases, diners expect to pay more for an item served in the evening meal period than for that same item served at a lunch period. Sometimes, this is the result of a smaller "luncheon" portion size, but in other cases the portion size, as well as service levels, may be the same in the evening as earlier in the day. This is true, for example, in buffet restaurants that charge a different price for lunch than they charge for dinner, perhaps expecting that guests on their lunch break will spend less time in the restaurant and thus will eat less. You must exercise caution in this area, however. Guests should clearly understand why a menu item's price changes with the time of day. If this cannot be answered to the guest's satisfaction, it may not be wise to implement a time-sensitive pricing structure.

DELIVERY METHOD

Today's restaurant guests have three basic delivery options when purchasing products from most restaurants. These are:

1. On-site dining
2. Pickup
3. Delivery

It is interesting to note that, by far, delivery is most expensive option for both diners and foodservice operators. With the delivery of meals, buyers pay delivery fees, and operators are charged a fee by the delivery company. As a result, many foodservice operators seek to encourage customer pickup of ordered items, and they do so by reducing prices for their carryout items.

Whether they are promoted as "takeout discounts," "pickup specials," or "to-go" deals, these promotions seek to reduce an operator's overall expense by minimizing delivery costs. In most cases, guests will respond positively to these efforts. In a survey, 58% of restaurant customers said an important reason they choose pickup is to save on the cost of a delivery fee, and 42% of restaurant customers said they pick up their meals because not having to tip a delivery person is important to them.[2]

[2] https://www.pymnts.com/restaurant-innovation/2023/most-consumers-drop-delivery-for-pickup-when-ordering-food/. Retrieved 10/31/2025.

The Move to Off-Premise Dining

The move to off-premise dining has caused many foodservice operators to reconsider the way they price their menu items. For example, Domino's is the largest pizza chain in the United States. One way they seek to encourage their guests to pick up their pizzas rather than have them delivered is via the use of their carryout deal program.

In this promotion, guests call in or use an ordering app to select products for pickup. The products are then sold at prices lower than the selling price for the same item as a delivered product.

This program is successful because it benefits both guests, who pay lower prices, and Domino's franchisees, who avoid charges from delivery company such as DoorDash® and Uber Eats.

LOCATION

Location can be a major factor in price determination. One needs look no further than America's many themed amusement parks, movie theaters, or sports arenas to see evidence of this. Foodservice operators in these locations are able to charge premium prices because they have, in effect, a monopoly on the food sold to visitors. The only all-night diner on the interstate highway exit is in much the same situation. Contrast that with an operator who is just one of 10 similar seafood restaurants on a restaurant row in a seaside resort town. In this case, it is unlikely that an operation will be able to charge, for similar items, prices significantly higher than its competitors.

It used to be said of restaurants that success was due to three things: location, location, and location! This may have been true before so many foodservice operations opened in the United States. There is, of course, no discounting the value of an excellent restaurant location, and location alone can indeed influence price in some cases. Location does not, however, guarantee long-term success. Location can be an asset or a liability. If it is an asset, menu prices may reflect that fact. If location is indeed a liability, menu prices may actually need to be lower to attract a sufficient clientele to ensure that the operation achieves its total revenue and profit goals.

SALES MIX

Of all the pricing-related factors addressed thus far, sales mix most heavily influences the menu pricing decision, just as guest purchase decisions will influence total product costs (see Chapter 5). Recall that sales mix refers to the specific menu items selected by guests. Managers can respond to this situation by employing a concept called price blending. **Price blending** refers to the process of pricing products with different individual cost percentages in a way that achieves a favorable overall product cost percentage.

The ability to blend prices is a useful skill and one that is well worth mastering. To illustrate, assume that you are the operations vice president for Texas Red's, a chain of upscale hamburger restaurants. Assume also that you hope to achieve an overall food cost of 40 percent in your units. For purposes of simplicity, assume that Figure 6.3 illustrates the three products you sell and their corresponding selling price if each is priced to achieve a 40 percent food cost.

In Chapter 5, you learned that the formula for computing food cost percentage is as follows:

$$\frac{\text{Cost of food sold}}{\text{Food sales}} = \text{Food cost \%}$$

FIGURE 6.3 Unblended Price Structure

Texas Red's Burgers			
Item	Item Cost	Desired Food Cost	Proposed Selling Price
Hamburger	$2.75	40%	$6.88
French fries	$0.65	40%	$1.63
Soft drinks (12 oz.)	$0.35	40%	$0.88
Total	$3.75	40%	$9.38

This formula can be worded somewhat differently for a single menu item without changing the formula's accuracy:

$$\frac{\text{Cost of a specific food item sold}}{\text{Food sales of that item}} = \text{Food cost \% of that item}$$

It is important to understand that when only a single menu item is sold, "Food sales of that item" equals its selling price. The principles of algebra allow you to rearrange the formula as follows:

$$\frac{\text{Cost of a specific food item sold}}{\text{Food cost \% of that item}} = \text{Food sales (selling price) of that item}$$

Thus, in Figure 6.3, the hamburger's selling price is:

$$\frac{\$2.75}{0.40} = \$6.88$$

Note that, in Figure 6.3, all products are priced to sell at a price that yields a 40 percent food cost. As a result, sales mix, that is, the individual menu selections of guests, would not affect this operation's overall food cost percentage. The sales mix resulting from this pricing strategy could, however, have very damaging results on your profitability. The reason is very simple. If you use the price structure indicated, your drink prices are too low. Most guests expect to pay far in excess of 88 cents for a soft drink at a QSR. You run the risk, in this example, of attracting many guests who are interested in buying only soft drinks at your restaurants. Your burger by itself, however, may be priced too high relative to your competitors who utilize price-blending. However, if you use the price-blending concept, and if you assume that each guest coming into your restaurants will buy a burger, French fries, and a soft drink, you can create a different menu price structure and still achieve your overall cost objective, as seen in Figure 6.4.

Note that, in this example, you actually achieve a total food cost slightly lower than 40 percent. Your hamburger's price is now less than $5.00 and more likely in line with your local competitors. Note also, however, that you have assumed each guest coming to Texas Red's will buy one of each item. In reality, of course, not all guests will select one of each item.

Some guests will not elect to buy fries, whereas others may stop in only for a soft drink. It is for this reason that popularity indexes discussed in Chapter 3 and sales mix records addressed in Chapter 5 are so critical. These sales-related histories let you know exactly what your guests are buying when they visit your operation. With this information, you can better develop your pricing strategy.

FIGURE 6.4 Blended Price Structure

Texas Red's Burgers			
Item	Item Cost	Proposed Food Cost %	Proposed Selling Price
Hamburger	$2.75	55.1%	$4.99
French fries	$0.65	26.1%	$2.49
Soft drinks (12 oz.)	$0.35	17.6%	$1.99
Total	$3.75	39.6%	$9.47

FIGURE 6.5 Sample Sales Mix Data

Texas Red's Burgers

Total Sales: $842.85 Guests Served: 100

Total Food Cost: $337.25 Food Cost %: 40.0%

Item	Number Sold	Item Cost	Total Food Cost	Selling Price	Total Sales	Food Cost %
Hamburger	92	$2.75	$253.00	$4.99	$459.08	55.1%
French fries	79	$0.65	$51.35	$2.49	$196.71	26.1%
Soft drink (12 oz.)	94	$0.35	$32.90	$1.99	$187.06	17.6%
Total		$3.75	$337.25		$842.85	40.0%

To illustrate how this works, assume that you monitored a sample of 100 guests who came into one of your Texas Red's units and found the results presented in Figure 6.5.

As you can see from Figure 6.5, you can use the price-blending concept to achieve your overall cost objectives if you have a good understanding of how many people buy each menu item. In this example, you have achieved the 40 percent food cost you sought. It matters little if the burger has a 55.1 percent food cost if the burger is sold in conjunction with the appropriate number of soft drinks and fries. Obviously, there may be a cost-related danger if your guests begin to order nothing but hamburgers when they come to your establishment. That, however, is unlikely. Again, the use of detailed sales histories to carefully monitor guest preferences will allow you to make price adjustments, as needed, to keep your overall costs and prices in line.

A word of caution regarding the manipulation of sales mix and price blending is in order. Because price itself is one of the factors that impact item popularity, a change in menu price may cause a change in customer buying habits. If, in an effort to reduce overall product cost percentage, you were to increase the selling price of soft drinks at Texas Red's, for example, you might find that a higher percentage of guests would elect *not* to purchase a soft drink. This could have the effect of actually increasing your overall product cost percentage because fewer guests would choose to buy the one item you sell that has an extremely low food cost percentage. Experienced managers know that understanding sales mix as well as the concept of price blending will have a major impact on their overall menu pricing philosophies and strategies.

Green and Growing!

You have now been introduced to at least 11 factors that may influence the prices you charge for the items you sell. In the future, the Level of Energy and Environmental Design (LEED) certification achieved by your operation may well constitute another such factor. The LEED rating system developed by the US Green Building Council (USGBC) evaluates facilities on a variety of standards. The rating system considers sustainability, water use efficiency, energy usage, air quality, construction and materials, and innovation. Currently, the maximum LEED score that can be achieved is 69 points, and operations scoring 26 or more points earn one of the following ratings:

Certified: 26–32 points

Silver: 33–38 points

Gold: 39–51 points

Platinum: 52–69 points

In many cases, when LEED certification is pursued, the initial construction cost of a building will be higher than the current industry standard. However, these costs are recovered by the savings incurred over time due to the lower-than-industry-standard operational costs that are typical of a LEED-certified building. In the foodservice business, these initial costs may also be mitigated by the fact that, increasingly, consumers are willing to pay more to dine in LEED-certified operations. In addition, LEED-certified buildings are healthier for workers and for diners. The LEED certification creates benefits for building owners, employees, and guests. Look for a continuing increase in its importance to your guests.

ASSIGNING MENU PRICES

The methods used to assign menu prices in foodservice operations are often as varied as the managers who utilize the methods. In general, however, menu prices in food and beverage operations have historically been determined on the basis of one of the following two approaches.

1. Product cost percentage
2. Product contribution margin

PRODUCT COST PERCENTAGE

This method of pricing is based on the idea that the cost of producing an item should be a predetermined percentage of the item's selling price. As was illustrated earlier in this chapter, if you have a menu item that costs \$3.50 to produce, and your desired cost percentage equals 40 percent, the following formula can be used to determine what the item's menu price should be:

$$\frac{\text{Cost of a specific food item sold}}{\text{Food cost \% of that item}} = \text{Selling price of that item}$$

or

$$\frac{\$3.50}{0.40} = \$8.75$$

Thus, given a \$3.50 food cost, the recommended selling price of this item would be \$8.75. If that item is sold for \$8.75, then a 40 percent food cost should be achieved for that item. A check of this work can also be done using the food cost percentage formula:

$$\frac{\$3.50}{\$8.75} = 0.40, \text{ or } 40\%$$

When management uses a predetermined or target food or beverage cost percentage to price menu items, it is stating its belief that product cost in relationship to selling price is of most importance.

Experienced foodservice managers know that a second method of calculating selling prices based on predetermined product cost percentage goals can be used. This method uses a pricing factor, or multiplier, which can be assigned to each desired food or beverage cost percentage. This factor, when multiplied by an item's portion cost, will result in a selling price that yields the desired product cost percentage.

Figure 6.6 details such a pricing factor table.

In each case, the factor is calculated by the following formula:

$$\frac{1.00}{\text{Desired product cost \%}} = \text{Pricing factor}$$

Thus, if a manager were attempting to calculate the pricing factor needed to price menu items while achieving a product cost of 40 percent, the manager's factor computation would be

$$\frac{1.00}{0.40} = 2.50$$

To return to our example, if you want to achieve a 40 percent food cost and if your item costs \$3.50 to produce, then the computation is:

$$\text{Food cost} \times \text{pricing factor} = \text{Menu price}$$

or

$$\$3.50 \times 2.5 = \$8.75$$

As can be seen, these two methods of arriving at the proposed selling price yield identical results. One mathematical formula simply relies on division, whereas the other relies on multiplication. The decision about which formula to use is completely up to you. With either approach, the selling price of an item will be determined with a goal of achieving a specified food or beverage cost percentage for each item sold.

FIGURE 6.6 **Pricing-Factor Table**

Desired Product Cost %	Pricing Factor
20	5.000
23	4.348
25	4.000
28	3.571
30	3.333
33⅓	3.000
35	2.857
38	2.632
40	2.500
43	2.326
45	2.222

HERE'S HOW IT'S DONE 6.1

To calculate the selling price of a menu item when applying the food cost percentage method, you must know two things:

1. The cost of the menu item
2. The food cost percentage you wish to achieve

When you know those two things, you can choose to calculate selling prices using either division or multiplication. You will always get the same answer. To illustrate, assume you know that an entrée costs \$4.50 to make. You wish to achieve a 25 percent food cost on the menu item.

Using division: To determine the proper selling price using division, you simply divide the cost of the menu item by the food cost percentage you wish to achieve. In this example:

$$\frac{\$4.50 \text{ menu item cost}}{25\% \text{ desired food cost}} = \$18.00 \text{ selling price}$$

Using multiplication: To determine the proper selling price using multiplication, you must first know the pricing factor that is associated with your desired food cost. In this example, the desired food cost is 25 percent, so the pricing factor to use (from Figure 6.6) is 4.0.

$$\$4.50 \text{ menu item cost} \times 4.0 \text{ pricing factor} = \$18.00 \text{ selling price}$$

If you don't have a pricing factor table readily available to you, you can easily calculate the pricing factor you should use simply by dividing your desired food cost percentage into 100%. In this example:

$$100\% \div 25\% \text{ desired food cost} = 4.0 \text{ pricing factor}$$

In many settings, such as all-you-can-eat buffets and banquet-style meals, foodservice managers cannot determine a single portion cost but rather must calculate their selling prices based upon their **plate costs.** A plate cost is simply the sum of all product costs included in a single meal (or "plate") served to a guest for one fixed price. In such settings, prices are determined in exactly the same manner as they are for an individual menu item.

To illustrate, assume that you manage the foodservice operation at an exclusive Country Club. The club specializes in wedding receptions. Assume further that a couple that will hold their reception at your facility has selected a meal from your banquet menu that resulted in your operation incurring the following per person plate costs:

Premeal reception	\$14.50
Dinner	\$21.00
Dessert	\$8.00
Beverages (nonalcoholic)	\$6.00
Total plate cost	\$49.50

Assume also that you sought to achieve a 40 percent overall product cost on each of your banquet meals. Utilizing the pricing factor method, the computation is:

$$\$49.50 \text{ plate cost} \times 2.5 \text{ pricing factor} = \$123.75$$

In this example, with a total plate cost of \$49.50 and a desired food cost of 40 percent, the selling price of your banquet meal would be \$123.75.

PRODUCT CONTRIBUTION MARGIN

Some foodservice managers prefer an approach to menu pricing that focuses on a menu item's contribution margin rather than its product cost percentage. **Contribution margin** (CM) is defined as the amount of money that remains after the product cost of a menu item is subtracted from the item's selling price. As a result, CM is the amount that a menu item "contributes" to pay for labor and all other expenses, and provide for a profit. Thus, if an item sells for \$8.75 and the product cost for this item is \$3.50, the CM is computed as follows:

$$\text{Selling price} - \text{Product cost} = \text{CM}$$

or

$$\$8.75 - \$3.50 = \$5.25$$

HERE'S HOW IT'S DONE 6.2

The CM is simply the amount of money that remains after the cost of making a menu item is subtracted from the item's selling price. To apply the CM method of pricing, managers use an easy two-step process.

Step 1: Determine the average CM required.

Step 2: Add the CM required to the item's product cost.

Step 1: Managers can determine the average CM they require based on the number of items to be sold or on the number of guests to be served. The process for each approach is identical. For example, to calculate CM based on the number of items to be sold, managers add their nonfood operating costs to the amount of profit they desire, and then divide the result by the number of items expected to be sold:

$$\frac{\text{Nonfood costs} + \text{Profit desired}}{\text{Number of items to be sold}} = \text{CM desired per item}$$

To calculate CM based on the number of guests to be served, managers divide all of their nonfood operating costs, plus the amount of profit they desire, by the number of expected guests:

$$\frac{\text{Nonfood costs} + \text{profit desired}}{\text{Number of guests to be served}} = \text{CM desired per guest served}$$

For example, if a manager's budgeted nonfood operating costs for an accounting period are $105,000, desired profit is $15,000, and the number of items estimated to be sold is 25,000, the manager's CM per item would be calculated as follows:

$$\frac{\$105{,}000 \text{ (nonfood costs)} + \$15{,}000 \text{ (profit desired)}}{25{,}000 \text{ (number of items to be sold)}} = \$4.80 \text{ CM desired per item}$$

Step 2: Managers complete this step by adding their desired CM per item (or guest) to the cost of making a menu item. For example, if, as in the aforementioned example, a manager's desired CM per item is $4.55 and a specific item's cost is $2.40, the item's selling price would be calculated as follows:

$$\$4.55 \text{ CM} + \$2.40 \text{ item cost} = \$6.95 \text{ selling price}$$

The CM method of pricing is popular because it is easy to use, and it helps ensure each menu item sold contributes to an operation's profits. When using the CM pricing approach, prices for menu items vary only due to their variations in product cost. When managers have accurate budget information about their nonfood costs, as well as realistic profit expectations, use of the CM method of pricing can be very effective.

When this approach is used, the formula for determining selling price is:

$$\text{Product cost} + \text{CM desired} = \text{Selling price}$$

Establishing menu price using this method is a matter of combining product cost with a predetermined desired CM. Thus, management's role here is to determine the desired CM for each menu item. When using this approach, you are likely to establish different CMs for various menu items or groups of items.

For example, in a restaurant where items are priced separately, entrées might be priced with a CM of $8.50 each, desserts with a CM of $3.25, and nonalcoholic drinks, perhaps, with a CM of $1.75. Those managers who rely on the CM approach to pricing do so in the belief that the average CM margin per item is a *more* important consideration in pricing decisions than is each item's product cost percentage.

PRODUCT COST PERCENTAGE OR PRODUCT CONTRIBUTION MARGIN

Although the debate over the "best" pricing method for food and beverage products is likely to continue for some time, you should remember to view pricing as an important process that has as its end goal the establishment of a desirable price/value relationship in the mind of your guests.

Regardless of whether the pricing method used is based on product cost percentage, CM, or even a completely different approach, the selling price selected must provide for a predetermined operational profit. For this reason, it is important that the menu is not priced so low that no profit is possible nor so high that you will not be able to sell a sufficient number of items to make a profit. In the final analysis, it is the customer who will eventually determine what your sales will be on any given item. Being sensitive to required profit as well as to your guests' needs, wants, and desires is critical to an effective pricing strategy.

Consider the Costs

"We have to lower our price, or we'll just get killed," said Hoyt Jones, director of operations for the seven-unit Binky's Sub shops. Binky's was known for its modestly priced, but very high-quality sandwiches and soups. Business and profits were good, but now Hoyt and Rachel, who served as Binky's Director of Marketing, were discussing the new $10.99 "Foot-Long Deal" sandwich promotion that had just been rolled out by their major competitor, an extremely large chain of sub shops that operated over 5,000 units nationally and internationally.

"They just decided to lower their prices to appeal to value-conscious customers," said Hoyt.

"But how can they do that and still make money?" asked Rachel.

"There's always a less expensive variety of turkey, ham, and cheese on the market," replied Hoyt. "They use lower-quality ingredients than we do. We charge $13.99 for our foot-long sub. That wasn't bad when they sold theirs at $11.99. Our customers know we are worth the extra two dollars. Now that they are at $10.99 . . . I don't know, but I think this is really going to hurt us in the market. We need to do something—fast."

1. How large a role do you believe cost likely played into the decision of this competitor to reduce its sandwich prices? Explain your answer.
2. Do you think the typical foodservice customer will consistently pay a higher price for better-quality food and beverage products? Give a specific example to support your answer.
3. Assume that you were the president of Binky's Subs. What steps would you instruct Hoyt and Rachel to take to address this specific pricing/cost challenge?

SPECIAL PRICING SITUATIONS

Some pricing decisions faced by foodservice managers call for a unique approach. In such cases, strategic pricing is used as a way to influence guests' purchasing decisions or to respond to particularly complex service situations. Areas that may require a manager's special pricing consideration include the following:

1. Coupons
2. Value pricing
3. Bundling
4. Dynamic pricing
5. Salad bars and buffets
6. Bottled wine
7. Craft beers
8. Alcoholic beverages at receptions and parties

COUPONS

Coupons are a popular way to vary an operation's menu prices. Essentially, there are two types of coupons in use in the hospitality industry. The first type offers guests a free menu item when they purchase another menu item. Thus, a coupon might state that if one entrée is purchased, a second entrée (usually of equal or lesser value) can be selected for no charge. This **Buy One Get One (BOGO)** approach has the effect of reducing by 50 percent the selling price of the couponed item. Similarly, a coupon that allows a guest who purchases an item at its regular price to buy a second item at 50 percent off has the effect of reducing by 25 percent the selling price of the coupon item.

When utilizing the second coupon type, one or more restrictions are placed on the coupon's use. For example, the coupon may only be accepted on a certain day, at a certain time of day, or the reduction in price granted by the coupon's use may be available only if the guest purchases a specific designated menu item. Examples include a coupon redeemable for $3 off a lunch entrée or $4 off a dinner entrée.

Recall that total revenue is calculated as price × number sold. Whether offered in a digital format or in hard copy form, redeemed coupons are popular with guests and have the effect of reducing an operation's selling prices in the hope that the total number of guests (number sold) will increase to the point that the operation's total revenue will also increase.

VALUE PRICING

Value pricing refers to the practice of reducing prices on selected menu items in the belief that, as in couponing, the total number of guests served will increase to the point that total sales revenue also increases. A potential danger with value pricing is that if guest counts do not increase significantly, total sales revenue may, in fact, decline rather than increase.[3]

Many credit the Wendy's restaurant chain with first establishing value pricing on a large scale, but currently its use is widespread, as is evident by the large number of $1.99 and $2.99 "value-priced" menu items for sale in the hospitality industry's QSR segment. Additional examples of value pricing include restaurant operations, such as Applebee's, which promotes its "half-price" appetizers menu every day, but the price discount is offered only during specific (late night) hours of operation and McDonald's "Value menu" that features a "Buy One, Add One for $1" deal on select breakfast, lunch, and dinner items.

BUNDLING

Bundling refers to the practice of selecting specific menu items and pricing them as a group (bundle) so that the single menu price of the group is lower than if the items in the group were purchased individually. The most common example is the combination meals offered by QSRs. In many cases, these bundled meals consist of a sandwich, French fries (or other side), and a drink. Bundled meals, often promoted as "combo meals" or "value meals," are typically identified by number for ease of ordering.

Bundled menu offerings are carefully designed to encourage guests to buy all of the menu items included in the bundle, rather than to separately purchase only one or two of the items. Bundled meals are typically priced very competitively, so that a strong value perception is established in the guest's mind.

---

[3]See: Hayes, D. K., and Huffman, V. L., "Value pricing: How low can you go?" *The Cornell Quarterly*, February, 1995, pp. 51–56.

DYNAMIC PRICING

Dynamic pricing is often thought of as a relatively new foodservice industry pricing approach, but in reality it is not. Essentially, dynamic pricing is a strategy that adjusts the price of an item based on real-time supply and demand, essentially raising prices during periods of high demand periods and reducing them during periods of low demand.

Many restaurants already have dynamic pricing in all but name, including "Happy Hour" at Dunkin and "Happier Hour" at Taco Bell. During off-peak hours, these restaurants offer massive discounts, including $1 beverages, to attract customers during an otherwise-slow time of the day. Experienced foodservice operators know that this form of dynamic pricing is simply a variation of the "Early Bird Special" and "Happy Hour" price reductions that have long been used in the industry.

SALAD BARS AND BUFFETS

The difficulty in establishing one set price for an all-you-care-to-eat salad bar or buffet is that the amount eaten and, thus, the total product costs incurred by an operation will vary from one guest to the next. For example, a person weighing 100 pounds will, most likely, consume fewer products from a buffet line than will a 300-pound person. The general restaurant industry practice, however, is that each of these two guests would pay the same price to go through the salad bar or buffet line.

Short of charging guests for the amount they actually consume (a product-weighing approach that has been tried by some commercial operators but with only limited success), when offering an all-you-care-to-eat salad bar or a buffet, a single selling price must be established. This price must be based on a known "average" plate cost for the diner who is given the all-you-can-eat option. This can be accomplished rather easily if product usage recordkeeping is accurate and timely.

Regardless of the self-selected items offered for sale, their usage must be accurately recorded. Consider the situation of Mei, the manager of Lotus Gardens, a Chinese restaurant where patrons pay one price but may return as often as they like to a buffet line. Mei finds that a form like the one presented in Figure 6.7 is helpful in recording both product usage and guests served.

Note that Mei uses the ABC method to determine her menu items. She does so because total food costs on a salad bar or buffet line are a function of the following:

- What is eaten
- How much is eaten

To establish her prices, Mei notes the amount of product she puts on the buffet to begin the dinner meal period (Beginning Amount), she adds any additional amounts placed on the buffet during the meal period (Additions), and then calculates the amount of usable product left at the end of the meal period (Ending Amount). From this information, Mei can compute her total product usage and her total product cost per guest.

Based on the data in Figure 6.7, Mei knows that her total product cost for dinner on January 15 was $522.98. She can then use the following formula to determine her buffet product cost per guest:

$$\frac{\text{Total buffet product cost}}{\text{Guests served}} = \text{Buffet product cost per guest}$$

or

$$\frac{\$522.98}{116} = \$4.51$$

FIGURE 6.7 Salad Bar or Buffet Product Usage

Unit Name: Lotus Gardens

Date: 1/15 (Dinner)

Item	Category	Beginning Amount	Additions	Ending Amount	Total Usage	Unit Cost	Total Cost
Sweet and Sour pork	A	6 lb.	44 lb.	13 lb.	37 lb.	$6.40/lb.	$236.80
Bean sprouts	B	3 lb.	17 lb.	2 lb.	18 lb.	$2.60/lb.	$46.80
Egg rolls	B	40 each	85 each	17 each	108 each	$0.96 each	$103.68
Fried rice	C	10 lb.	21.5 lb.	8.5 lb.	23 lb.	$1.60/lb.	$36.80
Steamed rice	C	10 lb.	30 lb.	6.5 lb.	33.5 lb.	$1.40/lb.	$46.90
Wonton soup	C	2 gal.	6 gal.	1.5 gal.	6.5 gal.	$8.00/gal.	$52.00
Total product cost							$522.98

Total product cost: $522.98

Guests served: 116

Cost per guest: $4.51

On this day, Mei had a per-guest portion cost (plate cost) of $4.51. She can use this information to establish a menu price that she feels is appropriate for helping her reach her profit goals. To illustrate, assume that Mei uses the product cost percentage approach to establishing menu price. She determines that a 25 percent food cost is her goal. Using the pricing factor table in Figure 6.6, Mei would use the following formula to establish her per-guest buffet price:

4.0 Pricing factor × $4.51 per-guest cost = $18.04 selling price

For marketing purposes, and to still achieve very close to her desired food cost percentage, Mei may well round her buffet selling price down to, say, $17.99 per person. The significant point to remember here is that the amount consumed by any individual guest is less important than the average amount consumed per guest. That is because it is the amount consumed by the average, or typical, guest that is used to establish the menu price.

It is also important to recognize that Mei's buffet cost per guest will likely vary somewhat each day. This is not a cause for great concern. Minor variations in product cost per guest should be covered adequately if the selling price is properly established. By monitoring buffet costs on a regular basis, however, Mei can be assured that she has control over her costs per guest and has developed an appropriate selling price.

Mei also knows that the secret to keeping selling prices low in a salad bar or buffet setting is to apply a variation of the ABC inventory approach (see Chapter 4). That is, A items, which are expensive, should be no more than approximately 20 percent of the total product available. The B items, which are moderate in price, should be about 30 percent of the item offerings, and C items, which are inexpensive, should be 50 percent of the offerings. Using this approach, a menu listing of items can be prepared to ensure that only items that stay within these predetermined ranges are offered to guests.

BOTTLED WINE

Few areas of restaurant industry pricing create more controversy than that of pricing wines by the bottle. The reason for this may be the incredible variance in cost among different vintages (years of wine production), as well as the wide range in quality of alternative wine offerings. If your foodservice operation will sell wine by the bottle, it is likely that you will have some wine products that appeal to value-oriented guests and other, higher-priced wines that are preferred by guests willing to pay higher prices for superior wines.

An additional element that affects wine pricing is the fact that many wines that are sold by the bottle in restaurants are also sold in retail grocery or liquor stores. Thus, guests may have a very good idea of what a similar bottle of wine would cost them if it were purchased in either of these locations. How you decide to price the bottled-wine offerings on your own menu will definitely affect your guest's perception of the price/value relationship offered by your operation.

Properly pricing wine by the bottle calls for skill and insight. Consider the case of Claudia, who owns and manages a casual-dining Armenian restaurant. Using the product cost percentage method of pricing, Claudia attempts to achieve an overall wine product cost in her restaurant of 25 percent. Thus, when pricing her wines and using the pricing factor table in Figure 6.6, Claudia multiplies the cost of each bottled wine she sells by four to arrive at her desired selling price. Following are the four wines she wants to sell and the costs and selling prices associated with each type:

Wine	Product Cost	Selling Price	Product Cost %
1	$8.00	$32.00	25%
2	$12.00	$48.00	25%
3	$30.00	$120.00	25%
4	$40.00	$160.00	25%

After reviewing the selling price range that results when using the product cost percentage method of pricing, Claudia decides that she would like to explore the CMs resulting from her wine list. She computes the CM (selling price – product cost CM) for each wine she sells and finds the following results:

Wine	Selling Price	Product Cost	Contribution Margin
1	$32.00	$8.00	$24.00
2	$48.00	$12.00	$36.00
3	$120.00	$30.00	$90.00
4	$160.00	$40.00	$120.00

Her conclusion, after evaluating the CM approach to pricing and what she believes to be her customers' perception of the price/value relationship she offers, is that she may be hurting sales of wines 3 and 4 by pricing these products too high, even though they are currently priced to achieve the same 25 percent product cost as wines 1 and 2. She also determines that the selling price of wine 1 may actually be too low because it yields only a $24.00 CM.

In the case of bottled wine, the CM approach to price can often be used to your advantage. Guests can be price conscious when it comes to bottled wine. When operators seek to achieve profits guests feel are inappropriate, bottled-wine sales may decline. After reviewing all of her price-related information, Claudia develops her own bottled wine pricing strategy as shown below:

Wine	Product Cost	Selling Price	Contribution Margin	Product Cost %
1	$8.00	$38.00	$30.00	21.1%
2	$12.00	$44.00	$32.00	27.3%
3	$30.00	$66.00	$36.00	45.5%
4	$40.00	$78.00	$38.00	51.3%

Note that, although the selling price has increased in the case of wine 1, it has been reduced for wines 2, 3, and 4. The CM is still higher for wine 4 than for wine 1. The difference, however, is not as dramatic as before. Product cost percentages have, of course, been altered due to the price changes Claudia is proposing. Note also that the **price spread**, defined as the range between the lowest and the highest-priced menu item, has been drastically reduced. Whereas the price spread was previously $32.00 to $160.00, it is now $38.00 to $78.00. This reduction in price spread may assist Claudia in selling more of her higher-priced wine because her guests may be more comfortable with the price/value relationship perceived under this new pricing approach.

Most often, pricing bottled wine only by the product cost percentage method is a strategy that results in reduced bottled-wine sales. In this special pricing situation, the best approach to establishing prices calls for evaluating both the product cost percentages and CMs resulting from bottled-wine sales. The careful monitoring of the bottled-wine sales mix she achieves will also be important for Claudia's future assessment of her proposed pricing strategy.

CRAFT BEERS

Restaurant managers face a variety of challenges when pricing their food and beverage products. This is especially the case when significant changes in customer preferences create heightened demand for specific new products. Nowhere is this challenge more apparent than in today's growth in consumer demand for craft beers.

In the United States, **craft beers** and ales are produced by small breweries or microbreweries; most commonly defined as those with an annual production of 6 million barrels or less. The demand for craft beers in the United States is ever increasing, but large numbers of customers who are intrigued by craft beers also have little familiarity with them. They may want to try craft beers, but they may also be hesitant to pay the higher prices these products command. As a result, the challenges surrounding the proper pricing and selling of craft beers are not related to the demand for the product, which is skyrocketing. Rather, the challenge is related to reference prices.

A **reference price** is simply the price perceived by customers to be the "normal" and "fair" price for a product or service. Buyers use a variety of information to establish their reference prices. This includes what they have paid for similar products in the past, the prices offered by other sellers for the same or similar products, and what they perceive as a price *that provides good value for the money they will spend*. Buyers then evaluate all other prices for the same item in comparison to their established reference price.

It is important to recognize that, in the United States and the many other countries where demand for newly created craft beers is strong, these products can and should be sold at prices significantly higher than those of more traditional beer products produced and distributed by mega-breweries. Doing so allows operators to recover their higher cost of purchasing craft beer products. More importantly, it allows operators to enjoy the benefits of increased profits that should come from selling high-demand products. It can also help create loyal patrons who return again and again to enjoy the unique experience of drinking a craft beer on tap or in a bottle.

But that can only be done if customers perceive prices to be fair. This can be a challenge when craft beers can and should be sold at 50 to 200 percent higher prices than consumers are accustomed to paying for regular beer. This challenge can be better met as managers recognize that, while craft beers are *not* the same as mass-distributed beers, not all their customers will know that. To help address that challenge, managers can teach servers and bartenders four things that are true about craft beers:

1. Their supply is limited:

   In most cases, buyers perceive higher value and are willingly pay more for products that are in short supply. That's one reason the price of diamonds is high and the price of sand is low. The available amount of a specific craft beer typically ranges from limited to extremely limited. For example, in the fall, many craft breweries offer a selection of pumpkin beers and hard ciders. These concoctions, flavored with spicy hints of clove and nutmeg, are incredibly popular, and the demand for them is heightened by the knowledge that in a few short weeks or months these beers will not be available until next year. Even for brews that are offered all year, the quantity of beer produced is extremely limited in comparison to beers made by large-scale breweries.

   For many guests, a simple and truthful explanation regarding the limited availability of a specific craft beer will go a long way toward increasing those guests' perception of an appropriate reference price.

2. They are specially made:

   Craft beers are *crafted* rather than mass-manufactured. There is an artisanal element of craft brewing that many customers appreciate. Craft brewers focus on the quality of their ingredients, like locally grown hops and seasonal fruits and nuts that give their craft brews distinct flavor and aromas. In contrast, many large macro breweries focus their time and creative energies on external marketing plans.

   Because of these distinct characteristics, servers who are made aware of each of the unique characteristics of an operation's craft beer offerings can emphasize those same characteristics to price-resistant customers. When they do, customers who appreciate creativity, taste experience, and quality in production techniques will most often be willing to pay higher prices for craft beers.

3. They are locally made:

   Increasing numbers of consumers recognize the importance of local industries on local economies. In many cases, consumers will prefer to buy locally, and from those they know, rather than from product manufacturers perceived as being far away and disconnected from the local economy.

4. Their ABV is higher:

   Almost all craft beers have a greater alcohol by volume (ABV) content than that of regular beers. Certainly, one reason customers drink any alcoholic beverage is that it includes alcohol. Some customers may not be aware that the ABV of mass-distributed beers is approximately 4 to 6 percent, while the ABV of craft beers is regularly around 6 to 8 percent. In some cases, it is as high as 12 percent or more! While all operators should serve alcohol responsibly, pointing out the actual ABV levels of craft beers can help justify their higher prices to some customers.

ALCOHOLIC BEVERAGES AT RECEPTIONS AND PARTIES

Pricing beverages for open-bar receptions and similar special events can be challenging, and the reason for this is very simple. Each consumer group can be expected to behave somewhat differently when attending an open-bar or hosted-bar function. Clearly, we would not expect the guests at a formal political fund-raising cocktail reception to consume as many drinks during a 1-hour reception as a group of fun-loving individuals celebrating their favorite team's victory in a championship football game.

Establishing a price per person in these two situations may well result in quite different numbers. One way to solve this problem is to charge each of the guests for what they actually consume. In reality, however, many party hosts want their guests to consume beverage products without having the guest pay for each drink. When this is the case, you could elect to charge the host on a per-drink-consumed basis or charge the host on a per-person (served), per-hour basis. When charging on a per-person, per-hour basis, you must have a good idea of how much the average attendee will consume during the length of the party or reception so that an appropriate price can be established.

To illustrate beverage pricing in a hosted-bar situation, assume that you are the food and beverage director at the Carlton House, a luxury hotel. Ms. Swan, a potential food and beverage guest, approaches you with the idea of providing a 1-hour champagne reception for 100 guests prior to an important dinner that she is considering booking at your facility. Ms. Swan would like all of the attendees to be able to drink as much champagne during the reception as they care to.

Ms. Swan's specific question is: "How much will I be charged for the champagne reception if 100 guests attend?" Clearly, an answer of "I don't know" or "It's impossible to say because it depends on how much they drink" is inappropriate. It is, of course, your business to know the answer to such questions, and you can know. If you are aware, from past events and records you have kept what the average consumption for a group of this type has been previously, you can establish an appropriate per-person price. To do so, records for this purpose must be maintained. Figure 6.8 is an example of one such management tool that can be used to provide the information you will need.

Note that the average consumption of any beverage product type can be recorded. In this example, assume that you had recently recorded the data from the Gulley wedding, an event very similar to the one requested by Ms. Swan. In this case, a wedding reception, which also requested champagne, was held for 97 guests. The product cost per guest for that event, based on your records in Figure 6.8, equaled $7.91.

Based on what you know about the drinking pattern of similar groups, you could now use either the product cost percentage method or the CM pricing method to establish your per-person champagne reception price.

For purpose of illustration, assume that you used the product CM approach to pricing alcoholic beverage receptions. Further assume that the CM you desire per person served is $15.00. Recall that the selling price using the CM formula is:

Product cost + CM desired = Selling price

In this example:

$7.91	(Per-person product cost)
+ $15.00	(CM desired)
= $22.91	(Selling price per person)

FIGURE 6.8 Beverage Consumption Report

Event: Gulley Wedding Date: 1/15/20xx

Unit Name: The Carlton Hotel

Beverage Type	Beginning Amount	Additions	Ending Amount	Total Usage	Unit Cost	Total Cost
Liquor A						
B						
C						
D						
E						
F						
G						
Beer A						
B						
C						
D						
E						
Wine A						
B						
C						
D						
Other: Champagne: A. Sparkling B. Sparkling Pink	 8 bottles 8 bottles	 24 24	 9 11	 23 21	 $16.00/btl. $19.00/btl.	 $368.00 $399.00
Total Product Cost						$767.00

Total Product Cost: $767.00

Guests Served: 97 Cost per Guest: $7.91

Armed with this historical information, as well as that from as many other similar events you have served in the past, you are better prepared to answer Ms. Swan's question: "How much will I be charged for my champagne reception if 100 guests attend?"

Monitoring your product costs and then assigning reasonable menu prices based on these costs is an important part of your job as a foodservice manager. You must be able to perform this task well. Increasingly, however, the cost of labor, in addition to the cost of food and beverage products, has occupied a significant portion of the typical foodservice manager's cost control efforts.

In fact, in some foodservice facilities, the costs of labor and employee benefits provided actually exceed that of the food and beverage products sold. Because labor cost control is so important, in the next chapter we will turn our attention to the unique set of skills and knowledge you must acquire to properly manage and control your operation's labor costs.

Technology Tools

In this chapter, you learned about the menu formats you most often encounter as a hospitality manager, as well as the factors affecting menu prices and the procedures used to assign individual menu item prices based on cost and sales data. The mathematical computations required to evaluate the effectiveness of individual menu items and to establish their prices can be complex, but there is a wide range of software products available that can help you to do the following:

1. Develop menus and cost recipes.
2. Design and print menu "specials" for meal periods or Happy Hours.
3. Compute and analyze menu items and overall food cost percentage.
4. Compute and analyze menu items and overall CM.
5. Price banquet menus and bars based on known product costs.
6. Evaluate the profitability of individual menu items.
7. Estimate future item demand based on past guest purchase patterns.
8. Assign individual menu item prices based on additional management-supplied parameters.

Menu analysis and pricing software is often packaged as part of a larger software program. Its importance, however, is great. It is an area that will continue to see rapid development in the future as software and app makers seek additional ways to improve their menu pricing-related products.

FUN ON THE WEB!

Advances in menu management software continue to occur rapidly. Increasingly, restaurateurs are looking for programs that will give them many options to choose from when designing their own sales tracking (and pricing) processes. To stay current with newly developed menu pricing software and apps, simply enter "menu item pricing software" in your favorite search engine and then review the results.

Apply What You Have Learned

Dominic Carbonne owns Hungry Henry's pizza, a four-unit chain of takeout pizza shops in a city of 60,000 people (with an additional 25,000 college students attending the local state university). Recently, a new chain of pizza restaurants has opened in town. The products sold by this new chain are of lower quality and use lower quantities of ingredients (cheese, meat, and vegetable toppings) and are also priced 25 percent less than Hungry Henry's equivalent-size pizza.

Dominic has seen his business decline somewhat since the new chain opened. This is especially true with the college students.

1. How would you evaluate the new competitor's pricing strategy?
2. What steps would you advise Dominic to take to counter this competitor?
3. Describe three specific strategies restaurants can use to communicate "high quality" rather than "lowest price" to their potential customers.

For Your Consideration

1. Historically, foodservice managers have established prices for their menu items based on the amount paid for the ingredients used to make the items. Today, labor costs in many operations equal or exceed the operations' food and beverage product costs. Do you think it would be a good idea if managers setting menu prices begin to deemphasize ingredient costs in favor of a greater emphasis on labor costs? Why or why not?
2. Consider your favorite foodservice operation today. Would it still be your favorite if it raised its menu prices by 5 to 10 percent per menu item? Why?

Key Terms and Concepts

The following are terms and concepts addressed in the chapter that are important for you as a manager. To help you review, define the terms below:

Tip-on (menu)
Standard menu
CEO (chief executive officer)
Menu specials
Selling price
Price/value relationship
Quick-service restaurant (QSR)
Fast casual (restaurant)
Casual (restaurant)
Fine dining (restaurant)
Ambience
Daily menu
À la carte (menu)
Price blending
Plate cost
Contribution margin
Buy one get one (BOGO)
Prix fixe (menu)
Cycle menu
Value pricing
Bundling
Dynamic pricing
Price spread
Craft beer
Reference price

Test Your Skills

You may download the Excel spreadsheets for the Test Your Skills exercises from the student companion website at www.wiley.com/go/dopson/foodandbeveragecostcontrol8e. Complete the exercises by placing your answers in the shaded boxes and answering the questions as indicated.

1. Carlita operates a takeout-only pizza parlor that specializes in four-cheese pizzas. Carlita offers four different pizza sizes, and she knows what it costs to make each size. Help her calculate her food cost percentage and CM for each size pizza.

Pizza Size	Product Cost	Selling Price	Food Cost %	Contribution Margin
Small	$3.60	$9.99		
Medium	$4.60	$11.99		
Large	$5.60	$13.99		
Extra large	$6.60	$15.99		

What is the food cost percentage of Carlita's small pizza?
Which pizza size do you think is Carlita's most profitable? Why?

2. Bill owns Bill's Burger Barn, and he offers three different combo meals. These are the Hamburger combo, Bacon Cheeseburger combo, and Chicken

Sandwich combo. Help Bill calculate the food cost percentage and the CM for each of his combo meals and then answer the questions that follow.

Combo 1	Selling Price	Number Sold	Item Cost	Total Cost	Food Cost %	Total Sales
Hamburger	$7.99	200				
Combo Items						
Hamburger		200	$1.50			
French fries (large)		200	$0.38			
Soft drink (16 oz.)		200	$0.20			
Total						
			Contribution margin per sale			

Combo 2	Selling Price	Number Sold	Item Cost	Total Cost	Food Cost %	Total Sales
Bacon cheeseburger	$8.99	160				
Combo Items						
Bacon cheeseburger		160	$2.05			
Onion rings		160	$0.59			
Soft drink (16 oz.)		160	$0.20			
Total						
			Contribution margin per sale			

Combo 3	Selling Price	Number Sold	Item Cost	Total Cost	Food Cost %	Total Sales
Chicken sandwich	$6.99	75				
Combo Items						
Chicken sandwich		75	$1.15			
French fries (large)		75	$0.38			
Soft drink (16 oz.)		75	$0.20			
Total						
			Contribution margin per sale			

Which combo meal has the lowest food cost percentage?
Which combo meal generates the largest CM per sale?

3. Tonekwa has priced her menu items using the product cost percentage method in the past. She has asked her evening shift manager to price new menu items, and she believes that he will feel more comfortable using the factor method to price the new items. Help Tonekwa convert her desired product cost percentages to factors. (Spreadsheet hint: Use the ROUND function for the "Factor" column to three decimal places.)

Pricing Factor Table	
Desired Product Cost %	**Pricing Factor**
18%	
21%	

Pricing Factor Table	
Desired Product Cost %	**Pricing Factor**
22%	
24%	
26%	
31%	
32%	
36%	
37%	
41%	
42%	
44%	
46%	

4. Karis Elli owns Blazing Fast Subs. Ericka Braelon heads her highly trained staff. They produce custom-ordered subs for guests in less than 3 minutes. Karis forecasts she will sell one sub sandwich to each of the 2,700 guests she estimates she will serve in the coming week.

   Karis uses the CM method of pricing. Based on the revenue and cost forecast she has started to put together in the following spreadsheet, help Karis fill in the information needed to determine her forecasted CM per guest for the coming week.

Blazing Fast Subs			
Guest count forecast		$	%
Revenue forecast		$23,500	100%
Breads	970		4.1%
Cheese	$1,195		5.1%
Meats	$		13.0%
Produce	$420		1.8%
Toppings and other food	$425		1.8%
Total cost of food		$	25.8%
Cost of labor		$	30.0%
Other operating expenses		$	32.0%
Profit		$	
Contribution margin per guest	$		

   Assume that Karis' revenue and cost forecast assumptions are correct. What will be her profit margin percentage? What will be her CM per guest?

   Assume also that Karis creates a new Scottish-style sub featuring Ruby Red smoked salmon and that it costs her $3.53 to make one. At what price should Karis sell the new Scottish-style sub?

5. Bess and David own two small diners in a mid-sized city in Oklahoma. Bess has primary responsibility for the diner in the suburbs and David has primary responsibility for the diner in the inner city. The menu items and product costs are the same in both diners, but the market in the inner city demands lower menu prices than that in the suburbs. So, Bess has set her desired product cost percentage at 35 percent, and David's desired product

cost percentage is 37 percent because he can't charge as much as Bess. Bess likes to use the product cost percentage method to price menu items and David likes to use the factor method. Help both of them determine their selling prices. (Spreadsheet hint: Use the ROUND function for "Pricing Factor" column to three decimal places.)

Bess and David's Diner–Suburbs (Bess)

Desired Product Cost Percentage: 35%

Product Cost Percentage Method

Menu Item	Product Cost	Desired Product Cost Percentage	Selling Price
Chicken Breast Dinner	$5.25		
Seafood Platter	$7.45		
Steak Dinner	$6.99		
Turkey Sandwich	$3.25		
Pork Chop	$4.45		
Hamburger	$2.50		
Cheeseburger	$2.75		
French Fries	$0.45		
Meat Loaf	$3.25		
Small Drink	$0.65		

Bess and David's Diner–Inner City (David)

Desired Product Cost Percentage: 37%

Factor Method

Menu Item	Product Cost	Pricing Factor	Selling Price
Chicken Breast Dinner	$5.25		
Seafood Platter	$7.45		
Steak Dinner	$6.99		
Turkey Sandwich	$3.25		
Pork Chop	$4.45		
Hamburger	$2.50		
Cheeseburger	$2.75		
French Fries	$0.45		
Meat Loaf	$3.25		
Small Drink	$0.65		

6. Frankie Marie owns Frankie's Bistro in a small town on the South Carolina coast. She has decided to price her menu items using the CM method. She has determined the following CMs for her food categories:

Contribution Margins:

Salads: $2.30
Entrées: $8.25
Desserts: $3.75
Drinks: $2.10

Help her price her menu items.

Contribution Margin Approach			
Menu Item	Product Cost	Desired Contribution Margin	Selling Price
Salads			
Dinner salad	$1.30		
Macaroni salad	$1.55		
Potato salad	$1.65		
Carrot and raisin salad	$1.40		
Bavarian salad	$1.60		
Entrées			
Liver and onions	$4.50		
Steak patty	$4.75		
Meatloaf	$5.85		
Chicken fried steak	$4.10		
Fried catfish	$4.35		
Chicken casserole	$4.25		
Turkey and dressing	$4.55		
Desserts			
Chocolate cream pie	$1.75		
Coconut cream pie	$1.75		
Pecan pie	$2.25		
Chocolate cake	$1.60		
Pudding	$0.90		
Jell-O	$0.60		
Carrot cake	$1.70		
Drinks			
Coffee	$0.55		
Tea	$0.65		
Soft drink	$0.42		

7. Gabriel Hinojosa owns Gabriel's Tex-Mex Restaurant, an extremely popular 250-seat establishment in a large California city. Gabriel has decided to offer a 4-hour Sunday brunch buffet for his guests because he thinks he can achieve a guest count of 625 during that time frame.

   Last Sunday, June 1, he offered the buffet for the first time, and he charged $21.99 per guest. However, he only served 400 people. He believes he could attract more guests if he offered the buffet at a lower price. He collected information on last Sunday's buffet product usage, and he used the ABC method to put his menu items into categories. His desired food cost percentage is 35 percent. Help him complete the buffet product usage report. (Spreadsheet hint: Use the ROUND function for "Cost per Guest" and "Desired Selling Price Based on Cost" to two decimal places.)

   After completing this analysis, what should be Gabriel's selling price? If he uses this new selling price and he serves 625 guests next Sunday, June 8, will his total revenue increase? If so, by how much?

Buffet Product Usage (Sunday, June 1): Gabriel's Tex-Mex Restaurant

Menu Item	Category	Unit	Beginning Amount	Additions	Ending Amount	Total Usage	Unit Cost	Total Cost
Steak Fajitas	A	lb.	20	60	6		$8.50	
Chicken Fajitas	A	lb.	15	70	10		$4.75	
Carne Asada	A	lb.	10	50	4		$6.25	
Cheese Enchiladas	B	lb.	2	80	15		$3.00	
Beef Enchiladas	B	lb.	3	60	10		$3.50	
Enchiladas Verde	B	lb.	1	70	8		$2.95	
Chili Rellenos	B	lb.	10	45	5		$3.75	
Tacos	C	Each	0	150	20		$0.85	
Bean Chalupas	C	Each	0	175	5		$0.65	
Tortilla Soup	C	gal.	2	10	4		$0.75	
Spanish Rice	C	lb.	5	70	12		$0.65	
Refried Beans	C	lb.	15	75	6		$0.65	
Sopapillas	C	Each	25	200	30		$0.45	
Total Product Cost								

Guests Served:	400	Desired Food Cost %:	35%
Total Product Cost:		Desired Selling Price Based on Cost:	
Cost per Guest:			

Revenues, June 1:		Projected Revenues, June 8:		Difference:	

8. JoAnna is the foodservice director at Reading Hospital. She has just started a new menu program to offer guests visiting hospitalized patients guest trays so the patients and their guests can eat together. She has been offering her guest-tray program for 1 month, and it has been very popular. She would like to know her average selling price per guest (check average) to see if she needs to change the price of her menu items. She would like to keep her average selling price above $16.99. Given the following information, and the fact that JoAnna priced each menu item with a $12.00 per-guest-tray CM, calculate her average selling price for the month.

Menu Item	Number Sold	Product Cost	Contribution Margin	Selling Price	Total Sales
Hunter's Chicken	2,560	$3.50			
Jambalaya	750			$15.75	
Grilled Salmon	1,200	$8.50			
Beef Tenderloin	500			$24.95	
Vegetarian Cheese Bake	1,210	$2.75			
Total					

Average Selling Price	

9. Stella operates a Cuban-themed food truck that sells sandwiches and drinks to office workers in the downtown area of a mid-sized city. Stella has been approached by the Grab-On coupon company to enter a partnership arrangement with them, whereby Grab-On would, on their own website, sell coupons good for $20.00 worth of Stella's food and beverage products.

The $20.00 coupons would sell for $10.00 each, and Grab-On would split these sales proceeds 50/50 with Stella. The coupons sold would have a 30-day expiration period, and Grab-On states that, in their experience, only 90 percent of the coupons sold will be redeemed prior to their expiration date.

Stella has already begun forecasting the next 30 days' sales and product expenses (without the coupon program), as well as her estimated total contribution margin. Recalculate Stella's projected financial results assuming her original forecast and the sale of 10, 100, and 500 Grab-On coupons.

Stella's Food Truck								
# of Coupons Sold	0		10		100		500	
Forecast	$	%	$	%	$	%	$	%
Noncoupon revenue	$15,000	100.0%	$15,000	99.67%	$15,000		$15,000	
Coupon revenue	$0	0%	$50	0.33%				
Total Revenue	$15,000	100.0%	$15,050	100.0%		100.0%		100.0%
Noncoupon product cost	$4,500	30.0%	$4,500	30.0%	$4,500	30.0%	$4,500	30.0%
Coupon product cost	$0	0.0%	$54	0.36%				
Total Product Cost	$4,500.00		$4,554					
Total Contribution margin		70.00%						

In addition to her product costs, what other factors should Stella consider prior to agreeing to partner with Grab-On in the sale of the coupons?

What will be the total increase in her CM if 500 coupons are sold?

Would you recommend to Stella that she agree to the proposed partnership?

10. Ming has recently inherited the Quick Wok restaurant started by her parents. It is located in a busy strip shopping area surrounded by many office complexes, but it is also near many QSRs. The Quick Wok has been successful because of the quality of its food, but Ming feels that it could do even better at lunchtime if she could create a value meal option to appeal to the price-conscious consumer. Because both a McDonald's and a Wendy's are within a quarter-mile of her store, she has determined that her own value meal menu item needs to be priced at $3.99. She creates a stir-fry dish that, when served with white rice, has a portion cost of 1.85 cents. Her beverages have a cost of 45 cents. The beverages already sell for $1.99 each, and she does not want to raise this price.

    She believes she could sell 75 of the new value meals per day if she offers the stir-fry dish at $3.99. In addition, she seeks an overall product cost percentage of 35 percent. From historical data she knows that 80 percent of her customers purchase a drink with their meals.

    Based on the information given, calculate the overall product cost percentage of the value meals and beverages. Would you advise Ming to go for it? Why or why not?

Value Meal and Beverage

Menu Item	Number Sold	Item Cost	Total Cost	Selling Price	Total Sales	Food Cost %
Stir-fry dish	75					
Beverage						
Total						

How many beverages must be sold in addition to the value meals if Ming is to achieve her target food cost percentage goal? Is this number feasible? Spreadsheet hint: You will have to arrive at the number of beverages sold by trial and error. Specifically, start with 60 and increase or decrease the number until you reach a total food cost % of 35 percent. This Test Your Skills problem is designed to show you how changes in the number of items (beverages) sold affect the total food cost percentage.

Number of Beverages to Achieve Target Overall Food Cost Percentage

Menu Item	Number Sold	Item Cost	Total Cost	Selling Price	Total Sales	Food Cost %
Stir-fry dish	75					
Beverage						
Total						

11. Jackson Daniels is the director of food and beverage at the Foxfire Country Club. In June, Jackson's club hosted three weddings. Each wedding featured a 4-hour hosted bar paid for by the bride and groom. The consumption data from each event is listed below. Complete the missing data in the report, and then help Jackson answer the questions that follow:

Wedding Date	Number of Guests Served	Beer Cost	Wine Cost	Spirit Cost	Total Cost	Cost per Guest
June 07		$700	$600	$1,100		
June 14	325	$850	$1,425			
June 21	400		$1,000	$2,250		
Total	975	$2,075		$5,150		

a. What do you think should be Jackson's best estimate of the cost to the Club of providing a 4-hour hosted bar at weddings?

b. If Jackson seeks to ensure a 25 percent beverage cost, what should be his selling price, per guest, for a 4-hour hosted bar?

c. Would you recommend Jackson charge half of the amount in the answer in part b. for a 2-hour hosted bar? Explain your answer.

12. Cindy Larson is the owner of the Baker's Choice bakeshop. Cindy is considering offering a "Tuesday Only" promotion, where her customers who buy a specific loaf of bread can get a second loaf of the same type at half price.

Cindy believes that approximately 20% of her bread buyers would take advantage of the promotion. She knows her contribution margin (CM) for each of the three loafs she sells at regular price but wants to know what her total revenue for each bread type and what her CM per loaf would be if she implements the promotion.

Help Cindy find that out by completing the promotion cost sheet she has started below.

Baker's Choice Promotion						
Bread Type	Regular Forecasted Sales	Estimated Sales with Promotion	Selling Price	Total Revenue	Cost per Loaf	CM Per Loaf with Promotion
French Loaf	120	144	7.99	$1,054.68	$2.45	
Sourdough	90		10.99		$3.50	
Seeded Rye	60		8.99		$2.80	

CHAPTER 7

Managing the Cost of Labor

OVERVIEW

This chapter explains the techniques foodservice managers can use to control their labor costs by establishing and monitoring labor cost standards. Factors that affect labor productivity, as well as methods for improving labor productivity, are presented. This chapter teaches you how to schedule employees based on established labor productivity standards, as well as how to compute a labor cost percentage and other measures of labor productivity most commonly used in the foodservice industry.

Chapter Outline

- Labor Expense in the Hospitality Industry
- Evaluating Labor Productivity
- Maintaining a Productive Workforce
- Measuring Current Labor Productivity
- Managing Payroll Costs
- Reducing Labor-Related Costs
- Technology Tools
- Apply What You Have Learned
- For Your Consideration
- Key Terms and Concepts
- Test Your Skills

LEARNING OUTCOMES

At the conclusion of this chapter, you will be able to:

- Identify the factors that affect employee productivity in foodservice operations.
- Develop labor standards and employee schedules used in a foodservice operation.
- Analyze and evaluate actual labor utilization.

LABOR EXPENSE IN THE HOSPITALITY INDUSTRY

You have learned that having the correct amount of food and beverage products available to serve your guests is important. Knowing how those products should be prepared and served is also vital. To see why, consider the case of Pauline. She manages the open-to-the-public cafeteria located in a large urban hospital. Both hospital staff and patients' visitors, who constitute the majority of her cafeteria guests, have good things to say about the quality of her food. They complain often, however, about the slowness of her cafeteria line, the soiled tables during the busy lunch hour, and the frequent running out of items at the self-serve sandwich station. Pauline often feels that she needs more employees. She knows, however, that her current staff is actually larger than it was a few years ago. She also knows that she now serves more guests each day than she has in the past. Her question is "Do I have the right number of employees scheduled to work, and at the right times, for the number of guests I am serving today?" Unfortunately for her, Pauline is so busy "helping" her employees get through the meal periods that there seems to be little time for thinking about and planning the strategies and techniques she needs to apply if she is to solve her labor-related customer service problems.

In years past, when labor was relatively inexpensive, Pauline might have responded to her need for more workers by simply hiring more employees. Today's foodservice manager, however, does not have that luxury. In an increasingly scarce and costly labor market, you must learn the supervisory skills needed to maximize the effectiveness of your staff and the cost control skills required to evaluate their efforts. That is because labor is a significant foodservice operating cost. In fact, in some foodservice establishments, the cost of labor actually exceeds the cost of their food and beverage products.

Today's competitive labor market indicates that, in the future, foodservice managers will likely find it even more difficult to recruit, train, and retain an effective team of employees. Therefore, the control of labor expenses takes on a greater level of importance than ever before. In some sectors of the foodservice industry, a reputation for long hours, poor pay, and undesirable working conditions has caused some high-quality employees to look elsewhere for more satisfactory careers. It does not have to be that way, and it is up to you to help ensure that in your organization it is not.

When labor costs are adequately controlled, management has the funds necessary to create desirable working conditions and pay wages that will attract the very best employees. In every service industry, better employees mean better guest service and, ultimately, better business profits.

LABOR EXPENSE DEFINED

Payroll is the term generally used to refer to the salaries and wages you will pay your employees. **Labor expense** includes salaries and wages, but it also includes other labor-related costs. These labor-related costs may include the following:

1. FICA (Social Security) taxes, including taxes due on employees' tip income
2. FUTA (Federal unemployment taxes)
3. State unemployment taxes
4. Workers' compensation
5. Group life insurance
6. Health insurance, including:
   Medical
   Dental

Vision
Disability
7. Pension/retirement plan payments
8. Employee meals
9. Employee training expenses
10. Employee transportation costs
11. Employee uniforms, housing, and other benefits
12. Vacation/sick leave/personal days
13. Tuition reimbursement programs
14. Employee incentives and bonuses

Not every operation will incur all of the costs listed. But some operations will have all of these and more. You can be sure, however, that regardless of the facility you manage, you will incur some labor-related expenses in addition to salaries and wages. The critical question you must answer is "How much should I spend on payroll and other labor expenses to deliver the quality of products and service that I feel is appropriate?" Before you can hope to answer that question, it is important that you understand well the individual components that make up payroll and total labor expense.

FUN ON THE WEB!

In the United States, the Patient Protection and Affordable Care Act (ACA) was signed into law in 2010. This health insurance–related law directly affects the benefit costs incurred by those foodservice organizations employing more than 50 full-time equivalent workers. Its actual requirements are ever evolving. To learn more about the ACA, and its current requirements, go to http://cms.gov/cciio/index.html.

PAYROLL

Payroll refers to the gross pay received by an employee in exchange for his or her work. That is, if an employee earns $20.00 per hour and works 40 hours per week for their employer, the gross paycheck (the employee's paycheck before any mandatory or voluntary deductions) would be $800 ($20.00 per hour × 40 hours = $800). This gross amount is considered a payroll expense.

If the employee earns a salary, that salary amount is also a payroll expense. A **salaried employee** generally receives the same income per week or month regardless of the number of hours worked. Thus, if a salaried employee is paid $1000 per week when they work a complete week, that $1000 is included in payroll expense. Salaried employees are actually more accurately described as **exempt employees** because their duties, responsibilities, and level of decisions make them exempt from the overtime provisions of the federal government's Fair Labor Standards Act (FLSA). Exempt employees do not receive overtime for hours worked in excess of 40 per week and are expected by most organizations to work the number of hours needed to fully do their jobs.

FUN ON THE WEB!

The designation of which workers can (and cannot) be considered exempt employees is governed by the US Department of Labor and the Fair Labor Standards Act. Minimum allowable salary levels are also determined by the Wage and Hour Division of this department. To learn more about wages that must be paid to exempt and to nonexempt employees, go to http://www.dol.gov/agencies/whd/fact-sheets/17a-overtime.

FIXED PAYROLL *VERSUS* VARIABLE PAYROLL

When you manage a foodservice facility, you must make choices regarding the number and types of employees you will hire to help you serve your guests. Some employees are needed simply to open the doors for minimally anticipated business. **Minimum staff** is the term used to describe the least number of employees, and thus the least number of payroll dollars, needed to operate a business. For example, in a small operation, this may include only one manager, one server, and one cook. The cost of providing payroll for these three individuals would be the operation's minimum staff payroll.

Suppose, however, that the operation anticipated much greater than minimum volume on a given day. The increased number of guests expected means that the operation may need more cooks and more servers, as well as cashiers, dish room personnel, and, perhaps, more supervisors to handle the additional workload. Clearly, these additional staff positions create a work group that is far larger than the minimum staff, but it is needed to adequately service the anticipated number of guests. In this scenario, payroll costs will increase.

Payroll costs may be fixed or variable. **Fixed payroll** most often refers to the amount an operation pays in salaries. This amount is typically fixed because it remains unchanged from one pay period to the next unless the salaried employee separates employment from the organization or is given a raise. On a profit and loss (Income) statement prepared according to the Uniform System of Accounts for Restaurants (USAR) these are considered to be "management" costs. **Variable payroll** consists primarily of those dollars paid to hourly employees. When using the USAR these are considered to be "staff" costs.

Variable payroll is the amount that should vary with changes in sales volume. Generally, as you anticipate increased volume levels in your facility, you may need to add additional hourly workers and, sometimes, additional salaried employees. The distinction between fixed and variable labor is an important one. As a manager, you may have little direct control over your fixed labor expense, whereas you will have nearly 100 percent control over variable labor expenses that are above your minimum staff levels.

LABOR EXPENSE

Unlike payroll expense, labor expense refers to the *total* of all costs associated with maintaining your workforce. Labor expense includes employee taxes and benefits costs and is always larger than payroll expense.

The actual amount of taxes and employee benefits paid for by a specific operation can vary greatly. Some expenses, such as payroll taxes and contributions to workers' unemployment and workers' compensation programs, are mandatory for all employers. Other benefit payments, such as those made for employee insurance and retirement programs, are voluntary and vary based on the benefits a business chooses to offer its employees. As employment taxes and benefit costs increase, an operation's labor expense can increase even if payroll expense remains constant.

Most foodservice operators have total control over their payroll expense. It is, therefore, often referred to as a "controllable" labor expense. Other labor expenses, such as taxes and some benefits, over which an operator has little or no control, are commonly called "non-controllable" labor expenses. In reality, however, you can exert some control over these non-controllable labor expenses, for example a foodservice manager who works very hard to ensure a well-trained workforce is working in a safe environment and thereby achieves a lower rate on workers' compensation, accident, and health insurance for their employees.

In this chapter, we deal primarily with payroll-related expenses. This is in keeping with the concept that these are the most controllable of labor-related expenses. To determine how much payroll is needed to operate your business, you must be able to determine how much work must be done and how much work each employee can perform. If too few employees are scheduled to work, poor service and reduced sales can result because guests may choose to go elsewhere in search of superior service levels. If too many employees are scheduled, payroll and other labor-related expenses will be too high, resulting in reduced profits. The best solution to this challenge is to know how many employees are needed given the estimated number of guests you will serve. To determine this number of employees, you must have a clear idea of the **productivity** of each of your employees. Productivity is the amount of work performed by an employee in a fixed period of time.

EVALUATING LABOR PRODUCTIVITY

There are many ways to assess labor productivity. In general, productivity is measured by calculating a **productivity ratio**:

$$\frac{\text{Output}}{\text{Input}} = \text{Productivity ratio}$$

To illustrate this formula, assume that a restaurant employs 4 servers and it serves 60 guests. Using the productivity ratio formula, the output is guests served and the input is servers employed, as follows:

$$\frac{\text{60 guests}}{\text{4 servers}} = \text{15 guests per server}$$

This formula states that, for each server employed, 15 guests can be served. The productivity ratio is 15 guests to 1 server (15 to 1) or, stated another way, 1 server per 15 guests (1/15).

There are several ways of defining foodservice output and input; thus, there are several types of productivity ratios. Some of these are presented later in this chapter.

All of these productivity ratios can help a manager determine the answer to the key question "How much should I spend on labor?" The answer, however, is even more complicated than it might seem at first. In the preceding example, you know that, on average, one of your servers can serve 15 guests. But how many guests will a slow server serve? How about your best server? How much do you pay the most productive server? Your least productive? Are you better off scheduling your best server if you anticipate 20 guests or should you schedule two of your slower servers? How can you help the slower server become an average or above-average server? At what cost? These are the types of questions that must be answered if you are to effectively manage your total payroll costs.

BACK OF THE HOUSE AND FRONT OF THE HOUSE PRODUCTIVITY

The effective control of labor requires managers to assess both their back of the house (see Chapter 4) and **front of the house** productivity levels. Front of the house refers to areas within a food operation that are open to public access. These areas typically include guest waiting and ordering areas, dining rooms, bar areas, and restrooms.

While each foodservice operation is unique, typical measures used to assess back of the house productivity can include the following:

- Number of covers (guest orders) completed per labor hour
- Number of guest checks (table orders) processed per labor hour
- Average guest check completion time (in minutes)

- Average time to process drive-thru orders
- Number of menu items produced per hour worked
- Number of improperly cooked items (mistakes) produced per hour worked

Examples of typical measures used to assess front of the house productivity of servers include the following:

- Number of guests (not tables) served per server hour worked
- Number of menu "specials" sold per server hour worked
- Number of errors (voided sales) produced per shift worked
- Average guest check size (per guest served)

Regardless of the productivity measures they choose to use, foodservice operators must develop their own methods for managing payroll because every foodservice unit is different. Consider, for example, the differences between managing payroll costs at a small, quick-service food kiosk located in a shopping mall and a large banquet kitchen located in a 1000-room convention hotel. Although the methods used to manage payroll costs may vary, it is always true that payroll costs can and should be managed.

MAINTAINING A PRODUCTIVE WORKFORCE

Before we address how managers calculate and utilize productivity ratios, it is important to understand the factors that make employees more productive because these factors directly affect employee productivity.

The following are 10 key employee-related factors that directly affect levels of employee productivity:

Ten Key Factors Affecting Employee Productivity

1. Employee selection
2. Training
3. Supervision
4. Scheduling
5. Breaks
6. Morale
7. Menu
8. Convenience food use *versus* scratch preparation
9. Equipment/tools
10. Service level desired

EMPLOYEE SELECTION

Choosing the right employee from the beginning is vitally important in developing a highly productive workforce. Good foodservice managers know that proper employee selection procedures go a long way toward establishing the kind of workforce that can be both efficient and effective. This involves matching the right employee with the right job. The process begins with the development of the job description.

FIGURE 7.1 Sample Job Description

Job Description

Unit Name: Thunder Lodge Resort **Position Title:** Room Service Delivery Person

PRIME TASKS:

1. Answer telephone/read texts to receive guest orders
2. Set up room service trays in steward area
3. Deliver trays to room, as requested
4. Remove tray covers upon delivery
5. Remove soiled trays from floors
6. Maintain guest check control
7. Balance room service cash drawer.
8. Clean room service setup area at conclusion of shift
9. Other duties, as assigned by supervisor
10.
11.
12.

Special Comments: Hourly rate excludes tips. Uniform allowance is $55.00 per week

Salary Range: $18.00–$22.25/hours Signature: Matt V.

JOB DESCRIPTION

A **job description** is a listing of the tasks that must be accomplished by the employee hired to fill a particular position. For example, in the case of a room service delivery person in a full-service hotel, the tasks might be listed as indicated on the job description shown in Figure 7.1.

A job description should be maintained for every position in every foodservice operation. From the job description, a job specification can be prepared.

JOB SPECIFICATION

A **job specification** is a listing of the personal characteristics needed to perform the tasks contained in a job description. Figure 7.2 shows the job specification that would match the job description shown in Figure 7.1.

As can be seen, this position requires a specific set of personal characteristics and skills. When a room service delivery person is hired, the job specification requirements (specs) must be foremost in management's mind. If the job specs do not exist or are not followed, it is likely that employees may be hired who are simply not able to be highly productive. Each employee hired must be able to do, or be trained to do, the tasks indicated in the employee's job description.

It will be your role to develop and maintain both job descriptions and job specifications for every position in your operation so employees will know what their jobs are and so that you know the characteristics your employees must have, or be

FIGURE 7.2 Sample Job Specification

Job Specification

Unit Name: Thunder Lodge Resort **Position Title:** Room Service Delivery Person

Personal Characteristics Required:

1. Good telephone skills; speaks clearly, speaks easily understood English
2. Ability to operate point-of-sale systems
3. Detail oriented
4. Pleasant personality
5. Discreet

Special Comments: Good grooming habits are especially important in this position, as employee will be a primary guest contact person.

Job Specification Prepared By: Matt V.

trained in, to do their jobs well. Managers selecting new employees have several important tools available to them. These include the following:

1. Applications
2. Interviews
3. Pre-employment testing
4. Background/reference checks

APPLICATIONS

The **employment application** is a document to be completed by the candidate for employment. It will generally list the name, address, work experience, and related information of the candidate. It is important that each employment candidate for each position be required to fill out (in person or online) an identical application, and that their applications be kept on file for each candidate who is ultimately selected to fill a vacant position.

FUN ON THE WEB!

To compete effectively in an increasingly tight labor market, foodservice operators should make it easy for potential employees to apply for a job. One good way to do this is to ensure that an operation has an easy-to-use online job application program.

An online job application is a form on the internet where applicants can communicate their skills and relevant experience for a specific job or position. Foodservice operators use online applications to help speed up the hiring process and increase the number of potential job candidates for the position.

Several technology companies offer online application services appropriate for small businesses. To learn more about the ways online application programs work and can be publicized to potential job applicants, enter "online job application programs for restaurants" in your favorite search engine and view the results.

INTERVIEWS

From the employment applications submitted, you will select some candidates for the personal interview process. It is important to realize that the types of questions that can be asked in the interview process are highly regulated by law. As a result, job interviews, if improperly performed, can subject an employer to significant legal liability. For example, if a candidate is not hired based on his or her answer to—or refusal to answer—an inappropriate question, that candidate may have the right to file a lawsuit.

What questions can and cannot be asked of potential employees? The Equal Employment Opportunity Commission (EEOC) suggests that all employers consider the following three issues when deciding whether to include a particular question on an employment application or in a job interview:

1. Does this question tend to screen out minorities or females?
2. Is the answer needed to judge this individual's competence for performance of the job?
3. Are there alternative, nondiscriminatory ways to judge the person's qualifications?

In all cases, questions asked both on the application and in the interview should focus on the applicant's ability to do a job, and nothing else.

FUN ON THE WEB!

Very few foodservice operators would intentionally ask a question of an applicant that is illegal. Sometimes, however, avoiding asking illegal questions can be tricky.

For example, assume an interview is taking place on a Sunday. The interviewer greets the applicant by saying "Welcome, how are you today?"

The applicant replies, "I'm doing great, I just got out of church!"

While it might seem reasonable for the interviewer to follow up with a "Good for you, What church do you attend?" query, and such a question might be considered normal and even friendly, it would also be illegal!

Asking an applicant about the church they attend is prohibited under Title VII of the Civil Rights Act because it could potentially make religion (and church attendance!) a factor in the interviewer's job offer decision making.

One good way to become more familiar with questions that are, and are not, appropriate to be asked in an interview is to watch videos on YouTube that address this key issue.

To learn about questions that can and cannot be asked of job applicants, go to the YouTube website, enter "illegal job interview questions to avoid asking" in the YouTube search bar, and then watch one or more of the posted videos addressing this important topic.

PRE-EMPLOYMENT TESTING

Pre-employment testing is a common way to help improve employee productivity. In the hospitality industry, pre-employment testing will generally fall into one of the following categories:

1. Skills tests
2. Psychological tests
3. Drug screening tests

Skills tests allow an applicant to show that he or she can complete a task. Example activities could include drink production for bartenders, food production tasks for cooks and chefs, and formula creation for those involved in using spreadsheet tools.

Psychological testing can include personality tests, tests designed to predict mental performance, or tests of mental ability.

Pre-employment drug testing is used by some organizations to determine if an applicant uses illegal drugs. When properly used, such testing is allowable in most states, and it can be a very effective tool for reducing insurance rates and potential employee liability issues.

BACKGROUND/REFERENCE CHECKS

Increasingly, hospitality employers are utilizing background checks prior to hiring employees in selected positions. Common verification points include the following:

- Name
- Social Security number
- Address history
- Dates of past employment and duties performed
- Education/training
- Criminal background
- Credit history

Background checks, like pre-employment testing, can leave an employer subject to litigation if the information secured during a check is false or is used in a way that violates federal or state employment laws. Employers are not allowed to discriminate among applicants when testing for drug or alcohol use. Neither educational background, demeanor, physical appearance, nor race may be used to determine if an individual applicant will be tested. Employers must treat all applicants for the same job in the same manner. In addition, if the test's results are improperly disclosed to third parties, it could violate the applicant's right to privacy.

Not conducting background checks on some positions can, however, subject the employer to potential litigation under the doctrine of **negligent hiring**, that is, a failure on the part of an employer to exercise reasonable care in the selection of employees. When background checks are performed, a candidate for employment should be asked to sign a consent form authorizing you to conduct the background check.

FUN ON THE WEB!

E-Verify, authorized by the Illegal Immigration Reform and Immigrant Responsibility Act of 1996 (IIRIRA), is a web-based system through which employers electronically confirm the employment eligibility of their new employees.

In the E-Verify process, employers create "cases" (individual employee records) based on information taken from an employee's Form I-9, Employment Eligibility Verification.

E-Verify then electronically compares that information to records available to the US Department of Homeland Security (DHS) and the Social Security Administration (SSA). The employer usually receives a response within a few seconds that either confirms the employee's employment eligibility or indicates the employee needs to take further action to complete the case.

E-Verify is administered by the SSA and US Citizenship and Immigration Services (USCIS). USCIS facilitates compliance with US immigration law by providing E-Verify program support, user support, training and outreach, and developing innovative technological solutions in employment eligibility verification.

The use of E-Verify is not mandatory (except in certain states), but its use is always a good idea.

To find out more about the E-Verify program and how it helps employers stay in compliance with the law, enter "Why use E-Verify?" in your favorite search engine and review the results.

Green and Growing!

Consumers are increasingly aware that when they support businesses committed to sustainability, their dollars make an impact socially and environmentally. As a result, influential and leading-edge thinkers concerned about society and the environment seek out companies that share their health, social, and environmental interests and priorities. So committed are they that they are, on average, willing to spend 20 percent more than the typical guest for products that conform to their values and lifestyle.

In a similar manner, environmentally conscious workers are increasingly aware that a company's care for the environment most often is also reflected in its care for its employees. As a result, companies espousing genuine commitment to the environment typically attract a more committed and, as a result, higher-quality staff. These workers tend to value health, the environment, social justice, personal development, and sustainable living. They want to contribute their efforts to companies that share those values, and their numbers are growing.

Do you have to be "certified" as green to attract these environmentally conscious employees? That is, do you have to be a "perfect" green facility? No. But you cannot just pretend that you are, either. Workers easily see beyond false claims of care, in part because they share information so freely and easily via social media. The more environmentally friendly, socially responsible, and healthy you are known to truly be, the easier it will be for prospective employees who share these goals to find you—because they *are* looking.

TRAINING

Perhaps no area under your control holds greater promise for increased employee productivity than effective training. In too many cases, however, training in the hospitality industry is poor or almost nonexistent. Highly productive employees are usually well-trained employees, and, frequently, employees with low productivity have been poorly trained. Every position in a foodservice operation should have a specific, well-developed, and ongoing training program.

Effective training will improve job satisfaction and instill in employees a sense of well-being and accomplishment. It will also reduce confusion, product waste, and loss of guests. In addition, supervisors find that a well-trained workforce is easier to manage than one in which employees are poorly trained.

Effective training begins with a good **orientation program**. An orientation program prepares a new worker for success in their job. The following list includes some of the concerns that most employees have when they start a new job. You should identify which items are relevant to your new employees and take care to provide them with information in each area, in either written or verbal form.

Potential Orientation Program Information

1. Payday
2. Annual performance review
3. Probationary period
4. Dress code
5. Telephone call, cell phone use policy
6. Smoking policy
7. Uniform allowance
8. Disciplinary system
9. Educational assistance
10. Work schedule
11. Mandatory meetings
12. Tip policy
13. Departmental transfers
14. Employee meal policy
15. Harassment policy
16. Lockers/ security
17. Jury duty
18. Leave of absence
19. Birth-related leave
20. Alcohol/drug policy
21. Employee assistance programs
22. Tardy policy
23. Sick leave policy
24. Vacation policy

25. Holidays and holiday pay
26. Overtime pay
27. Insurance
28. Retirement programs
29. Safety/ emergency procedures
30. Grievance procedures

Orientation and training programs need not be elaborate. They must, however, be consistent and continual. Hospitality companies can train in many different areas. Some training seeks to influence attitudes and actions, for example when training to prevent work-related harassment is presented. In other cases, training may be undertaken to assist employees with stress or other psychologically related job aspects.

In most cases, however, the training you will be responsible for as a unit manager is **task training**, the training used to ensure that an employee has the skills to meet productivity goals. The development of a training program for any task requires managers to:

1. Determine how the task is to be done.
2. Plan the training session.
3. Present the training session.
4. Evaluate the session's effectiveness.
5. Retrain at the proper interval.

DETERMINE HOW THE TASK IS TO BE DONE

Often, jobs can be done in more than one way. When management has determined how a task should be completed, however, that method should be made part of the training program and should be strictly maintained unless a better method can be demonstrated and successfully implemented. If this is not done, employees will find that anything goes, and product consistency, along with service levels, will vary tremendously.

PLAN THE TRAINING SESSION

Properly planning for a training session involves answering key questions:

1. Who should be trained?
2. Who should do the training?
3. Where should the training take place?
4. When should the training take place?
5. What tools, materials, or supplies are needed to conduct the training session?
6. What should be the length of the session(s)?
7. How frequently should the sessions occur?
8. How and where will the attendance and completion records regarding each training session offered be kept?

Good training sessions are the result of a need felt by management to train personnel matched with a management philosophy that training is important. Taking time to effectively plan training sessions is a good way for you to let employees know that you take the training process seriously.

Whether the training session is a self-paced interactive program delivered online, a hands-on demonstration, a group discussion, or a lecture-style presentation, time devoted to planning your training sessions is time well spent.

PRESENT THE TRAINING SESSION

Some managers feel that they have no time for training. But good management is about teaching, encouraging, and coaching others. As a result, you must find the time to train. Managers interested in the long-term success of their operations and employees will often set aside time each week to conduct training. Some managers maintain that all of the training in their unit must be on-the-job training. They feel that structured training either takes too long or is inappropriate. In nearly all cases, this is not correct and is a major cause of the rather low rate of productivity so prevalent in some parts of the hospitality industry.

The best training sessions are presented with enthusiasm and an attitude of encouragement. Make sure that training is presented not because employees "don't know" but because management wants them to "know more." Involve employees in the presentation of training exercises. Seek their input in the sessions. Ask questions that encourage discussion and always conclude the sessions on a positive note.

A brief, but effective, plan for each session could be to:

1. Tell the employees what you hope to teach them and why it is important.
2. Present the session.
3. Reemphasize the main training points and discuss why they are important.
4. Ask for questions to ensure trainee understanding.

EVALUATE THE SESSION'S EFFECTIVENESS

There is a saying in education that "if the student hasn't learned, then the teacher hasn't taught." This concept, when applied to hospitality training, implies that simply presenting a training session is not enough. Training should result in change. Either employees' learn, improve, and gain new skills, knowledge, or information, or they have not learned. To ensure learning occurred, you must evaluate your training sessions. This process can be as simple as observing an employee's behavior (to test skill acquisition) or as detailed as preparing written questions for employees to answer (to test knowledge retention and application).

Post-training evaluation should also be directed at how the sessions were conducted. Were they too long? Were they planned well? The evaluation of training is as important as its delivery. Both the content of the session and the delivery itself should be evaluated because employees who are well trained are more productive, are more highly motivated, and provide better service to guests.

RETRAIN AT THE PROPER INTERVAL

Humans, learn, unlearn, and relearn on a regular basis. The old telephone number you might have known so well 10 years ago may now be gone from your memory. Similarly, the friend or teacher's name you knew well at one time and felt you would never forget may be forgotten. In the same way, employees who are well trained in an operation's policies and procedures need to be constantly reminded and updated if their skill and knowledge levels are to remain high.

Performance levels can also decline because of a change in the operational systems you have in place or changes in equipment used. When this is true, you must retrain your employees. Training a workforce is one, if not the best, method of improving employee productivity. Effective training costs a small amount of time in the short run but pays off extremely well in dollars in the long run. Managers who have risen to the top in the hospitality industry have some specific characteristics and traits. Chief among these is their desire to teach and encourage their employees and, thus, get the best results from each and every one of them.

SUPERVISION

All employees require proper supervision. This is not to say that all employees want to be told what to do. Proper supervision means assisting employees in improving productivity. In this sense, the supervisor is a coach and facilitator who provides employee assistance. Supervising should be a matter of assisting employees to do their best, not just identifying their shortcomings. It is said that employees think one of two things when they see their boss approaching:

1. Here comes help!

   or

2. Here comes trouble!

For those supervisors whose employees feel that the boss is an asset to their daily routine, productivity gains are remarkable. Supervisors who only see their positions as one of power, or who see themselves as taskmasters, rarely maintain the quality workforce necessary to compete in today's competitive marketplace.

It is important to remember that it is the employee, not management, who services the guest. When supervision is geared toward helping employees, an operation's guests benefit and, as a result, the operation benefits. This is why, in most foodservice operations, it is so important for managers to be **on-the-floor**, in other words, in the dining or service area, during meal periods. Greeting guests, solving bottleneck problems, maintaining food quality, and ensuring excellent service are all tasks of the foodservice manager during the service period. When employees see that management is committed to providing high-quality customer service and is there to assist employees in delivering that level of service, worker productivity will improve.

SCHEDULING

Even with highly productive employees, poor employee scheduling by management can result in low productivity ratios. Consider the example in Figure 7.3, where management has determined a schedule for pot washers in a unit that is open for three meals per day.

In Schedule A, four employees are scheduled for 32 hours at a rate of $18.00 per hour. Pot washer payroll, in this case, is $576 per day (32 hours/day × $18.00/hour = $576 per day). Each shift, breakfast, lunch, and dinner, has two employees scheduled.

In Schedule B, three employees are scheduled for 24 hours. At the same rate of $18.00 per hour, payroll is $432 per day (24 hours/day × $18.00/hour = $432 per day). Wages, in this case, are reduced by $144 ($576–$432 = $144) per day, and further savings will be realized due to reduced employment taxes, benefits, employee meal costs, and other labor-related expenses. Schedule A assumes that the amount of work to be done is identical at all times of the day. Schedule B covers both the lunch and the dinner shifts with two employees but assumes that one pot washer is sufficient in the early-morning period as well as very late in the day.

When scheduling is done to meet projected demand, productivity ratios will increase. If production standards are to be established and monitored, management must do its job in ensuring that employees are scheduled only when needed to meet the sales or volume anticipated. Returning to our formula for computing the productivity ratio, Figure 7.4 assumes that 600 pots are to be washed, and as a result, shows the effect on the productivity ratio of different scheduling decisions.

Proper scheduling ensures that the correct number of part-time, full-time, or **full-time equivalent (FTE)** employees is available to do the necessary amount of work. An FTE is used to convert the number of hours worked by several part-time employees to the hours worked by one full-time employee (typically 40 hours per week or 8 hours per day). If too many employees are scheduled, productivity ratios will decline. If too few employees are scheduled, customer service levels may suffer, or necessary tasks may not be completed on time or as well as they should be.

Work in a foodservice operation tends to occur in peaks and valleys, and the foodservice manager is often faced with uneven demands regarding the number of employees needed. Different days of the week may also require different levels

FIGURE 7.3 Two Alternative Schedules

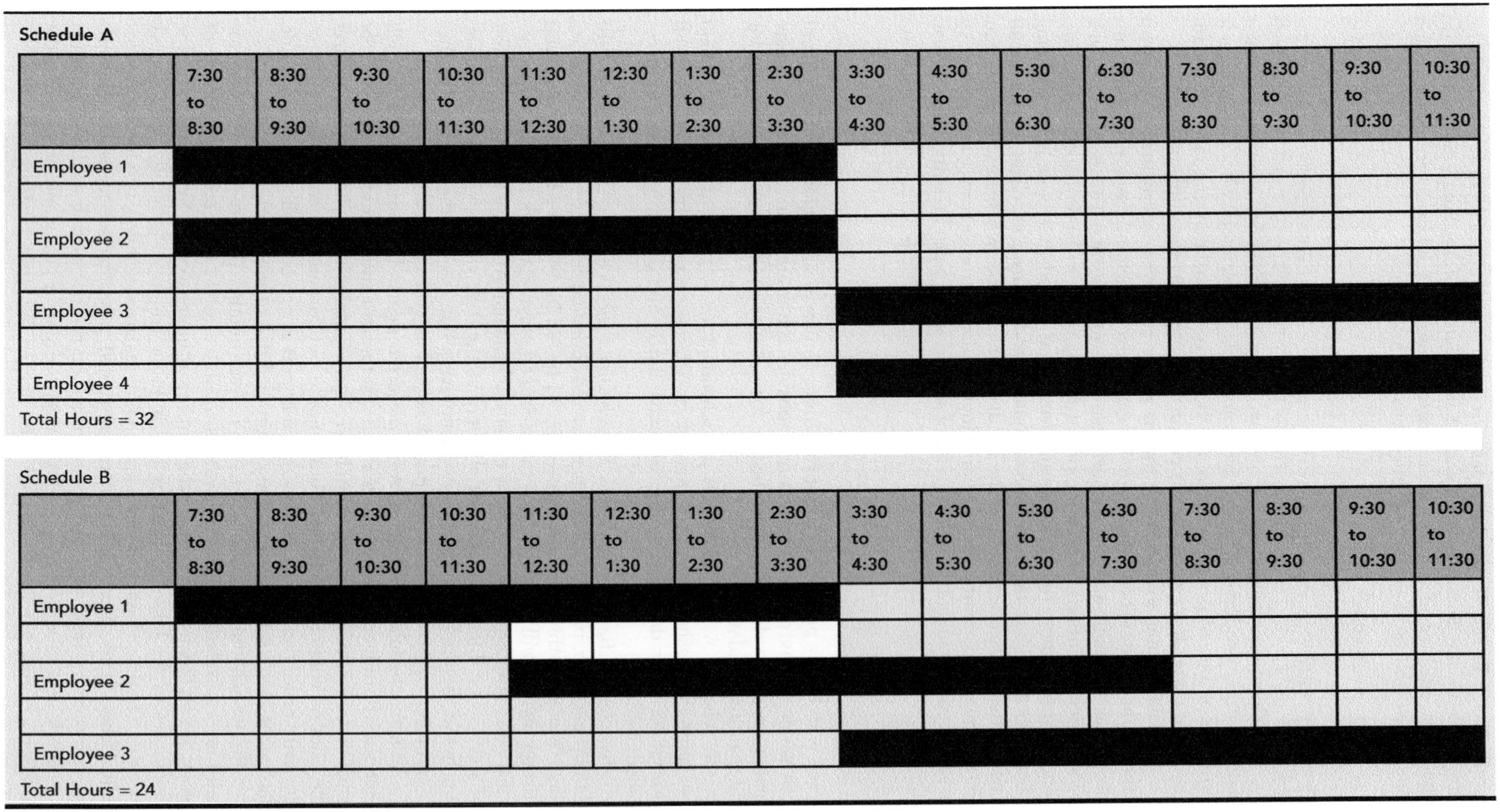

FIGURE 7.4 Effect of Scheduling on Productivity Ratios

Number of Pots to Be Washed	Number of Pot Washers Scheduled	Productivity Ratio
600	4	150 pots/washer
600	3	200 pots/washer

of staffing. In a restaurant located in hotel that caters to business travelers, for example, the slow period might be a weekend when most business travelers are at home rather than staying in the hotel.

In an upscale restaurant, on the other hand, the slow period may be during the week, with volume picking up on the weekends. For the weekend increase in business, the manager might hire part-time employees to handle the higher volume. To further complicate matters, some operations are faced with seasonal variations. In a college dining facility, the summers may be slow or the operation may even be closed, whereas a foodservice operation located in beach resort town may be extremely busy during that same time period.

Demand can also vary from hour to hour because people in the United States tend to eat in three major time periods. In restaurants doing strong lunch business, 5 cooks and 15 servers may be necessary. At 3:00 in the afternoon at the same restaurant, one cook and one server may find themselves with few guests to serve.

Scheduling efficiency during the day can often be improved through the use of the **split-shift**, a technique used to match individual employee work shifts with peaks and valleys of customer demand. In using a split-shift, the manager would, for example, require an employee to work a busy lunch period, be off in the afternoon, and then return to work for the busy dinner period.

Employee scheduling in the hospitality industry can be challenging. It is important, however, that it be done well. Some managers prefer to prepare work schedules using the pen and paper method. This traditional approach to making employee schedules does have its advantages. For one, it requires only a pen and a piece of paper to produce! Nor does this system require any familiarity with computers or software. Such a system, however, makes it harder to distribute the schedule to all workers and even harder to communicate modifications of it.

Increasingly, foodservice managers can utilize **cloud-based employee scheduling**. Cloud-based employee scheduling systems are designed to allow managers to put the right number of workers in the right shifts and on the right days. Most of these systems also include features that (with management's pre-approval) allow employees to pick up, drop, or swap shifts with other workers, giving staff more flexibility about when, and how much, they will work. This type of work flexibility is most often highly valued by employees and can contribute to reduced turnover rates caused by inconvenient worker scheduling.

As with any new computer program, the use of cloud-based employee scheduling does require a learning curve. Its use also assumes that all workers have Internet access via computer or smart device. Advantages of using cloud-based employee scheduling systems are many, however, and include the following:

- Recording and incorporating workers' time-off requests and scheduled vacations
- Instantaneously distributing the initial schedule to all employees
- Allowing any schedule changes to be promptly distributed 24/7
- Identifying and immediately notifying employees who have agreed to be available to replace workers who have called in sick or were no-shows

- Flagging for management any workers who have been scheduled at high-cost overtime wage rates
- Providing payroll metrics such as projected total worker hours scheduled, hours worked per employee, and total payroll costs

Regardless of the employee scheduling system used, productivity standards help the foodservice operator meet their goal of matching projected workload to the number of needed employees.

FUN ON THE WEB!

In the past, creating employee work schedules was one of a foodservice manager's most complex and time-consuming tasks. Balancing employee availability with requests for preferred shifts, time off, and vacations, while meeting operational staffing needs that could change daily (and doing so in a way all employees feel is fair!) was a real challenge. Fortunately, advances in technology allow today's managers to get much needed help in this complex area.

To see how one creative company's employee scheduling app saves foodservice managers time and money, while increasing employee satisfaction, enter "HotSchedules" in your favorite search engine. HotSchedules makes it very easy for managers to create cost-effective employee schedules, but it also makes it easy for employees to access their schedules and even trade shifts with their coworkers using their own mobile devices.

BREAKS

Most employees simply cannot work at top speed for 8 hours straight. They have both a physical and a mental need for breaks from their work. These breaks give them a chance to pause, collect their thoughts, converse with their fellow employees, and, in general, prepare for the next work session. Employees who are given frequent short breaks will outproduce those who are not given any breaks. It simply makes no sense, then, for management to behave as if they begrudge their staff the breaks that are so beneficial to the organization, as well as to the employee.

Federal law in the United States does not mandate that all employees be given breaks, but some states do. As a result, foodservice supervisors often must determine both the best frequency and the best length of designated breaks. In some cases, however, and especially regarding the employment of students and minors, both federal and state laws may mandate special workplace requirements for employees. As a professional manager, you will need to be familiar with these laws if they apply to your operation.

FUN ON THE WEB!

Visit the website of the department or agency responsible for enforcing youth employment laws in your state. To do so, enter "child labor laws" and the name of your state in your favorite search engine, then view the results to learn about any special laws or regulations that apply to minors working in that state.

MORALE

Employee morale is not often mentioned in discussions about controlling foodservice labor costs. Yet, as experienced managers will attest, it is impossible to overestimate the financial value of a highly motivated employee or crew. Although it is a truism that employees must motivate themselves, it is also true that effective managers can create an environment that makes it easy for employees to be motivated.

History is filled with examples of teams and groups who have achieved goals that seemed impossible but were accomplished because they were highly motivated. Professionally serving guests should be fun. It should be exciting. If this sense of fun and excitement can be instilled in foodservice employees, their work also becomes fun and exciting.

Work groups with high morale share common traits. In general, these groups work in an environment where:

1. A positive vision has been created.
2. The vision is constantly communicated to employees.
3. The vision is shared and embraced by both management and employees.

Creating a positive vision is nothing more than finding a "purpose" for the workforce. Any manager who communicates that the purpose for a pot washer is simply to clean pots cannot expect to have a fired-up, turned-on employee. Yet, pot washers can have high morale. They can be a critical part of management's overall purpose for the work crew. Consider, for purposes of illustration, some of the following techniques you could use to communicate a customer service vision to your pot washers, a group of workers not normally viewed as one having direct customer contact:

1. If your unit has been, as it should be, free from cases of foodborne illness, part of the credit goes to your pot-washing staff. Recognize them for this achievement on a quarterly basis.
2. Conduct regular pot-washing station inspections. Score the area for cleanliness on a scale of 1 to 100. Present each pot washer with a certificate when the area score exceeds 90. If it does not exceed 90, increase training until it does!
3. Recognize your best pot washer(s) at an annual employee recognition luncheon or dinner for:
   a. Best attendance
   b. Best productivity
   c. Cleanest work area
   d. Most improved
   e. Most thorough
   f. Most often in proper uniform
   g. Safest worker
4. Include a pot washer on your safety committee. Publicly recognize all safety committee members on a regular basis.
5. Make it a point to go to the pot-washing area on a daily basis to thank those employees for a job well done. Emphasize the importance of their efforts.
6. Encourage food production employees (your pot washers' internal customers) to take the time to thank the pot washers for their hard work and valuable contributions to the food production process.

If the purpose or vision that management proposes is only a financial profit for the organization or its owners, employees will rarely share the vision, even if it is strongly communicated. If management's vision, however, includes the vital importance of each employee's efforts, then each employee can share that vision. A shared purpose between management and employee is important for the development and maintenance of high morale. It is just not enough for management to feel that employees should be "glad to have a job." This type of attitude by management most often results in high employee turnover and lost productivity.

Employee turnover can be high in some sections of the hospitality industry. By some estimates, in the foodservice industry it exceeds 125 percent per year, with average "hard costs" (separation, replacement, and training) to replace hourly staff at \$2,305, managers (non-general managers) at \$10,518, and general managers at \$16,770.[1]

You can measure turnover in your own operation by using the following formula:

$$\frac{\text{Number of employees separated}}{\text{Number of employees in workforce}} = \text{Employee turnover rate}$$

For example, assume that you have a total of 50 employees in your foodservice operation. In the past year, you replaced 35 employees. Your turnover rate is computed as follows:

$$\frac{\text{35 employees separated}}{\text{50 employees in workforce}} = \text{70\% turnover rate}$$

Employee separation is the term used to describe employees who have either quit, been terminated, or in some other manner have separated themselves from the operation. The number of employees in the workforce refers to the average number of people employed by the operation during a given time period. It is computed by adding the number of employees at the beginning of the time period being evaluated to the number of employees at the end of the same period and dividing the sum by 2.

HERE'S HOW IT'S DONE 7.1

The formula used to calculate employee turnover rate is a simple one:

$$\frac{\text{Number of employees separated}}{\text{Number of employees in workforce}} = \text{Employee turnover rate}$$

The number of employees separated is the numerator in the formula. It simply refers to the number of employees that have left the organization during a defined accounting period, such as a month or a year.

The number of employees in workforce is the denominator in the formula. It represents the *average* number of workers employed in an organization. For that reason, managers calculating their employee turnover rates must identify the number of employees on staff at the beginning of the accounting period and the number on staff at the end of the accounting period. They then divide by two (2) to get the average number of employees on staff.

For example, in an organization that employed 40 people at the beginning of the year but, due to increased sales, employed 50 people at the end of the year, the calculation for the average number of employees in the workforce would be:

$$\frac{\text{40 (beginning or the period employees)} + \text{50 (endof the period employees)}}{2} = \text{45 number of employees in workforce}$$

If 18 employees were replaced during the year, the employee turnover rate for this operation would be:

$$\frac{18}{45} = \text{40\% turnover}$$

Using the *average* number of employees in the workforce when calculating an operation's turnover rate makes the turnover rate formula more accurate because it accounts for increases or decreases in the total number of staff members employed in the operation during the time period for which the employee turnover rate is being calculated.

[1] https://blackboxintelligence.com/news/state-of-restaurant-workforce-2024. Retrieved 10/31/2025.

Some foodservice operators prefer to distinguish between voluntary separation and involuntary separation. A **voluntary separation** is one in which the employee made the decision to leave the organization. This may be due, for example, to retirement, a better opportunity in another organization, or relocation to another town. An **involuntary separation** is one in which management has caused the employee to separate from the organization. This may have been, for example, terminating the employee because of poor attendance or violation of procedures or policy or as a result of a reduction in the workforce.

Some managers want to know the amount of turnover that is voluntary compared to that which was involuntary. For example, excessively high voluntary turnover rates may be a signal that wages and salaries are too low to keep your best employees. High involuntary turnover rates may mean that employee screening and selection techniques need to be reviewed and improved. If it is your preference, you can modify the turnover formula to create these two ratios:

$$\text{Involuntary employee turnover rate} = \frac{\text{Number of employees involuntarily separated}}{\text{Number of employees in workforce}}$$

$$\text{Voluntary employee turnover rate} = \frac{\text{Number of employees voluntarily separated}}{\text{Number of employees in workforce}}$$

Whether separation is involuntary or voluntary, turnover is expensive. In both cases, an operation will incur employee turnover costs that are both actual and hidden. Actual costs, such as those involved in advertising the position vacancy, interviewing, and new employee training time required, are easy to determine. The hidden costs are harder to quantify but can cost dearly in an operation with a high turnover rate. Increased china and glassware breakage may result as a new dishwasher learns the job. A new server may provide slower customer service, or a new cook may cause an increase in food waste or in the number of improperly prepared menu items.

Good foodservice managers regularly calculate and closely monitor their employee turnover rates. High turnover rates can mean trouble. Low turnover rates most often mean that employee morale is high and workers feel good about the operations that employ them.

MENU

A major factor in employee productivity is the foodservice operation's actual menu. The items that you select to serve on your menu will have a significant effect on your employees' ability to produce the items quickly and efficiently.

In most cases, the greater the number of menu items a kitchen is asked to produce, the less efficient that kitchen will be. Of course, if management does not provide the guest with enough choices, loss of sales may result. Clearly, neither too many nor too few menu choices should be offered. The question for management is "How many are too many?" The answer depends on the operation, the skill level of employees, and the level of menu item variety management feels is necessary to properly attract and service its guests.

Menus that continuously expand are costly in many ways. The quick-service restaurant (QSR) that elects to specialize only in hamburgers can prepare them quickly and efficiently. If the same restaurant decides to add pizza, tacos, salads, tofu wraps, and fried chicken strips, its management may find that not only are employees less productive but also guests are confused as to what the operation really is. Again, the dilemma management faces is how to serve an appropriately large variety of menu items, but not so many as to significantly reduce employee productivity.

Although the number of menu items produced is important, so is the type of item. Obviously, a small diner with one deep-fat fryer will have production problems if the day's specials are fried fish, fried chicken, and a breaded fried vegetable platter! Menu items must be selected to match the skill level of the employees as well as the equipment available to produce the menu items. Most foodservice operations change their menu fairly infrequently. Print and signage costs are often high, and restaurateurs are most often reluctant to radically change their product offerings. Thus, it is extremely important that the initial menu items selected by management are items that can be prepared efficiently and serviced well. If this is done, productivity rates will be high and so will guest satisfaction.

CONVENIENCE FOOD USE VERSUS SCRATCH PREPARATION

Few, if any, foodservice operators today make all of their menu items from scratch. Indeed, there is no real agreement among foodservice operators as to what "scratch cooking" or "on-site cooking" truly is. Canned fruits, frozen seafood, and ready-to-bake pastries are examples of foods that would not be available to many guests if it were not for the fact that they were processed to some degree before they were delivered to the foodservice operator's door. At one time, even presliced white bread was considered to be a convenience food by some in foodservice.

Other foods, such as canned cheese sauce, can be modified by an operator to produce a unique item. This can be done by the addition of special ingredients such as salsa or jalapeno peppers, according to the standardized recipe, with the intent of creating a unique product served only by that foodservice operation.

The decision of whether to make or buy a specific menu item most often involves two major factors. The first is, of course, product quality. In general, if an operation can make a product that is superior to the one it can buy from a supplier, it should produce that item. Product cost is the second factor, and it is also a major issue. It is possible that management determines that a given menu item can be made in-house and that it is a superior product. The cost of producing that item, however, may be so great that it is simply not cost-effective to do so. Fortunately, convenience products are becoming more quality-driven and less expensive due to advances in technology and increased competition among the major food suppliers.

To illustrate the impact on labor costs of the make *versus* buy decision, consider the situation that would arise if you managed a quick-service Mexican restaurant. One of the items on your menu is frijoles, cooked and seasoned pinto beans, which are mashed prior to serving.

Assume that you use 50 edible portion (EP) pounds of these beans per day. You can buy the seasoned beans in cans for $1.45 cents per pound. Your EP cost per pound with the canned product includes only the cost of the beans. In addition, you would incur the labor cost required to open the cans and heat the beans, as well as the cost of cooking fuel.

If you were to make the frijoles from scratch, your *food* cost would go down to 53 cents per pound; the $22.50 you would pay for dried pinto beans and $4.00 for seasonings = ($22.50 + $4.00)/50 pounds = $0.53 per pound.

In this example, it appears that making this menu item from scratch would save your operation money. The complete story, however, can only be viewed when you also consider the labor cost required to produce the frijoles.

Figure 7.5 details the hypothetical costs involved in the decision that you must make, assuming a usage of 50 pounds of product per day and a labor cost of $22.00 per hour to both cook the beans and clean up the production process.

As you can see in this example, you would experience a reduction in your cost of food if you made the frijoles from scratch but an *increase* from $22.00 to $66.00 in the cost of labor because your own employees must now complete the cooking process. In the case of frijoles, your decision may well be to purchase the convenience item, if it is of acceptable quality, rather than make it from scratch because the *overall* food and labor cost would be less.

FIGURE 7.5 Frijoles: 50 Pounds

Component	Cost of Convenience Product	Cost of Scratch Product
Beans (canned)	$72.50 ($1.45/lb.)	
Beans (dry)		$22.50 ($1.35/lb.)
Seasoning	$0	$4.00
Labor	$22.00 (1 hour)	$66.00 (3 hours)
Cooking fuel	$2.80	$8.80
Total cost	$97.30	$101.30
Cost per pound	$1.95	$2.03

Management, often in consultation with kitchen production staff, must address a large number of make-or-buy decisions. It is important to note, however, that these decisions affect both food *and* labor costs. One cannot generally achieve significant food cost savings for items without expending additional labor dollars. Conversely, when a manager elects to buy a convenience item, rather than make the item from scratch, food costs tend to rise, but labor costs most often should decline.

In general, labor productivity in your operation will rise when you elect to buy, rather than make from scratch, any item that you cannot produce efficiently. This may be due to the specialized skills required, as is the case with some purchased gourmet bakery items, or it could simply be a case of your supplier having the tools and equipment necessary to do a time-consuming task at a great savings to you, such as the case of buying a prechopped, frozen diced onion. It is important, however, that you do not fall into the trap of electing to buy more convenience-type items without reducing your labor expenditures. When that happens, you lose in terms of both higher food costs and higher-than-required labor costs.

EQUIPMENT/TOOLS

In most cases, foodservice productivity ratios have not increased as much in recent years as have those of other businesses. Much of this is due to the fact that we are a labor-intensive, rather than machine-intensive, industry. In some cases, equipment improvements have made kitchen work easier. Slicers, choppers, and mixers have replaced human labor with mechanical labor. However, robotics and automation are not yet a part of our industry in any major way. Nonetheless, it is critical for you to understand the importance of a properly equipped workplace and how that improves productivity. This can be as simple as understanding that a sharp knife cuts more safely, quickly, and better than a dull one or as complex as deciding which wireless system will be used to provide data and communication links to the 1,000 stores in a QSR chain. In either case, management must ask itself a fundamental question: "Am I providing my employees with the tools necessary to do their job effectively?" The key word in that question is "effectively." If the proper tools are provided but they are mounted at the wrong height, placed in the wrong location, or unavailable at the right time, the tools will not be used effectively. Similarly, if the proper tools are provided but employees are not adequately trained in their use, productivity gains will not occur. In all cases, it is part of your job to provide your employees with the tools they need to do their jobs quickly and effectively.

SERVICE LEVEL DESIRED

It is simply a fact that the average QSR employee can normally serve more guests in an hour than the fastest server at an exclusive fine-dining restaurant. The reason for this is quite obvious. In a QSR, it is speed, not total level of service rendered, that is of the utmost importance to guests. In a fine-dining operation, service is more elegant, and the personal service delivered is of a much higher level. Thus, when you vary service levels, you also vary employee productivity ratios.

In the past, foodservice managers focused very heavily on speed of service. Although that is still important today, many operators are finding that guests expect and demand higher levels of service than ever before. If this trend continues, one could expect that foodservice productivity levels will tend to go down. To prevent this from happening, foodservice operators will need to become very creative in finding ways to improve employee productivity in other areas (e.g., through training and improved morale) so that these "savings" can be used to provide the higher level of customer service demanded by today's sophisticated foodservice consumer.

Now that we have closely examined some key factors that directly impact employee productivity and have addressed what you can do to affect them, we return to the question of "How many employees are needed?" to effectively operate a foodservice unit. The key to answering that question lies in developing and applying productivity standards.

There are several measures of employee productivity used in the foodservice industry. You will learn next about the most commonly used measures and examine their weaknesses and strengths. In the final analysis, the best productivity measure for any unit you manage is the one that makes the most sense for your unique operation.

MEASURING CURRENT LABOR PRODUCTIVITY

There are a variety of ways managers measure productivity in the hospitality industry.

Six of the most commonly used are as follows:

1. Labor cost percentage
2. Sales per labor hour
3. Labor dollars per guest served
4. Guests served per labor dollar
5. Guests served per labor hour
6. Revenue per available seat hour (RevPASH)

LABOR COST PERCENTAGE

Perhaps the most commonly used measure of employee productivity in the foodservice industry is the **labor cost percentage**. In its simplest form it is computed as follows:

$$\frac{\text{Cost of labor}}{\text{Total sales}} = \text{Labor cost \%}$$

A labor cost percentage allows an operation's managers to measure the relative cost of labor used to generate a known quantity of sales. It is important to realize, however, that there are actually several ways managers could define "cost of labor." You should select the one that makes the most sense for your own operation.

As you have learned, an operation's total cost of labor includes wages, salaries, and benefits. To measure productivity, you may choose, for example, to include only hourly staff wages or staff wages and salary costs but not benefit costs. This approach makes sense if, for example, in your operation you have the ability to control only your payroll costs and not your total labor costs. Remember, however, that when comparing your labor cost percentage with those of other foodservice units, it is important that you make sure both your unit and the one you are comparing use the same formula for calculating the labor cost percentage. If you do not, your comparisons may not accurately reflect true differences between the units.

Controlling the labor cost percentage is extremely important in the foodservice industry because it is often used to assess the effectiveness of a manager. If labor cost percentage increases beyond what is expected, management will likely be held accountable by the operation's ownership.

Labor cost percentage is a popular measure of productivity, in part because it is so easy to compute and analyze. To illustrate, consider the case of Roderick, a foodservice manager in charge of a table-service restaurant in a year-round theme park. The unit is popular and has a $40 per guest check average. Roderick uses only payroll (staff wages and management salaries) when determining his overall labor cost percentage because he does not have easy access to the actual cost of taxes and benefits provided to his employees. Roderick's own supervisor considers these labor-related expenses to be non-controllable and thus beyond Roderick's immediate influence.

Roderick has computed his labor cost percentage for each of the last 4 weeks using the labor cost percentage formula. His supervisor has given Roderick a goal of 35 percent for the 4-week period. Roderick feels that he has done well in meeting that goal. Figure 7.6 shows Roderick's 4-week performance.

Using the labor cost percentage formula and the data in Figure 7.6, Roderick's labor cost percentage is calculated as follows:

$$\frac{\text{Cost of labor}}{\text{Total sales}} = \text{Labor cost \%}$$

$$\frac{\$29{,}330}{\$83{,}800} = 35\ \%$$

While Roderick did achieve a 35 percent labor cost for the 4-week period, Madeline, his supervisor, is concerned because she received many negative comments in week 4 regarding poor service levels in Roderick's unit. Some of these were even posted online and Madeline is concerned about the postings' potential impact on future visitors to the park's foodservice operations. When she carefully analyzes the numbers in Figure 7.6, she sees that Roderick far exceeded his goal of a 35 percent labor cost in weeks 1 through 3 and then reduced his labor cost to 27.9 percent in week 4.

FIGURE 7.6 Roderick's 4-Week Labor Cost % Report

Week	Cost of Labor	Total Sales	Labor Cost %
1	$14,200	$36,800	38.6%
2	$16,100	$43,000	37.4%
3	$14,516	$38,200	38.0%
4	$13,844	$49,600	27.9%
Total	**$58,660**	**$167,600**	**35.0%**

Although the monthly overall average of 35 percent is within budget, she knows all is not well in this unit. Roderick elected to reduce his payroll in week 4, and yet it is clear from the negative guest comments that, at a 27.9 percent labor cost, service to his guests suffered. That is, too few employees were on staff to provide the necessary guest attention. As you can see, one disadvantage of using an overall labor cost percentage is that it can hide daily or weekly highs and lows.

In Roderick's operation, labor costs were too high in the first 3 weeks, and too low in the last week, but he still achieved his overall target of 35 percent. Recall from the discussion about cost percentages in Chapter 1 that the total labor cost of 35 percent indicates that, for each dollar of sales generated, 35 cents was paid to the employees who assisted in generating those sales. In many cases, a targeted labor cost percentage is viewed as a measure of employee productivity and, to some degree, management's skill in controlling labor costs.

In addition to its tendency to mask productivity highs and lows, labor cost percentage has some additional limitations as a measure of productivity. Notice, for example, what happens to this measure of productivity if all of Roderick's employees are given a 5 percent raise in pay. If this were the case, Roderick's labor cost percentages for last month would be calculated as shown in Figure 7.7.

Note that labor now accounts for 36.8 percent of each sales dollar. It is important to realize that Roderick's workforce did not become *less* productive simply because they got a 5 percent increase in pay. Rather, the labor cost percentage changed due to a change in the price paid for labor. When the price paid for labor increases, labor cost percentage increases. When the price paid for labor decreases, labor cost percentage decreases. Because of this, using labor cost percentage alone to evaluate workforce productivity can sometimes be misleading.

For another example of the limitations of labor cost percentage as a measure of labor productivity, consider the effect on labor cost percentage of increasing selling prices. Return to the data in Figure 7.6 and assume that Roderick's unit raised all menu prices by 5 percent prior to the beginning of the month. Figure 7.8 shows how an increase of this size in his selling prices would affect his labor cost percentage.

Note that increases in selling prices (assuming no decline in guest count or changes in guests' buying behavior) will result in decreases in the labor cost percentage. Alternatively, lowering selling prices without increasing total revenue will result in an increased labor cost percentage.

Although labor cost percentage is easy to compute and widely used, it is difficult to use as a measure of productivity over time because it depends on labor dollars spent and sales dollars received for its computation. Even in relatively non-inflationary times, wages do increase and menu prices are adjusted. Both activities directly affect labor cost percentage, but not worker productivity. In addition, institutional foodservice settings, which often have no daily dollar sales figures to report, can find that it is not easily possible to measure labor productivity using labor cost percentage because they generally calculate and report guest counts or number of meals served rather than sales dollars earned.

FIGURE 7.7 Roderick's 4-Week Revised Labor Cost % Report (Includes 5% Pay Increase)

Week	Original Cost of Labor	5% Pay Increase	Cost of Labor (with 5% Pay Increase)	Total Sales	Labor Cost %
1	$14,200	$710.00	$14,910.00	$36,800	40.5%
2	$16,100	$805.00	$16,905.00	$43,000	39.3%
3	$14,516	$725.80	$15,241.80	$38,200	39.9%
4	$13,844	$692.20	$14,536.20	$49,600	29.3%
Total	**$58,660**	**$2,933.00**	**$61,593.00**	**$167,600**	**36.8%**

FIGURE 7.8 Roderick's 4-Week Revised Labor Cost % Report (Includes 5% Increase in Selling Price)

Week	Cost of Labor	Original Sales	5% Selling Price Increase	Sales (with 5% Selling Price Increase)	Labor Cost %
1	$14,200	$36,800	$1,840	$38,640	36.7%
2	$16,100	$43,000	$2,150	$45,150	35.7%
3	$14,516	$38,200	$1,910	$40,110	36.2%
4	$13,844	$49,600	$2,480	$52,080	26.6%
Total	**$58,660**	**$167,600**	**$8,380**	**$175,980**	**33.3%**

SALES PER LABOR HOUR

It has been said that the most perishable commodity any foodservice operator buys is the labor hour. When labor is not productively used, it disappears forever. It cannot be carried over to the next day, like an unsold head of lettuce or a slice of turkey breast. It is for this reason that some foodservice operators prefer to measure labor productivity in terms of the amount of sales generated for each labor hour used. This productivity measure is called **sales per labor hour**. The formula for computing this measure of labor productivity is as follows:

$$\frac{\text{Total sales}}{\text{Labor hours used}} = \text{Sales per labor hour}$$

In this formula, labor hours used is simply the sum of all labor hours paid for by the operation in a specific sales period. Consider Roderick's labor usage and the resulting sales per labor hour information presented in Figure 7.9.

In this example, sales per labor hour ranged from a low of $39.00 in week 1 to a high of $57.32 in week 4. Operators who compute sales per labor hour do so because they feel it is a more accurate measure of labor productivity than is the labor cost percentage.

Sales per labor hour will vary with changes in selling prices (as does the labor cost percentage), but it will not vary based on changes in the price paid for labor. In other words, increases and decreases in the price paid per hour of labor will not affect this productivity measure. On the negative side, however, sales per labor hour neglects to consider the hourly amount paid to employees per hour to generate the sales. As a result, a foodservice unit paying its employees an average of $24.00 per hour could, using this type of measure for labor productivity, have the same sales per labor hour as a similar unit paying $30.00 for each hour of labor used. Obviously, the manager paying $24.00 per hour has paid far less for an equally productive workforce if the sales per labor hour used are identical in the two units.

Many managers, particularly those managing commercial foodservice operations, like utilizing the sales per labor hour productivity measure because records on both the

FIGURE 7.9 Roderick's 4-Week Sales per Labor Hour

Week	Total Sales	Labor Hours Used	Sales per Labor Hour
1	$36,800	943.5	$39.00
2	$43,000	1,006.3	$42.74
3	$38,200	907.3	$42.10
4	$49,600	865.3	$57.32
Total	**$167,600**	**3,722.4**	**$45.02**

numerator (total sales) and the denominator (labor hours used) are readily available. However, depending on the recordkeeping system employed, it may be more difficult to determine total labor hours used than total labor dollars spent. This is especially true when large numbers of employees are paid by salary rather than by the hour. In most cases, the efforts of both salaried managers and hourly paid staff should be considered when computing an operation's overall sales per labor hour.

LABOR DOLLARS PER GUEST SERVED

Had Roderick preferred, he might have measured his labor productivity in terms of the labor dollars spent per guest served his operation. This productivity measure is called **labor dollars per guest served**. The formula for this measure is as follows:

$$\frac{\text{Cost of labor}}{\text{Guests served}} = \text{Labor dollars per guest served}$$

Using Roderick's data, the labor dollars per guest served computation would be as shown in Figure 7.10.

In this example, the labor dollars expended per guest served for the 4-week period would be computed as follows:

$$\frac{\$58{,}660}{4{,}190} = \$14.00$$

Using this measure of productivity, it is fairly easy to see why Roderick experienced guest complaints during the fourth week of operations. Note that in the first 3 weeks, he supplied his guests with more than $14.00 of guest-related labor costs per guest served, but in the fourth week that amount fell to less than $12.00 per guest.

As is the case with labor cost percentage, labor dollars per guest served is limited in that it will vary based on the price paid for labor. It is not, however, affected by changes in menu prices.

GUESTS SERVED PER LABOR DOLLAR

A variation on the formula of labor dollars per guest served is to reverse the numerator and denominator to create a new productivity measure, which is **guests served per labor dollar**. The formula for this measure of labor productivity is as follows:

$$\frac{\text{Guests served}}{\text{Cost of labor}} = \text{Guests served per labor dollar}$$

FIGURE 7.10 Roderick's 4-Week Labor Dollars per Guest Served

Week	Cost of Labor	Guests Served	Labor Dollars per Guest Served
1	$14,200	920	$15.44
2	$16,100	1,075	$14.98
3	$14,516	955	$15.20
4	$13,844	1,240	$11.16
Total	**$58,660**	**4,190**	**$14.00**

FIGURE 7.11 Roderick's 4-Week Guests Served per Labor Dollar

Week	Guests Served	Cost of Labor	Guests Served per Labor Dollar
1	920	$14,200	0.065
2	1,075	$16,100	0.067
3	955	$14,516	0.066
4	1,240	$13,844	0.090
Total	4,190	**$58,660**	0.072

Had Roderick wanted to use the guests served per labor dollar, his labor productivity data could have been calculated as is presented in Figure 7.11.

In this situation, Roderick served, for the 4-week average, a total of 0.072 guests for each labor dollar expended. As a measure of labor productivity, guests served per labor dollar spent has several advantages. First, it can be used by foodservice units, such as non-commercial institutions, which do not routinely record dollar sales figures. Also, this measure is relatively easy to compute because you are very likely to keep records, on a daily basis, of the total number of guests that have been served.

GUESTS SERVED PER LABOR HOUR

One good way to evaluate a table-service restaurant's labor efficiency is the number of guests that can be seated and served in a meal period. As **table turns** (the number of times a single table is reset and used for new guests) increases, labor efficiency increases. As a result, **guests served per labor hour** is another powerful and popular measure of labor productivity in table-service restaurants, QSRs, and fast casual restaurants.

The formula used to calculate guests served per labor hour is as follows:

$$\frac{\text{Guests served}}{\text{Labor hours used}} = \text{Guests served per labor hour}$$

The guests served per labor hour productivity measure is especially powerful because it includes neither dollar sales figures nor labor dollar expense in its computation. That means it is free from variations due to changes in menu selling prices and changes in the price paid for labor.

Guests served per labor hour is strong in its ability to measure productivity gains across time due to changes that are unrelated to selling prices or wages paid. It is also extremely useful in comparing similar operating units in areas with widely differing wage rates or menu item selling prices. Thus it is popular with multiunit corporations comparing operational units in diverse geographic regions. It is also useful in comparing dissimilar facilities with similar wages and selling prices because it can help identify areas of weakness in management scheduling, employee productivity, facility layout and design, or other factors that can affect labor productivity.

Had Roderick elected to evaluate his workforce productivity through the use of the guests served per labor hour formula, his data would look like that shown in Figure 7.12.

As the data shows, Roderick's guests served per labor hour ranges from a low of 0.975 guests per hour (week 1) to a high of 1.433 guests per hour (week 4). The average for the 4-week period is 1.126 guests served per labor hour (4,190 guests served/3,722.4 hours used = 1.126 guests per labor hour). Note that, in week 4, the number of guests served per worker hour used was nearly 50 % higher than

FIGURE 7.12 Roderick's 4-Week Guests Served per Labor Hour

Week	Guests Served	Labor Hours Used	Guests Served per Labor Hour
1	920	943.5	0.975
2	1,075	1,006.3	1.068
3	955	907.3	1.053
4	1,240	865.3	1.433
Total	4,190	3,722.4	1.126

in previous weeks, the result of which was poor guest service and the source of Roderick's increased customer complaints.

Managers who use guests served per labor hour as a measure of productivity generally do so because they like its focus on service levels and not merely costs. However, it may be more difficult and time-consuming to compute this measure of productivity because you must compute the number of labor hours used as well as make decisions on how to define a guest. For example, in an outdoor café, a guest who orders a cup of coffee is indeed a guest, but he or she requires much less service than one who consumes a full meal. Unless you decide differently, however, the guests served per labor hour productivity measure would treat these two guests in the same manner.

Consider the Costs

"You wanted to see me, sir?" said Francis to the clearly agitated guest seated at the six-top table in the corner of Chez Lapin, the upscale French Bistro-style restaurant that Francis managed.

"I've been waiting 10 minutes for my waiter to bring us our check. And as slow as he's been, it will probably take another 10 minutes to process my credit card. I just want to pay and leave. The food was fine, but this service is ridiculous!"

"I'm really sorry, sir. I'll find your server," replied Francis, as he glanced around the dining room. As he did, he noticed several unbussed tables that were littered with dirty dishes, as well as the host stand where the line of guests waiting to be seated hadn't gotten any smaller in nearly an hour.

When Francis entered the kitchen looking for the disgruntled guest's server, he was surprised to see several of the line cooks relaxing on the production line.

"How's it going back here tonight?" asked Francis, as he glanced around the kitchen.

"No problems. Just waiting for the orders to come in boss," replied Sasha, the sous chef in charge of the production line. "We're keeping up easily."

Assume that all of the workers at Chez Lapin are well trained and highly motivated:

1. Do you think the servers are likely doing their best to provide good service to the restaurant's guests? If you believe so, then why was the guest in this scenario unhappy?
2. What do you think is the cause of a consistently long line of waiting diners when there are numerous vacant, but unbussed, tables in the dining room?
3. The sous chef in this case said, "We're keeping up easily." Do you think that means they are being very efficient and, thus, very productive? Explain your answer.

REVENUE PER AVAILABLE SEAT HOUR (REVPASH)

The final assessment of productivity that you should know about is also one of the newest. Dr. Sherri Kimes[2] has developed and advocates the use of **revenue per**

[2] https://www.sherrikimes.com. Retrieved 5/25/2025.

available seat hour (RevPASH) as a way to measure the efficiency with which commercial restaurants manage their operations. RevPASH helps managers evaluate both how much guests purchase and how quickly they are served. It does so primarily by assessing the average sales per guest (check average) you learned about in Chapter 2 and the average duration of guests' dining experiences. Duration, in this case, is defined as the length of time customers occupy a seat or table.

Those managers who utilize RevPASH recognize that, from one perspective, a table-service restaurant does not sell products as much as it rents space. As a result, as the number of spaces (seats) is rented increases, the opportunity for sales increases as well. Thus, the foodservice operation that can maximize table turns will have more opportunities to sell than one that does not.

Although you will not generally have the ability to directly control the menu items your guests will purchase, the time it takes them to eat, or how long they linger after eating, RevPASH can give you a good idea of the speed at which your kitchen produces your menu items and the time it takes your service personnel to deliver them.

To calculate RevPASH, you must be able to identify the number of diners you serve each hour, as well as the amount these guests spend (revenue). Typically, this information is easily retrieved from your operation's point of sale (POS) system. In addition, you must know the number of dining room seats available to your guests as well as the number of hours those seats were available. The formula used to calculate this measure of productivity is as follows:

$$\frac{\text{Revenue}}{\text{Available seat hours}} = \text{Revenue per available seat hour (RevPASH)}$$

To illustrate, assume that Roderick's operation had 100 seats and that it was open for dinner from 5:00 P.M. to 10:00 P.M. Figure 7.13 shows the revenue his operation generated during each of the hours he was open last Friday.

Roderick calculates the total available seat hours in his operation simply by taking the number of seats available for guests and multiplying that number by the number of hours the seats are available: (available seats × hours of operation = available seat hours). Note that, on average, Roderick sells 80 percent of his seats at a RevPASH of $40.70.

His operation is most efficient from 7:00 P.M. to 10:00 P.M. as can be determined by a higher than average ($40.70) RevPASH for each of these hours. If

FIGURE 7.13 Revenue per Available Seat Hour (RevPASH)

				For: Last Friday Night
Hour	Available Seats	Guests Served	Revenue	RevPASH
4–5 P.M.	0	0	$0.00	$0.00
5–6 P.M.	100	25	$1,500.00	$15.00
6–7 P.M.	100	75	$3,700.00	$37.00
7–8 P.M.	100	100	$5,200.00	$52.00
8–9 P.M.	100	100	$5,150.00	$51.50
9–10 P.M.	100	100	$4,800.00	$48.00
10–11 P.M.	0	0	$0.00	$0.00
Total	500	400	$20,350.00	$40.70
% Seats Sold = 400/500 = 80%				

Roderick could add incentives to attract or move diners to earlier times (5:00 P.M. to 6:00 P.M.) or perhaps stay open later (10:00 P.M. to 11:00 P.M.), an increase in his RevPASH and his potential profit could be considerable.

Also, note that if Roderick could increase his seat turnover (i.e., if in 1 hour he served 120 guests in 100 seats), his operation's RevPASH could also increase significantly. It is important to realize that RevPASH does not directly utilize the price paid for labor in its evaluation of operational efficiency. Managers using this measure, however, gain valuable information about when their operations' labor needs are the greatest and then can use that information to efficiently schedule workers for each hour the operation is serving guests.

It is widely believed that work really does magically expand to meet (and often exceed!) the number of people available to do the job and so measures of productivity must be available to guide management in making labor productivity assessments. Figure 7.14 summarizes the six productivity measures addressed in this chapter and lists some advantages and disadvantages associated with each. You may

FIGURE 7.14 Productivity Measures Summary

Measurement	Advantages	Disadvantages
Labor cost% = $\frac{\text{Cost of labor}}{\text{Total sales}}$	1. Easy to compute 2. Most widely used	1. Hides highs and lows 2. Varies with changes in price of labor 3. Varies with changes in menu selling price
Sales per labor hour = $\frac{\text{Total sales}}{\text{Labor hours used}}$	1. Fairly easy to compute 2. Does not vary with changes in the price of labor	1. Ignores price per hour paid for labor 2. Varies with changes in menu selling price
Labor dollars per guest served = $\frac{\text{Cost of labor}}{\text{Guests served}}$	1. Fairly easy to compute 2. Does not vary with changes in menu selling price 3. Can be used by non-revenue-generating units	1. Ignores average sales per guest and, thus, total sales 2. Varies with changes in the price of labor
Guests served per labor dollar = $\frac{\text{Guests served}}{\text{Cost of labor}}$	1. Fairly easy to compute 2. Does not vary with changes in menu selling price 3. Can be used by non-revenue-generating units	1. Ignores average sales per guest and, thus, total sales 2. Varies with changes in the price of labor
Guests served per labor hour = $\frac{\text{Guests served}}{\text{Labor hours used}}$	1. Can be used by non-revenue-generating units 2. Does not change due to changes in price of labor or menu selling price 3. Emphasizes serving guests rather than reducing costs	1. Time-consuming to produce 2. Ignores price paid for labor 3. Ignores average sales per guest and, thus, total sales
RevPASH = $\frac{\text{Revenue}}{\text{Available seat hours}}$	1. Measures overall efficiency in seating and selling products to guests 2. Identifies most and least efficient serving periods 3. Assesses how much guests buy and how long it takes to serve them their selections	1. Most suitable for commercial operations 2. Varies with changes in menu selling price 3. Does not utilize the price paid for labor in its calculation 4. Requires a detailed data collection/reporting system

select one or more of the measures described previously or create your own measure. In most cases, it is recommended that you monitor your labor cost percentage (the easiest measure to compute) and at least one other measure of productivity if you are truly serious about controlling labor-related expenses.

SIX-COLUMN DAILY PRODUCTIVITY REPORT

Many operators, upon selecting one or more labor productivity measures, want to compute those measures on a daily, rather than on a weekly or monthly, basis. This can be done by using a six-column form similar to the one introduced in Chapter 5.

A six-column form for Roderick's restaurant sales and labor cost in week 1 is presented in Figure 7.15.

To estimate daily labor cost percentage (cost of labor/total sales = labor cost %), the amount in the "Cost of Labor Today" column is divided by the "Total Sales Today" column to create the "Labor Cost % Today", just as the amounts in the "To Date" columns are divided to create the "Labor Cost % To Date".

Roderick's daily labor cost percentage during week 1 ranged from a low of 31.2 percent (day 7) to a high of 47.8 percent (day 2). The labor cost percentage for the week was 38.6 percent. Again, you can see the effect of averaging highs and lows when using this measure of labor productivity.

Any of the six productivity-related measures can be calculated on a daily basis using a modification of the six-column form. Figure 7.16 details the method to be used to establish six-column forms for each of the six productivity measures presented in this chapter. When using the six-column report, it is important to remember

FIGURE 7.15 Six-Column Labor Cost %

Unit Name: Roderick's Date: 1/15–1/21

	Cost of Labor		Total Sales		Labor Cost %	
Weekday	Today	To Date	Today	To Date	Today	To Date
1	$1,600	$1600	$4,000	$4,000	40.0%	40.0%
2	$1,760	$3,360	$3,680	$7,680	47.8%	43.8%
3	$1,840	$5,200	$4,300	$11,980	42.8%	43.4%
4	$1,960	$7,160	$4,600	$16,580	42.6%	43.2%
5	$2,000	$9,160	$4,200	$20,780	47.6%	44.1%
6	$2,600	$11,760	$8,200	$28,980	31.7%	40.6%
7	$2,440	$14,200	$7,820	$36,800	31.2%	38.6%
Total	$14,200		$36,800		38.6%	

FIGURE 7.16 Six-Column Labor Productivity Form

Measure of Productivity	Columns 1 & 2	Columns 3 & 4	Columns 5 & 6
Labor cost % = Cost of labor / Total sales	Cost of Labor Today Cost of Labor To Date	Sales Today Sales To Date	Labor Cost % Today Labor Cost % To Date
Sales per labor hour = Total sales / Labor hours used	Sales Today Sales To Date	Labor Hours Used Today Labor Hours Used To Date	Sales per Labor Hour Today Sales per Labor Hour To Date
Labor dollars per guest served = Cost of labor / Guests served	Cost of Labor Today Cost of Labor To Date	Guests Served Today Guests Served To Date	Labor Dollars per Guest Served Today Labor Dollars per Guest Served To Date
Guests served per labor dollar = Guests served / Cost of labor	Guests Served Today Guests Served To Date	Cost of Labor Today Cost of Labor To Date	Guests Served per Labor Dollar Today Guests Served per Labor Dollar To Date
Guests served per labor hour = Guests served / Labor hours used	Guests Served Today Guests Served To Date	Labor Hours Used Today Labor Hours Used To Date	Guests Served per Labor Hour Today Guests Served per Labor Hour To Date
Revenue per available seat hour = Revenue / Available seat hours	Revenue (Sales) Today Revenue (Sales) To Date	Available Seat Hours Today Available Seat Hours To Date	RevPASH Today RevPASH To Date

that the To Date column value, on any given day, is always the sum of the values of all the preceding Today columns, including the current day.

DETERMINING COSTS BY LABOR CATEGORY

Because of the various approaches used to measure effectiveness, it is not surprising that many operators find a single measure of their labor productivity is insufficient for their needs. Consider the case of Otis, the owner and operator of a restaurant located in the mountains and called the Squirrel Flats Diner. This operation services both backpacking tourists and loggers taking a break from work at a nearby logging camp. Otis's sales last month were $100,000. His labor costs were $30,000.

Thus, his labor cost percentage was 30 percent ($30,000/$100,000 = 0.30). Otis, however, knows more about his labor cost percentage than this overall number alone tells him. Figure 7.17 shows the method Otis uses to compute his overall labor cost percentage.

Note that Otis divides his labor expense into four distinct labor subcategories. Production includes all those individuals who are involved with preparing the food Otis sells. Service includes all the servers and cashiers involved in delivering the food to guests and receiving guest payments. Sanitation consists of the individuals responsible for ware washing and after-hour cleanup of the establishment. Management includes the salaries of Otis's two supervisors.

FIGURE 7.17 Labor Cost % for Squirrel Flats Diner

Time Period: 1/1–1/31 Sales: $100,000

Labor Category	Cost of Labor	Labor Cost %
Production	$12,000	12%
Service	$9,000	9%
Sanitation	$3,000	3%
Management	$6,000	6%
Total	$30,000	30%

By establishing four labor categories, Otis has a better idea of where his labor dollars are spent than if only one overall figure had been used. Just as it is often helpful to compute food cost percentages by category (see Chapter 5), it is often helpful to calculate more than one labor cost percentage or other measure of labor productivity.

Notice in Figure 7.17 that the sum of Otis's four labor cost percentage subcategories equals the amount of his total labor cost percentage:

$$12\% + 9\% + 3\% + 6\% = 30\%$$

You may establish any number of labor subcategories that make sense for your own unique operation. Of course, you can apply any measure of labor productivity to these identified labor subcategories. The important points for you to remember when determining labor productivity measures by subcategory are to:

1. Be sure to include all the relevant data, whether it is amount spent, hours used, sales generated, or guests served.
2. Use the same method to identify the numerator and denominator for each category.
3. Compute an overall total to ensure that the sum of the categories is consistent with the overall total.

Keep these points in mind as you examine Figure 7.18, which details Otis's second measure of labor productivity. He has selected guests served per labor dollar as a supplement to his computation of labor cost percentage. Otis feels that this second measure helps him determine his effectiveness with guests without losing sight of the total number of dollars he spends on payroll expense. The formula he uses for this computation is as follows:

$$\frac{\text{Guests served}}{\text{Cost of labor}} = \text{Guests served per labor dollar}$$

Note that each labor category in Figure 7.18 yields a different guests served per labor dollar figure. As could be expected, when labor dollars in a category are low, guests served per labor dollar in that category are relatively high. This is clearly demonstrated in the Sanitation category of Figure 7.18. When labor dollars expended are high, as in the production category, guests served per labor dollar is lower.

With the exception of RevPASH, each of the measures of labor productivity presented in this chapter can be subcategorized in any logical manner that is of value to management. The purpose of computing numbers such as these is, of course, that they are valuable in developing staff schedules and estimating future payroll costs.

FIGURE 7.18 Guests Served per Labor Dollar for Squirrel Flats Diner

Guests Served: 25,000 | Time Period: 1/1–1/31

Labor Category	Cost of Labor	Guests Served per Labor Dollar
Production	$12,000	2.083
Service	$9,000	2.778
Sanitation	$3,000	8.333
Management	$6,000	4.167
Total	$30,000	0.833

For example, if Otis knows that next month he is projecting 30,000 guests and uses the data generated in Figure 7.18, he can estimate his labor costs for each subcategory. By following the rules of algebra and adding the word "estimated," the guests served per labor dollar formula can be restated as follows:

$$\frac{\text{Estimated number of guests served}}{\text{Guests served per labor dollar}} = \text{Estimated cost of labor}$$

From the data in Figure 7.18 and using the Production subcategory as an example, Otis estimates his production labor costs in a month where 30,000 guests are anticipated as follows:

$$\frac{\$30{,}000}{2.083} = \$14{,}402.30$$

Using this estimation method, Otis knows that his production-related costs for the month should be approximately $14,400. Using the same guests served per labor dollar formula, Otis can project and budget different labor costs for each of his labor subcategories, as well as his overall labor cost. Using this information, Otis can prepare an employee schedule that uses the planned for amount of payroll cost and no more.

MANAGING PAYROLL COSTS

Professional foodservice operators must manage their payroll costs. Essentially, the management of payroll costs is a four-step process, which includes the following:

Step 1. Determine productivity standards.
Step 2. Forecast sales volume.
Step 3. Schedule employees.
Step 4. Analyze results.

STEP 1: DETERMINE PRODUCTIVITY STANDARDS

The first step in controlling payroll costs is to determine productivity standards for your operation. A **productivity standard** is defined as a guideline that constitutes an appropriate productivity ratio in your foodservice operation. Thus, a productivity standard might be a particular labor cost percentage, a specific number of guests served per labor dollar expended, or any other predetermined productivity ratio you want to utilize. In other words, you must find the answers to the questions of how long it should take an employee to do a job and how many employees it takes to do the complete job.

A productivity standard, then, is simply management's expectation of the productivity ratio of each employee. Establishing productivity standards for every employee is an essential management task and the first step in controlling payroll costs.

Thus far, we have discussed some methods of measuring current productivity ratios based on historical sales and expense data. This tells you where you are in relation to productivity, but it does not identify where you *should* be. To illustrate this concept, assume that you have decided to buy and operate a franchise unit within a themed steakhouse chain. Currently, four units of the same chain exist in your immediate geographic area, which has been designated a district by the **franchisor**, the entity responsible for selling and maintaining control over the franchise brand's name.

The area franchisor representative who monitors labor productivity using the labor cost percentage has shared the following data with you, as shown in Figure 7.19.

The figures for units one to four represent labor cost percentages in those franchise steakhouse units located in your district. The district average is the unweighted mean (average) of those four units. Company average refers to the overall labor cost percentage in all of the units franchised by the steakhouse chain you are joining. Industry average refers to the average labor cost percentage reported by theme-style steakhouses of the type similar to the one you will own and operate.

Using the data presented in Figure 7.19, you could begin to establish your own desired productivity measures and goals. Your restaurant is not yet open, so you face the problem of not having historical data from your own unit to help you. Of course, this is always the case when opening new restaurants, but it is also true when significantly changing the theme, decor, or menu of an existing restaurant.

Using your own judgment, and the information you do have available, you could choose as your labor cost goal the lowest labor cost percentage (unit one), the highest (unit two), the district, company, or industry average, or even some other number not included in this listing. Any of these numbers could become your target, or ideal, labor cost percentage. Finally, you must use the data you have available to you, as well as your own insight, to establish an appropriate productivity standard for your operation. Productivity standards are typically based on the following types of information:

1. Unit history
2. Company average

FIGURE 7.19 Steakhouse Labor Cost % Summary

Unit Description	Labor Cost %
Unit 1	34.1%
Unit 2	35.5%
Unit 3	34.3%
Unit 4	35.2%
District average	34.8%
Company average	34.0%
Industry average	35.5%

3. Industry average
4. Management experience
5. A combination of two or more of the above

In this case, you might choose to begin with a 35 percent labor cost as one of your productivity standards. This figure is close to the district average (34.8 percent) and is likely an aggressive, but realistic, goal for your first year. As mentioned previously, you would likely want to choose at least one measure of productivity in addition to labor cost percentage to help you monitor your labor efficiency. In the years to come, you would want to effectively manage the factors that affect productivity, carefully monitor your actual productivity using the productivity measures you select, and then establish a productivity goal that is both realistic and attainable.

STEP 2: FORECAST SALES VOLUME

Sales volume forecasting (see Chapter 2), when combined with established labor productivity standards, allows you to determine the number of employees that should be needed to effectively service your guests. All foodservice units must forecast sales volume if they are to supply an adequate number of employees to service that volume. This forecasting may be done in terms of either sales dollars or number of guests to be served.

An important, but frequently misunderstood, distinction must be made between forecasting sales volume and forecasting the number of employees needed to service that volume. The distinction is simply that, as a manager, you will view guests coming to your operation in "block" fashion, that is, in groups at a time. Employees, on the other hand, are added to your schedule one individual at a time and, thus, will significantly affect productivity measures. A brief example will make this clear. Ted owns a small shop that sells only specialty coffees. His service staff productivity standard is one server for each 30 guests. Thus, he schedules his employees as shown in Figure 7.20.

When 30 or fewer guests are expected, Ted needs only one server on duty. In effect, each time a block of 30 new guests is added, Ted must add another server. Thus, for example, if Ted anticipates 20 guests on Monday and 30 guests on Tuesday, no change in staff scheduling from Monday to Tuesday is necessary. That is, an addition of 10 extra guests does not dictate the addition of another server. On the other hand, if Ted anticipates 30 guests one day and 40 the next, an additional staff member is required because Ted has introduced a new block of guests.

It does not matter how much of the new block actually arrives; Ted has staffed for all of it. He hopes, of course, that all or nearly the entire new block of guests will arrive because this will keep his cost per guest served low. If only a small portion of the block comes, Ted's cost per guest served will rise unless he takes some other action to reduce his labor-related expense.

FIGURE 7.20 Ted's Coffee Shop Staffing Guide

Number of Guests Anticipated	Number of Servers Needed
1–30	1
31–60	2
61–90	3
91–120	4

STEP 3: SCHEDULE EMPLOYEES

Forecasting sales volume is important to labor cost control because it begins to take management out of the past and present and allows them to project into the future. To illustrate how established productivity standards (Step 1) are combined with sales forecasts (Step 2) to develop employee schedules (Step 3), consider Darla, the foodservice director at Langtree College. Darla is responsible for operating both a residence hall feeding situation (Geier Hall) and an open snack bar/cafeteria (Lillie's Cafeteria and Snack Bar).

Geier Hall houses 1,200 students. Lillie's Cafeteria and Snack Bar is open to all students, staff, and faculty of the school. Darla is committed to controlling all of her labor-related expenses. As such, she has carefully monitored her past labor productivity ratios, those of other similar schools, and national averages. In addition, she has considered the facilities she operates, the skill level and morale of her workforce, and the impact of her aggressive training program on her employees' future productivity.

Because the Lillie's operation generates a daily sales figure, Darla has determined that productivity in this operation should be measured using a labor cost percentage and that it should be able to operate at a labor cost of 30 percent.

In the residence hall, where dollar sales figures are replaced with daily guest counts, she has decided that her preferred labor productivity measure will be guests served per labor hour. Her goal is a ratio of 30 to 1, that is, 30 guests served per labor hour expended. Darla may now develop her labor cost expense budget in terms of both dollars (Lillie's) and labor hours used (Geier Hall).

Figure 7.21 shows how Darla would establish her labor budget for Lillie's using her productivity standards, her sales forecast, and the labor cost percentage formula.

Recall that the labor cost percentage formula is calculated as follows:

$$\frac{\text{Cost of labor}}{\text{Total sales}} = \text{Labor cost \%}$$

If you include the words *forecasted*, *standard*, and *budget* and then apply the rules of algebra, the labor cost percentage formula can be restated as follows:

$$\text{Forecasted total sales} \times \text{Labor cost \% standard} = \text{Cost of labor budget}$$

Thus, the cost of labor budget amount in the preceding formula becomes Darla's forecasted, or targeted, labor expense. To determine, for example, her targeted labor expense budget for Lillie's in week 1, and assuming her sales forecast for Lillie's for the period was $6,550, Darla would compute her labor budget as follows:

$$\$6,550 \times 30\% = \$1,965$$

FIGURE 7.21 Labor Budget for Lillie's Cafeteria and Snack Bar Using Labor Cost %

Time Period	Forecasted Total Sales	Labor Cost % Standard	Cost of Labor Budget
Week 1	$6,550	30%	$1,965
Week 2	$6,850	30%	$2,055
Week 3	$6,000	30%	$1,800
Week 4	$8,100	30%	$2,430
Total	$27,500	30%	$8,250

Figure 7.22 illustrates how Darla would establish a budget for total number of labor hours to be used to service Geier Hall residents. Recall that, in this food facility, sales refers to the number of guests served rather than dollars, and this is reflected in Darla's sales forecast. Also remember that the guests served per labor hour formula is as follows:

$$\frac{\text{Guests served}}{\text{Labor hours used}} = \text{Guests served per labor hour}$$

If you include the words *forecasted*, *standard*, and *budget* and follow the rules of algebra, the guests served per labor hour formula can be restated as follows:

$$\frac{\text{Forecasted number of guests served}}{\text{Guests served per labor hour standard}} = \text{Labor hour budget}$$

Using week 1 in Figure 7.22 as an example, and assuming 20,000 meals would be served, the computation would be:

$$\frac{\text{20,000 forecasted guests}}{\text{30 guests served per labor hour standard}} = \text{666.7 labor hours required}$$

Note that in Figures 7.21 and 7.22, Darla could have varied her weekly productivity standard and still have produced a 4-week budget. In other words, on weeks when volume was high, she could have elected to reduce her desired labor cost percentage or increase the guests served per labor hour standard. This can be a logical course of action if the operator feels that increased volume can have the effect of reducing the cost percentage of fixed labor or can increase the number of guests served per labor hour by staff persons in specific positions, such as cashiers and managers. Experience tells Darla, however, that, in her case, a standard labor productivity ratio that remains unchanged across the 4 weeks is her best option.

From the labor budgets she developed, Darla can now schedule her production staff in terms of dollars to be spent for labor (Lillie's) and labor hours to be used (Geier Hall). She must be careful to schedule employees only when they are needed. To do this, she must forecast her volume in time blocks smaller than 1-day segments. Perhaps, in her case, volume should be predicted in three periods (breakfast, lunch, and dinner) in the residence hall and in 1- or 2-hour blocks in her snack bar/cafeteria.

To see how Darla might schedule her residence hall staff during week 1 of her 4-week projection, assume that Darla's weekly projection of sales volume for Geier Hall is as presented in Figure 7.23.

On any given day, Darla can match her volume projections with budgeted hours or dollars. To see exactly how she would use this information to determine her

FIGURE 7.22 Labor Budget for Geier Hall Using Guests Served per Labor Hour

Time Period	Forecasted Number of Guests Served	Guests Served per Labor Hour Standard	Labor Hour Budget
Week 1	20,000	30 guests/hour	666.7
Week 2	18,600	30 guests/hour	620.0
Week 3	18,100	30 guests/hour	603.3
Week 4	17,800	30 guests/hour	593.3
Total	74,500	30 guests/hour	2,483.3

FIGURE 7.23 Weekly Labor Hour Budget for Geier Hall

Day	Forecasted Number of Guests Served	Guests Served per Labor Hour Standard	Labor Hour Budget
1	3,000	30 guests/hour	100.0
2	2,900	30 guests/hour	96.7
3	2,900	30 guests/hour	96.7
4	2,850	30 guests/hour	95.0
5	3,000	30 guests/hour	100.0
6	2,700	30 guests/hour	90.0
7	2,650	30 guests/hour	88.3
Total	20,000	30 guests/hour	666.7

employees' schedules, let us examine day 1 in Figure 7.23, a day when Darla projects 3,000 guests (meals) served and 100 labor hours needed. She knows that she should "spend" no more than 100 total hours for labor on that day. She also knows, from recording her productivity ratio in the past (Figure 7.24), where she has spent her labor hours in prior time periods. Thus, Darla should invest approximately 60 hours (100 hours available × 0.60 average usage) for production employees, 30 hours for service employees, and 10 hours of management time for Monday if she is to stay within her labor budget while providing an adequate number of workers in each of the three categories of labor she monitors regularly.

Presented in a different way, Darla knows that, to stay within her labor goals, each labor category will have its own unique guests served per labor hour ratio, as shown in Figure 7.25.

FIGURE 7.24 Recap of Past Percentage of Total Usage by Category

Geier Hall	
Labor Category	**% of Total**
Production	60%
Service	30%
Management	10%
Total	100%

FIGURE 7.25 Recap of Guests Served per Labor Hour

Geier Hall			
Labor Category	**Forecasted Number of Guests Served**	**Labor Hour Standard**	**Budgeted Guests Served per Labor Hour**
Production	3,000	60	50
Service	3,000	30	100
Management	3,000	10	300
Total	3,000	100	30

FIGURE 7.26 Employee Schedule

Unit Name: Geier Hall | Date: Monday 1/1
Labor Category: Production | Shift: A.M. and P.M. | Labor Budget: 60 hours

Employee Name	Schedule	Hours Scheduled	Rate	Total Cost
Sally S.*	6:00 A.M.–2:30 P.M.	8	N/A	N/A
Tom T.*	6:30 A.M.–3:00 P.M.	8		
Steve J.*	8:00 A.M.–4:30 P.M.	8		
Lucy S.*	10:00 A.M.–6:30 P.M.	8		
Janice J.	7:00 A.M.–11:00 A.M.	4		
Susie T.	6:30 A.M.–10:30 A.M.	4		
Peggy H.	10:30 A.M.–1:30 P.M.	3		
Marian D.	2:00 P.M.–5:00 P.M.	3		
Larry M.*	11:00 A.M.–7:30 P.M.	8		
Jill D.*	1:00 P.M.–7:30 P.M.	6		
Total		60	N/A	N/A

*Includes 30-minute meal break.

It is important to recognize that, although guests served per labor hour varies by labor category, the overall total yields 30 guests per labor hour used, which is Darla's productivity standard. Darla's employee schedule for the production area on Monday might look as shown in Figure 7.26.

Because employee schedules are based upon the number of hours to be worked or dollars to be spent, an employee scheduling form similar to the one in Figure 7.26 can be an effective tool in a daily analysis of labor productivity. Because labor is purchased on a daily basis, it should be monitored on a daily basis. The labor schedule should even be adjusted as needed during the day. This constant adjustment is key to the quick-service industry's profitability because schedule modifications by good managers in this segment of the industry are implemented hourly, not daily! In other words, if customer demand is lower than expected, employees can be released from the schedule to reduce labor costs. If volume is higher than expected, additional employees should be available on an as-needed basis.

Some foodservice managers practice an **on-call** system in which employees who are off duty are assigned to on-call status. This means that these employees can be contacted by management on short notice to fill in for other employees who are absent. On-call employees may also be called in to work if customer demand suddenly increases. State laws vary regarding the compensation that must be paid to these on-call employees; thus, managers who use this method should know about existing state laws and their own company policies regarding the practice before it is implemented.

Other managers practice a **call-in** system, in which employees who are off duty are required to check in with management on a daily basis to see if the predicted sales volume is such that they may be needed. This is a particularly good way to make rapid changes in staffing because of unforeseen increases in projected sales volume while minimizing the payment of overtime wages. **Overtime wages** are those that must be paid at a higher than normal rate. Overtime wages should generally be held to a minimum and should require documented management approval before they are authorized.

In the case of Geier Hall, the Rate and Total Cost columns in Figure 7.26 are not computed because, in this unit, they are not part of the productivity measure (recall that guests served per labor hour relies on neither cost of labor nor sales revenue for its computation). This data would, of course, be used when determining the cost of the employee schedule for Lillie's snack bar because, in that unit, it is labor dollars, not hours used, that is used to calculate the productivity standard. In all cases, an employee schedule, reviewed on a daily basis, should be established for each operating unit, labor category, and individual worker if managers are to best match labor usage with projected volume.

Cost Control Around the World

Employee scheduling is directly affected by the laws, union contracts, and even the customs of the country in which a business operates. For example, Figure 7.27 details the amount of paid vacation earned by employees (who have worked for at least 1 year) in several different countries in which US foodservice companies typically do business.

Additional regulations that can directly affect management's scheduling of employees include local child labor laws, overtime wage payment requirements, and union contract terms. The best managers ensure that their employee schedules always comply with the laws and requirements in place in those countries in which their operations are located.

FIGURE 7.27 Minimum Annual Earned Vacation Time

Country	Vacation Time Earned
United Arab Emirates	30 days
France	30 days
United Kingdom	28 days
Austria	25 days
Denmark	25 days
Spain	22 days
Germany	20 days
Japan	20 days
Hong Kong	20 days
South Korea	15 days
Puerto Rico	15 days
South Africa	15 days
Turkey	12 days
Haiti	11 days
Costa Rica	10 days
Argentina	10 days
United States	*No national requirement. Two weeks (10 work days) is common but not mandatory.*

STEP 4: ANALYZE RESULTS

Darla has done a good job of using established labor standards and volume projections in building her employee schedule. To complete the job of managing labor-related expense, she should analyze her results.

Figure 7.28 details Lillie's actual and budgeted operating results for the first 4 weeks of the year.

To determine the percentage of budget, the following formula, introduced in Chapter 1, is used:

$$\frac{\text{Actual amount}}{\text{Budgeted amount}} = \text{\% of budget}$$

Note that total sales were somewhat less than budgeted (98 percent), whereas total labor cost dollars were somewhat higher than budgeted (102 percent). Consequently, labor cost percentage was somewhat higher than anticipated, in other words, a 31 percent actual result compared to a 30 percent budget percentage. Also notice that, when sales are projected perfectly (week 3) but labor dollars are overspent, the resulting labor cost percentage will be too high (33 percent). Conversely, when labor costs are exactly as budgeted (week 4), but sales volume does not reach its estimate, labor cost percentage will similarly be too high (32 percent).

It may seem that a 1 percent variation in overall labor cost percentage is insignificant. In fact, in this case it represents $185 ($8435 actual cost − $8250 budgeted cost = $185). However, if a foodservice company achieved sales of $30,000,000 per year, exceeding planned labor cost by 1 percent would represent a $300,000 ($30,000,000 × 0.01 = $300,000) per year cost overrun! Small percentages can add up. What constitutes a significant labor cost variation can only be determined by management. Darla may well want to review standard scheduling techniques with her supervisors because she exceeded her budgeted labor cost in two of the four weeks shown. In Darla's case, $185 may well be a significant budget variation.

When referring to labor costs, some foodservice operators use the term **standard labor cost**—that is, the labor cost needed to meet established productivity standards—rather than budgeted cost. If productivity standards are used to establish budgeted labor costs, then the two terms are synonymous. It is important to recognize that actual labor costs will often vary somewhat from standard labor costs unless guest counts and/or revenues can be predicted perfectly, which, of course, is rarely the case. We can, however, compare budgeted labor expense with actual results to determine if the reasons for the variation from budget are valid and acceptable to management. The complete process for establishing the labor schedule is summarized in Figure 7.29.

FIGURE 7.28 Labor Recap for Lillie's Actual *Versus* Budgeted Labor Cost

	Sales			Labor Cost			Labor Cost %	
Week	Budgeted	Actual	% of Budget	Budgeted	Actual	% of Budget	Budgeted	Actual
1	$6,550	$6,400	98%	$1,965	$1,867	95%	30%	29%
2	6,850	7,000	102%	2,055	2,158	105%	30%	31%
3	6,000	6,000	100%	1,800	1,980	110%	30%	33%
4	8,100	7,600	94%	2,430	2,430	100%	30%	32%
Total	$27,500	$27,000	98%	$8,250	$8,435	102%	30%	31%

FIGURE 7.29 Ten-Point Labor Schedule Checklist

- ☐ 1. Monitor historical operational data (or alternative data if historical data are not available).
- ☐ 2. Identify productivity standards.
- ☐ 3. Forecast sales volume.
- ☐ 4. Determine budgeted labor dollars or hours.
- ☐ 5. Divide monthly budget into weekly budgets.
- ☐ 6. Divide weekly budget into daily budgets.
- ☐ 7. Segment daily budget into meal period budgets.
- ☐ 8. Build schedule based on the budget.
- ☐ 9. Analyze service levels during schedule period.
- ☐ 10. Review and adjust productivity standards as needed.

HERE'S HOW IT'S DONE 7.2

In some cases, managers are interested in assessing the productivity levels of individual employees. This is the case, for example, when it is important to know the productivity levels of servers or bartenders. The use of modern POS systems makes such assessments easy. To illustrate, assume that a beverage manager's POS system requires bartenders to enter their own unique employee identification number when making a drink sale. The sales data from last night are summarized below:

Bartender	Sales	Hours Worked
#201	$2,800	4.5
#133	$2,450	3.0
#155	$1,500	2.5
Total	$6,750	10.0

The manager can now calculate each bartender's sales per hour generation using the following formula:

$$\frac{\text{Sales}}{\text{Hours worked}} = \text{Sales per hour worked}$$

The manager makes the calculations and finds the following:

Bartender	Sales	Hours Worked	Sales per Hour Worked
#201	$2,800	4.5	$622.22
#133	$2,450	3.0	$816.67
#155	$1,500	2.5	$600.00
Total	$6,750	10.0	$675.00

From the results of the manager's analysis, it is clear that bartender #133 outperformed the other two bartenders. The reasons why this could be so are varied, but it is easy to see that the productivity rates of these bartenders are not the same. Knowing that, the manager can determine if bartenders # 201 and #155 might benefit from additional training.

LABOR PRODUCTIVITY IN HIGH-VOLUME BARS

High-volume bars can provide significant operational profits when they are designed and managed properly. These bars may operate as stand-alone businesses or can be located in high-volume restaurants and hotels. High-volume bars can generate sales measured in the millions of dollars per year and may serve hundreds, or even thousands, of guests at a time all while employing extremely large numbers of servers and bartenders, all working at the same time. The following are some cost-related strategies managers can use to help ensure that high-volume bars are operated effectively.

Selected Labor Control Strategies for High-Volume Bars:

1. **Fully Stock Work Areas:** Every bartender's own work area must be fully stocked for high-volume drink production. Per-drink labor costs are reduced when drink production times are lessened. Fully stocking each bartender's work area with sufficient beverage products, juices, glassware, and ice helps ensure speedy drink production.
2. **Ensure Sufficient Adequate Pre-prep:** Many cocktail recipes require the use of flavorful mixers and are served with one or more attractive garnishes. When too few of such items are readily available in a high-volume bartender's work area, drink production times will go up, and guest satisfaction will surely go down. Because most of these items can be prepared ahead of time (and will easily retain their quality levels beyond one day), there should always be ample quantities of them pre-prepped and available at the beginning of each bartender's shift.
3. **Optimize Organization:** The use of sales histories (see Chapter 2) and beverage purchases records allows managers to know which beverage products sell best. These are the items that should be most readily within easy reach (one or two steps) of high-volume bartenders. Thus, the fastest selling items should be arranged in a way that provides bartenders with easiest access to them. If the operation is offering specials, is introducing new products, or notices changes in beverage item popularity, modification to the organization of the work area can be implemented.
4. **Make Work Areas Safe:** Beverage production areas can become slippery through drink spills and normal production activities. This is especially so in high-volume operations. Slippery floors cause professional bartenders to slow down their movements because they know the safety risks involved in slip-and-fall accidents. While all drink spills cannot be avoided, nonskid safety mats of the proper size and number should be placed on the floor in every work area to minimize the negative effect of such spills on bartenders' speed. Significant spills or bottle breakage should be addressed immediately by the proper staff member.

The Move to Off-Premise Dining

The decision by increasing numbers of guests to move their foodservice purchases to off-premise consumption formats has had, and will continue to have, a direct impact on restaurant labor costs. Interestingly, however, that impact is not the same across all operations. In some cases, the move to off-premise dining may increase an operation's overall labor cost expense and in other cases it may reduce it.

For example, in some traditional table-service restaurants the move to off-premise dining has resulted in a significantly reduced need for front-of-house employees such as hosts, waitstaff, table bussers, and dining room cleaners. The savings resulting from such staff reductions are often offset, in varying degrees, by an increased need for employees skilled in packaging foods and in managing on-site delivery systems (i.e., food pickup stations).

For those foodservice operators who elect to manage their own off-site meal delivery fleets, reductions in onsite staff needs are offset by payments required for delivery drivers and, in most cases, their associated vehicle costs.

The need for changes in current training programs will, perhaps, be one of a foodservice operator's most important labor cost-related issues. This is because the move to off-premise dining can result in completely new training needs.

For example, while fewer servers may need to be trained in the intricacies of high-quality table-service methods, increasing numbers of staff may need to be trained in the operation of offsite ordering apps, food packaging and holding procedures, and the management of interfacing of contactless payments systems such as Apple Pay, Google Wallet, and Samsung Pay.

Additionally, training in a chain restaurant's own mobile app may be required to understand how, for example, PayPal payments should be recorded and processed (in addition to traditional payment card processing such as Visa, Discover, American Express, and MasterCard).

REDUCING LABOR-RELATED COSTS

If through your analysis you find that your labor costs are too high, problem areas must be identified and corrective action must be taken. If the overall productivity of your work group cannot be improved, other action must be taken. The approaches managers take to reduce labor-related costs are different for fixed payroll costs than for variable payroll costs. Figure 7.30 indicates strategies managers can use to reduce labor-related expenses in each of these two categories. Notice that you can only decrease variable payroll expenses by increasing productivity, improving the scheduling process, eliminating employees, or reducing wages paid.

One too often ignored way of increasing employee productivity, and thus reducing labor-related expenses, is through employee empowerment. Employee empowerment results from a decision by management to fully involve employees in the decision-making process as far as guests and the employees themselves are concerned.

Empowerment refers simply to the fact that, whereas it was once customary for management to make all decisions regarding every facet of the operational aspects of its organization and present them to its employees as inescapable facts to be accomplished, employees can be given the power to get involved.

Employees can be empowered to make critical decisions concerning themselves and, most importantly, those concerning guests. Many foodservice employees work closely with guests. Many guest-related problems in the hospitality industry can be easily solved when employees are given the power to make it right for the guest. Management has found in many cases that, through a well-planned and consistently delivered training program, and by empowering their employees, they are nurturing a loyal and committed workforce that is more productive, is supportive of management, and is willing to go the extra mile for guests. Doing so helps build repeat sales and profits.

FUN ON THE WEB!

Monthly journals and daily Web briefings are important sources of management information related to managing employees. The Society for Human Resource Management is a great source of human resource management-related information. Visit their website. While you are there, browse "Topics and Tools." Also consider a daily reading of their "News and Trends" section.

FIGURE 7.30 Reducing Labor-Related Expense Percentage

Labor Category	Actions
Fixed	1. Improve productivity. 2. Increase sales volume. 3. Combine jobs to eliminate fixed positions. 4. Reduce wages paid to fixed-payroll employees.
Variable	1. Improve productivity. 2. Schedule appropriately to adjust to changes in sales volume. 3. Combine jobs to eliminate variable positions. 4. Reduce wages paid to variable employees.

Technology Tools

As labor costs continue to increase, and as labor cost management becomes increasingly important to the profitability of restaurateurs, the tools available to manage these costs have significantly improved and increased as well.

Current software programs and apps that will enable you to manage and control labor costs include those that:

1. Maintain employment records, such as:
   a. Required employment documents (e.g., applications, I-9s, W-2s)
   b. Pay rates
   c. Payroll tax data
   d. Earned vacation or other leave time
   e. Subcategories of labor data
   f. Benefits eligibility
   g. Training records
2. Present and record the results of online or computer-based training programs.
3. Compute voluntary and involuntary employee turnover rates by department.
4. Track employee days lost due to injury or accident.
5. Maintain employee availability records (e.g., requested days off, vacation, or mandated leave).
6. Develop employee schedules and interface employee schedules with time clock systems.
7. Monitor overtime wages costs.
8. Maintain job descriptions and job specifications.
9. Develop and maintain daily, weekly, and monthly productivity reports, including:
   a. Labor cost percentage
   b. Sales per labor hour
   c. Labor dollars per guest served
   d. Guests served per labor dollar
   e. Guests served per labor hour
   f. Optimal labor costs based on actual sales achieved
10. Interface employee scheduling software with artificial intelligence forecasting sales volume software in the POS system.

FUN ON THE WEB!

Innovation in management tools that aid in managing payroll costs is not always aimed toward larger companies. Intuit is a company that produces a popular payroll program for smaller operators as well as the widely used Quick Books accounting program. To review the payroll program, go to the Intuit website. When you arrive, choose "Quick Books," then select "Products and Services."

Apply What You Have Learned

Teddy Fields is the kitchen manager at the Tanron Corporation International Headquarters. The facility he helps manage serves 3,000 employees per day. Teddy very much needs an additional dishwasher. He is now interviewing Wayne, who is an excellent candidate with five years of experience and who is now washing dishes at the nearby Roadway restaurant.

Teddy normally starts his new dishwashers at $18.00 per hour. Wayne states that he currently makes $19.25 per hour; a rate that is higher than all but one of Teddy's seven current dishwashers. Wayne states that he simply will not leave his current job to take a pay cut.

1. If you were Teddy, would you hire Wayne at a rate higher than most of your current dishwashers? Why or why not?
2. Assume you answered "No" to question one above, what would you say to Wayne?
3. Assume you answered "Yes" to question one above, what would you say to your current dishwashing employees if in the future Wayne shared his pay information with them?

For Your Consideration

1. Some foodservice managers operate their facilities without the use of detailed job descriptions and job specifications. How would the lack of these key tools affect a manager's ability to recruit, hire, and retain highly qualified workers?
2. Salaried (exempt) employees are those whose are expected to work the number of hours required to effectively perform their jobs. In most cases, this exceeds 40 hours per week. Should there be an upper limit to the number of hours salaried employees are expected, or required, to work? Explain your answer.

Key Terms and Concepts

The following are terms and concepts addressed in the chapter that are important for you as a manager. To help you review, define the terms below:

Payroll
Labor expenses
Salaried employee
Exempt employees
Minimum staff
Fixed payroll
Variable payroll
Productivity
Productivity ratio
Front of the house
Job description
Job specification
Employment application
Skills test
Psychological testing
Pre-employment drug testing
Negligent hiring
Orientation program
Task training
On-the-job
On-the-floor
Full-time equivalent
Split-shift
Cloud-based employee scheduling
Employee separation
Voluntary separation
Involuntary separation
Labor cost percentage
Sales per labor hour
Labor dollars per guest served
Guests served per labor dollar
Table turn
Guests served per labor hour
Revenue per available seat hour (RevPASH)
Productivity standard
Franchisor
On-call
Call-in
Overtime wages
Standard labor cost
Empowerment

Test Your Skills

You may download the Excel spreadsheets for the Test Your Skills exercises from the student companion website at http://www.wiley.com/go/dopson/foodandbeveragecost-control8e. Complete the exercises by placing your answers in the shaded boxes and answering the questions as indicated.

1. Serrena owns a large gourmet coffee shop and bakery. In Serrena's operation, payroll costs for hourly employees and salaried employees are calculated separately. Serrena recorded the revenue her operation generated last month as well as the amount she spent for hourly staff and for salaried managers. Calculate Serrena's last month labor cost percentages for her hourly staff and for her salaried employees, as well as her total payroll cost and total payroll cost percentage.

	Payroll Last Month	Payroll %
Revenue	$290,000	
Cost of hourly staff	$59,000	
Cost of salaried managers	$13,000	
Total payroll		

2. Alli Katz is the General Manager of Gattino's Italian Kitchen, an upper-scale restaurant serving Northern Italian cuisine. Alli has a very large staff and wants to carefully monitor her front of the house and back of the house employee turnover rates, as well as the turnover rate for her entire operation for the time period January 1 to June 30 of this year.

   Using the following information, help Alli calculate the turnover rates her operation has experienced for the 6-month period, and then answer the questions that follow.

Employees	January 1 to June 30 Separations	Number of Employees on January 1	Number of Employees on June 30	Turnover Rate %
Back of House		45	49	
Involuntary separations	2			
Voluntary separations	5			
Front of House		32	30	
Involuntary separations	4			
Voluntary separations	1			
Total		77		

   a. What was Alli's back of the house involuntary employee turnover rate?

   b. What was Alli's front of the house voluntary employee turnover rate?

   c. What was overall employee turnover rate at Gattino's Italian Kitchen in the past 6 months?

3. Rosa is the manager of a French bistro in a large Midwest city. She has experienced high turnover with her hourly employees over the past several months because they say that she isn't paying competitive wages. More employees have threatened to leave if she doesn't give them a raise. She has determined that she can compete with local restaurants if she raises the hourly wage from $15.50 per hour to $16.50 per hour, a 6.5 percent increase.

   Rosa is concerned about what this will do to her labor cost percentage. Her current labor cost is 35 percent, and she feels that 38 percent is the highest labor cost ratio she can maintain and still make a profit. Using last month's data, help Rosa calculate the effect of a 6.5 percent increase in wages. Can she give the employees what they want and still make a profit?

Week	Original Cost of Labor	Raise in Dollars	Total Cost of Labor	Sales	Labor Cost %
1	$10,650			$27,600	
2	$12,075			$32,250	
3	$10,887			$28,650	
4	$10,383			$37,200	
Total					

4. Jennifer operates Joe Bob's Bar-B-Q Restaurant. She specializes in beef brisket and blackberry cobbler. Her operation is very popular. The following data are taken from her last month's operation. She would like to establish labor standards for the entire year based on last month's figures because she believes that a month represents a good level of both customer service and profitability for her operation. Jennifer has an average guest check of $12 and an overall average payroll cost of $8 per hour.

   a. Use Jennifer's last month's operating results to calculate the following productivity standards: labor cost percentage, sales per labor hour, labor dollars per guest served, guests served per labor dollar, and guests served per labor hour. (Spreadsheet hint: Use the ROUND function to two decimal places for "guests served per labor hour" because you will use it in part c.)

Operating Results for Joe Bob's		
Week	**Number of Guests Served**	**Labor Hours Used**
1	7,000	4,000
2	7,800	4,120
3	7,500	4,110
4	8,000	4,450
Total	30,300	16,680

Calculate:

Average guest check	$27
Average wage per hour	$17
Total sales	
Total labor cost	

Productivity Measurement	Productivity Standard
Labor cost percentage	
Sales per labor hour	
Labor dollars per guest served	
Guests served per labor dollar	
Guests served per labor hour	

   b. Jennifer has subdivided her employees into the following categories: meat production, bakery production, salad production, service, sanitation, and management. She wants to develop a sales per labor hour standard for each of her labor categories. She believes this will help her develop future labor budgets based on forecasted sales. Help Jennifer calculate this standard based on her current usage of labor hours.

Labor Category	% of Labor Hours Used	Labor Hours	Sales per Labor Hour
Meat production	25%		
Bakery production	15%		
Salad production	10%		
Service	20%		
Sanitation	20%		
Management	10%		
Total	100%		

c. Now that Jennifer has calculated her productivity standards, she would like to use them to develop a labor hours budget for each day next week. She has forecasted 8,000 guests and she wants to use the guests served per labor hour standard that was calculated in part a. Use this information to develop a labor hours budget for Jennifer.

Day	Forecasted Number of Guests Served	Guests Served per Labor Hour Standard	Labor Hours Budget
1	900		
2	925		
3	975		
4	1,200		
5	1,400		
6	1,600		
7	1,000		
Total	8,000		

5. Mikel owns Mikel's Steak House, a popular dining establishment just outside of town on a busy state highway. Mikel uses labor cost percentage as his productivity measure, but he has been calculating it only once per month. Since his monthly costs have been higher than he expected, Mikel has decided that he needs a daily measure of his labor cost percentage to better control his costs.

   a. Calculate Mikel's daily labor cost percentage using the six-column daily productivity report, which follows.

Six-Column Labor Cost Percentage

Unit Name: Mikel's Steak House **Date: 3/1–3/7**

	Cost of Labor		Sales		Labor Cost %	
Weekday	Today	To Date	Today	To Date	Today	To Date
1	$1,900		$5,040			
2	$2,240		$5,220			
3	$2,080		$5,440			
4	$2,200		$5,560			
5	$3,200		$7,130			
6	$3,400		$8,200			
7	$2,600		$7,820			
Total						

b. Mikel wants to keep his labor cost percentage at 37 percent. Given the results of his six-column daily productivity report for the first week of March, will he be able to achieve his labor cost percentage standard if he continues in the same manner for the remainder of the month? If not, what actions can he take to reduce both his fixed and his variable labor-related expenses?

6. Jeffrey operates a high-volume, fine-dining restaurant called the Baroness. His labor productivity ratio of choice is guests served per labor hour. His standards for both servers and bussers are as follows:

Servers = 12 guests per labor hour
Bussers = 25 guests per labor hour

On a busy day, Jeffrey projects the following volume in terms of anticipated guests. His projections are made in 1-hour blocks. Determine the number of labor hours Jeffrey should schedule for each job classification for each time period.

How often in the night should Jeffrey check his volume forecast to ensure that he achieves his labor productivity standards and, thus, is within budget at the end of the evening? (Spreadsheet hint: Format "Server Hours Needed" and "Busser Hours Needed" to one decimal place.)

Volume/Staff Forecasting for Saturday: The Baroness			
Time	**Forecasted Number of Guests Served**	**Server Hours Needed**	**Busser Hours Needed**
11:00–12:00	85		
12:00–1:00	175		
1:00–2:00	95		
2:00–3:00	30		
3:00–4:00	25		
4:00–5:00	45		
5:00–6:00	90		
6:00–7:00	125		
7:00–8:00	185		
8:00–9:00	150		
9:00–10:00	90		
10:00–11:00	45		
Total	1,140		

7. Steve is in trouble. He has never been a particularly strong labor cost control person. He likes to think of himself more as a "people person." His boss, however, believes that Steve must get more serious about controlling labor costs or he will make Steve an unemployed people person! Steve estimates his weekly sales and then submits that figure to his boss, who then assigns Steve a labor budget for the week. Steve's operating results and budget figures for last month are presented as follows.

Operating Results: Steve's Airport Deli									
									For Weeks 1–5
	Sales			Labor Cost			Labor Cost %		
Week	Budget	Actual	% of Budget	Budget	Actual	% of Budget	Budget	Actual	
1	$2,500	$2,250		$875	$900				
2	$1,700	$1,610		$595	$630				
3	$4,080	$3,650		$1,224	$1,300				
4	$3,100	$2,800		$1,085	$1,100				
5	$2,600	$2,400		$910	$980				
Total									

a. Compute Steve's percentage of budget figures for both sales and labor cost. Also, compute Steve's budget and actual labor cost percentages per week and for the 5-week accounting period.

b. Do you feel that Steve has significant variations from budget? Why do you think Steve's boss assigned Steve a lower labor cost percentage goal during week 3? How do you feel about Steve's overall performance? What would you do if you were Steve's boss? If you were Steve?

8. Megan Merkle is the manager of the Half Shell Oyster House. Megan is reviewing her labor cost percentages for the current week's accounting period and is doing some financial planning for the next period. She separates her labor cost into three categories: front of the house, back of the house, and management. Megan wants to know what the impact on her labor cost percentages will be in the next accounting period if she:

a. Gives all of her employees a 3 percent raise and experiences no decline in sales.

b. Increases her menu prices and thus achieves a 5 percent sales increase, but does not give any of her employees a raise.

c. Increases her menu prices, achieves a 5 percent sales increase, and also gives all of her employees a 3 percent raise.

Help Megan with her financial projections by completing the following chart and then answering the questions that follow.

Half Shell Oyster House								
	Current Period		Next Period					
			With 3% Wage Increase		With 5% Menu Price Increase		With 5% Menu Price Increase and 3% Wage Increase	
	$	%	$	%	$	%	$	%
Sales/Forecasted Sales	$85,000	100%	$85,000	100%		100%		100%
Cost of Labor								
Front of the house	$10,200		$10,506					
Back of the house	$12,750							
Management					$5,525	6.2%		
Total cost of labor	$28,475	33.5%						

a. What was Megan's back of the house labor cost percentage for the current period?

b. What will be Megan's front of the house labor cost percentage next period if she gives all of her employees a 3 percent raise and experiences no decline in sales?

c. What will be Megan's total labor cost percentage next period if she increases her menu prices, achieves a 5 percent sales increase, and also gives her employees a 3 percent raise?

9. Jordan is the new western regional manager for The Lotus House, an Asian buffet restaurant chain. Her territory consists of 12 stores in four states. Last week she received the following data from her stores. Compute Jordan's labor cost by store, by state, and for her region.

	Sales	Cost of Labor	Labor Cost %
California			
Store 1	$91,000	$34,500	
Store 2	$106,500	$38,750	
Store 3	$83,500	$31,500	
Total			
Oregon			
Store 1	$36,800	$12,250	
Store 2	$61,000	$18,750	
Store 3	$52,000	$17,500	
Total			
Washington			
Store 1	$47,500	$14,750	
Store 2	$46,500	$15,000	
Store 3	$45,500	$15,000	
Total			
Nevada			
Store 1	$53,000	$17,250	
Store 2	$56,000	$18,500	
Store 3	$55,100	$17,250	
Total			
Region			

Can Jordan compute the average labor cost percentage for her region by summing the labor cost percentages of the four states and dividing by four? Why or why not? How is the overall labor cost percentage for her region computed?

10. Ravi Shah is the food and beverage director at the St. Andrews Golf Course and Conference Center. The facility is a popular place for weddings, and Ravi finds that on many Friday and Saturday nights, the banquet space at St. Andrews is completely booked. That is good news, but Ravi also finds that Allisha, the one full-time (paid $20.00 per hour) employee he has utilized as a supervisor for the banquet area, is averaging 15 hours overtime per week.

    Ravi is considering three alternative courses of action:

    1. Maintain the status quo and pay Allisha for 55 hours per week.
    2. Create a salaried position, schedule the employee who holds the position 55 hours per week, and pay that individual $50,000 per year.
    3. Split the job into two part-time positions of 30 and 25 hours per week and pay these employees $22.50 per hour.

    Assume the following:

    - Overtime is paid at 1.5 times the normally paid rate.
    - The operation's benefit package for part-time employees is 20 percent of the wages paid to them.
    - The operation's benefit package for full-time employees is 35 percent of the wages paid to them.
    - All full-time employees receive 2 weeks of paid vacation per year.

    Which of these three courses of action will cost the facility the most money? The least? If you were Ravi, which of these alternatives would you implement? Why?

	Hours Worked	Pay per Hour	Pay per Week	Weeks in a Year	Pay Before Benefits	Benefits	Annual Pay with Benefits
Alternative 1							
Total							
Alternative 2							
Alternative 3							
Total							

11. Luis manages Havana, a 150-seat full-service restaurant featuring Cuban and other Caribbean-style menu items. His restaurant is open for dinner from 5:00 P.M. to 11:00 P.M. Luis was excited to read, in one of the industry publications to which he subscribes, an article explaining RevPASH. In the past, Luis used labor cost % to create his allowable labor cost budget. Luis now wants to calculate RevPASH, as well as continue to use labor cost % to establish his *hourly* allowable labor cost budget. He has created a form (below) to calculate each of these measures.

    Luis is a good manager. When his dining room is slower, he encourages his servers to aggressively sell appetizers and desserts to increase his check average. When the restaurant is very busy, he encourages his servers to stress the quick turn of tables to minimize guest wait times and maximize the number of guests that can be served. As a result, and based on his historical records, when he serves 100 or fewer guests per hour, the restaurant achieves a $20.00 per guest check average. When 101 to 150 guests are served per hour, the check average drops to $18.00 per guest. When over 150 guests are served per hour, Havana achieves a $16.00 per guest check average.

Consider Luis's forecast, shown in the following, of the number of guests he will serve this coming Friday night, and then calculate his forecasted RevPASH for each hour he will be open. Next, help him find out how much he can spend for labor for each hour he will be open. Assume that Luis's target is a 30 percent labor cost at all times. (Spreadsheet hint: Calculate Total Check Average *after* Total Revenue and Total Guests Served have been calculated.)

Havana Forecast for: This FRIDAY Night						
Hour	Available Seats	Guests Served	Check Average	Revenue	RevPASH	Allowable Cost Based on 30% Labor Cost
5–6 P.M.		50				
6–7 P.M.		100				
7–8 P.M.		150				
8–9 P.M.		175				
9–10 P.M.		125				
10–11 P.M.		75				
Total						
% Seats Sold =						

a. What percentage of his total seats available does Luis believe he will fill on Friday night? What overall check average does he estimate he will achieve?

b. What would be Luis's forecast for his hourly and overall RevPASH on this day?

c. What would be Luis's labor budget for each hour his restaurant will be open, as well as the total amount that could be spent for labor that night?

d. What are some specific steps Luis might take to improve his RevPASH on Fridays from 5–6 P.M.? From 8–9 P.M.?

12. Aaliyah Moghaddam is the owner of Shawarma Shack, a popular QSR. She employs a large number of part-time workers, most of which are high school and college students who work 10 to 25 hours per week. Aaliyah currently pays these workers an average wage of $14.75 per hour, the current minimum wage in the state in which her restaurant is located.

In its most recent session, the legislators in her state have proposed raising the state minimum wage by approximately 15 %, to $17.00 an hour. Assuming that such legislation is passed, and using her last month revenue and expenses, help Aaliyah determine by what percentage (rounded to the nearest tenth of one percent) she would need to increase her menu prices to maintain the same level of profitability for her operation that she currently enjoys, then answer the questions that follow.

Shawarma Shack	Under Current Minimum Wage	%	Under Proposed Minimum Wage	%	% Change
Revenue	$80,000	100.0%		100.0%	
Cost of food sold	$23,500	29.4%			
Minimum wage payroll	$14,000	17.5%			
Non-minimum wage payroll	$11,000	13.8%			
All other operating expenses	$18,500	23.1%			
Total expenses	$67,000	83.8%			
Profit	$13,000	16.3%	$13,000		

a. If Aaliyah wanted to maintain her current level of profitability, what alternative options would she have to increasing her menu prices?

b. Do you think that increasing her menu prices would likely cause her total revenue to go up, or go down? Explain your answer.

CHAPTER 8

Controlling Other Expenses

OVERVIEW

This chapter explains how to manage foodservice costs that are neither food, beverage, nor labor. These "other" costs can represent 10 to 20 percent, or even more, of an operation's sales revenue and must be controlled well if your financial goals are to be achieved in the operations you manage.

The chapter teaches how to identify the other expense costs you can directly control, as well as those considered to be non-controllable expenses. In addition, it details how to compute other expenses as a percentage of total sales and on a per-guest served basis.

Chapter Outline

- Other Expenses
- Controllable and Non-Controllable Other Expenses
- Fixed, Variable, and Mixed Other Expenses
- Monitoring Other Expenses
- Managing Other Controllable Expenses
- Examining Non-Controllable Expenses
- Technology Tools
- Apply What You Have Learned
- For Your Consideration
- Key Terms and Concepts
- Test Your Skills

LEARNING OUTCOMES

At the conclusion of this chapter, you will be able to:

- Classify other expenses as being either controllable or non-controllable.
- Categorize other expenses in terms of being fixed, variable, or mixed.
- Compute other expense costs in terms of percentage of sales and on a cost-per-guest basis.

OTHER EXPENSES

When you operate a foodservice business, in nearly all cases the sum of a foodservice operation's food, beverage, and labor expenses will represent the largest cost areas it will encounter. In fact, these costs are so important they are considered to be an operation's **prime costs**. In addition to food, beverage, and labor expenses, however, there are a large number of "other" operating expenses you must pay. These other expenses can account for a significant amount of the total cost of operating your foodservice unit. They include a variety of items. Some common examples include advertising, website maintenance, utility bills, repair of equipment, liability insurance, property taxes, and mortgage payments (if a building is owned by the operation) or rent/lease payments (if the building is owned by others).

Some of these other expenses are directly under the control of management. When following the Uniform System of Accounts for Restaurants (USAR) (see Chapter 1), these types of expense are reported on an operation's income (P&L) statement as **other controllable expenses**. Some other expenses, however, are not directly controlled by management and, as a result, these are reported on the income statement as **non-controllable expenses**.

Managing your other expenses will be just as important to your success as controlling your prime costs. Remember that the **profit margin** (the percentage of revenue remaining after all operating expenses have been paid) in many restaurants can be small, so the control of *all* operating costs is critically important. Even in those situations that are traditionally considered nonprofit, such as hospitals and educational institutions, all costs must be controlled because dollars that are wasted in the foodservice area will not be available for use in other important areas of the institution.

You must look for ways to control all of your expenses, but sometimes the environment in which you operate will influence some of your operating costs in positive or in negative ways. An excellent example of this is in the area of energy conservation and waste recycling. Energy costs are one of the other expenses examined in this chapter. In the past, serving water to each guest upon arrival in a restaurant was simply a **standard operating procedure (SOP)**. But the rising cost of energy and increased awareness of the effect on the environment of wasted resources caused most foodservice operations to implement a policy of serving water only on request rather than with each order.

Guests have found this change quite acceptable, and the savings in the expenses related to glass washing, equipment usage, energy, and cleaning supplies, as well as labor, are significant. In a similar manner, many operators today are finding that recycling fats and oils, cans, jars, and paper products can be good not only for the environment but also for their bottom line. Recycling these items reduces your cost of routine trash disposal, and, in some communities, the recycled materials themselves have a cash value.

CONTROLLABLE AND NON-CONTROLLABLE OTHER EXPENSES

It is helpful for managers to consider other expenses in terms of the costs being either controllable or non-controllable. To see why, consider the case of Benny, the owner of a neighborhood tavern. Most of Benny's sales revenue comes from the sale of beer, sandwiches, and his special pizza. Benny is free to decide on the monthly amount he will spend on advertising. Advertising expense, then, is under Benny's direct control and is considered to be a controllable other expense.

Some of his other expenses, however, are not under his direct control. One example is the licenses needed to operate his facility legally. The state in which Benny operates charges a liquor license fee to all those businesses that serve alcoholic beverages. If his state increases the liquor license fee, Benny will be required to pay the additional fee. In this situation, the alcoholic beverage license fee would be considered a non-controllable expense, that is, an expense that is outside of Benny's direct control.

As an additional example, assume that you own a franchised quick-service restaurant that features fried chicken strips. Your store is part of a nationwide chain of such stores. Each month, your store is charged a $500 advertising and promotion fee by the chain's regional headquarters. The $500 is used to purchase television and online advertising time for the chain. This $500 charge is a non-controllable operating expense because it is set by the franchisor, and it must be paid each month.

A controllable expense is one in which decisions made by the foodservice manager will have the effect of either increasing or reducing the expense. A non-controllable expense is one that the foodservice manager can neither increase nor decrease. Management has some control over controllable expenses but usually has little or no control over non-controllable expenses. Because this is true, managers should focus their attention primarily on controllable, rather than non-controllable, expenses. In all likelihood, it is your ability to manage your controllable other expenses that will determine your effectiveness in the control of other expense costs.

It is also important to recognize that the items categorized as other expenses can constitute almost anything in the foodservice business. If your restaurant is a floating ship, periodically scraping the barnacles off the boat is categorized as an other expense. If an operator is serving food to oil field workers in Alaska, heating fuel for the dining rooms and kitchen will be an other expense, and probably a very large one! If a foodservice company has been selected to serve food at the Olympics in a foreign country, airfares and lodging for its employees may be a significant other expense.

Each foodservice operation will have its own unique list of required other expenses. The USAR actually lists several hundred other expenses commonly incurred by foodservice operations. It is not possible, therefore, to list all imaginable expenses that could be incurred by every foodservice operator. The expenses that are incurred, however, should be recorded and reported in a meaningful way. For example, some restaurants will incur napkin, straw, paper cup, and plastic lid costs. All of these costs could be conveniently reported under a general grouping or cost category such as "Paper supplies and packaging."

Similarly, nearly all bars will incur expense for items such as stir sticks, paper coasters, tiny plastic swords, and small paper umbrellas. These costs might be combined and reported in a cost category such as "Bar utensils and supplies."

When cost groupings are used, they should make sense to the operator and should be specific enough to let the operator and others reading the operation's financial reports know exactly what is in the category. Although some operators prefer to make up their own other expense cost categories, the cost categories suggested in this textbook are those recommended in the USAR. The individual other expense cost categories generally recommended by the USAR are divided into two primary groups:

1. Other Controllable Expenses
2. Non-Controllable Expenses

Within each of these two major other expense groups, the USAR's recommended groupings for individual cost categories are as follows:

1. Other Controllable Expenses
   - Direct Operating Expenses
   - Music and Entertainment
   - Marketing
   - Utilities
   - General and Administrative Expenses
   - Repairs and Maintenance

2. Non-Controllable Expenses
   - Occupancy Costs
   - Equipment Rental
   - Depreciation and Amortization
   - Corporate Overhead (multiunit restaurants)
   - Management Fees
   - Interest Expense

Within each of the two primary other expense cost subcategories, an operation's individual other expenses can be even further detailed to aid managers in their reporting and decision-making. For example, the USAR suggests the following individual costs be reported under the "Other Controllable Expenses" subcategory of "Utilities."

Utilities

- Electricity
- Gas
- Heating Oil and Other Fuel
- Recycling Credits
- Trash Removal
- Water and Sewage

Similarly, those expenses that are non-controllable may be detailed within their individual cost categories. For example, the USAR suggests the following individual costs be reported under the "Non-Controllable Expenses" subcategory of "Occupancy Costs."

- Rent—Minimum or Fixed Amount
- Rent—Percentage Rent Amount
- Rent—Parking Spaces
- Ground Rent
- Common Area Maintenance
- Insurance on Building and Contents
- Other Municipal Taxes
- Personal Property Taxes
- Real Estate Taxes

One major purpose of the USAR is to give foodservice managers and business owners guidance on how to best report the individual other expenses their businesses incur. The reasons why this is important will become very clear in Chapter 9 (Analyzing Results Using the Income Statement) of this textbook.

FIXED, VARIABLE, AND MIXED OTHER EXPENSES

While it is important to identify all costs as either controllable or non-controllable, it is also important to understand that, as a foodservice manager, some of the other expenses amounts you will pay will stay the same each month, but others will vary. For example, if you elect to lease a building to house a restaurant you operate, your lease payment will likely require you to pay the same lease amount each month.

In some other instances, the amount you pay for an expense will vary based on the success of your business. For example, because each drive-thru guest will receive at least one paper bag for each order filled, the expenses a manager would incur for paper bags will increase as the number of drive-thru guests served increases and will decrease as the number of guests served decreases.

As an effective cost control manager, it is important to recognize the difference between expenses that are fixed and those expenses that vary as your revenue varies. A **fixed expense** is one that remains constant despite increases or decreases in sales volume. A **variable expense** is one that generally increases as sales volume increases and decreases as sales volume decreases. A **mixed expense** is one that has some properties of both a fixed and a variable expense.

To illustrate all three of these expense types, consider Jo Ann's Hot Dogs, a mid-size, freestanding restaurant located near a shopping mall. Jo Ann's sells upscale Chicago-style hot dogs. Assume that Jo Ann's average sales volume is $136,000 per month. Also assume that rent for her building and parking area is fixed at $8,000 per month. Each month, Jo Ann computes her rent as a percentage of total sales, using the other expenses cost percentage formula:

$$\frac{\text{Other expenses}}{\text{Total sales}} = \text{Other expenses cost \%}$$

In this case, the other expense category she is interested in looking at is rent; therefore, the formula becomes:

$$\frac{\text{Rent expense}}{\text{Total sales}} = \text{Rent expense cost \%}$$

Jo Ann has computed her rent expense percentage for the last 6 months. The results are shown in Figure 8.1.

Note that Jo Ann's rent expense *percentage* ranges from a high of 6.67 percent (February) to a low of 4.88 percent (May), yet it is very clear that rent itself was a fixed, or constant amount of $8,000 per month. Thus, rent, in this lease arrangement, is considered to be a fixed expense.

It is important to note that, although the dollar amount of her rent expense is fixed, her rent expense percentage declines as her volume increases. Thus, the rent payment as a percentage of sales or cost per item sold is not fixed. This is true because as sales volume increases, the number of guests contributing to rent expense also increases, thus it takes a smaller percentage amount of each guest's sales revenue to generate the $8,000 Jo Ann needs to pay her rent.

It makes little sense for Jo Ann to be concerned about the fact that her rent expense percentage varies monthly. If Jo Ann is comfortable with the 6-month

FIGURE 8.1 Jo Ann's Fixed Rent

For Period: 1/1–6/30

Month	Rent Expense	Sales	Rent %
January	$8,000	$121,000	6.61%
February	$8,000	$120,000	6.67%
March	$8,000	$125,000	6.40%
April	$8,000	$130,000	6.15%
May	$8,000	$164,000	4.88%
June	$8,000	$156,000	5.13%
6-Month average	$8,000	$136,000	5.88%

average rent percentage (5.88 percent), she is in control of, and managing, her other expense "Rent" category. If, however, Jo Ann feels that her rent expense percentage is too high, she has only two options. She must increase sales, and thereby reduce her rent expense percentage, or she must negotiate a lower monthly rental with her landlord.

When rent, or any type of other cost, is fixed, the expense as expressed by the percentage of sales will vary with sales volume. The total amount of the expense, however, does not change in response to changes in sales volume.

Some restaurant lease payment arrangements are not fixed but rather are based on the sales revenue an operator achieves in the leased facility. To illustrate, assume that Jo Ann has a lease arrangement of this type, requiring her to pay 5 percent of her monthly sales revenue as rent. If that were the case, Jo Ann's monthly lease payments would be completely variable, as shown in Figure 8.2.

Note that the dollar amount of Jo Ann's rent under this arrangement ranges from a low of $6,000 (February) to a high of $8,200 (May). The percentage of her sales revenue that is devoted to rent, however, remains 5 percent.

A third type of lease that is common in the hospitality industry illustrates the fact that some other expenses are mixed, that is, there is both a fixed and a variable component to this type of expense. Figure 8.3 demonstrates such a lease arrangement. In it, Jo Ann pays a flat rent amount of $5,000 per month plus 1 percent of total sales revenue.

FIGURE 8.2 Jo Ann's Variable Rent

For Period: 1/1–6/30

Month	Sales	Rent %	Rent Expense
January	$121,000	5.00%	$6,050
February	$120,000	5.00%	$6,000
March	$125,000	5.00%	$6,250
April	$130,000	5.00%	$6,500
May	$164,000	5.00%	$8,200
June	$156,000	5.00%	$7,800
6-Month average	$136,000	5.00%	$6,800

FIGURE 8.3 Jo Ann's Mixed Rent

For Period: 1/1–6/30

Month	Sales	Fixed Rent Expense	1% Variable Rent Expense	Total Rent Expense
January	$121,000	$5,000	$1,210	$6,210
February	$120,000	$5,000	$1,200	$6,200
March	$125,000	$5,000	$1,250	$6,250
April	$130,000	$5,000	$1,300	$6,300
May	$164,000	$5,000	$1,640	$6,640
June	$156,000	$5,000	$1,560	$6,560
6-Month average	$136,000	$5,000	$1,360	$6,360

Under this arrangement, a major portion ($5,000) of Jo Ann's lease is fixed, whereas a smaller amount (1 percent of revenue) is variable because the amount Jo Ann must pay each month varies based on her sales revenue. Mixed expenses of this type are common and often include items such as some energy costs, garbage pickup, some franchise fees, and other expenses where the operator must pay a base (fixed) amount and then additional amounts are paid as usage or sales volume increases.

In summary, the total dollar amount of fixed expenses does not vary with sales volume. The total dollar amount of variable expenses changes as volume changes. As a percentage of total sales, however, a fixed-expense percentage decreases as sales increase, and a variable-expense percentage should not change with changes in sales.

A mixed expense has both fixed and variable components; therefore, as sales increase, the mixed-expense percentage decreases while the total dollar amount of the mixed expense increases. Figure 8.4 shows how fixed, variable, and mixed expenses are affected by increases in an operation's sales volume.

A convenient way to remember the differences between fixed, variable, and mixed expenses is to consider the paper napkins in a napkin holder placed at the end of the line in a fast casual restaurant. The napkin holder is a fixed expense. One holder is sufficient whether you serve 10 guests at lunch or 100 guests. The napkins themselves, however, are a variable expense. As you serve more guests (assuming each guest takes one napkin), you will incur a greater paper napkin expense. The cost of the napkin holder and napkins, if considered together, would be a mixed expense.

For some very large restaurant chains, it makes sense to separate some mixed expenses into their fixed and variable components, whereas smaller operations may elect, as in the case of the napkin holder and napkins, to combine these expenses. The company you work for may make the choice of how to account for other expenses, or you may be free to record other expense costs in a manner you feel is best for your own operation.

Effective managers know they should not categorize controllable, noncontrollable, fixed, variable, or mixed costs in terms of being either "good" or "bad." Some expenses are, by their very nature, related to sales volume. Others are not.

It is important to remember that, in most cases, the goal of management is not to *reduce*, but rather to *increase* variable expenses in direct relation to increases in sales volume.

Expenses are required if you are to service your guests. In the example of the paper napkins, it is clear that management would prefer to use 100 paper napkins at lunch rather than 10. As long as the total cost of servicing guests is less than the amount charged to guests, increasing the number of guests served will increase variable other expenses, but it will increase profits as well.

Normal variations in expense percentages that relate only to whether an expense is fixed, variable, or mixed should not be of undue concern to management. It is only when a fixed expense is too high or a variable expense is out of line with expectations that management should act. This is called the concept of **management by exception**. That is, if the expense is within an acceptable range of variation, there is no need for management to intervene. You must take corrective action only when operational results are outside the range of acceptability. This approach keeps you from overreacting to minor variations in expense but allows you to monitor all important other expense-related activities.

FIGURE 8.4 Fixed-, Variable-, and Mixed-Expense Behaviors as Sales Volume Increases

Expense	As a Percentage of Sales	Total Dollars
Fixed expense	Decreases	Remains the same
Variable expense	Remains the same	Increases
Mixed expense	Decreases	Increases

MONITORING OTHER EXPENSES

When managing other expenses, two control and monitoring calculations can be helpful to you. They are as follows:

1. Other expense cost percentage
2. Other expense cost per guest

Each alternative can be used effectively in specific management situations so it will be helpful for you to understand both.

As you learned earlier in this chapter, the other expenses cost percentage is computed as:

$$\frac{\text{Other expenses}}{\text{Total sales}} = \text{Other expenses cost \%}$$

For example, in a situation where a restaurant you operate incurs an advertising expense of \$5,000 in a month, serves 10,000 guests, and achieves sales of \$78,000 for that same month, you would compute your advertising expense percentage for that month as follows:

$$\frac{\$5{,}000}{\$78{,}000} = 6.4\,\%$$

This other expense cost per guest is computed as:

$$\frac{\text{Other expense}}{\text{Number of guests served}} = \text{Other expense cost per guest}$$

Using the preceding formula, you would compute your advertising expense cost per guest in this example as follows:

$$\frac{\$5{,}000}{10{,}000 \text{ guests}} = \$0.50 \text{ per guest}$$

The formula used to calculate an other expense percentage requires that the other expense cost category be divided by total sales. In many cases, this approach yields useful management information. In some cases, however, this computation alone may not provide adequate information. When that is true, using the concept of other expense cost per guest can be very helpful.

HERE'S HOW IT'S DONE 8.1

The best managers know how to calculate their other expense costs as a percentage of sales as well as on a per-guest basis. To do either calculation, three pieces of information must be known:

1. Total sales
2. Amount of other expense
3. Number of guests served

To calculate other expense as a percentage of sales, use this formula:

$$\frac{\text{Other expense}}{\text{Total sales}} = \text{Other expense cost \%}$$

To calculate other expenses on a per-guest basis, use this formula:

$$\frac{\text{Other expense}}{\text{Number of guests served}} = \text{Other expense cost per guest}$$

To practice applying these two formulas, assume that a restaurant achieved the following operating results:

1. Total sales = $250,000
2. Amount of other expense = $3,750
3. Number of guests served = 50,000

In this example, the operation's other expense percentage is:

$$\frac{\$3{,}750 \text{ other expense}}{\$250{,}000 \text{ total sales}} = 1.5\% \text{ other expense cost}$$

In this example, the operation's other expense when calculated on a per-guest basis is:

$$\frac{\$3{,}750 \text{ other expense}}{50{,}000 \text{ guests served}} = \$0.075 \text{ per guest (or 7.5 cents per guest)}$$

To further illustrate this, consider the following example: Scott operates Chez Scot, an exclusive, fine-dining establishment in a suburban area of a major city. One of Scott's major other expenses is linen. He uses both linen tablecloths and linen napkins. Scott's partner, Joshua, believes that linen costs are a variable operating expense and should be monitored through the use of a linen cost percentage. In fact, says Joshua, records indicate that the operation's linen cost percentage has been declining over the past 5 months; therefore, current control systems must be working well. As shown in Figure 8.5, the operation's linen cost percentage has indeed been declining over the past 5 months.

Scott, however, is convinced that there are linen control problems. He has monitored linen expense on a cost-per-guest basis. His information is presented in Figure 8.6, and it validates Scott's concern. There is indeed a control problem in the linen area.

While Figure 8.5 would indicate a declining linen expense, Figure 8.6 clearly shows that linen cost per guest served has increased from $1.06 in January to a May high of $1.22.

Chez Scot is enjoying increased sales ($68,000 in January *vs.* $74,000 in May), but its guest count is declining (2,566 in January *vs.* 2,305 in May). Since the guest count per month is declining, the average sale per guest (check average) has obviously increased. This is a good sign because it indicates that each guest is spending more. The fact that fewer guests are being served per month should, however, result in a *decrease* in demand for linen and, thus, a decline in total linen cost. In fact, on a per guest basis, linen costs are up. Scott is correct to be concerned about possible problems in the area of linen control.

FIGURE 8.5 Chez Scot Linen Cost %

Month	Total Sales	Linen Cost	Cost %
January	$68,000	$2,720	4.00%
February	$70,000	$2,758	3.94%
March	$72,000	$2,772	3.85%
April	$71,500	$2,753	3.85%
May	$74,000	$2,812	3.80%
Total	$355,500	$13,815	3.89%

FIGURE 8.6 Chez Scot Linen Cost per Guest

Month	Linen Cost	Number of Guests Served	Cost per Guest
January	$2,720	2,566	$1.06
February	$2,758	2,508	$1.10
March	$2,772	2,410	$1.15
April	$2,753	2,333	$1.18
May	$2,812	2,305	$1.22
Total	$13,815	12,122	$1.14

Calculating an other expense cost per guest may also be useful in a situation where the foodservice manager normally receives no daily payments and thus generates no daily dollar sales figure. Consider, for example, a college residence hall dining situation where paper products such as cups, napkins, straws, and lids are placed on the serving line to be used by the students while eating their meals. Students utilizing this dining hall pay a per-semester dining fee rather than pay for each meal eaten.

In this situation, Juanita, the cafeteria manager, wonders whether students are taking more of these items than is normal. The problem is, of course, that she is not exactly sure what "normal" usage is when it comes to supplying paper products to her students. Juanita belongs to an industry trade association that asks its members to supply annual cost figures to a central location where the costs are tabulated and reported back to the membership. Figure 8.7 shows the tabulations for last year from five different colleges in addition to those of Juanita's school.

Juanita has compared her paper product cost per student for the year and has found it to be higher than at P. University and the University of T., but lower than O. University, C. State University, and A. State University. Juanita's costs appear to be in line in the paper goods area. If, however, Juanita hopes to reduce paper products cost per student even further, she could, perhaps, call or arrange a visit to either P. University or the University of T. to observe their operations and/or purchasing techniques.

The other expense cost-per-guest formula is of value any time management believes it can be helpful or when lack of a sales figure makes the regular computation of other expense cost percentage inappropriate.

FIGURE 8.7 Average Paper Product Cost

Institution	Cost of Paper Products	Number of Students	Paper Product Cost per Student
O. University	$140,592	8,080	$17.40
C. State University	$109,200	6,500	$16.80
P. University	$122,276	7,940	$15.40
University of T.	$184,755	11,300	$16.35
A. State University	$61,560	3,600	$17.10
5-University Average	$123,676.60	7,484	$16.53
Juanita's Institution	$77,220	4,680	$16.50

FIGURE 8.8 Six-Column Cost of Paper Products

Juanita's Institution						Date: 4/1–4/8
	Other Expense Cost		Number of Guests Served		Cost per Guest	
Weekday	Today	To Date	Today	To Date	Today	To Date
Monday	$145.50	$145.50	823	823	$0.18	$0.18
Tuesday	$200.10	$345.60	751	1,574	$0.27	$0.22
Wednesday	$417.08	$762.68	902	2,476	$0.46	$0.31
Thursday	0	$762.68	489	2,965	0	$0.26
Friday	$237.51	$1,000.19	499	3,464	$0.48	$0.29
Saturday	$105.99	$1,106.18	375	3,839	$0.28	$0.29
Sunday	0	$1,106.18	250	4,089	0	$0.27
Monday	$157.10	$1,263.28	841	4,930	$0.19	$0.26
Total	$1,263.28		4,930		$0.26	

While other expenses are often reported on an operation's monthly income statement, they can be monitored on a more frequent basis. Figure 8.8 presents a six-column form that can be used to track both daily and monthly cumulative cost-per-guest figures for paper products. It is maintained by inserting Other Expense Cost and Number of Guests Served in the first two sets of columns. The third set of columns, Cost per Guest, is obtained by applying the other expense cost per guest formula.

MANAGING OTHER CONTROLLABLE EXPENSES

Managers are responsible for reducing and controlling their operation's other expenses. Effective operators are those who are constantly on the lookout for ways to reduce unnecessary additions to any of their other expense categories. For most managers, the effective management of other expenses must include special attention paid to the six "Other Controllable Expenses" categories identified in the USAR. These are:

- Direct Operating Expenses
- Music and Entertainment
- Marketing
- Utilities
- General and Administrative Expenses
- Repairs and Maintenance

FIGURE 8.9 Examples of Direct Operating Expenses

Kitchen utensils	Fryer filters	Plates and dishes
Pots and pans	Cleaning supplies	Foils and wraps
Plants and flowers	Extermination services	Silverware
Phone charges	To-go boxes, bags, and packaging	Linens and linen rentals
Uniforms	Cups and glasses	Paper products

MANAGING DIRECT OPERATING EXPENSES

Direct operating expenses are those that are not specifically related to the sale of a single menu item. Rather, these are the expenses related to the overall operation of a foodservice business. It is important to remember that while most operations will incur some similar direct operating expenses, such as those required for cooking equipment and utensils, each foodservice manager will face their own unique set of direct operating expenses.

A restaurant with a liquor license will need to purchase glassware for serving cocktails, while a similar restaurant that does not serve alcohol would not. Every foodservice operation is unique.

Figure 8.9 lists some examples of the dozens of costs classified as "Direct Operating Expenses" by the USAR.

While monitoring the use of all of the direct operating expenses an operator incurs is important, many operators are increasingly paying special attention to those direct operating expenses related to the service of off-premise menu items.

It is important to note that there currently is no complete uniformity in how direct expenses related to the off-premise delivery of menu items are recorded. Some operators consider that the packaging costs required to service off-premise customers should be recorded as a COGS (see Chapter 5). Their rationale for doing so can be easily understood if one considers the off-premise delivery of a single pizza. Because the pizza cannot be delivered without a box, these operators consider the box to be just as integral an ingredient as the cheese and pepperoni that the pizza may contain!

Other operators consider packaging related to the off-premise delivery of menu items as a traditional "Paper goods" expense and thus record these costs as a "Direct Operating" expense.

If a foodservice operation is using a **third-party delivery service** in an effort to increase their sales and expand their market, then paper products costs as well as delivery fees might be classified as a "Marketing" expense. A third-party delivery service consists of a smartphone or computer application that creates a marketplace where customers can search and browse restaurant menus, place orders, and have them delivered to a location of the customer's choosing. In nearly all cases the guest orders are delivered by independent contractors who have been retained by the third-party delivery company retained by the company operating the third-party delivery app.

As a result, some operators incurring third-party delivery costs and service fees take the position that partnering with a third-party delivery company is primarily a market expansion strategy, and thus, those costs relating to these types of partnerships should be considered "Marketing" expenses rather than "Direct Operating" expenses.

Regardless of the decision on where to account for off-premise packaging and delivery costs, it is most important to recognize that these operating costs are real and thus they must be well-managed.

Green and Growing!

As foodservice guests continually move toward off-premise dining, the demand for takeout food containers and packaging has increased dramatically. Unfortunately, this increase in demand has also increased the use of single-use and non-sustainable packaging materials such as Styrofoam and plastics.

The use of Styrofoam containers is especially problematic because, when they are discarded, they break down into tiny microplastics which can contaminate the environment and be ingested by animals and fish. Plastic containers contribute significantly to plastic waste, leading to pollution of landfills and the oceans.

As a result of the negative impact resulting from increased potentially harmful single-use food container, in the United States increased state and local regulations have been developed aimed at reducing the use of Styrofoam and plastic single-use food containers. Because to-go packaging costs are normally recorded as a "Direct Operating Expense," controlling them is important.

Fortunately, today's foodservice operators can increasingly choose takeout packaging made from paper, sugarcane, bamboo, or wheat straw that is quite suitable for hot and cold foods and carryout meals, and poses no harm to the environment.

In addition to being socially aware and following newly imposed bans and regulations, the fact that increasing numbers of foodservice customers are choosing to spend their dollars at operations that demonstrate a commitment to bettering the environment means the use of eco-friendly packaging can actually result in increased sales! And some drinks-to-go operators are going even further, offering cash discounts or extra loyalty program reward points for customers who bring in their own reusable cups.[1]

The Move to Off-Premise Dining

Despite the many advantages gained by operators serving increased numbers of off-premise guests, the continual move to off-premise dining also creates challenges.

One significant challenge facing foodservice operators in the production and delivery of off-premise menu items is that not all items are suitable for delivery. Those foodservice operations that have menus specifically designed for off-premise consumption face fewer challenges in this area, but many traditional restaurants responding to the off-premise dining movement frequently find that not all of their dishes are designed to be transported well. The resulting packaging and delivery challenges faced by these operators must be addressed to ensure their menu items are not soggy, cold, or otherwise damaged in the delivery process. To cite just one example, a restaurant serving takeout lasagna may find few challenges related to the proper delivery of this item, but a restaurant featuring hot fudge sundaes as a dessert item may find that specific menu item virtually impossible to package and deliver in a way that ensures its quality.

A second cost-related challenge involves the packaging of off-premise menu items. In many cases utilizing those packaging materials appropriate for maintaining high levels of product quality can be quite costly. For example, the average cost of a paper cup appropriate for the delivery of a hot beverage ranges from 10 to 15 cents per cup, with lids adding another 5 to 10 cents per cup. Thus, for a foodservice operator selling a hot beverage at a menu price of $5.00, the cost of the cup and lid alone can range from 15 to 25 cents. The cost of a small bag used to package the coffee order can average 20 to 25 cents per bag (more if the bag has a customized printed logo), thus the operator's cup, lid, and bag cost alone can range from 35 to 50 cents, or 7 to 10 percent of the coffee's actual selling price!

[1]https://www.today.com/food/restaurants/starbucks-loyalty-program-changes-rcna209988. Retrieved 5/30/2025.

FUN ON THE WEB!

The increased movement to off-premise dining will no doubt continue in the foreseeable future. Thoughtful food-service operators will want to monitor this movement on a regular basis. Doing so may help identify creative ideas applicable to their own operations. To stay abreast of changes in off-premise delivery systems innovations offered in the industry, on a periodic basis go to your favorite search engine and enter "off-premise restaurant dining trends" in the search bar and review the results.

MANAGING MUSIC AND ENTERTAINMENT EXPENSES

Some foodservice operations incur significant music and entertainment expense. These venues typically have live music and the cost of securing the services of those professionals who provide the entertainment are recorded in the "Music and Entertainment" expenses section of these operations' income statements.

Figure 8.10 lists some examples of the costs classified as "Music and Entertainment Expenses" by the USAR.

FIGURE 8.10 Examples of Music and Entertainment Expenses

Complimentary beverage, music, and entertainers	Live music	Satellite TV services
Complimentary food, music, and entertainers	Muzak costs	Cable TV services
Entertainers – comedians	Piano rentals	Professional entertainers
Entertainers – magicians	Piano tuning	Licenses – music

FUN ON THE WEB!

If your restaurant plays background music, hosts live bands or Karaoke nights, or contracts with DJs to play recorded music, your facility will be required to pay artists' royalties for the music your guests will hear (or sing!). The USAR classifies these expenses as "Licenses – Music."

The following groups represent those artists and companies who produce music and must be paid if you use their content. Go to their websites to see how and why you will interact with them.

The American Society of Composers, Authors and Publishers (ASCAP). ASCAP is a membership association of over 660,000 US composers, songwriters, lyricists, and music publishers of every kind.

Broadcast Music, Inc. (BMI). BMI is an American performing rights organization that represents more than 800,000 songwriters, composers, and music publishers in all genres of music.

The Society of European Stage Authors and Composers (SESAC). SESAC's repertory, once limited to European and gospel music, has diversified to include today's most popular music, including R&B/hip-hop, dance, rock classics, country, Latina, contemporary Christian, jazz, and the television and film music of Hollywood.

MANAGING MARKETING EXPENSES

All foodservice operations must successfully serve their customers if they are to stay in business. Whether a foodservice operation's customers are called guests, students, patients, or any other title, satisfying those being served is essential for the continued operation of any successful foodservice. In fact, foodservice operators should consider their main job to be adding new customers while retaining current customers.

To ensure a steady flow of customers, foodservice operators must understand **marketing:** all the ways a foodservice operation communicates with current and potential guests. All of the costs directly related to marketing to guests are recorded in the "Marketing" expenses section of an operations' income statements.

Figure 8.11 lists some examples of the costs classified as "Marketing Expenses" by the USAR.

Marketing involves how a business informs potential guests about what it is offering for sale and, most importantly, why customers should buy from them. The major goal, or purpose, of effective marketing is to obtain and retain a *growing* base of satisfied guests.

THE 4 Ps OF MARKETING

With so many marketing-related expenses to consider many foodservice operators find it convenient to focus on the **4 Ps of marketing** to record marketing expenses properly. The 4 Ps of Marketing are a good way to categorize a business's marketing expense based on the products sold, places where they are sold, the prices at which they are sold, and the promotional efforts used to sell them. The 4 Ps of marketing are illustrated in Figure 8.12.

Product

In a foodservice operation, the marketing of the specific menu items sold and how they are sold most often involve expenses related to communicating special offers to guests. These expenses may be recorded as marketing, advertising, or promotions expenses.

Place

Place refers to the physical location where products are available for sale or delivery, and this can be very important to the success of a foodservice operation. The location of an operation may include its indoor dining areas and outdoor patio areas, and it also includes the curb appeal of the operation's exterior, including its lighting, furnishings, and exterior signage. Curb appeal refers to the general attractiveness of a foodservice operation when viewed from the outside by a potential customer.

FIGURE 8.11 Examples of Marketing Expenses

Advertising/marketing fees to franchisor	Complimentary food and beverage	Outdoor signage
Advertising – radio, TV	Online advertising	Loyalty program costs
Advertising – magazines and newspapers	Billboards	Frequent diner discounts
Website development	Civic and community projects	Direct mail costs
Website maintenance and fees	Advertising agency fees	Contributions and donations

FIGURE 8.12 The 4 Ps of Marketing

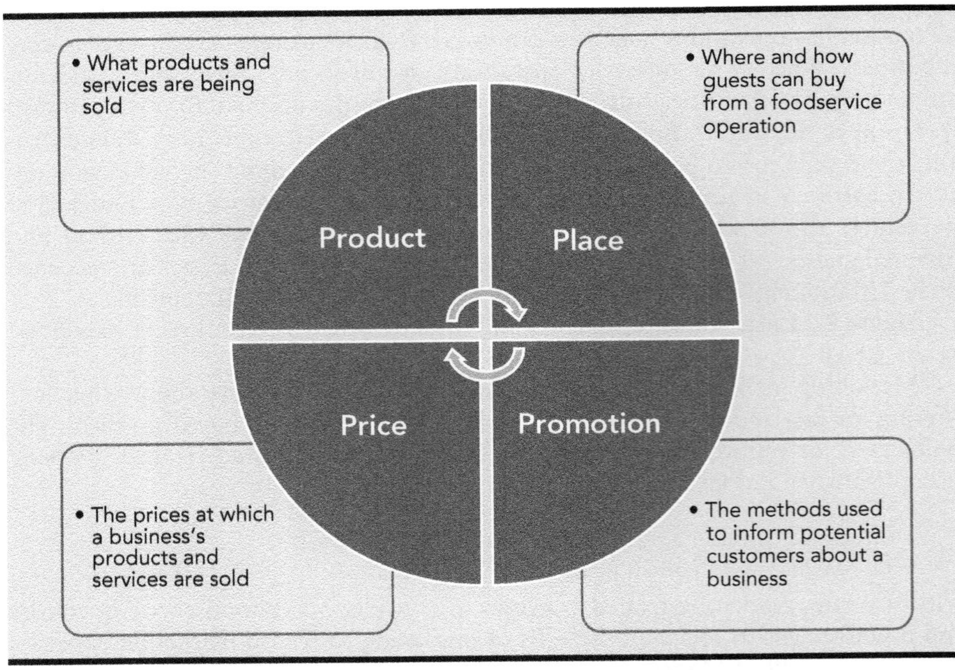

Minor expenses related to improving a facilities appearance are typically recorded in the "Repairs and Maintenance" section, rather than the "Marketing" section of an operation's income statement.

Major facility improvements and renovations are most often considered **capital expenditures**, which are facility expenses that benefit an operation for more than one year. Typically, these types of costs are depreciated over the useful life of the asset rather than charged as an expense in the month or year in which they were incurred.

Promotion

Promotion refers to all of the activities a foodservice operation uses to improve its sales or maintain its current sales level. Typical promotion expenses to be recorded are those directly related to advertising, personal selling, publicity, and public relations. These types of activities typically make up the bulk of marketing expenses recorded by most foodservice operations. These activities and their goals are summarized in Figure 8.13.

FIGURE 8.13 Promotion Activities and Goals Summary

Activity	Goal
Advertising	Delivering the marketing message in a cost-effective manner
Personal selling	Increasing customer counts, operational visibility, and guest check average through face-to-face interactions with customers
Promotions	Providing all guests (and especially loyal customers!) with special buying opportunities or rewards
Publicity	Increasing attention from the media at no cost to the operation
Public relations	Increasing name recognition and awareness of an operation in its own business community

FUN ON THE WEB!

Many foodservice operations use guest loyalty programs as an integral part of their marketing strategy. Essentially, these programs offer guests who already enjoy a foodservice operation's products and services some additional incentives to return often.

Typically, a customer loyalty program offers guests a discount, a free menu item, or specific merchandise as a reward for their frequent buying of qualifying purchases.

In the early days of foodservice rewards programs, customers were typically given a punch card to record the number of times they visited or the total value of their purchases over time. Today, specially designed customer loyalty program software does away with the need for such cards. Since they are easy to initiate and maintain today, nearly every commercial foodservice operator should utilize a customer loyalty program.

To review specific guest loyalty program software and apps appropriate for use in a foodservice operation, enter "restaurant customer loyalty program software" in your favorite search engine and view the results.

Price

While price is considered one of the four Ps of marketing, it is important to recognize that reductions in price are typically reflected in increased costs of product sold (COGS), rather than expensed as a marketing cost.

For example, if an operation decides to offer BOGO (see Chapter 6) entrees on a specific Friday night and the average entree sells for $20, the "cost" of such a promotion will be a $20 reduction in revenue and an increase in COGS. The costs related to communicating this special offer to customers, however, will be recorded as a marketing cost.

The Move to Off-Premise Dining

While it has long been a good idea for foodservice operations to have attractive and active proprietary websites, the move to off-premise dining has made the development and regular maintenance of an operation's website even more critical.

While a website's initial development costs are **capitalized** (i.e., recorded as an asset rather than an expense), ongoing website maintenance costs are expensed as incurred. Some operators prefer to record these as direct operating expenses, others as marketing expenses, and still others as administrative and general expenses.

The actual content on a proprietary website can vary widely based on the needs and interests of the operation and its ownership. In some cases, a website offers extremely limited information consisting only of the operation's name, address, and phone number (**NAP**), and perhaps a PDF of its menu(s). In other cases, a proprietary website may include a large number of features and functions.

Some content and features might be considered mandatory because guests expect every foodservice operation's website to address them, and these include:

- ✓ NAP
- ✓ Operating hours
- ✓ Driving directions
- ✓ Link to Google maps (to show an operation's map location)
- ✓ Menu items offered
- ✓ Menu item prices (optional unless online ordering is available)
- ✓ Payment forms accepted
- ✓ Links to related social media sites (if any)

Online ordering is not yet considered a mandatory content area for all operations. However, many foodservice operators believe this feature will soon be required on the foodservice's website (or an app). Designated payment apps are becoming more necessary because guests prefer using apps on mobile devices. The reason: it is easier for them to just tap on an icon located on their mobile screens instead of looking for a website with web browsers.

While the future of takeout and delivery ordering by guests is still being written, operators learned during the COVID-19 pandemic that they absolutely need a way for guests to place orders without being on-site as they do so. As a result, they found that an online ordering function coupled with online payments can help promote sales, despite the expense related to maintaining these increasingly robust websites.

FUN ON THE WEB!

Few foodservice managers have the technical skills required to develop and manage both their websites and social media presence. Fortunately, a number of companies provide these services. To see the variety of service options offered by these companies, enter "restaurant website management" in your favorite search engine and view the results.

MANAGING UTILITY EXPENSES

Utility services costs are an extremely important category of other expenses related to food and beverage operations. To produce their menu items, serve, and clean up, restaurants typically use thousands of gallons of water, consume significant amounts of natural gas (generally used for cooking and water heating), and utilize a large number of **kilowatt hours (kwh)** (the measure of electrical usage) each month.

Like food and labor costs, energy usage costs can be controlled. This process starts by understanding just where your restaurant uses its energy. Although the heating and cooling costs incurred by a restaurant in Alaska will be different than those of a restaurant in Arizona, the usage pattern shown in Figure 8.14 is a typical one.

Your utility costs can (and should) be controlled. Learning and then teaching your staff about the information in Figure 8.15 is a good way to start the process.

FUN ON THE WEB!

ENERGY STAR® is a joint program of the Environmental Protection Agency (EPA) and the Department of Energy (DOE). Its goal is to help consumers, businesses, and industry save money and protect the environment by adopting energy-efficient products and practices.

The ENERGY STAR label identifies top- performing, cost-effective products, homes, and buildings. The Department of Energy has also produced the "ENERGY STAR Guide for Cafes, Restaurants and Institutional Kitchens" to help foodservice operators identify ways to save energy and water in their businesses, boost their bottom lines, and help protect the environment. This resource also contains tips on how to upgrade equipment and highlights best practices that can positively impact a business's daily operations.

To learn more about foodservice specific money-saving energy-related tips, enter "ENERGY STAR Guide for Cafes, Restaurants and Institutional Kitchens" in your favorite search engine and review the results.

FIGURE 8.14 Typical Restaurant's Energy Consumption Pattern

Activity	Energy Use %
Cooking	32%
Heating (building)	19%
Cooling (building)	18%
Heating (water)	13%
Refrigeration	11%
Lighting	6%
Administrative	1%
Total	100%

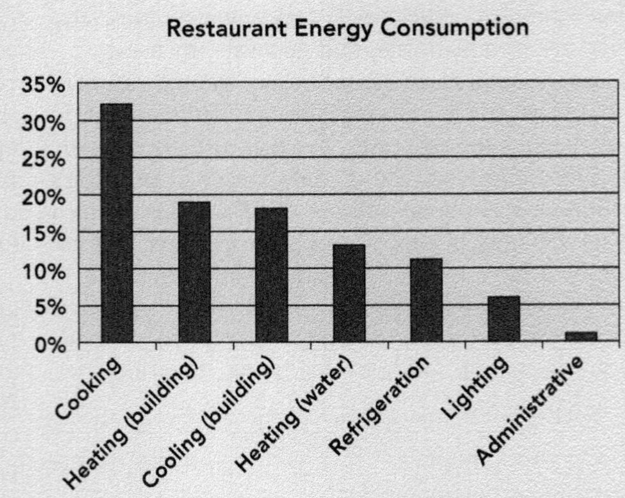

FIGURE 8.15 Ten Commonsense Energy Tips for Restaurateurs

1. Turn It Off
   - Turn off lights, cooking equipment, and exhaust fans when they are not being used.
   - Activate the standby mode for office equipment, in-house computers, and printers to effectively put these pieces of equipment to sleep when not in use.
2. Keep It Closed
   - Keep refrigerator doors closed.
   - Keep outside doors, if any, to the kitchen closed to minimize heating and cooling loss.
3. Turn It Down
   - Set air-conditioning units at 76°F (24.5°C) for cooling.
   - Set heating systems at 68°F (20°C) for heating.
   - Reduce the temperature of your hot water heater (where appropriate).
   - Adjust heating/cooling temperature settings when you close your operation for the night.
4. Vent It
   - Use ceiling fans to help recirculate dining room air.
   - Retrofit exhaust hoods with both low- and high-speed fans in dish room, food preparation, and cooking areas.
5. Change the Bulbs
   - Replace incandescent bulbs with fluorescent. They use 75 percent less electricity and last 10 times as long.
   - Install photocell light sensors (motion detectors) where appropriate (storage areas and the like) to activate lighting only when needed.
6. Watch the Water
   - Run dishwashers only when they are full.
   - Replace/repair leaking faucets immediately.
   - Insulate all hot water pipes.
   - Install water-saver spray nozzles in dish areas.
7. Cook Right
   - Stagger preheat times for equipment to minimize surcharges for high-energy use.
   - Bake during off-peak periods.
   - Idle cooking equipment (between meal periods) at reduced temperatures where appropriate.
8. Seal It
   - Caulk and weather-strip cracks and openings around doors, windows, vents, and utility outlets.
   - Check freezer, refrigerator, and walk-in seals and gaskets for cracks or warping. Replace as needed.
9. Maintain It
   - Change air filters on a regular basis (monthly during peak heating and cooling seasons).
   - Clean grease traps on ventilation equipment.
   - Clean air conditioner and refrigeration condenser/evaporator coils at least every 3 months.
   - Oil, lube, clean, and repair equipment as needed to maximize operating efficiency.
10. Get Help
    - Take advantage of any advisory services offered by your local utility company and governmental agencies.
    - Talk to your **heating, ventilation, and air-conditioning (HVAC)** repair person for tips on minimizing energy and maintenance costs with your particular HVAC system. It's like getting a free energy management consultant!

Consider the Costs

"Wow, this is ridiculous. I know it was hot last month, but I never expected our electric bill to look like this!" said Felix Unger, the manager of the Italian Garden restaurant.

Felix was talking to Myron Bollinger, the restaurant's owner. Each month Felix and Myron reviewed the operation's P&L. The previous month had been a busy one, but the weather was unseasonably hot. As a result, the operation's electric bill was nearly 40 percent higher than the prior year. It was also much higher than the amount that was budgeted for the month's utility bill.

Because the utility bill was so high, the restaurant's other expense percentage for utilities had cut into the operation's profits in a significant way.

"Well, it certainly was hot last month," said Myron. "And our cooling costs were way up, but what can we really do about it?"

1. Assume you were Felix. Where would you go to find information about reducing HVAC costs in particularly high-cost months?
2. What specific steps might you take in the future to help reduce energy costs in your operation during very warm periods?
3. How would you best communicate these steps to your staff?

MANAGING GENERAL AND ADMINISTRATIVE EXPENSES

General and administrative (G&A) expenses are (in most, but not all cases) those other expenses that stay the same regardless of the levels of sales achieved by a foodservice operation. Despite their typically constant nature, these somewhat fixed expenses are considered controllable when utilizing the USAR.

Figure 8.16 lists some examples of the dozens of costs classified as "General and Administrative" expenses by the USAR.

The specific G&A expenses incurred by a foodservice operation will depend greatly on the type of operation, but of increasing importance to nearly all operators are those expenses related to credit card fees.

It is important to recognize that in 1976 the US Federal government issued a ban on charging buyers more for using credit cards to pay for their purchases. The ban stated that "no seller in any sales transaction may impose a surcharge on a cardholder who elects to use a credit card in lieu of payment by cash, check or similar means."[2] But, in 1984 Congress opted to allow the ban to expire.

FIGURE 8.16 **Examples of General and Administrative Operating Expenses**

Computer/PC software	Membership dues	Educational seminars
Food and beverage taxes	Office supplies	Legal fees
Burglar alarms	Payroll service	Licenses
Help wanted ads	Postage and stamps	Telephone charges
Internet service	Cash register supplies	Credit card fees

[2] https://microbiz.com/credit-card-surcharges-and-cash-discounts/#:~:text=In%201976%2C%20these%20efforts%20paid,%2C%20check%20or%20similar%20means.%E2%80%9D. Retrieved 5/30/2025.

Today, merchants such as foodservice operators are, in most states, allowed to pass on credit card fees to those guests who use payment cards to pay for their purchases and many operators have elected to do so. These fees typically range from 2 to 4 percent of customer's total bill, so the fees can be significant.

The pros and cons of charging guests for the use of credit cards to pay for their purchases is beyond the scope of this text, but it is important to recognize that credit card fees can be a significant contributor to total G&A expenses and thus should be carefully monitored.

FUN ON THE WEB!

Electronic payments for restaurant purchases have become so commonplace that some foodservice operations are even opting to go cashless. This is just one example of how advances in technology directly impact the other expense costs operators incur for collecting guest payments. To monitor the evolving pros and cons of such an approach, enter "cashless restaurants" in your favorite search engine and view the results.

MANAGING REPAIRS AND MAINTENCE EXPENSES

Repairs and maintenance expenses are those that are required for a foodservice operation to maintain its equipment and property in good working condition. All foodservice operations will incur repairs and maintenance expenses due to the normal wear and tear that occurs when equipment and facilities are utilized to create sales.

Figure 8.17 lists some examples of the many of costs classified as "Repairs and Maintenance Expenses" by the USAR.

Any skilled craftsman knows that keeping work tools clean and in good working order will make the tools last longer and perform better. The same is true for tools and equipment in foodservice facilities. Proper care of mechanical equipment not only prolongs its life but also reduces operational costs. As prices for water, gas, and other energy sources needed to operate facilities continue to rise, you must implement a facility repair and maintenance program that seeks to discover and treat minor equipment and facility problems before they become major problems.

One way to help ensure that costs are as low as possible is to use a competitive-bid process before awarding contracts for services you require. For example, if you hire a company to service your dish machine monthly, it is a good idea to annually seek competitive bids from other vendors. This can help to reduce your costs by ensuring that the vendor you select has given you a price that is competitive with other service providers in the area. For general maintenance contracts in areas such as the repair of kitchen or for mechanical equipment, elevators, or grounds

FIGURE 8.17 Examples of Repairs and Maintenance Expenses

Air conditioning repairs	Building repairs	Carpet and rug repair
Kitchen equipment repairs	Computer repairs	Landscaping costs
Roof repairs	Dishwasher repairs	Painting and decorating costs
Sidewalk repairs	Electrical and mechanical repairs	Refrigerator and freezer repairs
Sign repairs	Furniture repairs	Window repairs

FIGURE 8.18 Equipment Inspection Report

Unit Name: Your Restaurant — Time Period: 1/1–1/31

Item Inspected	Inspection Date	Inspected By	Action Recommended
A. Refrigerator #6	1/1	D.H.	Replace gasket
B. Fryer	1/7	D.H.	Inspected, no maintenance needed
C. Ice Machine	1/9	D.H.	Drain, de-lime
D.			
E.			

maintenance, it is a good idea to get bids for these services at least once per year. This is especially true if the dollar value of the services contract is large.

To help control equipment maintenance and repair costs, air-conditioning, plumbing, heating, and refrigerated units should be inspected at least yearly, and kitchen equipment, such as dishwashers, slicers, and mixers, should be inspected at least monthly for preventive maintenance purposes. A form similar to the one shown in Figure 8.18 is useful in this process.

Some foodservice managers operate facilities that are large enough to employ their own full-time maintenance staff. If this is the case, make sure these employees have copies of the operating and maintenance manuals of all equipment. These documents can prove invaluable in the reduction of equipment and facility-related operating and repair costs.

FUN ON THE WEB!

Recordkeeping is one of the most important aspects of a preventive maintenance program. To see examples of software programs developed to help you record and plan preventive maintenance activities for your equipment, enter "restaurant equipment maintenance software" into your favorite search engine and review the results.

Cost Control Around the World

In a large number of cases, those who operate American chain restaurants outside the United States will experience equipment repair and replacement costs that are higher than their US-based counterparts. That's because, in many cases, a franchised restaurant's popularity is tied closely to the type of equipment or facility it requires for the production of its most popular menu items. When these key facility-related pieces of equipment are missing because repair service or spare parts are not readily available in foreign countries, the results can be disastrous.

For example, imagine a McDonald's with a US-built fryer that needs a replacement part, a Dairy Queen with a malfunctioning "Blizzard" machine, or a Rain Forest Café with a defective sound system. You can easily see how the character (and popularity) of those restaurants would be significantly and negatively altered if one or more of these critical pieces of equipment were unavailable for a significant amount of time.

In many cases, foodservice operators will find it more difficult to build, service, and maintain their physical facilities in foreign countries. This is especially true if the company has not identified a dependable, cost-effective, and local service representative for the building they occupy or the major equipment they utilize.

FUN ON THE WEB!

Finding reliable service and repair for foodservice equipment is always an issue for managers. This is especially the case for those operators whose restaurants are located in less populated areas of the world or in areas where service on popular equipment is hard to obtain. Fortunately, some quality companies do offer virtually worldwide sales and service. To view one such company, go to the Hobart corporation website.

When you arrive at the site, click on "Find Service" to view a list of the international sales and service centers maintained by this high-quality foodservice equipment and consulting company.

EXAMINING NON-CONTROLLABLE EXPENSES

Some foodservice operating costs are considered non-controllable because, in most cases, the manager of an operation does not have the ability to influence these costs. As well, these costs do not typically vary with sales levels.

Owners, including those managers who are owners of their operations do, however, have some control over these costs. Also, in some cases the costs must be recorded according to specific standards established by state or federal taxing authorities.

The basic format for an income statement prepared according to USAR recommendations includes three categories that are considered to be non-controllable expenses. These are:

1. Occupancy costs
2. Equipment leases
3. Depreciation and Amortization

OCCUPANCY COSTS

Occupancy costs include the cost of renting buildings and land, interest on loans used to purchase land and buildings (Note: mortgage payments for land and buildings are not listed on the income statement), property taxes, and insurance on an operation's **fixed assets**.

Fixed assets are those assets such as land, building, furniture, fixtures, and equipment (FF&E) that are purchased for long-term use and are not intended to be sold to customers. Occupancy costs can vary considerably between similar foodservice operations. Since the owners of a business most often control these costs, they are only infrequently under the direct control of those managers actually operating the business.

If your operation's occupancy costs are too high because of unfavorable rent or lease arrangements or due to excessive debts and required loan repayments, you may face extreme difficulty in generating operating profits.

Food, beverage, labor, and other controllable operating expenses can only be managed to a point; beyond that, efforts to reduce costs excessively can result in decreased product quality, reduced guest service levels, and lower guest satisfaction. If occupancy costs are unrealistically high, no amount of effective cost control can help save the operation's profitability.

The Move to Off-Premise Dining

One of the most lasting impacts of the COVID-19 pandemic and the move to off-premise dining has been the growing trend toward building new restaurants that are smaller than those built in the past. Increasingly, those owners who are considering building new foodservice units are opting to build facilities that greatly reduce, or even eliminate, indoor dining. Some are building **ghost kitchens,** which are operations that prepare meals solely for delivery, without any traditional indoor or outdoor dine-in space.

Owners doing so cite reduced expense for land (because lots can be smaller) reduced building costs for the smaller units and lowered local property taxes because their smaller buildings have a lower assessed taxable value. All of these factors combine to reduce an operation's total occupancy costs.

According to the National Restaurant Association (NRA) 2025 off-premise Restaurant Trends nearly 75 percent of all restaurant traffic now happens off-premises, meaning that almost three out of four restaurant orders are taken to go. Takeout is the most frequent off-premises method, followed by drive-thru and delivery.[3]

This significant change in buyer behavior continues to have a significant impact on many aspects of the foodservice business, and the move to smaller units is one of the most significant.

EQUIPMENT LEASES

Equipment lease expense results from the leasing or renting of equipment used in a foodservice operation. Common examples include charges for the leasing of POS systems, beverage dispensing equipment, and ice machines, as well as other specialty equipment.

The owners of foodservice operations often look to leasing rather than buying because it allows them to conserve their cash for other purposes, and to receive favorable tax treatment on their lease payments (because the amount of their lease payments is deducted as a normal monthly business expense. The amount of lease payments is subtracted from taxable income prior to calculating taxes owed (thus reducing taxes due). Leasing may also help ensure that operators do not end up with obsolete equipment when newer, better, and more efficient equipment can be obtained through a new lease.

[3] https://www.prnewswire.com/news-releases/from-trend-to-transformation-off-premises-dining-now-essential-for-restaurant-consumers-operators-302429894.html. Retrieved 5/30/2025.

DEPRECIATION AND AMORTIZATION

Depreciation and amortization are special cases of non-controllable other expenses. This is because these two types of expense are considered to be **non-cash expenses**. A non-cash expense is one that reduces an operation's taxable income, but does not involve cash coming into, or going out of, the business. Examples of non-cash expenses include **depreciation** and **amortization**.

Depreciation is the process of allocating the cost of equipment and other depreciable assets based on the projected length of their useful life. For example, if a piece of foodservice equipment is purchased, and it has an expected useful life of 10 years, then that asset would be depreciated over 10 years. The legal rules about how depreciation is actually calculated vary based on the type of asset purchased.

Amortization is the process of spreading the cost of intangible assets over those asset's useful life. Examples of intangible assets include goodwill, patents, customer lists, copyrights, and trademarks.

The individual entries in the Non-Controllable Expenses section of a USAR income statement are summed to yield an operation's Total Non-Controllable Expenses.

To illustrate the profit variability caused by non-controllable expenses, assume two foodservice operators manage virtually identical units. Both units generate $1 million in sales and have $250,000 remaining after subtracting all controllable expenses.

One operation has total non-controllable expense of $100,000 and generates a restaurant operating income (all of an operation's revenue minus its controllable and non-controllable expenses) of $150,000 ($250,000 after controllable expenses income – $100,000 non-controllable expenses = $150,000 restaurant operating income).

The second operation has total non-controllable expense of $200,000 and generates a restaurant operating income of $50,000 ($250,000 after controllable expenses income – $200,000 non-controllable expenses = $50,000 restaurant operating income). In this example, it is likely that both operators are doing good jobs managing their businesses even though the total restaurant operating income of the first unit is three times that of the second unit!

It is important to recognize that controllable and non-controllable other expenses in an operation can range from 10 to 20 percent, or even more, of the unit's total sales. As a result, experienced managers know that, although "other expenses" are sometimes considered to be minor expenses when compared to food and labor, they are extremely important to overall operational profitability.

In this and the previous chapters of this textbook you have learned a great deal about the ways foodservice operators manage and control their food, labor, and other expense costs. In the next chapter you will learn how foodservice managers report, and then analyze, their actual operating results.

Technology Tools

Those expenses that are related to neither food nor labor were introduced in this chapter. Depending upon the specific foodservice operation, these costs can represent a significant portion of the operation's total expenses. As a result, controlling these costs is just as important as controlling food- and labor-related costs.

It is important to recognize that, in most cases, an operation's technology-related acquisition and maintenance costs will be accounted for in its various "other expenses" categories. Some of the technology-related products and services that can be purchased to assist in this area include applications that relate to:

1. Assessing and monitoring utilities cost
2. Minimizing energy costs via the use of motion-activated sensors

3. Managing equipment maintenance records
4. Tracking marketing costs/benefits
5. Menu and promotional materials printing hardware and software
6. Website development and maintenance
7. Social media presence expense
8. Providing guest-facing (guest accessible) Wi-Fi access
9. Analysis of all other expense costs on a cost percentage or cost-per-guest basis
10. Comparing building/contents insurance costs across alternative insurance providers
11. Mobile ordering for menu item e-payment and pickup.
12. Software programs designed to assist in the preparation of the income statement, balance sheet, and statement of cash flows.

The unique needs of an individual foodservice operation will influence their specific other expense costs.

Apply What You Have Learned

"Well, I don't think it's complicated at all," said Amir, the assistant manager of the Hampton Deli. "The delivery fees have to be considered a cost of sales. It's no different than the ingredients we use to make our sandwiches. Delivery fees are a cost we incur when we make a sale. So, we should record them as a cost of sales."

Amir was talking to Peggy Richards, the owner of the Hampton Deli, a popular New York-style delicatessen located near a major college campus, and with Anna Rodgers, the deli's manager.

Peggy's restaurant features traditional deli fare like corned beef and pastrami sandwiches, homestyle soups, and traditional cheesecakes. Peggy has just entered an agreement with DineInDash, a third-party delivery service. The meeting occurring now was to discuss how to best record and account for the delivery fees that would be charged.

"Well, that doesn't make any sense to me," said Anna. "if we started our own delivery service, then we would charge the labor account for the cost of the people making the deliveries. So even though they're not our direct employees, I think the delivery fees have to be a labor cost."

"I don't know about that Anna," said Peggy. "if we delivered ourselves, we certainly wouldn't charge the cost of the vehicles and fuel to our labor account. Those costs would be placed in our G&A expense category for sure. And really, the reason I made the decision to go with DineInDash was to make more sales. Seems to me their fees should be a marketing cost. Guests will go to their website, see our information, and place their order. How is that any different than seeing an ad in a magazine or the campus newspaper and then coming to visit us? If we were going to record the cost of an ad, we would record it as a marketing cost, wouldn't we?"

1. Do you think delivery fees in this scenario should be recorded as a COGS, labor cost, or as some other type of "other expense"? Explain your answer.
2. If you were advising Peggy, where would you tell her to record these costs?
3. Whatever her final decision, why would it be important for Peggy to record these delivery costs in the same manner every time an income statement for her operation was produced?

For Your Consideration

1. The costs associated with making technology purchases are increasing rapidly in many foodservice operations. What are some questions you think foodservice managers should consider, or ask their technology vendors, before they decide to make significant investments in technology upgrades?
2. For chain-affiliated restaurants, monthly franchisee fees can represent 5 to 10 percent, or more, of total sales. Why do you think many restaurant owners and investors are willing to pay these monthly fees, when independent operators do not incur such costs?

Key Terms and Concepts

The following are terms and concepts addressed in the chapter that are important for you as a manager. To help you review, define the terms below:

Prime costs
Other controllable expenses
Non-controllable expenses
Profit margin
Standard operating procedure (SOP)
Fixed expense
Variable expense
Mixed expense
Management by exception
Third-party delivery service
Marketing
Curb appeal
Capital expenditure
Capitalized
Name, address, and phone number (NAP)
Kilowatt-hours (kwh)
Heating, ventilation, and air-conditioning (HVAC)
Occupancy costs
Fixed assets
Ghost kitchen
Non-cash expense
Depreciation
Amortization

Test Your Skills

You may download the Excel spreadsheets for the Test Your Skills exercises from the student companion website at http://www.wiley.com/go/dopson/foodandbeveragecostcontrol8e. Complete the exercises by placing your answers in the shaded boxes and answering the questions as indicated.

1. Sophia owns and operates a restaurant featuring lobster rolls and fried cod sandwiches. She is evaluating her revenue and expenses for last month. Help Sophia assess her operation's profit and other expense performance by completing the following revenue and expense worksheet.

Sophia's Restaurant	Amount	Percentage
Sales	$160,000	100%
Total cost of food	$50,000	31.3%
Total cost of labor	$44,000	27.5%
Other controllable expenses	$22,000	
Non-controllable expenses	$10,000	
Profit		

2. Gene Monteagudo is the owner of Margaritas-To-Go, a drive-thru only operation that sells various flavors and sizes of takeout Margaritas. Each single-serving Margarita sold requires the use of one plastic cup, a lid, and a wrapped straw. Gene is interested in knowing his per-serving beverage packaging costs. Using the information presented in the chart below, help Gene calculate his per-serving beverage packaging cost for a 16-ounce product sale. Please round your calculated answers to the nearest tenth of a cent.

Gene's 16-ounce Beverage Sale Packaging Cost			
Product	**Case Price**	**Units per case**	**Cost per unit**
16-ounce cup (plastic)	$98.50	1,000	$
16-ounce lid	$38.00	1,000	$
Straws (wrapped)	$11.00	500	$
Total cost per sale			$

3 John owns and operates the End Zone Steakhouse. The restaurant's "Other Expenses" operating cost categories are presented below. Place an "X" in the Controllable column for those operating expenses that John can control. Place an "X" in the Non-controllable column if he cannot directly control the cost.

Other Expenses	Controllable	Non-controllable
Real estate tax		
Menu printing		
Professional musicians		
Interest on long-term debt		
Charitable donations		
Cleaning supplies		
Flowers and decorations		
Licenses and permits		

4. Crystal Tate is the manager of a restaurant that is connected to her city's water and sewage system. Crystal is trying to forecast next month's water and sewage bill. Crystal's operation has a 2-inch water inlet pipe, and she forecasts that next month her operation will consume 130 units of water (1 unit = 100 cubic feet = approximately 750 gallons) per day.

The city charges Crystal $15.97 per month for her water inlet pipe and $0.58 for the first 10 units of water her operation uses each month. The city charges Crystal $2.25 for each unit used in excess of 10 units used each month. The city also charges a $3.15 per unit sewage treatment charge for all units used. Help Crystal estimate her next month's water bill by completing the worksheet she has begun.

Crystal's Water and Sewage Bill Forecast			
Item Charge	**Item Cost**	**Units**	**Total**
2-in. inlet line	$15.97	1	
Water Usage			
Units 1–10	$0.58	10	$5.80

Crystal's Water and Sewage Bill Forecast			
Item Charge	**Item Cost**	**Units**	**Total**
Units 11 and above	$2.25		
Wastewater/sewage	$3.15		
Month total forecast			

What will be Crystal's forecast of her water and sewage bill for next month?

5. Shanna operates a lounge in an extremely popular downtown convention hotel. The hotel regularly operates around the 80 percent occupancy mark, and its lounge, Luigi's, is very often filled to capacity. On weeks when business at the hotel is slower, Shanna attempts to build local sales by scheduling a variety of popular bands to play on the stage. She must select one band to play on Saturday night, 6 weeks from now, when the hotel is not busy. She has kept records of the costs and sales volume of the last four bands she has booked.

   a. Compute both band expense percentage and cost per guest served. Based on the cost percentage of the bands, which one should Shanna select for booking?

Expense Percentage and Cost per Guest Served—Bands						
Unit Name: Luigi's Lounge						
Date	**Band**	**Band Expense**	**Lounge Sales**	**Cost %**	**Number of Guests Served**	**Cost per Guest Served**
1/1	Tiny and the Boys	$1,400	$11,400		1,425	
2/1	Shakin' Bill and the Billfolds	$1,900	$12,250		1,980	
3/1	La Noise	$2,000	$12,000		2,005	
4/1	The Hoppers	$2,000	$10,250		2,100	

   b. Would your answer change if you knew Shanna charged a $20.00 cover charge to enter the lounge on the nights she has a band, and that the cover charge is reported separately from the lounge sales? If so, which band would you choose?

Date	Band	Number of Guests Served	Cover Charge per Guest	Total Cover Charges	Lounge Sales	Total Sales
1/1	Tiny and the Boys	1,425			$11,400	
2/1	Shakin' Bill and the Billfolds	1,980			$12,250	
3/1	La Noise	2,005			$12,000	
4/1	The Hoppers	2,100			$10,250	

6. Marjorie manages a 200-seat, white-tablecloth restaurant in an upscale neighborhood. Since her guests expect her tablecloths and napkins to be really white, she sends her linens to a local laundry service daily. The laundry service

charges her by the piece. She wants to keep track of her laundry cost per guest to see if she can use the information to control her laundry costs better. Help her complete her six-column cost-per-guest report. She has budgeted $0.60 per guest on average. How is she doing at controlling her laundry costs?

Six-Column Cost per Guest—Laundry Service						
Unit Name: Marjorie's						Date: 5/1–5/7
	Laundry Service Cost		Number of Guests Served		Cost per Guest	
Weekday	Today	To Date	Today	To Date	Today	To Date
1	$225		400			
2	$204		375			
3	$200		350			
4	$240		425			
5	$275		450			
6	$300		500			
7	$230		410			
Total						

7. Stella Daniels is a franchisee in the Bluebird Burgers franchise system. Stella wants to forecast the franchise fees she will likely pay next year.

Bluebird Burgers charges its franchisees a $500-per-month inspection fee, royalty fees of 6.5 percent of gross sales, and a national advertising fee of 2.25 percent of gross sales.

Last year Stella's operation generated $1,235,575 in gross sales. Stella would like to forecast the fees she will pay if her next year's gross sales decline by 2.5 percent, as well as the fees she will pay if her gross sales increase by 2.5, 5.0, 7.5, 10.0, and 12.5 percent. Help Stella forecast her next year franchise fees under each of these gross sales assumptions, then answer the questions that follow.

	Franchise Fees	Last Year's Revenue and Fees	Next Year Revenue Change					
			−2.5%	2.5%	5.0%	7.5%	10.0%	12.5%
Revenue		$1,235,575						
Monthly inspection fee	$500							
Royalty fee	6.50%							
Marketing fee	2.25%							
Total franchise fees								
Total franchise fee %								

a. What was the total amount of franchise fees Stella paid last year?
b. What was the total percentage (rounded to two places) of her gross sales Stella paid in franchise fees last year?
c. What is the total percentage of her gross sales (rounded to two places) Stella will pay in franchise fees next year if her sales increase 10.0 percent?

d. If her sales actually increase 10.0 percent next year, will the Total Franchisee Fee percent of sales Stella will pay her franchisor be higher or lower than it was last year? Why?

8. Josiam operates the foodservice at Springdale Valley school system. He has just been informed by City Power, the electrical company in his area, that the rate per kilowatt hour (kwh) for the school system's kitchens will be rising from $0.085 per kwh to $0.092 per kwh beginning next academic year (September). Based on last year's bill, what was each kitchen's electricity usage?

Assuming no operating changes, how much more will be spent next year? What are some of Josiam's best resources for discovering ways to limit electricity usage in the kitchens?

School	Electricity Cost Last Year	Number of kwh Used Last Year	Cost per kwh Estimate Next Year	Estimated Electricity Cost for Next Year
Springdale Elementary	$6,800.00		0.092	
Jefferson Elementary	$7,650.00		0.092	
Clinton Middle School	$10,200.00		0.092	
Tri-Valley High School	$12,750.00		0.092	
Total	$37,400.00		0.092	

9. Enrique has located the perfect spot for his restaurant. It is a 3,000 square foot location in the local shopping mall. The mall's manager has given him the following monthly lease options:

Option 1: Pay a flat fee of $2.50 per square foot per month.
Option 2: Pay a flat fee of $4,000 per month plus 5 percent of food sales.

Enrique estimates that his sales for the coming year will be as shown as follows. Calculate the monthly lease amounts under each option.

Which lease option should Enrique choose? Why?

		Option 1			Option 2		
Month	Sales Forecast	No. of Square Feet	Flat Fee per Square Foot	Monthly Lease $	Flat Fee per Month	Five % of Sales	Monthly Lease $
Jan	$65,000						
Feb	$55,000						
Mar	$65,000						
April	$70,000						
May	$80,000						
June	$70,000						
July	$70,000						
Aug	$85,000						

(continued)

		Option 1			Option 2		
Month	Sales Forecast	No. of Square Feet	Flat Fee per Square Foot	Monthly Lease $	Flat Fee per Month	Five % of Sales	Monthly Lease $
Sept	$90,000						
Oct	$95,000						
Nov	$110,000						
Dec	$135,000						
Total							

10. Libbey Hocking is the owner of the Hummingbird, an all-organic restaurant featuring fresh salads and a variety of vegetarian entrée dishes. As part of a dining room redesign, she is replacing all of the glassware in her 100-seat restaurant. Libbey would like to purchase 40 dozen glasses. Her glassware vendor has offered her similarly styled glassware at three different quality levels. The highest-quality glassware would cost Libbey $50.00 per dozen. The average life expectancy of these glasses is 1,000 uses before they either break or chip. A lower-priced, mid-quality glass sells for $35.00 per dozen and has an expected life of 750 uses. The least expensive glasses sell for $26.00 per dozen and have an expected life of 500 uses. Help Libbey get more information to assess her best purchase choice by completing the following product cost comparison worksheet. (Spreadsheet hint: Format the "Per Use Cost" column to five decimal places).

Hummingbird's Glassware Purchase Worksheet							
Product Durability	Price per Dozen	Number of Dozens	Total Cost	Total Number of Glasses	Per Glass Cost	Estimated Uses per Glass	Per Use Cost
Highest							
Middle							
Lowest							

a. Based on cost per use only, which quality glass should Libbey purchase?
b. What nonpurchase price factors might influence Libbey's choice of glassware?
c. If you were Libbey, which product alternative would you select? Explain your answer.

11. Braelon Frank is opening the first Papa Frank's pizza. Braelon is considering whether to deliver his takeout pizzas himself or contract with a popular third-party delivery company to make the deliveries for him. The third-party app delivery company is offering to deliver Frank's pizzas for 26 percent of the pizzas order's selling price. Braelon estimates his average pizza delivery order will generate $40.00 in revenue.

While Braelon knows there are many issues to be considered when making the decision to self-deliver his pizza or to contract with the third-party delivery company, he would like to know his relative costs, including his labor cost, if he should decide to self-deliver his pizzas.

Help Braelon better understand his delivery alternatives by completing the worksheet he has begun below. Then answer the questions that follow.

Papa Frank's Pizza	Self-Delivery	Third-Party Delivery	Difference (Net)
Sales per delivery	$40.00	$40.00	
Number of orders	25	25	
Gross sales	$1,000.00		
Third-party fees @ 26%	$0		
Wages ($18.25 per hour for 7 hours)		$0	
Vehicle costs ($3 per order)		$0	
Total wages and vehicle costs		$0	
Total delivery cost per $1,000 sales			

a. On a per-order basis, will it cost Braelon more to deliver his pizzas himself or to have a third-party delivery company make the deliveries?
b. What additional factors might cause Braelon to prefer the self-delivery option?
c. What additional factors might cause Braelon to prefer the third-party delivery option?

12. Vicki Font operates Casa Vieja, a Puerto Rican-themed restaurant in a mid-sized Florida city. Vicki has been approached by a third-party delivery company that wants to sign up for their services. The sales representative for the third-party delivery companies estimates that Casa Vieja's sales will increase by 10 to 25 percent if she agrees to list her operation with them. Vicki has decided to estimate the impact of such a potential sales increase on her restaurant's profitability.

    Vicki is preparing her profit comparison using the following assumptions:

    - ✓ 30 percent food cost on all sales
    - ✓ 26.5 percent delivery fees on all off-premise sales
    - ✓ 6 percent off-premise packaging cost on all off-premise sales
    - ✓ 4 percent management costs for on-premise sales, with no change for any additional off-premise sales
    - ✓ 24 percent staff cost on all sales
    - ✓ 13 percent of management + staff costs for employee benefits

    Help Vicki complete her cost comparison on the spreadsheet below and answer the following questions:

    1. What would be Vicki's net income if third-party delivery sales create a 10 percent increase in her total sales?
    2. What would be Vicki's net income if third-party delivery sales create a 25 percent increase in her total sales?
    3. What would you advise Vicki to do?

Vicki's Forecast	No Third-Party Sales	Plus 10% Third-Party Sales	Plus 25% Third-Party Sales
SALES			
Food sales	$2,000,000	$2,000,000	$2,000,000
Third-party delivery sales	$0	$200,000	$500,000
Total Sales	$2,000,000		
COST OF SALES			
Food @ 31%	$620,000		
Delivery fees @ 26.5%			
Takeout packaging @ 6%			
Total Cost of Sales	$620,000		
LABOR			
Management	$90,000	$90,000	$90,000
Staff @ 24%	$480,000		
Employee benefits @ 13%	$74,100		
Total Labor	$644,100		
PRIME COST	$1,264,100		
OTHER CONTROLLABLE EXPENSES			
Direct operating expenses	$80,000	$80,000	$80,000
Marketing	$100,000	$100,000	$100,000
Utilities	$20,000	$20,000	$20,000
Administrative & general expenses	$40,000	$40,000	$40,000
Repairs & maintenance	$10,000	$10,000	$10,000
Total Other Controllable Expenses	$250,000		
CONTROLLABLE INCOME	$485,900		
NON-CONTROLLABLE EXPENSES			
Occupancy costs	$20,000	$20,000	$20,000
Equipment leases	0	0	$0
Depreciation & amortization	$10,000	$10,000	$10,000
Total Non-Controllable Expenses	$30,000	$30,000	$30,000
RESTAURANT OPERATING INCOME	$455,900		
Interest expense	$24,000	$24,000	$24,000
INCOME BEFORE INCOME TAXES	$431,900		
Income taxes @ 24%	$103,656		
NET INCOME	$328,244		

CHAPTER 9

Analyzing Results Using the Income Statement

OVERVIEW

This chapter explains what you will do to analyze the cost-effectiveness of your operation. It teaches you how to read and use the income statement, a financial document that is also known as the profit and loss (P&L) statement. The chapter explains techniques used to analyze your sales volume as well as your food, beverage, labor, and other expenses. Finally, the chapter shows you how to review the income statement to best analyze your operation's profitability.

Chapter Outline

- Introduction to Financial Analysis
- Uniform System of Accounts
- Income Statement (USAR format)
- Analysis of Sales/Volume
- Analysis of Food Expense
- Analysis of Beverage Expense
- Analysis of Labor Expense
- Analysis of Other Expenses
- Analysis of Profits
- Technology Tools
- Apply What You Have Learned
- For Your Consideration
- Key Terms and Concepts
- Test Your Skills

LEARNING OUTCOMES

At the conclusion of this chapter, you will be able to:

- Prepare an income (P&L) statement using the USAR format.
- Analyze sales and expenses using the income statement.
- Evaluate a foodservice operation's profitability using the income statement.

INTRODUCTION TO FINANCIAL ANALYSIS

Far too many foodservice managers find that they collect information, fill out forms, and enter and receive numbers from their point of sale (POS) or accounting system with little regard for what they should actually do with all of the data they have collected.

Some have said that managers often make poor decisions because they lack information, but when it comes to the financial analysis of a hospitality operation, the opposite is usually true. Foodservice managers often find themselves awash in numbers! It is an important part of your job to sift through all of this information and select for analysis those numbers that can shed light on exactly what is happening in your operation. Among other things, three important things you will want to know are as follows:

- How much money did we take in?
- How much money did we spend?
- How much profit did we make?

This information is necessary to effectively operate your business. It also may be required by groups that are directly or indirectly involved with the financial operation of your facility. For example, local, state, and federal financial records relating to taxes and employee wages must be submitted to the appropriate governmental authorities on a regular basis. In addition, records showing the financial health of an operation may need to be submitted to new suppliers or vendors to establish an operation's credit worthiness.

If a foodservice operation has both operating partners and investors, those partners and investors will certainly require accurate and timely updates that focus on the financial health of the business. Owners, stockholders, and investment bankers may all have an interest in the day-to-day effectiveness of an operation's management. For each of these groups, and, of course, for the foodservice organization's own managers, a systematic and accurate examination of operational efficiency on a regular basis is essential.

As a professional foodservice manager, you will be very interested in examining your operation's cost of doing business. Documenting and analyzing sales, expenses, and profits is sometimes called **cost accounting**, but a more appropriate term for the process is **managerial accounting**, a term that reflects the importance managers place on this process. In this textbook, we use the term managerial accounting when referring to documenting, analyzing, and managing information related to an operation's sales, expenses, and profits.

It is important for you to be aware of the difference between **bookkeeping**, the process of simply recording and summarizing financial data, and managerial accounting, the actual analysis of that data. As an example, a POS system can be programmed to provide data about food and beverage sales per server. Management can, if it wishes, track by shift the relative sales effort of each service employee. If this is done with the goal of examining the data to increase the training needed by less productive servers or to reward the most productive ones, the POS system has, in fact, added information that is valuable and has assisted in the unit's operation.

If, on the other hand, information describing server effectiveness is dutifully recorded on a daily basis, filed away or sent to a regional office, and is then left to collect dust, the POS system has actually harmed the operation by taking management's time away from the more important task of running the business. It has converted the manager's role from that of skilled cost analyst to one of a bookkeeper (record keeper) only. This type of situation must be avoided because, as a professional manager, you need to maximize your effectiveness by being in the production area or dining room during high-service periods rather than in the office catching up on your paperwork.

Bookkeeping is essentially the summarizing and recording of data. Managerial accounting involves the summarizing, recording, and, most important, the analysis and use of that data. As a professional foodservice manager, you are also a managerial accountant!

You do not have to be a **certified management accountant** (CMA) or a **certified public accountant** (CPA) to analyze data related to foodservice revenue and expense. This is not meant to diminish the value of accounting professionals. It is important to recognize, however, that it is the professional foodservice manager, not an outside accounting expert, who is most qualified to assess the effectiveness of a foodservice team in controlling costs while providing the service levels desired by management. The analysis of operating data, a traditional role of the accountant, must also be part of your role as a foodservice manager. The process is not complex, and it is one of the most fun and creative aspects of a foodservice manager's job.

It is important to recognize that a good foodservice manager is, in fact, a manager first and not an accountant. It is also important, however, for you to be able to read and understand financial information and be able to speak intelligently and confidently with the many parties outside your operation who will read and use the information produced by you and your accountants.

It is also important to know that, by federal law, it is an operation's owners and managers (not its accountants) that will be called upon to verify the accuracy of the financial data they report. In the past, some managers in some industries have fraudulently reported their financial information. As a direct result, the United States Congress, in 2002, passed the **Sarbanes–Oxley Act** (SOX). Technically known as the Public Company Accounting Reform and Investor Protection Act, the law provides criminal penalties for those found to have committed accounting fraud.

SOX covers a whole range of corporate governance issues, including the regulation of those who are assigned the task of verifying a company's financial health. Ultimately, the US Congress determined that a company's implementation of proper financial reporting techniques is not merely good business, it is the law, and violators are subject to fines or even prison terms!

FUN ON THE WEB!

The 2002 Sarbanes–Oxley Act became law to help rebuild public confidence in the way corporate America governs its business activities. The Act has far-reaching implications for the tourism, hospitality, and leisure industry. To examine an overview of its provisions, go to http://www.sec.gov/about/laws.shtml

When you arrive at the site, select "Sarbanes-Oxley Act of 2002".

UNIFORM SYSTEM OF ACCOUNTS

Financial reports related to the operation of a foodservice facility are of interest to management, stockholders, owners, creditors, governmental agencies, and, often, the general public. To ensure that this financial information is presented in a way that is both useful and consistent, a **uniform system of accounts** has been established for many areas of the hospitality industry. In Chapter 1, you learned, for example, that the National Restaurant Association and its partners, have developed a Uniform System of Accounts for Restaurants (USAR).

Uniform systems of accounts also exist for hotels, clubs, nursing homes, schools, and hospitals. Each uniform system seeks to provide a consistent and clear manner in which managers working in those areas can record their sales, expenses, and the overall financial condition of their organizations. Sales categories, expense classifications, and methods of computing relevant ratios are included in the uniform systems of accounts. These uniform systems are often made available by the national trade or professional associations involved with each hospitality segment.

It is important to note that the uniform systems of accounts are guidelines, not a mandated methodology. Small foodservice operations, for example, may use the

USAR in a slightly different way than will large operations. In all cases, however, operators who use the uniform system of accounts "speak the same language," and it is truly useful that they do so. If operators prepared financial records in any manner they wished, it is unlikely that the many external audiences who must use them could properly interpret the reports. As a result, it is important that managers secure a copy of the applicable uniform system of accounts for their operations and then become familiar with its basic formats and reporting principles.

FUN ON THE WEB!

The USAR is a product of the National Restaurant Association. Hotels have their own uniform system of accounts. The Uniform System of Accounts for the Lodging Industry (USALI) is updated regularly. It is now in its 12th edition. Published by the Hospitality Financial and Technology Professionals (HFTP) organization, find out more about reporting foodservice results in hotels by visiting the their USALI website at https://usali.hftp.org.

FUN ON THE WEB!

Hospitality managers can learn what they need to know about reporting and analyzing their financial results in books devoted specifically to that subject. One of the very best books available is *Managerial Accounting for the Hospitality Industry*.

You can review the newest edition of the book at www.wiley.com. Enter the title in the search bar to explore the book's table of contents and, if you wish, to buy the book.

INCOME STATEMENT (USAR FORMAT)

The income statement is a summary report that describes the sales achieved, the money spent on expenses, and the resulting profit or loss generated by a business in a specific time period. Despite its formal title as the "Statement of Income and Expense," it is popularly referred to as the income statement, the profit and loss statement, or the P&L, the shortened title introduced in Chapter 1.

The P&L is a manager's most important tool for cost control. Properly designed and used P&Ls show the revenue and expenses of a business at a level of detail deemed best by the owners of the business and that conforms to the appropriate uniform system of accounts (e.g., for restaurants, hotels, clubs, or other industry segment).

P&Ls show the operating profits of a business. Interestingly, however, the word "profit" does not actually appear on a P&L prepared according to USAR recommendations! This is because the word "profit" can mean different things to different managers; therefore, the P&L statement can be somewhat confusing if you are not familiar with it. To see why, assume that someone asked you how much money you made on your last job. You could answer by telling the amount of gross earnings you achieved (your pay before taxes), or you could just as accurately answer by telling your take-home pay (your after-tax earnings). In either case, you would be accurately communicating the amount that you "earned," but the two answers you would provide would be very different numbers. In a similar manner, some business operators consider profit to be what they earn before they pay income taxes, whereas others reserve the term "profit" for their after-income tax earnings.

A precise definition of exactly what is meant by the term "profit" must be established for each P&L statement if it is to be helpful to its readers. In all cases, however, one purpose of the P&L statement is to identify **net income**, which is the amount of money remaining after all operating expenses of a business have been paid.

Figure 9.1 details 2 years of P&L statements presented in the USAR recommended format for Joshua's Inc., a foodservice complex that includes a cocktail

FIGURE 9.1 Joshua's Income Statement (P&L)[1]

Joshua's Inc.
Last Year *versus* This Year

	Last Year	%	This Year	%
SALES				
Food	$1,891,011	82.0%	$2,058,376	81.0%
Beverage	$415,099	18.0%	$482,830	19.0%
Total Sales	**$2,306,110**	**100.0%**	**$2,541,206**	**100.0%**
COST OF SALES				
Food	$712,587	37.7%	$767,443	37.3%
Beverages	$94,550	22.8%	$96,566	20.0%
Total Cost of Sales[2]	**$807,137**	**35.0%**	**$864,009**	**34.0%**
LABOR				
Management	$128,219	5.6%	$142,814	5.6%
Staff	$512,880	22.2%	$571,265	22.5%
Employee benefits	$99,163	4.3%	$111,813	4.4%
Total Labor	**$740,262**	**32.1%**	**$825,892**	**32.5%**
PRIME COST	**$1,547,399**	**67.1%**	**$1,689,901**	**66.5%**
OTHER CONTROLLABLE EXPENSES				
Direct operating expenses	$122,224	5.3%	$132,143	5.2%
Music & entertainment	$2,306	0.1%	$7,624	0.3%
Marketing	$43,816	1.9%	$63,530	2.5%
Utilities	$73,796	3.2%	$88,942	3.5%
General & administrative expenses	$66,877	2.9%	$71,154	2.8%
Repairs & maintenance	$34,592	1.5%	$35,577	1.4%
Total Other Controllable Expenses	**$343,611**	**14.9%**	**$398,970**	**15.7%**
CONTROLLABLE INCOME	**$415,100**	**18.0%**	**$452,335**	**17.8%**
NON-CONTROLLABLE EXPENSES				
Occupancy costs	$120,000	5.2%	$120,000	4.7%
Equipment leases	–	0.0%	–	0.0%
Depreciation & amortization	$41,510	1.8%	$55,907	2.2%
Total Non-controllable Expenses	$161,510	7.0%	$175,907	6.9%
RESTAURANT OPERATING INCOME	**$253,590**	**11.0%**	**$276,428**	**10.9%**
Interest Expense	$86,750	3.8%	$84,889	3.3%
INCOME BEFORE INCOME TAXES	**$166,840**	**7.2%**	**$191,539**	**7.5%**
Income Taxes	$65,068	2.8%	$76,616	3.0%
NET INCOME	**$101,772**	**4.4%**	**$114,923**	**4.5%**

[1]Income Statement format is based on the USAR, 8th ed., with the inclusion of income taxes and net income.
[2]All percentages are computed as a percentage of Total Sales except Cost of Sales line items, which are based on their respective category sales.

lounge, two dining areas, and banquet space. Joshua prepares his P&L statement for the calendar year period of January 1 through December 31.

In some operations, a fiscal year (see Chapter 4) is used to prepare P&Ls. Fiscal year reporting may be used in seasonal resorts, non-commercial facilities such as schools and colleges, and any other situation in which it makes good sense for the operation to do so.

P&Ls can be prepared for any accounting period that is helpful to management. As a result, managers may prepare annual, quarterly, monthly, or even weekly P&Ls.

As can be seen in Figure 9.1, last year Joshua's operation generated $2,306,110 in total sales and achieved a net income of $101,772. This year, when the corporation generated total sales revenue of $2,541,206, Joshua achieved a net income of $114,923. The question Joshua must ask himself, of course, is, "How good is this performance?"

HERE'S HOW IT'S DONE 9.1

Most restaurant owners are interested in knowing their operations' ability to generate cash available for spending. A restaurant's earnings before interest, taxes, depreciation, and amortization (**EBITDA**) is a measure that business owners can use to determine their net cash (net operating) income.

EBITDA is a unique number because it does not include the noncash operating expenses of interest, taxes, depreciation, and amortization. Payments for these items are listed on the income statement, or elsewhere on a business's financial statements, but they are not related to the day-to-day core operation of the business. For example, the interest paid on debts is listed on the P&L, but this expense is related to how the business was financed, and not the ability of the business to generate sales or profits. Similarly, income taxes due are listed on the P&L, but these do not affect how a business's managers operate on a daily basis.

Calculating EBITDA from a USAR-formatted income statement is an easy two-step process. Using the *This Year* data shown in Figure 9.1, the two steps are as follows:

Step 1: Identify Restaurant Operating Income

Restaurant Operating Income = $276,428

Step 2: To the Restaurant Operating Income amount, add back the amount listed for Depreciation & Amortization.

Restaurant Operating Income	$276,428
+Depreciation and Amortization	$55,907
= EBITDA	$332,335

In this example, Joshua's EBITDA for this year = $332,335.

Many owners view EBITDA as a good way to assess the earning power of their businesses and to compare their own operations with other similar operations that have different debt levels or depreciation amounts. But it is also important to recognize the very real limitations of EBITDA. Interest and taxes are indeed actual business expenses that must be paid. In addition, depreciation and amortization should reflect the decline in the real value of a business's assets. Each of these may be "noncash" expenses, but the expenses are real!

In general, most owners would agree that EBITDA is one good way to help assess the cash-generating ability of a restaurant, but it should not be the only metric used to make that assessment.

It is important to note that each foodservice operation's P&L could be formatted slightly differently than the P&L shown in Figure 9.1 even when it follows the USAR. All income statements, however, typically will take a very similar, rather than identical, approach to reporting revenue and expenses.

Note that, although the detail is much greater, the layout of Joshua's P&L is very similar in structure to the abbreviated P&L presented as Figure 1.4 in Chapter 1. Both statements list revenue first, then expenses, and finally the difference between the revenue and expenses figures. If this number is positive, it represents a profit. If expenses exceed revenue, a loss, represented by a negative number or a number in brackets, is shown. Operating at a loss is often referred to as operating "in the red" or "shedding red ink," based on an old practice of using black ink to record profits and red ink to record debts or losses. Regardless of the color of the ink, operating at a loss can cause an operator to shed a few tears!

To help ensure that your operation does not produce a loss, you need to know some important components of the USAR income statement format. The USAR can best be understood by dividing it into four major sections: (1) sales, (2) prime cost, (3) other controllable expenses, and (4) non-controllable expenses.

Referring to Figure 9.1, the Sales category includes the revenue generated by all of the food and beverage purchases made by Joshua's customers. The Prime Cost (see Chapter 8) section of Joshua's P&L details the amount that was spent for food and beverages as well as the total amount spent for labor. Prime cost is clearly listed on the P&L because it is an excellent indicator of management's ability to control cost of sales and labor, which you have learned are the two largest expenses in most foodservice operations.

The prime cost concept is important because when prime costs are excessively high, it makes it very difficult to generate a sufficient amount of profit in a food-service operation. This can be so even when controllable and non-controllable other expenses are kept well in line. While each restaurant operation is different, prime costs in the range of 60 to 70 percent are common for full-service operations, while prime costs below 60 percent are most common in quick-service restaurants and fast casual operations.

The Other Controllable Expenses categories for Joshua's include entries for:

Direct operating expenses
Music and entertainment
Marketing
Utilities
General and administrative expenses
Repairs and maintenance

The Non-controllable Expenses categories listed on Joshua's P&L include:

Occupancy costs
Equipment leases
Depreciation and amortization

Although interest expense and income taxes are not included in the non-controllable expenses section of the USAR format P&L, they are commonly regarded as non-operating and non-controllable expenses, therefore these costs can be assessed within the fourth section of the income statement.

Note that, when following the USAR, the foodservice manager arranges the four main sections of a P&L in order from most controllable to least controllable by the foodservice manager. Knowing what the four main sections of the income statement contain allows you to focus on those things over which you have the most control as a foodservice manager. When you focus on these controllable areas, you can better manage your time and make the most out of your efforts to control your costs.

Managers most often choose **restaurant operating income (ROI)**, the operating income subtotal before interest expense is subtracted, or **income before income taxes** (the operating income after interest expense is subtracted, but before income taxes are subtracted) as the most relevant indicators of management performance. This is because most managers do not have interest expenses or taxes in their direct span of control.

As you know, the government controls taxes; to paraphrase Benjamin Franklin, the only sure things in life are death and taxes. So, the foodservice manager has little control over the amount of money "Uncle Sam" and local taxing authorities get every year. Regardless of the specific income subtotal being considered, managers know that number directly reflects their ability to manage their operations.

Note also that each revenue and expense category in Figure 9.1 is expressed both in terms of its dollar amount and its percentage of total sales. All ratios are calculated as a percentage of total sales except the following:

- Food costs are divided by food sales.
- Beverage costs are divided by beverage sales.

When calculating food and beverage cost percentages, food expense is divided by food sales, and beverage expense is divided by beverage sales. Food costs and beverage costs are among the most controllable items on the income statement; thus, it helps Joshua to separate these sales and costs out of the aggregate and evaluate these individual expense areas carefully.

Notice also that Joshua's accountant presents this year's P&L statement along with last year's P&L statement in a side-by-side comparison because this can help Joshua assess changes in his costs and analyze trends in his business.

Another aspect of the uniform system of accounts that you should know about is the supporting schedule. The income statement as shown in Figure 9.1 is an **aggregate statement**. This means that all details associated with the sales, costs, and profits of the foodservice establishment are summarized on the P&L statement. Although this summary gives the manager an overall look at the performance of the operation, detailed expenses within each cost category are not included directly on the statement. These details can be found in **supporting schedules**. Each line item on the income statement should be accompanied by a supporting schedule that outlines all of the information that the manager needs to know about that category to operate the business successfully.

For example, in Figure 9.1, Direct Operating Expenses could have an accompanying supporting schedule that details costs incurred for uniforms, laundry and linen, china and glassware, silverware, and the like. These expenses can also be broken down by their percentage of total direct operating expenses. In addition, the supporting schedule should have a column in which notes can be made about the costs it includes.

Figure 9.2 is an example of a supporting schedule that could accompany Joshua's Income Statement (Figure 9.1). Note that the Total Direct Operating Expenses, $132,143 in the Direct Operating Expense Schedule, matches exactly with this year's Direct Operating Expenses as listed on the income statement.

The type of information and the level of detail that are included on supporting schedules are left up to the manager, based on what is appropriate for the manager's operation. The schedules are used to collect the information needed to give additional detail about sales or costs and to determine problem areas and potential opportunities for improving each item on the income statement.

The P&L statement is only one of several documents an owner will use to help evaluate the success of a business. The P&L statement alone, however, yields important information that is critical to the development of future management plans and budgets.

The analysis of P&L statements is a fun and very creative process if basic procedures are well understood. In general, managers who seek to discover all that their P&L will tell them undertake the following areas of analysis.

1. Sales/volume
2. Food expense
3. Beverage expense
4. Labor expense
5. Other controllable and non-controllable expense
6. Profits

Using the data from Figure 9.1, each of these areas will be reviewed in order.

FIGURE 9.2 Direct Operating Expenses Schedule

Type of Expense	Expense	% of Direct Operating Expenses	Notes
Uniforms	$13,408	10.15%	
Laundry and linen	40,964	31.00%	
China and glassware	22,475	17.01%	Expense is higher than budgeted because China shelf collapsed on March 22
Silverware	3,854	2.92%	
Kitchen utensils	9,150	6.92%	
Linen	2,542	1.92%	
Cleaning supplies	10,571	8.00%	
Paper supplies	2,675	2.02%	
Bar uniforms	5,413	4.10%	
Menus and wine lists	6,670	5.05%	Expense is lower than budgeted because the new wine supplier agreed to print the wine lists free of charge.
Exterminating	1,803	1.36%	
Flowers and decorations	9,014	6.82%	
Licenses	3,604	2.73%	
Total direct operating expenses	$132,143	100.00%	

ANALYSIS OF SALES/VOLUME

As addressed earlier in this book, foodservice operators can measure their sales in terms of either dollars generated or number of guests served. In both cases, an increase in sales volume is usually to be desired. A sales increase or decrease must, however, be analyzed carefully if you are to truly understand the revenue direction of your business.

STEPS IN COMPUTING SALES MEASURES

Consider the sales portion of Joshua's P&L statement, as detailed in Figure 9.3. Based on the data from Figure 9.3, Joshua can compute his overall sales increase or decrease using the following steps:

Step 1. Determine sales for this accounting period.

Step 2. Calculate the following: this period's sales minus last period's sales.

Step 3. Divide the difference in Step 2 by last period's sales to determine the sales percentage variance (increase or decrease).

FIGURE 9.3 Joshua's P&L Sales Comparison

Sales	Last Year	% of Sales	This Year	% of Sales
Food sales	$1,891,011	82.0%	$2,058,376	81.0%
Beverage sales	$415,099	18.0%	$482,830	19.0%
Total sales	$2,306,110	100.0%	$2,541,206	100.0%

FIGURE 9.4 Joshua's P&L Sales Variance

Sales	Last Year	This Year	Variance	Variance %
Food sales	$1,891,011	$2,058,376	$167,365	+ 8.9%
Beverage sales	$415,099	$482,830	$67,731	+16.3%
Total sales	$2,306,110	$2,541,206	$235,096	+10.2%

For Joshua, the sales percentage variance is as indicated in Figure 9.4. To illustrate the steps outlined for calculating sales variance using Total Sales as an example, we find:

Step 1. $2,541,206

Step 2. $2,541,206 – $2,306,110 = $235,096

Step 3. $235,096 ÷ $2,306,110 = 10.2%

The P&L statement shows that Joshua has achieved an overall increase in total sales of 10.2 percent. There are, however, several ways Joshua's total sales could have increased in the current year. These are:

1. More guests were served at the same check average
2. The same number of guests was served, but at a higher average sale per guest (**check average**)
3. More guests were served and at a higher check average
4. Fewer guests were served, but at a higher check average

Each of these four possibilities can lead to different conclusions about the direction of the business. To examine which of the possibilities was responsible for the total sales increase, Joshua may need to apply a sales adjustment technique.

To see why, assume that Joshua raised his prices for food and beverage by 5 percent at the beginning of the year. If this was the case, and he wishes to determine his actual sales increase, he must first consider the impact of that 5 percent menu price increase. The procedure he would use to adjust his sales variance to include the menu price increases is as follows:

Step 1. Increase the prior-period (last year) sales by the amount of the price increase.

Step 2. Subtract the result in Step 1 from this period's sales.

Step 3. Divide the difference in Step 2 by the value of Step 1.

Thus, in our example, Joshua would follow these steps to determine his real (after selling price adjustments) sales increase.

In the case of Total Sales, the procedure would be as follows:

$$\textbf{Step 1.}\quad \$2,306,110 \times 1.05 = \$2,421,415.50$$

$$\textbf{Step 2.}\quad \$2,541,206 - \$2,421,415.50 = \$119,790.50$$

$$\textbf{Step 3.}\quad \$119,790.50 \div \$2,421,415.50 = 4.95\%$$

FIGURE 9.5 Joshua's P&L Sales Comparison with 5 Percent Menu Price Increase

Sales	Last Year	Adjusted Sales (Last Year × 1.05)	This Year	Variance	Variance %
Food sales	$1,891,011	$1,985,561.55	$2,058,376	$72,814.45	+ 3.67%
Beverage sales	$415,099	$435,853.95	$482,830	$46,976.05	+10.78%
Total sales	$2,306,110	$2,421,415.50	$2,541,206	$119,790.50	+ 4.95%

Figure 9.5 details the results that are achieved if this 5 percent adjustment process is completed for all sales areas. Joshua's total sales figure would be up this year by 4.95 percent if he adjusted it for a 5 percent menu price increase. Thus, although Joshua's overall sales did increase significantly this year, approximately half of the increase was due to increased menu prices rather than to additional guest purchases.

There is still more, however, that the P&L can tell Joshua about his sales. If he has kept accurate guest count records, he can compute his average sales per guest (check average) amount. With this information, he can determine whether his sales are up because he is serving more guests or because he is serving the same number of guests, but each one is spending more per visit or because some of both has occurred.

In fact, if each guest is spending quite a bit more per visit, Joshua may have even experienced a *decrease* in total guest count yet still achieved an increase in total sales. If this were the case, he would want to know about it because it may be quite unrealistic to assume that revenue will continue to increase over the long run if the number of guests visiting the establishment is actually declining.

OTHER FACTORS INFLUENCING SALES ANALYSIS

In some foodservice operations, special factors must be taken into consideration before sales revenue can be accurately analyzed. To illustrate, consider the situation you would face if you owned a restaurant across the street from a professional basketball arena. If you were to prepare a P&L that compared sales from this March to sales generated last March, the number of home basketball games in March for this professional team in each of the two periods (which could vary) would have to be determined before you could make valid conclusions about guest count increases or decreases because the number of home games would very likely directly affect your sales levels.

Similarly, if a foodservice facility is open only Monday through Saturday, the number of these operating days contained in two given monthly accounting periods may be different for the facility. When this is the case, percentage increases or decreases in sales volume are more usefully based on average daily sales, rather than the total sales figure.

To illustrate this, consider Hot Dogs!, a hot-dog stand that operates in the city center of a large metropolitan area. It is open only on Monday through Friday. In October of this year, the stand was open for 21 operating days. Last year, however, because of the number of weekdays and weekend days in October, the stand operated for 22 days. Figure 9.6 details the comparison of sales for the stand, assuming no increase in menu selling prices this year when compared with last year's selling prices.

FIGURE 9.6 Hot Dogs! Sales Data

	Last Year	This Year	Variance	Variance %
Total sales (October)	$17,710.00	$17,506.00	–$204	–1.2%
Number of operating days	22 days	21 days	1 day	
Average daily sales	$805.00	$833.62	$28.62	+3.6%

At first glance, it would appear that October sales this year are 1.2 percent *lower* than last year, but in reality, average daily sales are up 3.6 percent! Are sales for October up or down? Clearly, the answer must be qualified in terms of monthly or daily sales. For this reason, effective foodservice managers must be careful to consider all of the relevant facts before making determinations about an operation's overall sales direction.

Managers must consider a number of possible impacting factors when evaluating changes in revenue. These can include the number of operating meal periods or days; changes in menu prices, guest counts and check averages, and holidays and special events contained in the accounting period. Only after carefully reviewing all details of this type can you truly assess the degree to which your operation's sales are increasing or decreasing.

ANALYSIS OF FOOD EXPENSE

In addition to sales analysis, P&L statements provide information about other areas of operational interest. For the effective foodservice manager, the analysis of food expense is a matter of major concern.

Figure 9.7 details the food expense portion of Joshua's P&L as detailed in the Cost of Sales: Food Expense Schedule.

It is important to remember that the numerator of the food cost percentage equation is cost of food sold, whereas the denominator is total food sales, rather than total food and beverage sales.

With total cost of food sold this year of \$767,785 and total food sales of \$2,058,376, the total food cost percentage is 37.3 percent (\$767,785 ÷ \$2,058,376 = 37.3%). A food cost percentage can be computed in a similar manner for each subcategory of food. For instance, the food cost percentage for the category Meats and seafood costs for this year would be computed as follows:

$$\frac{\text{Meats and seafood costs}}{\text{Total food sales}} = \text{Meats and seafood cost \%}$$

or

$$\frac{\$344{,}063}{\$2{,}058{,}376} = 16.7\%$$

FIGURE 9.7 **Joshua's P&L Cost of Sales: Food Expense Schedule**

	Last Year	% of Food Sales	This Year	% of Food Sales
Food sales	\$1,891,011	100.0%	\$2,058,376	100.0%
Cost of sales: Food				
Meats and seafood	\$297,488	15.7%	\$343,063	16.7%
Fruits and vegetables	\$94,550	5.0%	\$127,060	6.2%
Dairy	\$55,347	2.9%	\$40,660	2.0%
Baked goods	\$16,142	0.9%	\$22,870	1.1%
Other	\$249,060	13.2%	\$233,790	11.4%
Total cost of food sales	\$712,587	37.7%	\$767,443	37.3%

FIGURE 9.8 Joshua's P&L Variation in Food Expense by Category

Category	Last Year %	This Year %	Variance %
Meats and seafood	15.7	16.7	+1.0
Fruits and vegetables	5.0	6.2	+1.2
Dairy	2.9	2.0	−0.9
Baked goods	0.9	1.1	+0.2
Other	13.2	11.4	−1.8
Total cost of food sold %	37.7	37.4	−0.4

At first glance, it appears that Joshua has done well for the year and that his total cost of goods sold expense has declined 0.4 percent, from 37.7 percent overall last year to 37.3 percent this year. This is true.

Closer inspection of Figure 9.7, however, indicates that, although the categories Dairy and Other showed declines, Meats and seafood, Fruits and vegetables, and Baked goods showed increased expense.

Figure 9.8 shows the actual differences in food cost percentage for each of Joshua's food categories. Although it is true that Joshua's overall food cost percentage is down by 0.4 percent (37.7 − 37.3 = 0.4), the variation among categories is quite significant. It is clearly to his benefit to subcategorize food products so that he can watch for fluctuations within and among groups rather than merely monitor his increase or decrease in overall food costs. Without such a breakdown of categories, he will not know exactly where to look if his food costs get too high.

Sometimes food costs rise because the purchase prices of needed ingredients rise. An additional potential area of loss is simply carelessness in following standardized recipes (see Chapter 3). In other cases, food costs rise because too much food is held in inventory, resulting in excessive product loss or theft.

It would be helpful for Joshua to determine how appropriate the inventory levels are for each of his product subgroups so that he can adjust the inventory sizes if needed. To do this, Joshua must be able to compute his food inventory turnover.

Cost Control Around the World

You have learned that the USAR contains suggestions for preparing income statements primarily for businesses operated in the United States. However, income statements prepared in other countries can vary in format from the income statements prepared according to USAR recommendations. One good example of this relates to the concept of profits. In the United Kingdom, for example, income statements are routinely prepared in a way that clearly indicates an operation's **gross profit**, often referred to as "gross operating profit" or GOP for short. GOP is calculated as follows:

Total sales − Total cost of sales = Gross operating profit (GOP)

Many operators feel GOP is a key number because it assesses food and beverage sales and those food- and beverage-related costs that can and should be directly controlled by the manager on a daily basis. Earlier editions of the USAR actually included GOP calculations, but the most recent edition does not. In many other countries, however, GOP remains a key entry on the income statement.

The proper preparation of income statements and understanding the concept of gross profit are both good examples of the expanded knowledge many foodservice managers must have when they are responsible for businesses operating in countries that establish their own preferred accounting systems used for reporting a foodservice unit's operating results.

FOOD INVENTORY TURNOVER

Inventory turnover refers to the number of times the total value of an operation's inventory has been purchased and replaced in an accounting period. Each time the cycle is completed once, we are said to have "turned" the inventory.

For example, if you normally keep $100 worth of oranges on hand at any given time and last month's usage of oranges was $500, you would have replaced your orange inventory five times that month. The formula used to compute a food inventory turnover is as follows:

$$\frac{\text{Cost of food consumed}}{\text{Average inventory value}} = \text{Food inventory turnover}$$

Note that it is cost of food consumed (see Chapter 5), rather than cost of food sold, that is used as the numerator in this ratio. This is because all food usage should be tracked so that you can better determine what is sold, wasted, spoiled, pilfered, and provided to employees as employee meals.

Inventory turnover is a measure of how many times an operation's inventory value is purchased and sold to guests. In the foodservice industry, we seek high inventory turnover rates. If a 5 percent profit is made on the sale of an inventory item, we would like to sell (turn) that item as many times per year as possible. If, in this example, the item were sold from inventory only once per year, then only one 5 percent profit would result. If the item turned 10 times, however, a 5 percent profit on each of the 10 sales would result. But managers must ensure that a high inventory turnover is caused by increased sales and not by increased food waste, food spoilage, or employee theft.

Joshua can compute his overall food inventory turnover or he can compute inventory turnovers for each of his food categories. To compute his inventory turnover for each food category, Joshua must first establish his average inventory value for each food category. The average inventory value is computed by adding the beginning inventory value for a food category for an accounting period to the ending inventory value of the food category for the same period, and then dividing by 2, using the following formula:

$$\frac{\text{Beginning inventory value} + \text{Ending inventory value}}{2} = \text{Average inventory value}$$

From his inventory records, Joshua creates the data recorded in Figure 9.9.

To illustrate the computation of average inventory value, note that Joshua's Meats and Seafood beginning inventory for this year was $16,520, whereas his ending inventory for that category was $14,574. His average inventory value for

FIGURE 9.9 Joshua's P&L Average Inventory Values

Inventory Category	This Year Beginning Inventory	This Year Ending Inventory	Average Inventory Value
Meats and seafood	$16,520	$14,574	$15,547
Fruits and vegetables	$1,314	$846	$1,080
Dairy	$594	$310	$452
Baked goods	$123	$109	$116
Other	$8,106	$9,196	$8,651
Total	$26,657	$25,035	$25,846

FIGURE 9.10 Joshua's P&L Food Inventory Turnover

Inventory Category	Cost of Food Consumed	Average Inventory	Inventory Turnover
Meats and seafood	$343,063	$15,547	22.1
Fruits and vegetables	$127,060	$1,080	117.6
Dairy	$40,660	$452	90.0
Baked goods	$22,870	$116	197.2
Other	$233,790	$8,651	27.0
Total/average	$767,443	$25,846	29.7

that category is $15,547 ([$16,520 + $14,574] ÷ 2 = $15,547). All other categories and the total average inventory value are computed in the same manner.

Now that Joshua has determined the average inventory values for his food categories, he can compute the inventory turnovers for each of these. As you recall from Chapter 5, cost of food consumed is identical to cost of food sold when no reduction is made in cost as a result of employee meals. That is the case at Joshua's facility because he charges employees full price for menu items that he sells to them as employee meals. Therefore, employee meals are included in regular food cost because a normal food sales price is charged for these meals. Figure 9.10 shows the result of his computations using the food inventory turnover formula for this year.

To illustrate, Joshua's Meats and seafood inventory turnover is 22.1 ($343,063 ÷ $15,547 = 22.1). That is, Joshua purchased, sold, and replaced his meat and seafood inventory, on average, just over 22 times this year. Note that all other food categories and the total inventory turnover are computed in the same manner. Note also that in categories such as Fruits and vegetables, Dairy, and Baked goods, the turnovers are very high, reflecting the perishability of these items. Joshua's overall inventory turnover is 29.7.

If Joshua's *target* inventory turnover for the year was 26 times (one turnover every 2 weeks), then he should investigate why his actual inventory turnover is higher. It could be because of his increase in sales, which is a good sign of his restaurant's performance, or it could be due to excessive waste, pilferage, or spoilage.

Alternatively, if his inventory turnover was less than 26 times per year, and his target was 26 times, the difference could be due to lower sales or overstocking of his inventory. Joshua could be buying excess inventory at a bulk discount (which is an intentional management decision to save money) or he might be purchasing too much due to poor inventory management. He should use inventory turnover analysis to help him determine how he can most effectively control his product costs in the future.

ANALYSIS OF BEVERAGE EXPENSE

Joshua's P&L (Figure 9.1) shows beverage sales for this year of $482,830. With total sales of $2,541,206, beverages represent 19 percent of Joshua's total sales ($482,830 ÷ $2,541,206 = 19%). Also from Figure 9.1, beverage costs for this year are shown as $96,224; thus, Joshua's beverage cost percentage, which is computed as cost of beverages divided by beverage sales, is 19.9 percent ($96,224 ÷ $482,830 = 19.9%).

To completely analyze the beverage expense category, Joshua would compute his beverage cost percentage, compare that to his planned or targeted beverage expense, and compute a beverage inventory turnover rate using the same inventory turnover formula he utilized for his food products.

If an operation carries a large number of rare and expensive wines, its overall beverage inventory turnover rate will be relatively low. Conversely, those beverage operations that sell their products such as beer and spirits primarily by the glass are likely to experience inventory turnover rates that are quite high. The important concept here is to compute the turnover rates at least once per year (or more often if needed) to gauge whether inventory sizes should be increased or decreased.

High beverage inventory turnovers accompanied by frequent product outages may indicate inventory levels that are too low, whereas low inventory turnover rates and many slow-moving inventory items may indicate the need to reduce beverage inventory levels.

Figure 9.1 shows that last year's beverage cost percentage was 22.8 percent (\$94,550 ÷ \$415,099 = 22.8%). This year's beverage cost is 19.9 percent. This, at first glance, would indicate that beverage costs have been reduced, not in total dollars spent, because sales were higher this year than last year, but in percentage terms. In other words, a beverage cost percent of 22.8 last year *versus* 19.9 percent this year indicates a 2.9 percentage point overall reduction (22.8% – 19.9% = 2.9%).

Assume, for a moment, however, that Joshua raised drink prices by 10 percent this year over last year. Assume also that Joshua pays, on average, 5 percent more for beverages this year compared with last year. After making those assumptions, is his beverage operation more efficient this year than last year, less efficient, or the same? To determine the answer to this important question in the beverage or any other expense category, Joshua must make adjustments to both his sales and his cost figures. Similar to the method for adjusting sales, the method for adjusting expense categories for known cost increases is as follows:

Step 1. Increase the prior-period expense by the amount of the cost increase.

Step 2. Identify the appropriate sales figure, remembering to adjust prior-period sales, if applicable.

Step 3. Divide costs determined in Step 1 by sales determined in Step 2.

Thus, in our example, Joshua's beverage expense last year, adjusted for this year's costs, would be \$94,550 × 1.05 = \$99,277.50.

His beverage sales from last year, adjusted for this year's menu prices, would be \$415,099 × 1.10 = \$456,608.90.

His last year's beverage cost percentage, adjusted for increases in this year's costs and selling prices, would be computed as follows:

$$\frac{\$99{,}277.50 \text{ adjusted beverage costs}}{\$456{,}608.90 \text{ adjusted beverage sales}} = 21.7\%$$

In this case, Joshua's real cost of beverage sold has, in fact, declined this year, although not by as much as he had originally thought. That is, a 21.7 percent adjusted cost for last year *versus* a 19.9 percent cost for this year equals a reduction of 1.8 percent, not 2.9 percent as originally determined (21.7% – 19.9% = 1.8%).

In nearly all cases, food and beverage expense categories must be adjusted in terms of both their menu selling prices and costs if truly accurate comparisons are to be made over longer periods of time. For example, when, in the long ago past, foodservice managers purchased hamburger for \$0.59 per pound, it is important to recall that the quarter-pound hamburger they made from it may have sold for \$0.59 also! The 25 percent resulting product cost percentage is no different from today's operator paying $7.00 a pound for ground beef and selling the resulting quarter-pound burger for $7.00.

It is not possible to compare efficiency in food and beverage usage from one time period to the next unless you are making that comparison in equal terms. As product costs increase or decrease and as menu prices change, they may result in changes in food and beverage expense percentages. As an effective manager, you must determine if variations in product usage and cost percentages are caused by real changes in your operation or simply by differences in the price you pay for your products or by changes in your selling (menu) prices.

ANALYSIS OF LABOR EXPENSE

From Figure 9.1, it is interesting to note that, although the total dollars Joshua spent on labor increased greatly from last year to this year, his labor cost percentage increased only slightly. This was true in the Management and Staff categories as well as the Employee benefits category. Recall that whenever your labor costs are not 100 percent variable, increasing sales volume will help you decrease your overall labor cost percentage, even when the total dollars you spend on labor will increase. The reason for this is simple. When total dollar sales volume increases, fixed labor cost percentages will decline. In other words, the dollars paid for fixed labor will consume a smaller percentage of your total revenue. Thus, as long as any portion of total labor cost is fixed (management's salaries, for example), increasing volume will have the effect of reducing an operation's overall labor cost percentage.

Variable labor costs increase with sales volume increases, but the percentage of revenue they consume should stay relatively constant. When you combine a declining percentage (fixed labor cost) with a constant percentage (variable labor cost), you should achieve a reduced overall percentage, although your total labor dollars expended will increase as sales increase.

It is important to note that serving additional guests will most often cost an operation additional labor dollars. That, in itself, is not a bad thing. In most food-service situations, you want to serve more guests. If labor expenses are controlled properly you will find that an increase in the number of guests served, and the sales generated by these additional guests, will result in increased total labor costs required to service those additional guests. You must be careful, however, to always ensure that increased labor costs are at the appropriate level for increases achieved in sales volume.

Remember, too, that declining costs of labor are not always a sign that all is well in a foodservice operation. Declining costs of labor may be the result of significant reductions in the number of guests served. If, for example, a foodservice facility produces poor-quality products and/or provides poor service, guest counts can be expected to decline, as would the cost of labor required to service those guests who do remain. Labor dollars spent would decline, but it would be an indication of improper, rather than effective, management. An effective foodservice manager seeks to achieve reductions in operational expense because of improved operational efficiencies, not because of reduced sales.

Figure 9.11 details the labor cost portion of Joshua's P&L.

Note that all the labor-related percentages he computes are based on his total sales, that is, the combination of his food and beverage sales. This is different from computing expense percentages for food and beverages because food cost percentage is determined using food sales as the denominator and beverage cost percentage computations use beverage sales as the denominator. Thus, Joshua's

FIGURE 9.11 Joshua's P&L Labor Cost

Labor Cost	Last Year	% of Total Sales	This Year	% of Total Sales
Management	$128,219	5.6%	$142,814	5.6%
Staff	$512,880	22.2%	$571,265	22.5%
Employee benefits	$99,163	4.3%	$111,813	4.4%
Total labor cost	$740,262	32.1%	$825,892	32.5%

total management, staff, and employee benefits expense percentage for this year is computed as follows:

$$\frac{\text{Total labor cost}}{\text{Total sales}} = \text{Labor cost \%}$$

or

$$\frac{\$825{,}892}{\$2{,}541{,}206} = 32.5\%$$

A brief examination of the labor portion of Joshua's P&L would indicate an increase in both dollars spent for labor and labor cost percentage. Just as adjustments must be made for changes in sales and food and beverage expenses before valid expense comparisons can be made, so too adjustments must be made for changes, if any, in the price an operator pays for labor.

In Joshua's case, assume that all employees were given a **cost of living adjustment (COLA)**, or pay raise, of 5 percent at the beginning of this year. Recall from Chapter 7 that this, coupled with an assumed 10 percent menu price increase, would indeed have the effect of changing the operation's overall labor cost percentage, even if labor productivity did not change.

From Figure 9.11, Joshua can see that his actual labor cost percentage increased from 32.1 percent last year to 32.5 percent this year, an increase of 0.4 percent (32.5% − 32.1% = 0.4%). To adjust for the changes in the cost of labor and his selling prices, if these indeed occurred, Joshua uses assessment techniques similar to those previously detailed in this chapter. Thus, based on the assumption of a 5 percent increase in the cost of labor and a 10 percent increase in the average menu selling price, he adjusts both sales and cost of labor, using the same steps as those employed for adjusting food or beverage cost percentage, and he computes a new labor cost for last year as follows:

Step 1. Determine sales adjustment: $2,306,110 × 1.10 = $2,536,721

Step 2. Determine total labor cost adjustment: $740,262 × 1.05 = $777,275.10

Step 3. Compute adjusted labor cost percentage: $777,275.10 ÷ $2,536,721 = 30.6%

As can be seen, last year Joshua's P&L would have indicated a 30.6 percent labor cost percentage *if* he had operated under this year's increased labor costs and this year's selling prices. This is certainly an area that Joshua would want to investigate. The reason is simple. If he were exactly as efficient this year as he was last year, his sales were constant, and he assumed a 10 percent menu price increase and a 5 percent labor cost increase, Joshua's cost of labor for this year should have been computed as follows:

This year's sales × Last year's adjusted labor cost % = This year's projected labor cost

or

$2,541,206 × 0.306 = $777,609.04

Put in another way, if Joshua were just as efficient with his labor this year as he was last year, he would have expected to spend 30.6 percent of sales for labor this year, given his 5 percent payroll increase and 10 percent menu price increase.

In actuality, Joshua's labor cost was $48,282.96 higher ($825,892 actual labor cost this year − $777,609.04 projected labor cost this year = $48,282.96). A variation in labor cost this large should be of concern to Joshua, and he should examine it closely.

Increases in payroll taxes, benefit programs, and employee turnover can all affect overall labor cost percentage. Although, in this example, we assumed that employee benefits (including payroll taxes, insurance, etc.) increased at the same 5 percent rate as did salaries and wages, these are, of course, different expenses and may increase at rates higher or lower than salary and wage payments to employees. Indeed, one of the fastest-increasing labor-related costs for foodservice managers today is the cost of health insurance benefit programs. This is especially the case with the passage of the Patient Protection and Affordable Care Act (the **Affordable Care Act**), the federal law mandating, in some situations, that foodservice employers provide health-care insurance for their employees (see Chapter 7). In addition, many States and even cities have moved in recent years to increase the minimum wages required to be paid to foodservice workers.

Controlling and evaluating total labor cost will be an important part of your job as a hospitality manager. In fact, many managers feel it is even more important to control labor costs than product costs because, for many of them, labor and labor-related benefit costs comprise a larger portion of their operating budgets than do food and beverage product costs.

Consider the Costs

"I'm just saying that we are already reporting our daily labor cost percentage to the owners, and now they also want to know this every day. It just seems like more paperwork to me," said Raul Sullivan, the assistant manager of the New York Sliders restaurant.

Raul was talking to Karis, the restaurant's manager. Karis had just informed Raul that the restaurant's owners wanted her to submit a daily "Average Drive-thru Ticket Time Report" along with the day's revenue and labor cost percentage reports.

Along with on-site dining, the New York Sliders drive-thru was seeing an increasing amount of business. However, there have been some complaints about slow service. In response, the restaurant's owners requested that Karis calculate and report an average drive-thru ticket time. With the POS system in use, ticket times began when a guest's order was placed and ended when the guest left the drive-thru window with their ordered items.

One reporting feature of the POS system was that it automatically calculated the average drive-thru ticket time for all drive-thru orders, and this was the information the operation's owners now wanted to review each day.

Assume you were Karis.

a. Why do you think this operation's owners are now requesting this new employee productivity report?

b. How important do you believe the information contained in this report will help your labor cost control efforts, and your efforts to optimize guest service in the drive-thru?

ANALYSIS OF OTHER EXPENSES

An analysis of Other Expenses should be performed each time the P&L is produced. Figure 9.12 details the Other Expenses listed on Joshua's P&L statement.

Joshua's Other Expenses consist of both Other Controllable Expenses and Non-controllable Expenses, (see Chapter 8) and he must review these carefully.

In Joshua's operation, Other Expenses have increased from last year's levels. To see why, each individual expense category must be reviewed. Note that his Repairs and maintenance expense category is higher this year than it was last year. This is one area in which he both expects and approves a cost increase. It is logical to assume that kitchen repairs will increase as a kitchen ages. In that sense, a kitchen is much like an automobile. As the auto ages, it maintenance costs will typically increase. Even with a good preventative maintenance program, Joshua does not expect an annual decline in kitchen repair expense. In fact, he would be somewhat surprised and concerned should this category be smaller this year than in the previous year because his sales were higher and likely caused more wear and tear on his one-year-older kitchen equipment.

In the same way, General and administrative expenses include charges for credit card commissions and fees (see Chapter 11) and the increase in this expense category likely reflects both an increase in Joshua's sales and an increase in the number of noncash (credit) sales made compared to the number of cash sales made.

A complete analysis of Other Expenses could prove difficult for Joshua if he is not sure how his results compare with other operations in his area or with operations of a similar nature. For comparison purposes, he is able to use industry trade publications to get national averages on Other Expense categories. One helpful source Joshua can use is an annual publication, *Restaurant Operations Data Abstract*, published by the National Restaurant Association (it can be ordered through the National Restaurant Association's website).

For operations that are a part of a franchise group or corporate chain, unit managers can most often receive comparison data from district and regional managers, who can supply performance information from similar units operating in the city, region, state, and nation.

FIGURE 9.12 Joshua's P&L Other Expenses

	Last Year	% of Total Sales	This Year	% of Total Sales
Other Controllable Expenses				
Direct operating expenses	$122,224	5.3%	$132,143	5.2%
Music & entertainment	$2,306	0.1%	$7,624	0.3%
Marketing	$43,816	1.9%	$63,530	2.5%
Utilities	$73,796	3.2%	$88,942	3.5%
General & administrative expenses	$66,877	2.9%	$71,154	2.8%
Repairs & maintenance	$34,592	1.5%	$35,577	1.4%
Total other controllable expenses	$343,611	14.9%	$398,970	15.7%
Non-controllable Expenses				
Occupancy costs	$120,000	5.2%	$120,000	4.7%
Equipment leases	$0	0%	$0	0%
Depreciation & amortization	41,510	1.8%	$55,907	2.2%
Total non-controllable expenses	$161,510	7.0%	$175,907	6.9%

The Move to Off-Premise Dining

The continued move by customers to off-premise dining affects foodservice operators in many ways. Of significant importance is the way that it impacts "Other Expenses" in a foodservice operation.

While every foodservice operation is unique, many operators are finding that their "Other Expenses" costs are increasing significantly in the following areas:

1. **Direct Operating Expenses:** When preparing an income statement using USAR, Direct Operating Expenses is the cost category used to record the cost of to-go boxes and containers, paper and packaging, and disposable food containers including disposable cups and straws. As increasing numbers of off-premise guest sales require these packaging items, the cost of supplying them will increase as well.
2. **Marketing:** Most foodservice industry professionals will agree that a restaurant's proprietary website is more important than ever. Guests increasingly search the Internet as they select restaurants. These searches include location, menu offerings, menu prices, and guest reviews.

   For many off-premise diners, websites must also include an easy method of ordering and paying for their purchases. This requires an operation's website to be interfaced (digitally connected) with its ordering and payment acceptance systems. Because of their increasing importance, the cost of maintaining a vibrant and impactful website has increased. Routine maintenance and upgrading of a website is typically recorded as an operation's "Marketing" expense. As a result, this cost category has increased significantly for many foodservice operators.
3. **Credit card fees:** The fees involved in an operation accepting credit cards, debit cards, and other electronic payment forms (e.g., Apple Pay, Google Pay, Pay Pal, and the like) are recorded as "credit card fees" in an operation's General and Administrative account. As fewer and fewer customers pay with cash, these payment processing fees have increased, and are quite likely to continue to do so in the future.

ANALYSIS OF PROFITS

As can be seen in Figure 9.1, "profits" for Joshua's Inc. is represented by the "Net Income" figure located at the bottom of the income statement. In fact, because of its location on the income statement, some managers refer to this number as "the bottom line."

In addition to being evaluated on the basis of their ability to generate profit dollars, foodservice managers are often evaluated on their ability to achieve targeted profit margins (see Chapter 8). A profit margin represents the amount of profit generated on each dollar of sales. Profit margin represents that portion of a dollar's sale that is returned to the operation in the form of profits. For that reason, profit margin is also referred to as **return on sales (ROS)**. This year, Joshua's profit margin is 4.5 percent. Thus, for each dollar of sales generated at Joshua's, 4.5 cents of profit was made. Joshua calculates his profit margin using the following formula:

$$\frac{\text{Net income}}{\text{Total sales}} = \text{Profit margin}$$

or

$$\frac{\$114{,}923}{\$2{,}541{,}206} = 4.5\%$$

For the foodservice manager, perhaps no number is more important than profit margin (ROS). This percentage is often considered to be the most telling indicator of a manager's overall effectiveness at generating revenues and controlling costs in line with forecasted results.

Although it is not possible to state what a "good" ROS figure should be for all restaurants, restaurant industry averages for ROS (depending on the specific restaurant segment) range from 1 percent to over 20 percent.

Some operators prefer to use ROI (see Figure 9.1) as the numerator for their ROS calculations instead of Net Income. This is because, in most cases, an operation's interest expense is a cost established by the operation's owners, not its managers. As a result, amounts paid for interest and the taxes paid by a business are not truly reflective of a manager's ability to generate a profit.

Regardless of the manner in which profits are viewed, and profit margins are calculated, it is most important to recognize that foodservice operators ultimately bank "dollars" and not "percentages." Clearly, a manager whose operation achieves a 15 percent ROS on revenues of $2,000,000 is going to generate more profits than a manager who achieves a 20 percent ROS on revenues of $500,000. Because that is true, effective managers should be concerned about both the dollar amount of profit they generate and their ROS percentage.

Joshua's ROS results of this year ($114,923 or 4.5 percent of sales) represent an improvement over last year's figure of $101,772, or 4.4 percent of total sales. Thus, he has shown improvement both in the dollar amount of his net income and in the size of net income as a percentage of total sales.

HERE'S HOW IT'S DONE 9.2

In most businesses, increased sales should mean increased profits. But it is important to recognize that sales increases will likely mean increased costs as well. For example, serving more restaurant guests will result in greater sales, but it will also result in increased food and labor costs. In some cases, additional revenue that comes from excessively discounting menu prices could actually cause an increase in sales, but a reduction in operating profits. As a result, business owners want to know exactly how much an increase (or decrease) in sales affects their bottom lines.

Flow-through is the measure used to identify how well a foodservice operation converts *additional* sales dollars into additional ROI.

Calculating flow-through from a USAR-formatted income statement is an easy, three-step process when you utilize the flow-through formula. That formula is as follows:

$$\frac{\text{ROI this year} - \text{ROI last year}}{\text{Sales this year} - \text{Sales last year}} = \text{Flow-through}$$

For example, to calculate flow-through using the data shown in Figure 9.1, the three steps are as follows:

Step 1: Subtract last year's ROI from this year's ROI:

$$\$276{,}428 - \$253{,}590 = \$22{,}838$$

Step 2: Subtract last year's sales from this year's sales:

$$\$2{,}058{,}376 - \$1{,}891{,}011 = \$167{,}365$$

Step 3: Divide the results from Step 1 by the result from Step 2:

$$\frac{\$22{,}838}{\$167{,}365} = 13.6\%$$

In this example, for each additional dollar of sales generated above last year's sales level, Joshua's operation achieved 13.6 cents (13.6 percent) of additional ROI. This is a full 2.7 cents higher than his overall ROI average per dollar (13.6 − 10.9 = 2.7). Calculating an operation's flow-through is one good way to help assess how revenue increases directly impact ROI, and now you know how to do it!

In Chapter 2, you learned how to calculate a sales variance and a sales variance percentage. The variance amount and percentage variance between last year's net income and this year's net income can be calculated using similar formulas. Joshua can compute the variance in his net income by subtracting net income last year from net income this year as follows:

$$\text{Net income this year} - \text{Net income last year} = \text{Net income variance}$$

or

$$\$114{,}923 - \$101{,}772 = \$13{,}151$$

Joshua can then compute his net income percentage variance as follows:

$$\frac{\text{Net income this year} - \text{Net income last year}}{\text{Net income last year}} = \text{Percentage variance}$$

or

$$\frac{\$114{,}923 - \$101{,}772}{\$101{,}772} = 12.9\%$$

An alternative, and shorter, formula for computing the percentage variance is as follows:

$$\frac{\text{Variance}}{\text{Net income last year}} = \text{Percentage variance}$$

or

$$\frac{\$13{,}151}{\$101{,}772} = 12.9\%$$

Yet another way to compute the percentage variance is to use the math shortcut of subtracting 1.00, as follows:

$$\frac{\text{Net income this year}}{\text{Net income last year}} - 1.00 = \text{Percentage variance}$$

or

$$\frac{\$114{,}923}{\$101{,}772} - 1.00 = 12.9\%$$

or

$$1.129 - 1.00 = .129 = 12.9\%$$

How much of this improvement is due to improved operational methods *versus* increased sales will depend, of course, on how much Joshua actually did increase his sales relative to increases in his costs. Monitoring selling price, guest count, sales per guest (check average), operating days, special events, and actual operating costs is necessary for accurate profit comparisons. Without knowledge of each of these areas, the effective analysis of profits becomes a risky proposition.

HERE'S HOW IT'S DONE 9.3

Sometimes managers simply want to know the difference between two numbers. For example, they may want to know the difference between the amount of money spent this year for food and the amount of money spent for food last year, or between the amount of profit earned this year and the amount of profit earned last year. Calculating the difference, or variance, between any two numbers simply requires the use of subtraction.

Sometimes, however, managers know they will gain more information when they calculate the percentage variance between two numbers. Calculating a *percentage variance* is a two-step process that is easy to do once you learn how.

To illustrate, assume that you wish to calculate the percentage variance between last year's profits and this year's profits. Last year's profit was $300,000. This year's profit is $310,500. The two-step process for calculating variance percentage in this example is as follows:

Step 1: Subtract last year's number from this year's number. In this example, that would be:

$310,500 − $300,000 = $10,500

Step 2: Divide the difference calculated in Step 1 by last year's number. In this example, that would be:

$10,500 ÷ $300,000 = 0.035 or 3.5%

That tells the manager this year's profit is 3.5 percent *higher* than last year's profit.

Or you can use a shortcut by dividing this year's number by last year's number and subtracting 1.00. In this example, that would be:

($310,500 ÷ $300,000) − 1.00 = 3.5%

The identical two-step process works when a number from this year is *smaller* than a number from last year. To illustrate, assume that last year's profit was $312,000 and profit this year is $297,600. The two-step process for calculating variance percentage in this case is:

Step 1: Subtract last year's number from this year's number. In this example, that would be:

$297,600 − $312,000 = −$14,400 (note this is a negative number!)

Step 2: Divide the difference calculated in Step 1 by last year's number. In this example, that would be:

−$14,400 ÷ $312,000 = −0.046 or −4.6%

That tells the manager that this year's profit was 4.6 percent *lower* than last year's profit.Using the short cut again, you can divide this year's number by last year's number and subtract 1.00. In this example, that would be:

($297,600 ÷ $312,000) − 1.00 = −4.6%

With a little practice you will be able to calculate and assess the percentage difference between any two numbers with ease!

Green and Growing!

Some foodservice operators think that adopting a sustainable, or "green," operations viewpoint inevitably means putting planet before profit. They would argue that businesses can survive only if they put profits first so that is where their focus should be. In the long run, of course, an organization's purpose must include a focus on both profit and planet. It is true that an unprofitable business will not stay in business long. It is also true that "planet friendly" management yields many positive financial outcomes for businesses, as well as for the health of the local communities these businesses count on to support them.

For example, the simple act of buying local (to minimize transportation costs and environmental impact) produces an additional reward: community connection. Buying local creates relationships with those who produce food and keeps money flowing through a local economy, resulting in a healthier community.

There are additional positive outcomes from the local food movement. For example, as more hospitality businesses use local, seasonal, sustainably raised food, more fresh fruits and vegetables may be consumed, resulting in positive nutrition and health impacts within the community. The long-term result may be a reduction in health-care costs. Perceptive foodservice operators now clearly recognize that profits, planet, and people all benefit from an operation's green commitment.

Technology Tools

This chapter introduced the concept of management analysis as it relates to sales, expenses, and profits. In this area, available software and apps are increasing rapidly and there are many tools available to assist you. The best of these tools can help you:

1. Analyze sales trends over management-established time periods.
2. Analyze food and beverage costs.
3. Analyze labor costs.
4. Analyze other expenses.
5. Analyze profits.
6. Evaluate the financial productivity of individual servers, day parts, or other specific time periods established by management.
7. Integrate sales enhancing customer relationship management programs with POS data.
8. Interface accounting systems with POS system data.
9. Compare actual operating results to budgeted results.
10. Suggest revisions to future budget periods based on current operating results.
11. "Red flag" areas of potential management concern regarding financial performance.
12. Compare the operating results of multiple profit centers within one location or across several locations.

Apply What You Have Learned

Terri Settles is a registered dietitian (RD). She supervises five hospitals for Maramark Dining Services, the management company the hospitals have selected to operate their foodservice departments. Her company produces a monthly and annual income statement for each hospital.

1. Discuss five ways in which income statements can help Terri better do her job.
2. What would a hospital do with profits or surpluses made in the foodservice area?
3. What effect will profit or loss have on the ability of Terri's company to continue to manage the foodservices for these hospitals?

For Your Consideration

1. Most managers want their P&Ls to indicate their operating results in both dollar amounts and as percentages. What are some specific reasons why managers can find it helpful to see their operating costs expressed as a percentage of total revenue?
2. Some managers prefer to produce P&Ls that compare their operating results to those of a prior accounting period. Other managers prefer to produce P&Ls that compare their operating results to their forecasted (or budgeted) results. Which of these two ways do you think would be best? Why do you think so?

Key Terms and Concepts

The following are terms and concepts addressed in the chapter that are important for you as a manager. To help you review, define the terms below:

Cost accounting
Managerial accounting
Bookkeeping
Certified management accountant (CMA)
Certified public accountant (CPA)
Sarbanes–Oxley Act (SOX)
Uniform system of accounts
Net income
EBITDA
Restaurant operating income (ROI)
Income before income taxes
Aggregate statement
Supporting schedules
Check average
Gross profit (GOP)
Inventory turnover
Cost of living adjustment (COLA)
Affordable Care Act
Return on sales (ROS)
Flow-through

Test Your Skills

You may download the Excel spreadsheets for the Test Your Skills exercises from the student companion website at http://www.wiley.com/go/dopson/foodandbeveragecost-control8e. Complete the exercises by placing your answers in the shaded boxes and answering the questions as indicated.

1. Jacob operates a Sicilian-themed restaurant that does a significant amount of carryout pizza business. Jacob's P&L has been designed to indicate the sales levels achieved in his dining room, carryout business, and bar. The revenue Jacob achieved this month and last month is indicated on the partial P&L shown as follows. Calculate Jacob's total sales for this month and last month as well as the variances and percentage variances in each of his revenue categories, then answer the questions that follow.

Jacob's Restaurant				
	Last Month	This Month	Variance	Variance %
Sales				
Dining room	$169,000	$172,500		
Carryout	$30,000	$31,000		
Bar	$27,000	$25,500		
Total sales				

a. Did Jacob's total sales this month increase or decrease when compared to last month's sales?

b. In what revenue producing area(s) did Jacob's sales increase?

c. In which sales area did Jacob experience the greatest dollar increase?

d. In which sales area did Jacob experience the greatest percentage increase?

e. In which sales area did Jacob's sales not improve?

2. Marylou Wishing operates Marylou's Shrimp boat. Her restaurant features a daily "All You Can Eat" shrimp and catfish special along with other Gulf coast seafood specialties. Mary Lou is evaluating her sales for February.

   Last year was a Leap year and thus, in February last year she was open 29 days. This year is not a leap year, so Marylou was open 28 days in February. Help Marylou better understand whether her sales are improving or declining by completing the chart below, and then answering the questions that follow:

Marylou's Shrimp Boat					
Revenue	**Operating Days**	**Last Year**	**This Year**	**$ Variance**	**% Variance**
Monthly revenue (Feb)		$568,000	$553,800	$	
Daily revenue last year	29	$			
Daily revenue this year	28		$	$	

   a. What was the total amount of Marylou's February dollar sales variance this year when compared to last year? What was the variance percentage?

   b. Are Marylou's daily sales increasing, declining, or staying the same when compared to February last year?

3. Roscco Burnett is the manager of a New York style delicatessen. Roscco's boss has established performance targets that determine the amount of bonus Roscco will receive each month. Roscco's performance is evaluated in five separate categories. Roscco receives a 2 percent bonus for each of the five targets he achieves. He earns a 2 percent bonus when:

   a. His food cost is 30 percent or less.

   b. His total cost of sales is 31 percent or less.

   c. His labor cost is 22 percent or less for the "Staff" category.

   d. His total labor cost is 35 percent or less.

   e. His prime costs are less than 63 percent.

   If he achieves all five of his targets, the maximum bonus Roscco can achieve is 10 percent of his monthly salary. Review the targets established for Roscco's performance, complete Roscco's P&L Sales, and Prime Cost sections, and then answer the question that follows.

Roscco's P&L Sales and Prime Cost		
	This Month	**%**
Sales		
Food	$71,500	
Beverage	9,250	
Total sales		
Cost of Sales		
Food	$22,250	
Beverage	2,200	
Total cost of sales		

(continued)

Roscco's P&L Sales and Prime Cost		
	This Month	%
Labor		
Management	$4,750	
Staff	16,252	
Employee benefits	6,225	
Total labor cost		
Prime Cost		

What percentage of his monthly salary will Roscco receive as a bonus this month?

4. Ethan Bischoff is the owner of the City-Grow Garden, a healthy foods restaurant that offers unique menu items from around the world. Ethan has just created his P&L showing this year's final operating results. Ethan wants to compare the labor portion of his P&L to his last year's results, as well as to the results he had originally forecasted for the year. Help Ethan complete the comparisons, and then answer the questions that follow:

City-Grow Garden						
Labor Analysis	This Year		Last Year		This Year Forecast	
	Actual	%	Actual	Variance %	Actual	Variance %
Total sales	$1,255,100	100.0%	$1,147,850	9.3%	$1,200,000	4.6%
Management	$52,087		$51,125	1.9%		
Staff	$229,056	18.3%	$227,325	0.8%	$231,900	−1.2%
Employee benefits			$67,815	1.8%	$72,345	−4.6%
Total labor cost	$350,173				$356,395	

a. What was Ethan's labor cost percentage last year?

b. What was Ethan's labor cost percentage forecast for this year?

c. Do Ethan's P&L results for this year indicate an improvement in his labor force productivity level over last year's level? Explain your answer.

5. Aafreen is the owner of a restaurant that features Pakistani and American cuisine. She operates on a calendar year and prepares a monthly P&L as well as a Year to Date (YTD) P&L. Aafreen is preparing her P&L for April. She has entered the amounts and calculated her percentages for the month. She has also entered her YTD revenue and expense amounts. Help Aafreen complete her P&L by calculating her missing YTD percentages.

Income Statement, April 20XX				
	April	%	YTD	%
Sales				
Food	$85,200	78.7%	$331,200	
Liquor	$7,150	6.6%	$27,500	
Beer	$5,100	4.7%	$20,500	
Wine	$10,850	10.0%	$23,250	
Total sales	$108,300	100.0%	$402,450	
Cost of Sales				
Food	$24,150	28.3%	$101,750	
Liquor	$1,875	26.2%	$6,400	
Beer	$1,430	28.0%	$4,850	
Wine	$4,150	38.2%	$8,750	
Total cost of sales	$31,605	29.2%	$121,750	
Labor				
Management	$7,125	6.6%	$28,500	
Staff	$19,925	18.4%	$79,500	
Employee benefits	$4,170	3.9%	$14,900	
Total labor	$31,220	28.8%	$122,900	
Prime Cost	$62,825	58.0%	$244,650	

Income Statement, April 20XX				
	June	%	YTD	%
Other Controllable Expenses				
Direct operating expenses	$6,850	6.3%	$22,450	
Music & entertainment	$475	0.4%	$1,850	
Marketing	$1,820	1.7%	$7,250	
Utilities	$3,125	2.9%	$12,450	
General & administrative expenses	$4,950	4.6%	$18,150	
Repairs & maintenance	$2,875	2.7%	$5,250	
Total other controllable expenses	$20,095	18.6%	$67,400	
Controllable income	$25,380	23.4%	$90,400	
Non-controllable Expenses				
Occupancy costs	$8,850	8.2%	$35,400	
Equipment leases	$1,725	1.6%	$6,900	
Depreciation & amortization	$2,125	2.0%	$8,500	
Total non-controllable expenses	$12,700	11.7%	$50,800	
Restaurant operating income	$12,680	11.7%	$39,600	
Interest expense	$1,100	1.0%	$4,400	
Income before income taxes	$11,580	10.7%	$35,200	
Income taxes	$2,300	2.1%	$7,040	
Net income	$9,280	8.6%	$28,160	

6. Faye manages Faye's Tea Room in a small suburban town. She sells gourmet food and a variety of teas. This year, Faye increased her selling prices by 5 percent and she increased her wages by 10 percent. Faye's condensed P&L follows. Help her calculate her variance and variance percentage change from last year to this year. Use her adjusted sales and labor cost to provide a more accurate picture of her performance this year.

Increase in Selling Prices					
Increase in Wages					
Faye's Condensed P&L					
	Last Year	**Adjusted Sales and Labor (for Last Year)**	**This Year**	**Variance**	**Variance %**
Sales	$1,865,000		$2,315,000		
Cost of food	$615,450		$717,650		
Cost of labor	$540,850		$671,350		
Other expenses	$428,950		$486,150		
Total expenses					
Profit					

7. a. Rudolfo owns Rudolfo's Italian Restaurante. He wants to compare last year's costs to this year's costs on his food expense schedule to see how he performed in each food category. Help Rudolfo complete his schedule.

Rudolfo's Food Expense Schedule				
	Last Year	**% of Food Sales**	**This Year**	**% of Food Sales**
Food Sales	$2,836,517		$3,087,564	
Cost of food sold				
Meats and seafood	$386,734		$445,982	
Fruits and vegetables	$122,915		$165,178	
Dairy	$71,951		$52,858	
Baked goods	$20,985		$29,731	
Other	$323,778		$303,927	
Total cost of food sold				

b. In addition to calculating the food cost percentage for each of his food categories, Rudolfo wishes to calculate his inventory turnover for the year. Help Rudolfo complete his inventory turnover table below. Assume that cost of food sold, part a, and cost of food consumed, part b, are the same for Rudolfo's.

Rudolfo's inventory turnover target for this year was 32 times. Did he meet his target? If not, what may have caused this?

Rudolfo's Food Inventory Turnover					
Inventory Category	This Year's Beginning Inventory	This Year's Ending Inventory	Average Inventory Value	Cost of Food Consumed	Inventory Turnover
Meats and seafood	$21,476	$17,489		$445,982	
Fruits and vegetables	$1,708	$1,015		$165,178	
Dairy	$772	$372		$52,858	
Baked goods	$160	$131		$29,731	
Other	$10,538	$11,035		$303,927	
Total					

8. Captain Jaymal Davies is director of club operations for five military bases in Florida. He has just received year-end income statements for each base. Information from the revenue and labor portion of those statements is shown as follows. Jaymal wants to use the current year's data to create next year's budget. Assume that Jaymal is happy with his labor productivity in each unit and that both wages and revenue in each will increase 2 percent next year. Calculate how much Jaymal should budget for revenue and labor in each unit. Also, calculate what Jaymal's labor cost percentage will be for each unit if he meets his budget. (Spreadsheet hint: Use the ROUND function to one decimal place for Projected Labor Cost %.)

Projected Increase in Wages	
Projected Increase in Revenues	

	This Year's Results		Next Year's Budget		
Location	This Year's Labor Cost	This Year's Revenue	Projected Labor Cost	Projected Revenue	Projected Labor Cost %
Pensacola	$285,000	$980,500			
Daytona	$197,250	$720,000			
Fort Myers	$235,500	$850,250			
Tampa	$279,750	$921,750			
Miami	$1,190,250	$3,720,000			

9. Ron MacGruder is senior vice president of acquisitions for Yummy Foods. Yummy Foods is a large multinational foodservice company. It owns over 2,000 restaurants. Among its famous brands are a chain of pizza parlors, a Mexican carryout group, a chain of fried chicken stores, and a large group of fried fish stores. Ron is constantly on the lookout for growing foodservice concepts that could be purchased at a fair price and added to the Yummy Foods group. One such concept that is currently available for sale is a small but expanding group of upscale Thai restaurants called Bow Thais. The current owners wish to sell the 17-unit chain and retire. They proudly point to the fact that both their sales and profits have increased in each of the last 3 years. Revenue has more than doubled this year compared to 2 years ago. Profits have risen from just $600,000 2 years ago to

over $1,000,000 this year. In addition, the Bow Thais chain consisted of 6 units 2 years ago, 12 units last year, and 17 units this year.

Develop a summary P&L for each of Bow Thais' last 3 years showing revenue, expense, and profit. Next compute a per-unit revenue, expense, profit, and profit margin level for each of the 3 years for which you have data.

Would you advise Ron to buy the company? Why or why not?

	Revenue	Expenses	Profit	No. of Units	Revenue per Unit	Expenses per Unit	Profit per Unit	Profit Margin %
2 Years Ago	$6,500,000		$600,000					
Last Year	$9,900,000		$800,000					
This Year	$13,500,000		$1,010,000					

10. Elena Visten is the manager of the Hoffbrau House, a German-themed restaurant that features steaks and roasted Bavarian-style pork dishes. Elena wants to compare this year's operating results to last year's results. Complete the 2-year comparison she has begun and then answer the questions that follow.

Hoffbrau House				
			Difference	
	This Year ($)	Last Year ($)	$	%
Sales				
Food	3,499,939	3,025,618		
Beverage	773,228	622,649		
Total sales				
Cost of sales				
Food	1,151,165	997,622		
Beverages	144,849	132,370		
Total cost of sales				
Labor				
Management	299,913	269,262		
Staff	771,206	628,277		
Employee benefits	167,720	138,828		
Total labor				
Prime cost				
Other controllable expenses				
Direct operating expenses	198,215	171,114		
Music & entertainment	11,436	3,228		
Marketing	95,295	61,342		
Utilities	133,413	103,314		
Administrative & general expenses	106,731	93,628		
Repairs & maintenance	53,366	48,429		
Total other controllable expenses				

Hoffbrau House				
			Difference	
	This Year ($)	Last Year ($)	$	%
Controllable income				
Non-controllable expenses				
Occupancy costs	180,000	168,000		
Equipment leases	0	0		
Depreciation & amortization	83,861	58,114		
Total non-controllable expenses				
Restaurant operating income				
Interest expense	127,334	121,450		
Income before income taxes				
Income taxes	313,850	261,315		
Net income				

a. One of Elena's goals this year was to increase sales by 15 percent when compared to last year. Did Elena meet that goal?

b. Another of Elena's goals this year was to increase her operation's net income by 8.5 percent when compared to last year. Did Elena meet that goal?

c. What was the operation's EBITDA this year?

11. Basil Bakal is the newly appointed food and beverage director at Telco Industries. Telco creates and markets software apps developed for use with Apple products. The company has 500 employees and operates its own cafeteria and executive dining room, where it offers daily free lunches to all employees.

Basil's employee cafeteria serves between 375 and 425 lunches per day. Approximately 50 more meals per day are served in the executive dining room. Basil has created his own modified version of a P&L for use in his operation. Calculate the percentages of meals served in the employee cafeteria and executive dining room and the costs per meal served and then answer the questions that follow.

a. How much more did it cost (cost of sales per meal) Basil to serve a meal in the executive dining room than it did in the employee cafeteria? Why do you think that would be so?

b. Basil's modified P&L combines all labor costs when calculating cost per meal served. Why do you think he elected not to allocate labor costs between the two serving areas? How could he do so?

c. Assume that you were on the board of directors of Telco. How would you decide how much more money you should allocate to Basil's area next year to account for rising food prices? Who would you expect to provide you with the information you need to make an informed decision about the appropriate size of this increase?

Telco Industries Food Services Department		
	Meals Served Last Year	% of Total Meals Served
Number of Meals Served		
Cafeteria	104,250	
Executive dining room	12,150	
Total served		
	Total cost $	Per meal cost $
Cost of Sales		
Cafeteria	$248,750	
Executive dining room	$48,450	
Total cost of sales		
Labor		
Management	$73,332	
Staff	$171,108	
Employee benefits	$61,110	
401(k) match	$25,350	
Total labor		
Prime Cost		
Other Controllable Expenses		
Direct operating expenses	$113,515	
Utilities	$96,000	
General & administrative expenses	$46,669	
Repairs & maintenance	$21,510	
Total other controllable expenses		

12. Lucir manages a full-service restaurant located in a large urban area. The restaurant's owner wants to know how well Lucir did this year at generating sales, controlling costs, and providing a profit. The owner promised Lucir that he would give her a raise if she increased return on sales (profit margin) by at least 1 percent. Complete Lucir's P&L.

    Should Lucir receive a raise?

Lucir's P&L				
	Last Year	%	This Year	%
Sales				
Food	$2,647,415		$2,675,889	
Beverage	$498,119		$965,660	
Total sales				
Cost of Sales				
Food	$855,104		$1,074,420	
Beverages	$104,005		$115,879	
Total cost of sales				

Lucir's P&L				
	Last Year	%	This Year	%
Labor				
Management	$192,330		$204,227	
Staff	$576,989		$581,260	
Employee benefits	$118,996		$122,994	
Total labor				
Prime Cost				
Other Controllable Expenses				
Direct operating expenses	$146,669		$145,357	
Music & entertainment	$2,767		$8,386	
Marketing	$52,579		$69,883	
Utilities	$88,555		$97,836	
Administrative & general expenses	$80,252		$78,269	
Repairs & maintenance	$41,510		$39,135	
Total other controllable expenses				
Controllable Income				
Non-controllable Expenses				
Occupancy costs	$144,000		$132,000	
Equipment leases	$0		$0	
Depreciation & amortization	$49,812		$61,498	
Total non-controllable expenses				
Restaurant Operating Income				
Interest expense	$104,100		$93,378	
Income Before Income Taxes				
Income taxes	$235,146		$343,150	
Net Income				

CHAPTER 10

Planning for Profit

OVERVIEW

This chapter shows you how to analyze your menu so you can identify which menu items contribute the most to your profits. In addition, it teaches you how to determine the sales dollars and guest volume levels you must achieve to break even and to generate a profit in your operation. Finally, the chapter shows you how to establish an operating budget and presents techniques you can use to monitor your effectiveness in staying within your budget.

Chapter Outline

- Financial Analysis and Profit Planning
- Menu Analysis
- Cost/Volume/Profit Analysis
- The Budget
- Developing the Budget
- Monitoring the Budget
- Technology Tools
- Apply What You Have Learned
- For Your Consideration
- Key Terms and Concepts
- Test Your Skills

LEARNING OUTCOMES

At the conclusion of this chapter, you will be able to:

- Analyze a menu for profitability.
- Prepare a cost/volume/profit (break-even) analysis.
- Establish a budget and monitor performance to the budget.

FINANCIAL ANALYSIS AND PROFIT PLANNING

In addition to analyzing the profit and loss (P&L) statement (see Chapter 9), you should also undertake a thorough study of three additional areas of analysis that will assist you in planning for profit:

1. Menu analysis
2. Cost/volume/profit (CVP) analysis
3. Budgeting

Menu analysis examines the profitability of the individual menu items you sell. CVP analysis examines the sales dollars and volume required by your foodservice unit to avoid an operating loss and make a profit. The process of budgeting allows you to plan your next year's operating results by estimating your future sales, expenses, and profits and then developing a budgeted, or projected, P&L statement.

Some foodservice operators "hope for profit" instead of "plan for profit." Although hoping is an admirable attitude when playing the lottery, it does little good when managing a foodservice operation. Smart managers know that preplanning is the key to achieving the cost and profit goals that will keep them in business.

MENU ANALYSIS

Effective managers want to know the answer to a basic operational question: "How does the sale of a particular menu item contribute to the overall success of my operation?" The answer to such a question can sometimes be provided by applying mathematics, but numbers are only one component of menu analysis. There are others to consider.

Consider the case of Danny, who operates a successful family restaurant, called The Mark Twain, in rural Tennessee. The restaurant has been in his family for three generations. One item on the menu is mustard greens with scrambled eggs. It does not sell often, but both mustard greens and eggs are ingredients in other, more popular items. Why does Danny keep the item in a prominent spot on the menu? The answer is simple and has little to do with finance. The menu item was Danny's grandfather's favorite. As a thank-you to his grandfather, who started the business and inspired Danny to become service and guest oriented, the menu item survives every menu reprint.

Menu analysis is about more than just analyzing numbers. It involves marketing, sociology, psychology, and oftentimes a good deal of emotion. Remember that guests respond best not to weighty financial analyses, but rather to menu item descriptions, placement on the physical menu, price, and current popularity. Although the financial analysis of a menu is indeed done "by the numbers," you must realize that while those numbers are an important part, they are merely one part of the complete menu analysis process.

The Move to Off-Premise Dining

Menu items to be sold for off-premise consumption must be carefully chosen because the precise time between order pickup and item consumption is not known in advance. For guests picking up menu items and transporting them to a nearby location the time between pickup and consumption may be very short. Alternatively, on a busy night when a guest order has been prepared for pickup and delivery by a third-party delivery service, the length of time items will be held before consumption may be relatively long.

Menu item selection for takeaway menu items is important because guests always equate an operation's quality standards with the products they receive when they are picked up or delivered. Some menu items can be held for a relatively long time with no reduction in quality. Examples include pizza, fried chicken, deli sandwiches, and undressed salads. Some menu items, however, do not travel well. Examples are many deep-fried items (including French fries!), ice cream-based items, sauced items such as Eggs Benedict, souffles, and delicate fish dishes.

Each foodservice operation is different. Operators should carefully consider whether items that deteriorate when not quickly consumed should be offered on their off-premise menus. Troublesome items that do not reflect the quality of the entire menu should be omitted in favor of items that can be transported while preserving food quality, texture, and flavor.

For the serious foodservice manager, the analysis of a menu deserves special study. Many components of the menu, such as item pricing, menu layout, design, and **menu copy** (the written descriptions of menu items), all play an important role in the overall success of a foodservice operation. This is true whether an operation presents its menu on paper, on a manual or digital menu board, or via a QRS code (see Chapter 4).

If you investigate the menu analysis methods that have been widely used by professional foodservice managers in recent times, you will find that each method seeks to perform a menu analysis using one or more of the following operational variables.

- Food cost percentage
- Item popularity
- Contribution margin (CM)
- Selling price
- Variable expenses
- Fixed expenses

Although there are many variations, three of the most popular systems of menu analysis are shown in Figure 10.1. They represent the three major philosophical approaches to menu analysis.

The **matrix analysis** referenced in Figure 10.1 provides a method for comparisons between menu items. The use of a matrix allows menu items to be placed into categories based on whether they are above or below overall menu item averages for factors such as food cost percentage, popularity, and CM.

Each approach to menu analysis has its proponents and detractors, but an understanding of each approach will help you as you attempt to develop your own philosophy of menu analysis.

FIGURE 10.1 Three Methods of Menu Analysis

Method	Variables Considered	Analysis Method	Goal
1. Food cost %	a. Food cost % b. Popularity	Matrix	Minimize overall food cost %
2. Contribution margin	a. Contribution margin b. Popularity	Matrix	Maximize contribution margin
3. Goal value	a. Contribution margin % b. Popularity c. Selling price d. Variable cost % e. Food cost %	Algebraic equation	Achieve predetermined profit % goals

FOOD COST PERCENTAGE

Menu analysis that focuses on food cost percentage is the oldest and most traditional method used. When analyzing a menu using the food cost percentage method, you are seeking menu items that have the effect of minimizing your overall food cost percentage. The rationale for this is that a lowered food cost percentage leaves a higher percentage of the sales dollar to be spent for labor and other operational expenses.

A criticism of the food cost percentage approach is that, when it is used, items that have a higher food cost percentage may be removed from the menu in favor of items that have a lower food cost percentage, but when purchased by guests, these lower food cost percentage items may also contribute fewer dollars to overall profit.

To illustrate the use of the food cost percentage menu analysis method, consider the case of Maureen, who operates a steak and seafood restaurant near the beach in a busy resort town. Maureen sells seven items in the entrée section of her menu. The items and information related to their popularity, selling price, portion cost, and CM (see Chapter 6) for a one-week period in January are presented in Figure 10.2.

The numbers in the rows of Figure 10.2 are calculated as follows:

Individual Menu Item Rows:

Number Sold and Selling Price are obtained from the point-of-sale system.

Total Sales = Number Sold × Selling Price

Item Cost (food only) is obtained from the item's standardized recipe cost sheet.

Total Cost = Number Sold × Item Cost

Item CM = Selling Price − Item Cost

Total CM = Number Sold × Item CM

Food Cost % = Total Cost/Total Sales *or* Item Cost/Selling Price

Total Row:

The totals in the "Number Sold," "Total Sales," "Total Cost," and "Total CM" columns are calculated by simply summing the numbers in those columns.

Average (Mean) Row:

To calculate the values in the "Mean (Average)" row (see Chapter 2), divide the sum of the column by the number of menu items.

FIGURE 10.2 Maureen's Menu Analysis Worksheet

Date: 1/15–1/21

Menu Item	Number Sold	Selling Price	Total Sales	Item Cost	Total Cost	Item CM	Total CM	Food Cost %
Strip Steak	73	$35.95	$2,624.35	$16.16	$1,179.68	$19.79	$1,444.67	45.0%
Coconut Shrimp	121	$33.95	$4,107.95	$10.18	$1,231.78	$23.77	$2,876.17	30.0%
Grilled Tuna	105	$35.95	$3,774.75	$14.38	$1,509.90	$21.57	$2,264.85	40.0%
Chicken Breast	140	$27.95	$3,913.00	$6.14	$859.60	$21.81	$3,053.40	22.0%
Lobster Stir-Fry	51	$43.95	$2,241.45	$22.40	$1,142.40	$21.55	$1,099.05	51.0%
Scallops/Pasta	85	$29.95	$2,545.75	$7.18	$610.30	$22.77	$1,935.45	24.0%
Beef Medallions	125	$31.95	$3,993.75	$11.82	$1,477.50	$20.13	$2,516.25	37.0%
Total	700		$23,201.00		$8,011.16		$15,189.84	34.5%
Average (Mean)	100	$34.24		$12.61		$21.63		35.5%
Weighted Average		$33.14		$11.44		$21.70		34.5%

Weighted Average Row:
To calculate the values in the "Average/Weighted Average" row (see Chapter 2), do the weighted average calculations.

Weighted Average	Calculation
Weighted average selling price	Total sales/total number sold = $23,201/700 = $33.14
Weighted average item cost	Total cost/total number sold = $8,011.16/700 = $11.44
Weighted average item CM	Total CM/total number sold = $15,189.84/700 = $21.70
Weighted average food cost %	Total cost/total sales = $8,011.16/$23,201.00 = 34.5%

To analyze her menu using the food cost percentage method, Maureen separates her items based on two characteristics:

1. Food cost percentage
2. Popularity (number sold)

Maureen's overall food cost is 34.5 percent, so she considers any individual menu item with a food cost percentage above 34.5 percent to be high in food cost percentage, whereas any menu item with a food cost below 34.5 percent will be considered low in food cost percentage.

In a similar manner, with a total of 700 entrées served in this accounting period and seven possible menu choices, each menu item would sell 700/7, or 100 times, if all were equally popular. Given that fact, Maureen determines that any item that sold more than 100 times during this week's accounting period will be considered high (above average) in popularity, whereas any item selling less than 100 times will be considered low (below average) in popularity. Having made these determinations, Maureen produces a matrix labeled as follows:

		Popularity	
		Low	High
Food Cost %	High	1 High food cost % Low popularity	2 High food cost % High popularity
	Low	3 Low food cost % Low popularity	4 Low food cost % High popularity

Based on the number sold and food cost percentage data in Figure 10.2, Maureen can classify her menu items in the following manner:

Square	Characteristics	Menu Item
1	High food cost % Low popularity	Strip Steak, Lobster Stir-Fry
2	High food cost % High popularity	Grilled Tuna, Beef Medallions
3	Low food cost % Low popularity	Scallops/Pasta
4	Low food cost % High popularity	Coconut Shrimp, Chicken Breast

Note that Maureen can place each menu item in one, and only one, of the four matrix squares. Using the food cost percentage method of menu analysis, Maureen would like as many menu items as possible to fall within square 4. These items have the characteristics of being low in food cost percentage but high in popularity (guest acceptance). Thus, both coconut shrimp and chicken breast have below-average food cost percentages and above-average popularity.

When developing a menu that seeks to minimize food cost percentage, items in the fourth square are highly desirable because the low food cost percentage of the individual items helps keep the operation's overall food cost percentage low and the items sell very well. These items are, of course, kept on the menu. They should be well promoted and have high menu visibility. With these items, managers should take care not to develop any new menu items with a higher food cost percentage that are similar enough in nature that they could detract from the sales of these lower-cost items.

The characteristics of the menu items that fall into each of the four matrix squares are unique and, thus, should be managed differently. Because of this, items that fall into each individual square require a special marketing strategy, depending on their matrix location. These strategies can be summarized as shown in Figure 10.3.

It can be quite effective to use the food cost percentage method of menu evaluation. It is fast, easy to calculate, and time-tested. Remember that if you achieve too high a food cost percentage, you run the risk that not enough percentage points will remain to generate a profit on your sales.

In spite of its popularity, however, the food cost percentage menu analysis method has limitations. For example, most foodservice operators would say it is better to achieve a 20 percent food cost on a menu item than a 40 percent food cost. Consider, however, a chicken dish that sells for $20.00 and costs you just $4.00 to make. This item yields a 20 percent food cost ($4.00 food cost ÷ $20.00 selling price = 20% food cost), and there is $16.00 ($20.00 – $4.00 = $16.00) remaining to pay for the labor and other expenses of serving this guest.

FIGURE 10.3 Analysis of Food Cost Matrix Results

Square	Characteristics	Problem	Marketing Strategy
1	High food cost % Low popularity	Marginal due to both high product cost and lack of sales	**a.** Remove from the menu. **b.** Consider if the item's method of preparation is unpopular. **c.** Survey guests to determine current wants regarding this item. **d.** If this is a high-contribution-margin item, consider reducing price and/or portion size.
2	High food cost % High popularity	Marginal due to high product cost menu	**a.** Increase price. **b.** Reduce prominence on the menu. **c.** Reduce portion size. **d.** "Bundle" the sale of this item with one that has a lower cost and, thus, will provide a better (lower) overall food cost %.
3	Low food cost % Low popularity	Marginal due to lack of sales	**a.** Relocate on the menu for greater visibility. **b.** Take off the regular menu and run as a special. **c.** Reduce menu price. **d.** Eliminate other unpopular menu items in order to increase demand for this one.
4	Low food cost % High popularity	None	**a.** Promote well and frequently. **b.** Increase visibility on the menu.

Compare that sale to a sale made to a guest buying a steak for $40.00 that costs you $16.00 to make. Your food cost percentage would be 40 percent ($16.00 food cost ÷ $40.00 selling price = 40% food cost). In this case, however, there would be $24.00 ($40.00 – $16.00 = $24.00) remaining to pay for the labor and other expenses of serving this guest. Clearly in this example it would be better to sell more steak than chicken, despite the steak's higher food cost percentage. In this example, the steak item has a larger CM than the chicken item. For this reason, many operators prefer to analyze their menus' overall CM. It is also the reason many operators are fond of saying "You bank dollars, not percentages!"

CONTRIBUTION MARGIN

When analyzing a menu using the CM approach, the operator seeks to produce a menu that maximizes the menu's overall CM. Recall from Chapter 6 that CM is defined as the amount of money that remains after the product (food or beverage) cost of a menu item is subtracted from that item's selling price.

HERE'S HOW IT'S DONE 10.1

The contribution margin (CM) for an individual menu item is the amount of money remaining from the item's selling price *after* the product cost of making the item has been subtracted. The formula for calculating a CM for a menu item is as follows:

Item selling price – Item product cost = Item contribution margin

To illustrate, if a menu item sells for $20 and it cost $4.00 to make it, the item's CM is $8.00. In this example, the menu item's contribution margin is calculated as follows:

$20.00 Selling price – $4.00 Product cost
= $16.00 Contribution margin

Remember also that when calculating the CM of a menu item, it is important to include the cost of the item as well as all of that item's accompaniments, such as sauces, bread, salads, side dishes, and garnishes, that are also presented to the guest when the menu item is ordered.

The cost of these accompaniments can be significant and must be included in the calculation of the item's CM. For example, if the item that sells for $20.00 costs $4.00 to make and is served with a side vegetable that costs $1.00 to produce, the true CM of the item would be calculated as follows:

$20.00 Selling price – $4.00 Product cost
– $1.00 Side vegetable cost = $15.00 contribution margin

CM is the amount that you will have available to pay for your labor, your controllable and non-controllable other expenses, and to keep as a profit. Thus, from Figure 10.2, if an item on Maureen's menu, such as strip steak, sells for $35.95 and the product cost for the item is $16.16, the CM for the item would be computed as follows:

Selling price – Item cost = Contribution margin

or

$35.95 – $16.16 = $19.79

When CM is the driving factor in analyzing a menu, the two variables used for the analysis are CM and item popularity.

To illustrate the use of the CM approach to menu analysis, the data in Figure 10.2 are again used. In this case, Maureen must again separate her items based on high or low popularity. Doing so gives the same results as those obtained when using the food cost percentage method; thus, any item that sells 700/7, or 100 times, or more is considered to be a high-popularity item. Any menu choice selling less than 100 times would be considered low in popularity.

Some menu analysis experts maintain that any menu item that achieves at least 70 percent of the average number sold be considered a high-popularity menu item, whereas those that sell below 70 percent of the average should be considered low popularity. However, to illustrate the CM menu analysis procedure, we will be using the average (mean) number of items sold (in this case, 100 items sold) for our calculations.

When computing average CM for the entire menu, Maureen must follow two steps. First, to determine the total CM for the menu, the following formula is used:

$$\text{Total sales} - \text{Total item (product) costs} = \text{Total contribution margin}$$

or

$$\$23{,}201.00 - \$8{,}011.16 = \$15{,}189.84$$

Next, because 700 total menu items were sold, you can determine the average CM per item using the following formula:

$$\frac{\text{Total item contribution margin}}{\text{Number of items sold}} = \text{Average contribution margin per item}$$

or

$$\frac{\$15{,}189.84}{700} = \$21.70$$

You develop the CM matrix in much the same way as the food cost percentage matrix. In this case, average item popularity is 100 and average item CM is $21.70.

The matrix is developed as follows:

		Popularity: Low	Popularity: High
Contribution Margin	High	1 High contribution margin Low popularity	2 High contribution margin High popularity
	Low	3 Low contribution margin Low popularity	4 Low contribution margin High popularity

Maureen now classifies her menu items according to the contribution margin matrix in the following manner:

Square	Characteristics	Menu Item
1	High contribution margin Low popularity	Scallops/Pasta
2	High contribution margin High popularity	Coconut Shrimp, Chicken Breast
3	Low contribution margin Low popularity	Strip Steak, Lobster Stir-Fry
4	Low contribution margin High popularity	Grilled Tuna, Beef Medallions

Again, when completing this analysis, each menu item can be placed in one, and only one, matrix square. Using the CM method of menu analysis, Maureen would like as many of her menu items as possible to fall within square 2, that is, high CM and high popularity.

From her analysis, Maureen knows that both coconut shrimp and chicken breast yield a higher than average CM. In addition, these items sell very well. Just as Maureen seeks to give high menu visibility to items with low food cost percentage and high popularity when using the food cost percentage method of menu analysis, she would seek to give that same visibility to items with high CMs and high popularity when using the CM approach.

The menu items that fall into each of the squares require a special marketing strategy, depending on its square location. These strategies are summarized as shown in Figure 10.4.

A frequent, and legitimate, criticism of the CM approach to menu analysis is that it tends to favor high-priced menu items over low-priced ones because higher-priced menu items, in general, tend to have the highest CMs. Over the long term, this can result in sales techniques and menu placement decisions that tend to put in the guest's mind a higher check average than the operation may warrant or desire. This may not be desirable at all!

Alternatively, the successful menu strategy related to the many quick-service restaurants (QSRs) that offer $2.99, $3.99, or $5.99 value menu items is certainly not one of maximizing CM, but rather is one of seeking to increase total revenue.

Some users of the CM method of menu analysis refer to it as **menu engineering** and classify the squares used in the analysis with colorful names. The most common of these are "Plow horses" (square 1) because these items have high CMs but are less popular, "Stars" (square 2), because these items are popular and have high CMs, "Dogs" (square 3) because these have low CMs and are not popular, and "Puzzles" or sometimes "Challenges" (square 4) because these menu items are highly popular but return lower than average CMs to the operator.

The selection of either food cost percentage or CM as a menu analysis technique is really an attempt by the foodservice operator to answer the following questions:

- Are my menu items priced correctly?
- Are the individual menu items selling well enough to warrant keeping them on the menu?
- Is the overall profit margin on my menu items satisfactory?

FIGURE 10.4 Analysis of Contribution Margin Matrix Results

Square	Characteristics	Problem	Marketing Strategy
1	High contribution margin Low popularity	Marginal due to lack of sales	a. Relocate on menu for greater visibility. b. Consider reducing selling price.
2	High contribution margin High popularity	None	a. Promote well. b. Increase prominence on the menu.
3	Low contribution margin Low popularity	Marginal due to both low contribution margin and lack of sales	a. Remove from menu. b. Consider offering as a special occasionally, but at a higher menu price.
4	Low contribution margin High popularity	Marginal due to low contribution margin	a. Increase price. b. Reduce prominence on the menu. c. Consider reducing portion size.

The use of either the matrix-based food cost or CM menu analysis systems to answer these questions is, the authors of this book believe, overly simplistic for use in today's increasingly sophisticated foodservice operations. This is especially so now that labor costs (which are not considered in either approach) make up an increasingly large proportion of most operation's prime costs (see Chapter 8).

Because of the limitations of matrix analysis, neither the matrix food cost nor the matrix CM approach is truly effective in analyzing menus. This is the case because, mathematically, the axes on the matrix are determined by the average (mean) food cost percentage, CM, or sales level (popularity). When this is done, some items will *always* fall into the less desirable categories. This is so because, in matrix analysis, high food cost percentage, for instance, really means food cost percentage above that operation's average, or arithmetic mean. Obviously, then, some items must always fall above and below the average regardless of their contribution to operational profitability. Eliminating the poorest items only shifts other items into undesirable categories.

To illustrate this significant drawback to matrix analysis, consider the following example. Assume that Homer, one of Maureen's competitors, sells only four items, as follows:

Homer's No. 1 Menu	
Item	**Number Sold**
Beef	70
Chicken	60
Pork	15
Seafood	55
Total	200
Average (mean) sold	50 = (200/4)

After conducting a matrix analysis, Homer may elect to remove the pork item because its sales level is below the average of 50 items sold. If Homer adds turkey to the menu and removes the pork, he could get the following results:

Homer's No. 2 Menu	
Item	**Number Sold**
Beef	65
Chicken	55
Turkey	50
Seafood	30
Total	200
Average sold	50(200/4)

As can be seen, the turkey item drew sales away from the beef, chicken, and seafood dishes and did not increase the total number of menu items sold. In this case, it is now the seafood item that falls below the menu average. Should that item be removed because its sales are now below average? Clearly, this might not be wise. Removing the seafood item might serve only to draw sales from the remaining items to the seafood's replacement item. The same type of result can occur when you use a matrix to analyze food cost percentage or CM. As someone once stated, half of us are below average in everything. Thus, the use of the matrix approach, because of its average-based format, inherently forces some menu items to be below average.

An additional and increasingly relevant criticism of both the food cost percentage and CM analysis methods relates to the fact that the results of both are determined only by the price operators pay for food or beverage products. As increasing numbers of operators find they pay more for labor to produce their menu items than for the ingredients to make them, some managers question any menu analysis system that ignores labor and other important costs.

How, then, can an operator answer important questions related to menu price, sales volume, and overall profit margin? One answer is to avoid the overly simplistic and ineffective matrix analysis and employ a more powerful and effective method of menu analysis called goal value analysis.

FUN ON THE WEB!

Some companies have developed menu engineering software designed to help managers analyze their menus. To view the menu analysis offerings of current software producers, enter "menu engineering software" in your favorite search engine and review the results.

GOAL VALUE ANALYSIS

Goal value analysis was introduced by Dr. David Hayes and Dr. Lynn Huffman in an article titled "Menu Analysis: A Better Way" (Hayes & Huffman, 1985), published by the highly respected hospitality journal *The Cornell Quarterly*.

Ten years later, at the height of what was known as the "value pricing" debate (i.e., extremely low pricing strategies used to drive significant increases in guest counts), the effectiveness of goal value analysis was again proved in a second article, "Value Pricing: How Low Can You Go?" (Hayes & Huffman, 1995), which was also published in *The Cornell Quarterly*.

Goal value analysis uses the power of an algebraic formula to replace less sophisticated menu averaging techniques. Before the widespread introduction of computerized spreadsheet programs, some managers found the computations required to use goal value analysis challenging. Today, however, such computations are easily made using standard computerized spreadsheet programs.

The advantages of goal value analysis are many, including ease of use, accuracy, and the ability to simultaneously consider more variables than is possible with two-dimensional matrix analysis. Mastering goal value analysis can help you design menus that are more effective, popular, and, most importantly, profitable.

Goal value analysis is used to evaluate each menu item's food cost percentage, CM, and popularity. Unlike the two previous matrix-based analysis methods introduced, however, it also includes the analysis of the menu item's nonfood variable costs as well as its selling price.

Returning to the data in Figure 10.2, we see that Maureen has an overall food cost percentage of 34.5 percent. In addition, she served an average of 100 guests per menu item offered and achieved an entrée check average of $31.14. If we knew about Maureen's overall fixed and variable expenses (see Chapter 8), we would know a great deal more about the profitability of each of Maureen's menu items. Goal value analysis evaluates that profitability information.

One difficulty encountered when seeking to evaluate a menu item's profitability is that the assignment of nonfood variable costs to individual menu items can be a challenge. The issue is complex. It is very likely true, for example, that different items on Maureen's menu require differing amounts of labor to prepare. For instance, the strip steak on her menu is purchased precut and vacuum-sealed. Its preparation simply requires opening the steak package, seasoning the steak, and placing it on a broiler. The lobster stir-fry, on the other hand, is a complex dish that requires cooking and shelling the lobster, cleaning and trimming the vegetables, then preparing the item when ordered by quickly cooking the lobster, vegetables, and a sauce in a wok. Thus, the variable labor cost of preparing the two dishes is very different.

It is assumed that Maureen responds to these differing labor-related costs by charging a higher price for a more labor-intensive dish and a lower price for one that is less labor intensive. Other dishes require essentially the same amount of labor to prepare; thus, their variable labor costs have less impact on individual menu item's selling prices. Because that is true, for analysis purposes, most operators find it convenient to assign variable costs to individual menu items based on their *average* labor and other variable costs. For example, if labor and other variable costs are 30 percent of total sales, all menu items may be assigned that same variable cost percentage of their selling price.

For the purpose of her goal value analysis, Maureen determines her total variable costs. These are all the costs that vary with her sales volume, excluding the cost of the food itself. She computes those variable costs from her P&L statement (see Chapter 9) and finds that they account for 30 percent of her total sales. Using this information, Maureen assigns a variable cost of 30 percent of selling price to each menu item.

Having compiled the information in Figure 10.2, Maureen can use the algebraic goal value formula to create a specific target, or goal value, for her entire menu, and then use the same formula to compute the goal value of each individual menu item.

Menu items that achieve goal values higher than that of the menu's overall goal value will contribute greater than average profit percentages. As the goal value for an item increases, so too does its profitability percentage. Assuming that Maureen's average food cost percentage, average number of items sold per menu item, weighted average selling price (check average), and average variable cost percentage all meet the overall profitability goals of her restaurant, each individual menu item's goal value can be analyzed to assess that item's contribution to profits. The goal value formula is as follows:

$$A \times B \times C \times D = \text{Goal value}$$

where

A = 1.00−Food cost %
B = Item popularity
C = Selling price
D = 1.00−(Variable cost %+Food cost %)

Note that A in the preceding formula is actually the CM *percentage* of a menu item. D is the *percentage* amount available to fund fixed costs and provide for a profit after all variable costs (including food costs) are accounted for.

Maureen uses this formula to compute the goal value of her overall menu and finds that:

A	× B	× C	× D	= Goal Value
(1.00 = 0.345)	× 100	× $33.14	× [(1.00 − (0.30 + 0.345)]	= Goal Value
			or	
0.655	× 100	× $33.14	× 0.355	= 770

According to this formula, any menu item whose goal value equals or exceeds 770 will achieve profitability that equals or exceeds that of Maureen's overall menu. The computed goal value carries no unit designation; that is, it is neither a percentage nor a dollar figure because its primary purpose is to serve as a numerical score or target.

Figure 10.5 shows the results of Maureen's goal value analysis of her menu.

It is important to recognize that every menu will have items that are more or less profitable than others. In fact, some operators develop items called **loss leaders.** A loss leader is a menu item that is priced very low, sometimes even below its preparation cost, for the purpose of drawing large numbers of guests to the operation and thus increasing total revenue. If, for example, Maureen's is the only operation in town that serves outstanding lobster stir-fry, that item may, in fact, contribute to the overall success of the operation by drawing people who will buy it, whereas others dining with them may order other menu items that are more profitable.

The accuracy and value of goal value analysis and similar menu analysis systems are well documented. Used properly, this is a convenient way for management to evaluate the profitability, sales volume, and pricing characteristics of menu items. Because all of the values needed for the goal value formula are readily available, management need not concern itself with puzzling through endless decisions about item replacement.

Items that do not achieve the targeted goal value tend to be deficient in one or more of the key areas of food cost percentage, popularity, selling price, or variable cost percentage. In theory, all menu items have the potential of reaching the goal value. Management may, however, determine that some menu items can best serve the operation as loss leaders, an approach illustrated by the continued use of "value" menu items by leading chains in the QSR segment.

Very sophisticated users of the goal value analysis system can, as suggested by Lendal Kotschevar in *Management by Menu*, 4th ed. (view at www.wiley.com), modify the goal value formula to increase its accuracy and usefulness even more. In the area of variable costs, a single menu item might be assigned a low, medium, or high variable cost. If overall variable costs equal 30 percent, for example, management may choose to assign a variable cost of 25 percent to those items with lower labor costs attached to them, 30 percent to those items with average labor costs, and 35 percent or more to items with even higher costs. This adjustment affects only the D variable of the goal value formula and can be made quite easily. Similarly, managers can use the Goal Value formula to estimate, in advance, how potential changes in selling price, and portion size (food cost percentage) may affect an individual menu's contribution to profits.

It is at this point that managers must recognize, however, that a purely quantitative approach to menu analysis is neither practical nor desirable. Menu analysis and pricing decisions are always a matter of experience, skill, insight, and educated predicting because it is challenging to know in advance how changing the characteristics of any one menu item may affect the sales mix (see Chapter 5) of that item and the remaining items.

FIGURE 10.5 **Goal Value Analysis Results**

Rank	Menu Item	A	B	C	D	Goal Value
1	Chicken Breast	(1 − 0.22)	140	$27.95	[1 − (0.30 + 0.22)]	1,465
2	Coconut Shrimp	(1 − 0.30)	121	33.95	[1 − (0.30 + 0.30)]	1,150
3	Scallops/Pasta	(1 − 0.24)	85	29.95	[1 − (0.30 + 0.24)]	890
4	Beef Medallions	(1 − 0.37)	125	31.95	[1 − (0.30 + 0.37)]	830
	Overall Menu Goal Value	***(1 − 0.345)***	***100***	***$33.14***	[1 − (0.30 + 0.345)]	***770***
5	Grilled Tuna	(1 − 0.40)	105	35.95	[1 − (0.30 + 0.40)]	680
6	Strip Steak	(1 − 0.45)	73	35.95	[1 − (0.30 + 0.45)]	360
7	Lobster Stir-Fry	(1 − 0.51)	51	43.95	[1 − (0.30 + 0.51)]	208

COST/VOLUME/PROFIT ANALYSIS

Many foodservice operators find that some accounting periods are more profitable than others. Often, this is because sales volume is higher or costs are lower during certain periods. The ski resort that experiences tremendous sales during the ski season but has a greatly reduced volume or may even close during the summer season is a good example. Profitability, then, can be viewed as existing on a graph similar to Figure 10.6.

The x axis in Figure 10.6 represents the number of **covers** sold (guests served or units sold) in a foodservice operation. The y axis represents the costs/revenues in dollars. The Total Revenues line starts at zero because if no covers are sold, no dollars are generated. The Total Costs line starts farther up the y axis because fixed costs are incurred even if no covers are sold. The point at which the two lines cross (intersect) is called the **break-even point**.

At the break-even point, operational expenses are exactly equal to sales revenue. Stated in another way, when sales volume in your operation equals the sum of your total variable and fixed costs, your break-even point has been reached. Below the break-even point, costs are higher than revenues, so losses occur. Above the break-even point, revenues exceed costs, so profits are made.

Most operators would like to know their break-even point on a daily, weekly, or monthly basis. In effect, by determining the break-even point, the operator is answering the question: "How much sales volume must I generate before I begin to make a profit?"

Beyond the break-even point, you will want to answer another question: "How much sales dollars and volume must I generate to make my *desired* profit level?" To answer this question, you must conduct a **cost/volume/profit (CVP) analysis**. A CVP analysis helps predict the sales dollars and volume required to achieve desired profit (or break-even) based on your known costs.

You can perform a CVP analysis either by constructing a CVP graph or by arithmetical calculation. Although there are advantages to both methods, the arithmetical calculation is typically the most accurate. You can make CVP calculations to determine the dollar sales volume required to break-even and/or achieve a desired profit, or to determine the number of guests (covers) that must be served to break-even.

FIGURE 10.6 Cost/Volume/Profit Graph

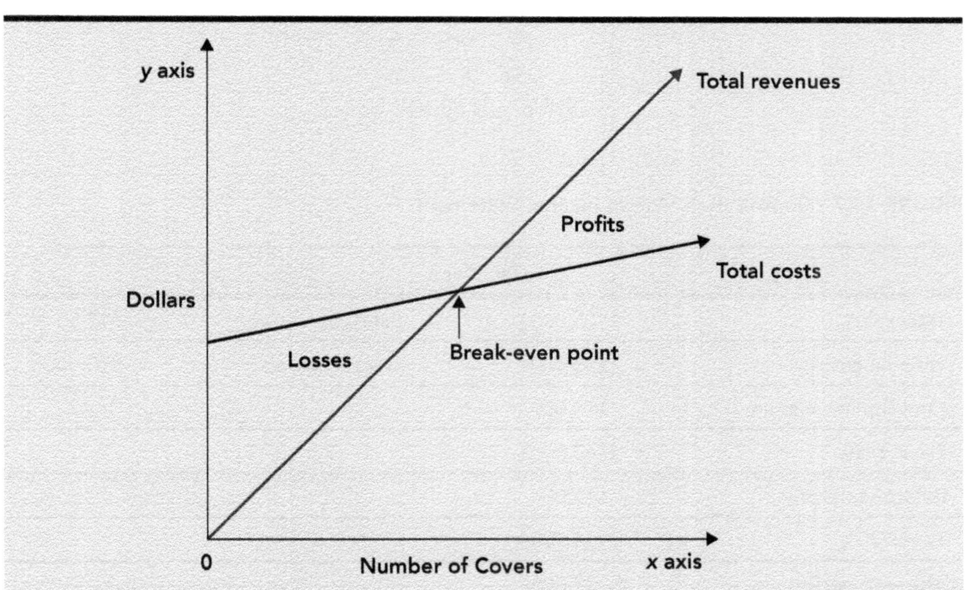

To see how a CVP analysis is conducted, consider the case of Jennifer, who operates an Asian-themed restaurant. Based on her income statement and sales records from last month, Jennifer has converted her P&L statement to a **contribution margin income statement**, as shown in Figure 10.7. A CM income statement simply shows P&L items in the categories of total sales, variable costs, CM, fixed costs, before-tax profit, taxes, and after-tax profit.

As you have learned, foodservice expenses can generally be classified as either fixed or variable. Of course, some expenses have both a fixed and a variable component and, thus, in reality, are mixed costs (see Chapter 8). For the purpose of engaging in a CVP analysis, however, it is necessary for the operator to assign costs to either a fixed or a variable category, as Jennifer has done. In addition, her **contribution margin for the overall operation** is defined as the dollar amount that contributes to paying an operation's fixed costs *and* providing for a profit.

CM for overall operation at Jennifer's is calculated as follows:

Total sales − Variable costs = Contribution margin for overall operation

or

$125,000 − $50,000 = $75,000

Jennifer can also view her CM income statement in terms of per-unit (guest) and percentage sales, variable costs, and CM, as shown in Figure 10.8.

Notice the box in Figure 10.8 that includes per unit (guest) and percentage calculations for sales per guest (selling price, [SP]), variable costs (VC), and contribution margin (CM). Note also that fixed costs are not calculated as per unit or as a percentage of sales. This is because fixed costs do not vary with sales volume increases or decreases.

To calculate these numbers, the following steps are taken:

Step 1. Divide total sales, variable costs, and CM by the number of guests served (units) to get per-guest (per-unit) values:

$$\frac{\text{SP}}{\text{Unit}} = \frac{\$125{,}000}{10{,}000\ \text{Units}} = \$12.50$$

$$\frac{\text{VC}}{\text{Unit}} = \frac{\$50{,}000}{10{,}000\ \text{Units}} = \$5.50$$

$$\frac{\text{CM}}{\text{Unit}} = \frac{\$75{,}000}{10{,}000\ \text{Units}} = \$7.50$$

$$\frac{\text{SP}}{\text{Unit}} - \frac{\text{VC}}{\text{Unit}} = \frac{\text{CM}}{\text{Unit}}$$

$$\$12.50 - \$5.00 = \$7.50$$

FIGURE 10.7 Contribution Margin Income Statement

Jennifer's			
Total sales	$125,000	Sales per guest	$12.50
Variable costs	$50,000	Guests served	$10,000
Contribution margin	$75,000		
Fixed costs	$60,000		
Before-tax profit	$15,000		
Taxes (40%)	$6,000		
After-tax profit	$9,000		

FIGURE 10.8 Contribution Margin Income Statement with per-Unit and Percentage Calculations

Jennifer's				
Total sales	$125,000			
Variable costs	$50,000	**Per Unit (Guest)**		**Percentage**
Contribution margin	$75,000	SP	$12.50	100%
Fixed costs	$60,000	VC	$5.00	40%
Before-tax profit	$15,000	CM	$7.50	60%
Taxes (40%)	$6,000	Guests served = 10,000		
After-tax profit	$9,000			

Step 2. Divide VC/Unit by SP/Unit, and CM/Unit by SP/Unit to get percentage values:

$$SP\% = 100\%$$

$$VC\% = \frac{\$5.00}{\$12.50} = 40\%$$

$$CM\% = \frac{\$7.50}{\$12.50} = 60\%$$

$$SP\% - VC\% = CM\%$$

$$100\% - 40\% = 60\%$$

After Jennifer's P&L statement has been converted to a CM income statement and per-unit values and percentages have been calculated, she can determine her operational break-even point and the sales required to achieve her desired profit. She wants to do this based both on dollar sales and on the number of covers (guests) required to be served.

To determine the dollar sales required to break even, Jennifer uses the following formula:

$$\frac{\text{Fixed costs}}{\text{Contribution margin \%}} = \text{Break-even point in sales}$$

or

$$\frac{\$60,000}{0.60} = \$100,000$$

Thus, Jennifer must generate $100,000 in sales per month before she begins to make a profit. At a sales volume of less than $100,000, she is operating at a loss. In terms of the number of guests that must be served to break-even, Jennifer uses the following formula:

$$\frac{\text{Fixed costs}}{\text{Contribution margin per-unit (per guest)}} = \text{Break-even point in guests served}$$

or

$$\frac{\$60,000}{\$7.50} = 8,000 \text{ guests (covers/units)}$$

Now assume that Jennifer has decided that next month she will plan for \$12,000 in after-tax profits. To determine sales dollars and needed guests (covers/units) to achieve her after-tax profit goal, Jennifer uses the following formula:

$$\frac{\text{Fixed costs} + \text{Before-tax profit}}{\text{Contribution margin \%}} = \text{Sales dollars to achieve desired after-tax profit}$$

Jennifer knows that her after-tax-profit goal is \$12,000, but the preceding formula calls for before-tax profit. To convert her after-tax profit to before-tax profit, Jennifer must compute the following:

$$\frac{\text{After-tax profit}}{(1.00 - \text{Tax rate})} = \text{Before-tax profit}$$

or

$$\frac{\$12{,}000}{(1.00 - 0.40)} = \$20{,}000$$

Now that Jennifer knows her before-tax profit goal is \$20,000, she can calculate the sales dollars needed to achieve her targeted after-tax profit as follows:

$$\frac{\text{Fixed costs} + \text{Before-tax profit}}{\text{Contribution margin \%}} = \text{Sales dollars to achieve desired after-tax profit}$$

or

$$\frac{\$60{,}000 + \$20{,}000}{0.60} = \$133{,}333.33$$

Thus, Jennifer must generate \$133,333.33 in sales to achieve her desired after-tax profit of \$12,000. In terms of calculating that must be served to make her desired profit, Jennifer uses the following formula:

$$\frac{\text{Fixed costs} + \text{Before-tax profit}}{\text{Contribution margin per guest (cover/unit)}} = \text{Guests to be served to achieve desired after-tax profit}$$

or

$$\frac{\$60{,}000 + \$20{,}000}{\$7.50} = \$10{,}666.67 \text{ guests (covers/units), round up to } 10{,}667 \text{ guests}$$

Notice that the number of guests was rounded up from 10,666.67 to 10,667. You *must always* round the number of guests *up* because (1) a guest (person) does not exist as a fraction and (2) it is better to slightly overstate the number of guests to achieve break-even or desired profits than to understate the number.

Also note that once you round the number of guests up to 10,667, you should adjust the total sales dollars to reflect this. Thus, $10{,}667 \times \$12.50 = \$133{,}337.50$ sales dollars to achieve the desired after-tax profit is more accurate than \$133,333.33, which you calculated using CM percentage. The variation in these two sales dollar levels is due to rounding. This difference is minimal and may not even warrant adjustment unless an exact sales dollar amount is required based on the number of guests served.

When calculating sales and covers needed to achieve break even and determine desired after-tax profits, you can easily apply the proper formula if you remember that:

1. CM percentage is used to calculate sales dollars.
2. CM per unit is used to calculate sales volume in units (guests or covers).

Once you fully understand the CVP analysis concept, you can predict any sales level for break even or after-tax profits based on your selling price, fixed costs, variable costs, and CM. You can also make changes in your selling prices and costs to improve your ability to break even and achieve your desired profit levels. This is where the menu pricing and cost control concepts addressed in this book come into play. As you make changes in your control areas, you will be able to manage your operation efficiently so that losses can be prevented and planned profits can be achieved.

FUN ON THE WEB!

The direct relationship between goal value analysis and CVP was first presented by Dr. Lea Dopson (L. Dopson (2004), in "Linking Cost-Volume-Profit Analysis with Goal Value Analysis in the Curriculum Using Spreadsheet Applications," *The Journal of Hospitality Financial Management* 12(1), 77–80).

These two forms of analysis are strategically linked, as follows:

Cost/Volume/Profit Analysis	Goal Value Analysis
Food cost % from contribution margin income statement	Food cost % goal
Guests served to achieve desired after-tax profit	Total average number of covers per menu item goal
Selling price	Selling price goal
Labor and other variable cost % from contribution margin income statement	Variable cost % goal

If you are a reader with a keen interest in how to better identify the profitability of individual menu items and their relationship to an operation's break-even analysis, enter "linking cost-volume-profit analysis with goal value analysis in the curriculum using spreadsheet applications" in your favorite search engine and review the results.

MINIMUM SALES POINT

Every foodservice operator should know his or her break-even point. The concept of minimum sales point is related to the break-even concept. The **minimum sales point (MSP)** is the sales volume required to justify keeping an operation open during a given period of time. At sales levels *below* the MSP, the operation is not operating profitably. At sales levels *above* the MSP, enough revenue is being taken in to make a contribution to profits. The information needed to calculate the MSP is as follows:

1. Food cost percentage
2. Minimum payroll cost needed for the time period
3. Variable cost percentage

Fixed costs are eliminated from the calculation because, even if the volume of sales in an extended opening period equals zero, fixed costs still exist and must be paid. To illustrate, consider the situation of Richard, who is trying to determine whether he should close his steakhouse at 10:00 P.M. or 11:00 P.M.

Richard wants to identify the sales volume necessary to justify staying open the additional hour. He can make the needed calculation because he knows that his food cost equals 40 percent, his minimum labor cost to stay open for the extra hour equals $150, and his variable costs (taken from his P&L statement) equal 30 percent.

In calculating MSP, his Food Cost % + Variable Cost % is called his **minimum operating cost**. Richard applies the MSP formula as follows:

$$\frac{\text{Minimum labor cost}}{1.00 - \text{Minimum operating cost}} = \text{MSP}$$

or

$$\frac{\text{Minimum labor cost}}{1.00 - (\text{Food cost \%} + \text{Variable cost \%})} = \text{MSP}$$

In this case, the computation would be as follows:

$$\frac{\$150}{1.00 - (0.40 + 0.30)} = \text{MSP}$$

or

$$\frac{\$150}{1.00 - 0.70} = \text{MSP}$$

or

$$\frac{\$150}{0.30}$$

thus

$$\$500 = \text{MSP}$$

If Richard can achieve total sales volume of $500 in the 10:00 P.M. to 11:00 P.M. time period, he should stay open. If this level of sales is not possible, he should close the operation at 10:00 P.M. Richard can also use MSP to determine the hours his operation is most profitable. Of course, some managers may not have the authority to close their operations, even when remaining open is not very profitable. Corporate policy, contractual hours, promotion of a new unit, competition, and other factors must all be considered before the decision should be made to modify operational hours.

THE BUDGET

In most managerial settings, you will be responsible for preparing and maintaining a budget (see Chapter 1) for your foodservice operation. This budget, or financial plan, will detail the operational direction of your unit and your expected financial results. The techniques used in managerial accounting will show you how close your actual performance conformed to your budget and will provide you with the information you need to make any needed changes to your operational procedures or budget process. This process will help ensure that your operation achieves the financial goals you have for it.

It is important to note that the budget should not be a static (unchanging) document. Rather, it should be modified and fine-tuned as managerial accounting presents data about sales and costs that affect the direction of the overall operation. For example, if you own a restaurant featuring Thai food, and you find that a major competitor in your city has closed its doors, you may quite logically determine that you want to increase your estimate of the number of guests who will come to your operation. This would, of course, affect your projected sales revenue, your costs, and, in most cases, your profitability. Not to do so might allow you to meet and even exceed your original sales and profit goals but would ignore a significant event that very likely will affect the financial results of your restaurant.

In a similar manner, if you are the manager of a New York–style delicatessen specializing in salads, sliced meat sandwiches, and related side items, and you find through your purchase orders that the price you pay per pound for the high-quality beef briskets used to make your corned beef sandwiches has greatly increased since last month, you must adjust your budget, or you will find that you have no chance of staying within your food cost guidelines. Again, the point is that budgets should be closely monitored through the use of managerial accounting, which includes the thoughtful analysis of changes in operating data.

Just as the P&L tells you about your past performance, the budget is developed to help you predict and achieve your future goals. The budget tells you what must be done if predetermined cost and profit objectives are to be met. In this respect, you are attempting to modify the profit formula, as presented in Chapter 1. With a well-thought-out and attainable budget, your profit formula would change from:

$$\text{Revenue} - \text{Expenses} = \text{Profit}$$

to

$$\text{Budgeted revenue} - \text{Budgeted expenses} = \text{Budgeted profit}$$

To prepare a budget and stay within it helps ensure attainable results. Without such a plan, you must guess how much to spend and how much sales you should anticipate. The effective foodservice operator builds their budget, monitors it closely, modifies it when necessary, and achieves desired results. Yet, many operators do not develop a budget. Some say that the process is too time-consuming. Others feel that a budget, especially one shared with the entire organization, is too revealing.

Budgeting can also cause conflicts. This is true, for example, when dollars budgeted for new equipment must be used for either a new kitchen stove or a new beer-tapping system. Obviously, in this example, a kitchen manager and a bar manager may hold different points of view on where these funds would best be spent!

Despite the fact that some operators avoid using them, budgets are extremely important. The advantages for having and using a budget are summarized in Figure 10.9.

FIGURE 10.9 Advantages of Budgeting

1. It is the best means of analyzing alternative courses of action and allows management to examine alternatives prior to adopting a particular one.
2. It forces management to examine the facts regarding what must be done to achieve desired profit levels.
3. It provides a standard for comparison, which is essential for good controls.
4. It allows management to anticipate and prepare for future business conditions.
5. It helps management to periodically carry out a self-evaluation of the organization and its progress toward its financial objectives.
6. It provides a communication channel whereby the organization's objectives are passed along to its various departments.
7. It encourages department managers who have participated in the preparation of the budget to establish their own operating objectives and evaluation techniques and tools.
8. It provides management with reasonable estimates of future expense levels and serves as an instrument for setting proper menu prices.
9. It identifies time periods in which operational cash flows may need to be supplemented from other sources.
10. It communicates realistic financial performance to owners, investors, and the operation's managers.

Budgeting is best done by the entire management team, for it is only through participation in the process that the whole organization will feel compelled to support the budget.

Foodservice budgets typically can be considered as one of three main types:

1. Long-range budget
2. Annual budget
3. Achievement budget

LONG-RANGE BUDGET

The **long-range budget** is most often prepared for a period of 3 to 5 years into the future. Although it is not highly detailed, it does provide a long-term view about where the operation should be going. It is particularly useful in those cases where extensive remodeling or the opening of additional operational units may increase sales volume and accompanying expenses.

To illustrate, assume that you are preparing a budget for a corporation you manage. Your corporation has entered into an agreement with an international franchise company to open 45 sub shops in strip shopping centers across the United States and Canada. You will open a new store approximately every month for the next four years. To properly plan for your revenue and expense in the coming four-year period, a long-range budget for your company will be necessary.

ANNUAL BUDGET

The annual, or yearly, budget is the type most operators think of when the term "budget" is used. As it states, the **annual budget** is for a one-year period or, in some cases, one season. This would be true, for example, in the case of a children's summer camp that is open and serving meals only while school is out of session and campers are attending; or in the case of a ski resort that opens in late fall/early winter but closes in the spring when the snow melts.

It is important to remember that an annual budget need not follow a calendar year. In fact, the best time period for an annual budget is the one that makes sense for the operation. A college foodservice director in the United States, for example, would want a budget that covers the typical time period of a school year, that is, from the fall of 1 year through the spring of the next. For a restaurant whose owners have a fiscal year (see Chapter 4) that is different from a calendar year, the annual budget may coincide with either the fiscal year or the calendar, as the owners prefer.

It is also important to remember that an annual budget need not consist of 12 one-month periods. Although many operators prefer one-month budgets, some prefer budgets consisting of 13 separate 28-day periods (see Chapter 1). Other managers may prefer to use quarterly (three-month) or even weekly budgets to plan for revenues and costs throughout the budget year.

ACHIEVEMENT BUDGET

The **achievement budget** is always of a shorter range, perhaps a month or a week. It provides current operating information and, thus, assists in making current operational decisions. A weekly achievement budget might, for example, be used to predict the number of gallons of milk needed for this time period, the number of vehicles to be serviced through the drive-thru on Friday night, or the number of servers to be scheduled on Tuesday night.

DEVELOPING THE BUDGET

Some managers think it is very difficult to establish a budget, and so they simply do not do it. Creating a budget is not that complex. You can learn to do it and do it well. To establish any type of budget, you need to have the following information:

1. Prior-period operating results (if available)
2. Assumptions of next-period operations
3. Goals
4. Monitoring policies

To examine how prior-period operating results, assumptions of next-period operations, and goals drive the budgeting process, we will consider the case of Levi, who is preparing the annual foodservice budget for his 100-bed health-care facility. This section addresses prior-period operating results, assumptions, and goals, and the final section looks closely at monitoring the budget.

PRIOR-PERIOD OPERATING RESULTS

Levi's facility serves three patient meals per day to an average of 80 patients, and he serves approximately 100 additional meals per day to staff and visitors. His department is allotted a fixed dollar amount by the facility's administration for each meal he serves.

His operating results for last year are detailed in Figure 10.10. Patient and additional meals served were determined by actual count. Revenue and expense figures were taken from Levi's income (P&L) statement at the year's end. It is important to note that Levi must have this information if he is to do any meaningful profit planning. Foodservice unit managers who do not have access to their past operating results are at a tremendous managerial disadvantage. Levi has his operational summaries and the data that produced them. Because he knows how he has operated in the past, he is now ready to proceed to the assumptions section of the planning process.

FIGURE 10.10 Levi's Last-Year Operating Results

Patient Meals Served: 87,600 Revenue per Meal: $7.74
Additional Meals Served: 36,500
Total Meals Served: 124,100

	Amount		Percentage	
Total Department Revenue		$960,000		100%
Cost of Food	$384,000		40%	
Cost of Labor	$307,200		32%	
Other Expenses	$172,800		18%	
Total Expense		$864,000		90%
Profit		$96,000		10%

ASSUMPTIONS OF NEXT-PERIOD OPERATIONS

If Levi is to prepare a budget with enough value to serve as a useful guide and enough flexibility to adapt to a changing environment, he must factor in the assumptions he and others feel will affect his operation in the coming year.

Although each management team will arrive at its own conclusions given the circumstances of its own operation, in this example, Levi makes the following assumptions regarding next year:

1. Food costs will increase by 3 percent.
2. Labor costs will increase by 5 percent.
3. Other expenses will rise by 10 percent due primarily to a significant increase in utility costs.
4. Revenue received for all meals served will be increased by no more than 1 percent.
5. Resident occupancy averaging 80 percent of his facility's capacity will remain unchanged.

Levi makes these assumptions after discussions with his facility administrators, suppliers, and labor union leaders, after reviewing his own records, and, most importantly, by his knowledge of the operation itself.

As an operator, Levi predicts items 1, 2, and 3 above by himself, whereas his own supervisor has given him input about items 4 and 5. Given these assumptions, Levi can establish realistic operating goals for next year.

In the commercial sector, when arriving at realistic future performance assumptions, operators must also consider new or reduced levels of competition, changes in traffic patterns, variations in product costs, menu prices, and national food trends. At the highest level of foodservice management, assumptions regarding the acquisition of new units or the introduction of new products directly affect the budgeting process.

Green and Growing!

Budgeting for utility costs is one of a foodservice operator's biggest challenges. This is due to both the instability of energy prices and the impact of the weather on usage. Both of these factors should lead cost-conscious and planet-conscious managers to consider all possible strategies for reducing energy usage. For many operators, these strategies should include the following:

1. Investigating the installation of smart lighting systems that automatically turn off lights when storage areas are vacant.
2. Replacing all incandescent lighting with LED lights or with an appropriate type of electric discharge lamp (such as fluorescent, mercury vapor, metal halide, or sodium) wherever possible.
3. Using dual-flush, low-flow, or waterless toilets to reduce water waste.
4. Installing low-flow faucet aerators on all sinks to cut water usage by as much as 40 percent; from a standard 4 gallons per minute to a cost-saving 2.5 gallons a minute.
5. Implementing an effective preventive maintenance program for all cooking equipment, including frequent and accurate temperature recalibrations.
6. Reducing waste disposal costs by implementing effective source reduction plans as well as pre- and post-production recycling efforts.

The benefits to implementing strategies such the preceding are twofold. They will improve an operation's bottom line and reduce its carbon footprint. Those results are good news for a business's P&L, and they directly benefit the business's current, as well as its future, customers.

ESTABLISHING OPERATING GOALS

Given the assumptions he has made, Levi considers his operating goals for the coming year. He will develop these goals (targets) for each of the following areas:

1. Meals served
2. Revenue
3. Food costs
4. Labor costs
5. Other expenses
6. Profit

MEALS SERVED

Given the assumption of no increase in patient occupancy, and in light of his results from last year, Levi budgets to prepare and serve 87,000 patient meals (80 patients × 3 meals per day × 365 days = 87,600 meals). He feels, however, that he can increase his visitor and staff meals somewhat by being more customer-service driven and by offering a wider selection of lower calorie and lighter items on the facility's cycle menu.

He decides, therefore, to raise his goal for additional meals from the 36,500 served last year to 40,000 for the coming year. Thus, his budgeted total meals to be served will equal 127,600 meals (87,600 patient meals + 40,000 visitor and staff meals = 127,600 total meals).

REVENUE

Levi knows that his total revenue is to increase by only 1 percent. His revenue per meal served can be estimated as $7.74 × 1.01 = $7.82. With 127,600 meals estimated to be served, Levi will generate $997,832 (127,600 × $7.82 = $997,832) if he meets his meals-served budget.

HERE'S HOW IT'S DONE 10.2

When preparing budget forecasts, a manager often must project either an increase or a decrease in a known sales or expense amount. For example, assume that a manager spent $80,000 for utilities last year. The manager is preparing next year's budget and believes that next year utility costs will *increase* by 5 percent. This manager can forecast next year's utility cost in two different ways that will yield the same result.

The first way requires the manager to multiply last year's utility costs times the 5 percent projected increase and then *add* the resulting number to last year's utility costs. It is a two-step process:

Step 1: Calculate the amount of the projected increase:

$80,000 × 0.05 = $4,000

Step 2: Add the amount of the projected increase to last year's cost to arrive at next year's estimated cost:

$80,000 + $4,000 = $84,000

The second way to arrive at the same number is faster because it requires only one step. That is so because the manager in this example is actually projecting next year's utility cost to be 105 percent (100% + 5% = 105%) of this year's cost. When using this approach, the one-step calculation is as follows:

Step 1

$80,000 × 1.05 = $84,000

What if the manager assumes that the amount of sales or expense will decrease? When projecting a sales or expense *decrease*, the same math principles apply. To illustrate, assume that the manager will implement cost savings approaches projected to *reduce* total utility costs by 5 percent. Using the two-step approach, the calculation is as follows:

Step 1: Calculate the amount of projected decrease:

$80,000 × 0.05 = $4,000

Step 2: Subtract the amount of the projected decrease from last year's cost to arrive at next year's estimated cost:

$80,000 − $4,000 = $76,000

The faster way to arrive at the same projected decreased amount requires only one step. That is possible because the manager is actually projecting next year's utility cost to be 95 percent of this year's cost (100% − 5% = 95%).

When using this approach, the one-step calculation is as follows:

Step 1

$80,000 × 0.95 = $76,000

Both the two steps and the one-step approach yield the same answer.

Choose the method you like best!

FOOD COSTS

Levi is planning to serve more meals, so naturally he expects to spend more on food. In addition, he assumes that this food will cost him, on average, 3 percent more than it did last year. To determine a food budget, Levi computes the estimated food cost for 144,200 meals as follows:

1. Last year's food cost per meal = Last year's cost of food/Total meals served = $384,000/$124,100 = $3.09 per meal
2. Last year's food cost per meal + 3% Estimated increase in food cost = $3.09 × 1.03 = $3.18 per meal
3. $3.18 × 127,600 Meals to be served this year = $405,768 Estimated cost of food this year

LABOR COSTS

Levi is planning to serve more meals next year, so he also expects to spend more on the labor needed to prepare and serve the extra meals. In addition, he assumes that this labor will cost, overall, 5 percent more than last year. To determine a labor budget, Levi computes the estimated labor cost for 127,600 meals to be served as follows:

- Last year's labor cost per meal = Last year's cost of labor/Total meals served = $307,200/124,100 = $2.48 per meal
- Last year's labor cost per meal + 5% Estimated increase in labor cost = $2.48 × 1.05 = $2.60 per meal this year
- $2.60 × 127,600 Meals to be served this year = $331,760 Estimated cost of labor this year

OTHER EXPENSES

Levi assumes a 10 percent increase in other expenses and, thus, they are budgeted as last year's amount plus an increase of 10 percent. Thus, $172,800 × 1.10 = $190,080.

Based on his assumptions about next year, Figure 10.11 details Levi's budget summary for next year.

FIGURE 10.11 Levi's Budget for Next Year

Patient meals budgeted: 87,600 Budgeted revenue per meal: $7.82
Visitor and staff meals budgeted: 40,000
Total meals budgeted: 127,600

	Amount		Percentage	
Total budgeted revenue		$997,832		100.0%
Budgeted expenses				
Cost of food	$405,786		40.7%	
Cost of labor	$331,760		33.2%	
Other expenses	$190,080		19.0%	
Total expense		$927,626		$93%
Profit		$70,206		7%

Cost Control Around the World

To illustrate one of the many ways in which the global economy affects financial reporting in the foodservice industry, consider the challenges and opportunities facing Inez Magana. Inez is the Southern Region Director of Operations for El Pollo Salsa, a restaurant chain with units in the United States, Caribbean, Mexico, Brazil, and other South American countries. She oversees 150 company-owned stores. One of Inez's most important tasks is to accurately prepare income and expense forecasts and annual budgets for her restaurants.

It is a fundamental principle of accounting that the financial performance of a business should be reported in an identifiable **monetary unit**. A monetary unit is simply defined as a specific and recognized currency denomination. Typically, financial statements are prepared in the currency of the country in which the business is based. Thus, in Inez's case, the US dollar is the monetary unit used for preparing many, but not all, of her financial forecasts.

At first, it might appear easy to ensure that financial summaries are prepared in the currency denomination used by the company whose records are being reported. But, in reality, fulfilling the monetary unit reporting principle can be quite complex. Consider that Inez operates restaurants internationally and her restaurants' sales and expenses are reported as follows:

1. Sales and expenses in her American restaurants are reported in US dollars.
2. Sales and expenses in her Mexican restaurants are reported in Mexican pesos.
3. Sales and expenses in her Brazilian restaurants are reported in Brazilian reals.
4. Sales and expenses in her Dominican Republic restaurants are reported in Dominican Republic pesos.
5. Sales and expenses in her Bolivian restaurants are reported in Bolivian bolivianos.
6. Sales and expenses in her Aruban restaurants are reported in Aruban florins.

Because the **exchange rate** (the relative price of one currency expressed in terms of another currency), can change daily, the ability to move easily back and forth among international currencies when reporting financial results for restaurants operating in various countries is just one of the many skills international foodservice managers must master to do their jobs well.

FUN ON THE WEB!

Managers who must prepare budgets using monetary units other than US dollars can take advantage of many free-to-use currency calculators. These calculators allow managers to enter amounts in any international currency and have those amounts automatically converted to their US dollar equivalent. Alternatively, they can use the calculators to enter a US dollar amount and have that amount converted to a chosen international currency.

To review one or more of these helpful tools, enter "currency converter" in your favorite search engine, then choose a site offering a free converter and try it yourself!

PROFIT

Note that the increased costs Levi will experience, when coupled with his minimal revenue increase, will cause his profit to fall from $96,000 for last year to a projected $70,206 profit for the coming year. If this is not acceptable, Levi must either increase his revenue beyond his original assumption or look to his operation for ways to reduce costs.

Levi has now developed concrete guidelines for his operation. Since his supervisor approves his budget as submitted, Levi is now ready to implement and then monitor his new budget.

Modern foodservice managers have monitoring tools available to them today that simply did not exist in the past. One of the most important of these tools is cloud storage. A service manager can use cloud storage to maintain and manage important data such as their operations' budgets. But the use of cloud storage for keeping a business's records comes with both pluses and minuses.

Pluses of cloud storage include the following:

1. Data can be stored remotely and accessed by multiple individuals at the same time.
2. Files can be shared with others via e-mail links to data rather than e-mail attachments of the actual data.
3. Files can be accessed anywhere an Internet connection is available.
4. Files cannot be inadvertently lost or deleted, as they can be if held on only one computer device.

Minuses of cloud storage include the following:

1. Cloud storage providers charge a fee for their services.
2. Changes to stored files changes can be "lost" if not carefully saved by users in a predetermined and properly labeled folder.
3. Lack of an Internet connection results in the inability to access needed files.
4. Data safety, security, and privacy concerns increase with the use of cloud-based storage.

Regardless of whether important budget-related (and other) information is maintained in the cloud or on premises, managers must always monitor their budgets carefully.

MONITORING THE BUDGET

An operational plan has little value if management does not use it. In general, the budget should be regularly monitored and analyzed in each of the following three key areas:

- Revenue
- Variable Expense
- Profit

REVENUE ANALYSIS

If revenue falls below projected levels, the impact on profits can be substantial. Simply put, if revenues fall far short of projections, it may be difficult or even impossible for you to meet your profit goals. If revenue consistently exceeds your projections, the overall budget must be modified or the expenses required to produce the increased sales will soon exceed the originally predicted amounts. Effective managers compare their actual revenue to budgeted revenue on a regular basis.

Increases in operational revenue should result in proportional total dollar increases in variable expense budgets. Fixed expenses, because they do not vary with volume, need not be adjusted for increases (or decreases) in revenue. For those foodservice operations with more than one meal period, the best monitoring of budgeted sales volume means monitoring each meal period.

Consider the case of Rosa, the night (P.M.) manager of a gourmet coffee shop. She feels that she is busier than ever, but Lois, the shop's owner, maintains that there can be no increase in Rosa's labor budget because the overall coffee shop sales volume is exactly in line with budgeted projections Lois approved at the beginning of the year.

Figure 10.12 shows the complete story of the sales volume generation at the coffee shop after the first 6 months of their fiscal year. Note that the year is half (50 percent) completed at the time of this analysis.

Based on the sales volume she generates, Rosa should indeed have an increase in her labor budget for the evening operating period. The amount of business she is generating in the evenings is substantially higher than budgeted. Note that she is half-way through her budget year but has already generated 71 percent of the annual revenue forecasted by the budget. This, however, does not mean that the labor budget for the entire coffee shop should be increased. In fact, the labor budget for the A.M. shift should likely be reduced because those dollars would be better used to purchase the labor needed to serve guests in the evening.

Some foodservice operators relate revenue to the number of seats they have available in their operation. As a result, they sometimes budget based on **sales per seat**, the total revenue generated by a facility divided by the number of available seats in the dining area(s). The size of a foodservice facility affects both total investment and operating costs, therefore this can be a useful metric. The formula for the computation of sales per seat is as follows:

FIGURE 10.12 Coffee Shop Annual Revenue Budget Summary

Time Period: First 6 Months

Operating Period	Annual Budget	Actual to Date	% of Budget
A.M.	$480,500	$166,698	35%
P.M.	$350,250	$248,677	71%
Total	$830,750	$415,375	50%

$$\frac{\text{Total sales}}{\text{Available seats}} = \text{Sales per seat}$$

To illustrate, assume that, if Rosa's coffee shop has 60 seats, her P.M. sales per seat thus far this year would be calculated as follows:

$$\frac{\$248{,}677 \text{ Total sales}}{60 \text{ Available seats}} = \$4{,}144.61 \text{ Sales per seat}$$

The A.M. sales per seat, given the same number of seats, would be computed as follows:

$$\frac{\$166{,}698 \text{ Total sales}}{60 \text{ Available seats}} = \$2{,}778.30 \text{ Sales per seat}$$

As you can see, Rosa's sales per seat at night are much higher than that of her A.M. counterpart.

Some commercial foodservice operators relate revenue to the number of square feet their operations occupy. These operators budget revenues based on a **sales per square foot** basis. This is a very common approach in strip shopping centers and malls where occupation costs are determined in large degree by the number of square feet a foodservice operation leases from the space's owner. Sales per square foot is also an important factor for those operators who are building new foodservice operations. This is because as the number of square feet in an operation increases, its construction and operating costs increase as well. As a result, the physical size of a foodservice operations is an important factor in its profitability

The computation of sales per square foot is quite similar to that used in Rosa's sales per seat calculation. The formula for sales per square foot is as follows:

$$\frac{\text{Total sales}}{\text{Total square footage}} = \text{Sales per square foot}$$

Regardless of how it is measured, when sales volume is lower than originally projected, management must seek ways to increase revenue or to reduce costs. As stated earlier, one of management's main tasks is to generate guests, whereas the staff's main task is to service these guests to the best of their ability.

The Move to Off-Premise Dining

As the move to off-premise dining continues, many foodservice operators constructing new units are increasingly reducing the size of those units. This is especially true in the QSR segment of the industry, where increasing number of sales come via drive-thru windows, delivery, or guest pickup. As a result, when constructing new units many QSR operators are substantially reducing the sizes of their on-premise dining rooms, and some are even eliminating them completely. These actions can significantly increase an operation's resulting sales per square foot.

As well, new operations are increasingly trading on-premise seating areas for expanded on-premise third-party pickup areas. This has been done in response to the great volume of business achieved through third-party delivery apps. Other QSR operators are removing self-serve beverage stations from dining areas because, as on-premise diners reduce in numbers, these beverage stations are used less and less.

The costs savings from building smaller units can be substantial. Due to rising real estate costs in many desirable areas, smaller units have become especially attractive for investors and owners. In addition, smaller units require less labor to operate and have reduced utility (e.g., heating and cooling) costs, both of which can also increase unit profitability.

FUN ON THE WEB!

It has often been said that the foodservice manager's most important job is to retain and maintain a growing customer base. With the advent of the Internet as a tool, guests increasingly use to choose those operations they will frequent, and the trend toward off-premise dining, it is essential that foodservice operators stay up to date on the best marketing and promotion techniques for their own specific operations.

Fortunately, some high-quality publications addressed this very topic. One of the best is the book *Successful Marketing in Foodservice Operations*, written by Dr. David Hayes and Dr. Jack Ninemeier. The book is published by John Wiley.

To view the book's outline of topics, go to the Wiley website (www.wiley.com) and enter "marketing in foodservice operations" in the search bar to review this valuable restaurant marketing aid.

VARIABLE EXPENSE ANALYSIS

Effective foodservice managers are careful to monitor operational expense because costs that are too high or too low may be cause for concern. Just as it is not possible to estimate future sales volume perfectly, it is also not possible to estimate future expense perfectly because, as you have learned, some expenses will vary as sales volume increases or decreases.

To know that an operation spent $1,800 for fruits and vegetables in a given week becomes meaningful only if we know what the sales volume for that week was. Similarly, knowing that $900 was spent for variable labor during a given lunch period can be analyzed only in terms of the amount of sales achieved in that same period.

In Chapter 9 (see Figure 9.4), you learned that managers can compare their operations' performance in a current accounting period with the results achieved in a prior period. In a similar manner, managers can compare their actual performance in an accounting period with the results they had planned to achieve. This comparison process is important if managers are to do the things they must do to ensure that their budgets are met. Accurate assessments of variable expenses can be complex for a number of inter-related reasons, including the following:

- Actual revenue achieved may be higher than budgeted
- Actual revenue achieved may be lower than budgeted
- Actual variable expenses incurred may be higher than budgeted
- Actual variable expenses incurred may be lower than budgeted

To help them make accurate variable expense assessments quickly, operators can take some simple steps that will help them relate actual revenue generation to their actual food (and other) variable costs and cost estimates.

For example, assume that Marion is a foodservice manager who estimates next year's sales in his operation to be $900,000. He created a budget with a targeted food expense of $335,800, as shown in Figure 10.13.

In the coming year, Marion wishes to know whether his monthly food purchases and usage are in line with his annual budget. For example, Marion knows that if his sales exceed his estimates, his food-related costs will increase. But he wants to make sure that any increases (or decreases) in his cost of food reflect changes in revenue, not inefficiencies in his operation.

Because he understands the use of cost percentages, he can do that easily for his food (and any other variable) expense by using a three-step method:

FIGURE 10.13 Marion's Next-Year Revenue Estimate and Food Expense Budget

Revenue	$900,000
Food Expense	
Meats	$133,200
Fish/Poultry	$73,000
Produce	$53,000
Dairy	$40,000
Groceries	$36,600
Total Food Expense	$335,800

Step 1. Calculate the percentage of revenue budgeted for each food category.

Step 2. Compute percentage costs to sales for each food category using actual food usage and actual sales levels (taken from an operation's P&L).

Step 3. Compare actual monthly food expense percentages to budgeted levels and take corrective action if necessary.

To illustrate, assume that, in the first month of the new year, Marion achieves $78,000 in revenue. The three-step process to assess his variable food expenses at that level of revenue would be as follows:

Step 1. Determine the percentage of annual revenue budgeted for each food category. For example, the calculation to determine the budgeted meat cost percentage would be as follows:

$$\frac{\text{Budgeted meats cost}}{\text{Budgeted revenue}} = \text{Budgeted meats cost percentage}$$

or

$$\frac{\$133{,}200}{\$900{,}000} = 14.8\%$$

Marion's One-Year Food Expense Budget

Revenue	$900,000	
Food expense		
Meats	$133,200	14.8%
Fish/poultry	$73,000	8.1%
Produce	$53,000	5.9%
Dairy	$40,000	4.4%
Groceries	$36,600	4.1%
Total	$335,800	37.3%

Step 2. Compute costs to sales percentages for each food category using actual food usage and actual sales levels. In this example, Marion achieved the following results in the first month of his budget period.

Marion's First Month Actual Revenue and Food Expense Results

Revenue	$78,000	
Food expense	$	%
Meats	$12,090	15.5%
Fish/poultry	$7,020	9.0%
Produce	$4,290	5.5%
Dairy	$3,432	4.4%
Groceries	$3,120	4.0%
Total	$29,952	38.4%

Step 3. Compare actual monthly food expense percentages to budgeted levels and take corrective action if necessary.

Revenue	$78,000		
Food expense	**Actual $**	**Actual %**	**Budgeted %**
Meats	$12,090	15.5%	14.8%
Fish/poultry	$7,020	9.0%	8.1%
Produce	$4,290	5.5%	5.9%
Dairy	$3,432	4.4%	4.4%
Groceries	$3,120	4.0%	4.1%
Total	$29,952	38.4%	37.3%

Marion can now compare his budgeted expense with his actual performance. In this example, Marion's food cost percentage exceeds his budget by 1.1 percent (38.4% actual−37.3% budgeted = 1.1%). It is important to recall that, while variable expense totals will increase when sales exceed budgeted levels and decrease when sales fall short of budgeted levels, the percentage of sales allotted to a variable expense should not vary.

A careful examination of Marion's actual food expense in month 1 indicates two areas in which actual costs exceed expected levels (Meats and Fish/Poultry), two in which actual costs were less than budgeted (Produce and Groceries), and one in which actual costs were equal to budgeted costs (Dairy). Marion can now direct his attention to investigating why his costs were out of line in the key areas of meats and fish/poultry.

Just as Marion used percentages to compare his actual food usage to his budgeted usage, he can use them to compare his actual variable labor cost to budgeted labor costs using the same three-step method. To assess his variable labor usage, the three steps would be as follows:

Step 1. Calculate the percentage of revenue budgeted for each variable labor category (e.g., cooks, servers, buss staff, and the like).

Step 2. Compute percentage costs to sales for each variable labor category using actual labor usage and actual sales levels (taken from an operation's P&L).

Step 3. Compare actual monthly variable labor expense percentages to budgeted levels and take corrective action if necessary.

Consider the Costs

Sofia Lancaster is the registered dietitian (RD) responsible for all dietary services at Parker Meridian Memorial Hospital. Her operation consists of two departments. The first, and largest, is patient feeding. It consists of the tray line staff and the majority of her food production staff.

The second department is the public cafeteria, which includes special dining areas for hospital staff and a large dining area for visitors. Each year, she submits to her supervisor an annual labor expense budget broken down by month for each of her two departments.

On June 1, and after 4 months of consideration and planning, Sofia and her staff were excited to implement a new public cafeteria menu. The response from hospital visitors was excellent. The dining room staff reported that there were many positive comments about the foods selected for the new menu, and the production staff reported a 25 percent increase in the amount of food prepared for cafeteria service. Sofia had only budgeted for a 5 percent sales increase because of the new menu, so she was very happy.

On July 7, when Sofia's assistant, Jason, brought her the financial reports for the month of June, he was concerned. "I think we are in big trouble, Boss," he said. "We spent a lot more on cafeteria labor last month than we budgeted!"

1. What do you think is the cause of Sofia's labor budget overage in the cafeteria?
2. Do you think that going over budget in this situation is a good thing or a bad thing? Why?
3. If you were Sofia, would you re-examine your cafeteria labor budget for future operating periods? Why or why not?

PROFIT ANALYSIS

As business conditions change, changes in the budget are to be expected. This is because budgets are based on a specific set of assumptions and if these assumptions change, so too will the budgets. Assume, for example, that you budgeted $1,000 in January for snow removal from the parking lot attached to the restaurant you own in New York state. If unusually severe weather causes you to spend $2,000 for snow removal in January instead, the assumption (normal levels of snowfall) was incorrect, and the budget will be incorrect as well.

HERE'S HOW IT'S DONE 10.3

Typically, managers will find it very easy to monitor the fixed expense portions of their budgets. In most cases, the amount to budget (and spend) for a fixed expense in one accounting period is calculated as follows:

$$\frac{\text{Budgeted amount to spend in 1 year}}{\text{Number of accounting periods in 1 year}} = \text{Amount to spend in 1 accounting period}$$

Thus, if an operation makes an annual budget, and operates 12 months in the year, the manager would divide all annual fixed costs by 12 to determine the estimated monthly cost. If, however, a manager utilizes a 28-day accounting period, and, for example, the lease on her facility is $90,000 per year, the amount budgeted for each accounting period would be:

$$\frac{\$90,000}{13} = \$6,923.08$$

This is so because there are thirteen 28-day accounting periods in 1 year.

The monitoring of fixed costs amounts is easy, but managers should recall that fixed costs *percentage of sales* estimates will vary with changes in sales levels. When sales levels are higher than budgeted, fixed cost percentages will be lower than budgeted. When actual sales levels are lower than budgeted, fixed cost percentages will be higher than budgeted.

Ultimately, budgeted profit levels must be realized if an operation is to provide adequate returns for owner and investor risk. Consider the case of James, the operator of a foodservice establishment with excellent sales but below-budgeted profits. James budgeted a 5 percent profit on $2,000,000 of sales, and thus, a $100,000 profit ($2,000,000 sales × 0.05 profit = $100,000) was planned.

At year's end, however, James achieved $2,000,000 in sales but only $50,000 in profit, or 2.5 percent of sales ($50,000 profit ÷ $2,000,000 sales = 2.5%). If the operation's owners feel that $50,000 is an adequate return for their investment, James's professional services may be retained. If they do not, he could lose his job, even though the operation is profitable.

Remember that your goal is not merely to generate a profit, but rather to generate budgeted profit. In most cases, you will be rewarded only when you meet this goal because a primary purpose of management is to help generate the profits needed to continue the business. Budgeting for, and monitoring, these profits is a fundamental step in the process.

If profit goals are to be met, safeguarding the incoming money, or the operational revenue you generate, is very critical. It is to that task that we turn our attention in the next chapter because the proper collection of money and accounting for guest payments is one of the final and most important steps in a successful food and beverage cost control system.

Technology Tools

This chapter introduced the concepts of conducting individual menu item analysis and identifying a break-even point for your operation. When this break-even point is known, an effective operational budget can be produced. The chapter concluded with a discussion of the importance of developing and monitoring a budget. Although menu analysis software is often packaged as part of a larger program and is somewhat limited, software designed for overall break-even analysis and budgeting is readily available. Software and apps available to you today can help you:

1. Evaluate item profitability based on:
   a. Food cost percentage
   b. Popularity
   c. Contribution margin
   d. Selling price
2. Conduct menu matrix analysis.
3. Perform break-even analysis.
4. Budget revenue and expense levels.
5. Budget profit levels.
6. Assemble and produce budgets based on days, weeks, months, years, or other identifiable accounting periods.
7. Conduct regular performance to budget analysis for both revenue and costs.
8. Maintain performance to budget histories.
9. Blend budgets from multiple profit centers (or multiple units) into single centers.
10. Perform budgeted cash flow analysis.

For commercial operators, it is simply not wise to attempt to operate an effective foodservice unit without a properly priced menu and an accurate budget that reflects accurate sales and expense estimates.

Apply What You Have Learned

"All I am saying is we should use one or two sentences to describe each wine we sell. That way our guests will know more about the exact type of wine they are buying and what entrees go best with the wines they choose," said Latisha, the wine steward at the Gaylord Bistro, a casual restaurant noted for its chef's imaginative entrée preparations. "It might also make their ordering faster."

"And what I am saying" replied Ralphie, the operation's dining room manger, "is that with the length of our beer menu, and the two-page specialty cocktail menu we already have, if we add even one or two sentences to describe each wine we sell the beverage menu is going to become massive."

Latisha and Ralfie were talking to Shivansh, the operation's general manager. All were members of the Gaylord Bistro's menu development team.

"Well," said Shivansh, "our current wine list includes about 50 different choices. I agree it's a good idea to tell guests as much as possible about what they're ordering, but my initial thought is that maybe this will actually slow down our dining room service as guests will spend a lot of extra time just reading our menus."

1. Assume you were Shivansh. After giving it some thought, would you likely agree with Latisha that detailed menus speed guest ordering or would you maintain your initial opinion that they will slow service?
2. How can operators balance their guests' desire to know as much as possible about what they are ordering with the practical risk of potentially producing food or beverage menus that are too long and unwieldy to be effective?

For Your Consideration

1. Managers should take action when a significant difference exists between actual revenue (or expense), results, and their budgeted (forecasted) results. How large should a difference be before it should be considered significant?
2. Managers of multi-unit properties within the same chain can compare operating results across the units they manage. What do you think are some reasons financial performance and profitability could vary among operating units serving identical menu items at identical prices?

Key Terms and Concepts

The following are terms and concepts addressed in the chapter that are important for you as a manager. To help you review, define the terms below:

Menu copy
Matrix analysis
Menu engineering
Goal value analysis
Loss leaders
Covers
Break-even point
Cost/volume/profit (CVP) analysis
Contribution margin income statement
Contribution margin (CM) for overall operation
Minimum sales point (MSP)
Minimum operating cost
Long-range budget
Annual budget
Achievement budget
Monetary unit
Exchange rate
Sales per seat
Sales per square foot

Test Your Skills

You may download the Excel spreadsheets for the Test Your Skills exercises from the student companion website at http://www.wiley.com/go/dopson/foodandbeveragecost-control8e. Complete the exercises by placing your answers in the shaded boxes and answering the questions as indicated.

1. Ray Lee is the manager of a casual seafood restaurant. Ray wants to do a Goal Value analysis of his entrée items and then use the Goal Value to assess the profitability of each of his 20 entrée items.

   Ray's check average is $25.95, his food cost is 32%, and his non-food variable costs are 40%. For this analysis, Ray is estimating that he will serve 3,000 guests.

   Calculate Ray's Goal Value target and then answer the question that follows.

   Ray's Goal Value Target

A	B	C	D	Goal Value
		$25.95		

   Rounded to the nearest whole number, what will be Goal Value against which Ray will compare each of his menu items?

2. Isabelle Elias is the managing owner of Barney's Gourmet Burgers. Isabelle sells a variety of Wagu beef burgers, as well as assorted sides such as French fries, onions rings, and deep-fried pickles. Isabell has created a condensed P&L detailing last year's financial results, and she is now forecasting her sales and expenses for next year.

   For next year Isabelle expects her menu price increases and marketing efforts to produce a 6 percent increase in sales. She predicts her food cost percentage will remain unchanged but, due to raises and the additional sales she will generate, she predicts the total amount she will spend for labor will increase by 5 percent.

   She predicts her other controllable expenses will increase by $30,000 and her non-controllable expenses will remain unchanged. Help Isabelle forecast the before tax profit she will generate next year by completing her condensed P&L below, then answer the question that follows.

Barney's Gourmet Burgers	This Year	%	Next Year	%
Total sales	$1,500,000	100%		100%
Cost of sales	$450,000	30.0%		
Labor costs	$480,000	32.0%		
Prime costs	$930,000	62.0%		
Other controllable expenses	$275,000	18.3%		
Non-controllable expenses	$150,000	10.0%		
Total costs	$1,355,000	90.3%		
Before tax profit	$145,000	9.7%		

   What will be Isabelle's before tax profit next year if her forecast is accurate?

3. Anna Farris is a District Manager for the Fun Time chain of pizzerias. Her new promotion to Regional Manager means she must now select one of the current unit managers in her district to fill her old District Manager job.

   Anna would like to promote the manager whose unit generated the most revenue per square foot on an average week last year in their current operation.

   Using last year's revenue figures, help Anna complete the weekly revenue per square foot worksheet she has begun, and then answer the questions that follow.

Fun Time Pizza District Results: Anna's District				
Unit Location	**Last Year's Revenue**	**Average Weekly Revenue**	**Unit Square Footage**	**Average Weekly Revenue per Square Foot**
Chippewa Falls	$1,027,000		780	
Altoona	$1,391,000		1,100	
Fall Creek	$1,586,000		1,550	
Cadot	$807,000		1,200	
Augusta	$1,651,000		1,400	
Clair	$1,235,000		1,225	

   1. Which unit's manager achieved the highest weekly revenue?
   2. Which unit's manager achieved the highest revenue per square foot?
   3. Based solely on their ability to generate per square foot revenue, which unit's manager should Anna promote?

4. Bart Masterson is considering leasing a 2,450 square foot free-standing restaurant site that has the potential to add drive-thru service to its currently 100 seat on-premise dining and bar areas. Bart has plans to open an operation featuring Cuban and other Caribbean-style foods. The owner of the space wants an annual rent of $22.50 per square foot the location.

   Bart feels he can make a good profit in this proposed location if his space rental cost is 6 percent or less of his gross sales. Help Bart complete the worksheet he has begun to determine the annual sales volume his operation must generate to achieve a 6 percent space rental cost in this proposed location and then answer the questions that follow.

Bart's Proposed Space Rental					
Annual space cost per square foot	$22.50	Monthly		Annual	
# of square feet	2,450	$	%	$	%
Total revenue			100%		100%
Space rental cost					6.0%

   1. At what monthly total revenue would Bart's space rental cost equal 6 percent of his sales?
   2. At what annual total revenue would Bart's space rental cost equal 6 percent of his sales?

5. Boniso operates Boniso's Mexican Restaurant in a rural area in West Texas. He has worked hard at setting up cost control systems, and he is generally happy with his overall results. However, he is not sure if all of his menu

items are providing profitability for his restaurant. He decides to use food cost matrix and contribution margin matrix analyses to study each of his menu items.

a. Complete his menu analysis worksheet.

Menu Analysis Worksheet

Menu Item	Number Sold	Selling Price	Total Sales	Item Cost	Total Cost	Item Contribution Margin	Total Contribution Margin	Food Cost %
Fajita plate	147	$25.90		$9.84				
Enchilada dinner	200	$19.90		$6.96				
Menudo	82	$13.90		$3.48				
Mexican salad	117	$15.90		$4.78				
Chalupa dinner	125	$17.90		$5.02				
Burrito dinner	168	$19.90		$6.50				
Taco dinner	225	$11.90		$3.10				
Total								
Weighted average								

b. Using the results of Boniso's menu analysis worksheet (in part a), fill in the appropriate average food cost, popularity, and contribution margin in the blanks below. Then, place the menu items in the appropriate squares in the matrices.

Food Cost Matrix

High Food Cost % (Above ____%)		
Low Food Cost % (Below ____%)		
	Low Popularity (Below ____ sales)	High Popularity (Above ____ sales)

Contribution Margin Matrix

High Contribution Margin (Above $ ____)		
Low Contribution Margin (Below $ ____)		
	Low Popularity (Below ____ sales)	High Popularity (Above ____ sales)

6. Eunice manages a Thai restaurant in a primarily Asian section of a major West Coast city. She is interested in determining dollar sales and number of guests needed to break even and to generate her desired profits. Her check average (selling price) is $16.00, her variable cost per unit (guest) is $6.60, and her fixed costs are $170,000.

a. Complete the following grid and determine her before-tax profit.

	Per Unit (Guest)	Percentage
SP		
VC		
CM		

Fixed costs	$170,000.00
Desired after-tax profit	$24,000.00
Tax rate	40%
Before-tax profit	

b. Using the information from part a, calculate the following. (Spreadsheet hint: Use the ROUNDUP function for "Rounded up—Break-even point in guests served" and "Rounded up—Guests served to achieve desired after-tax profit.")

Break-even point in sales dollars			
Break-even point in guests served		Rounded up =	
Sales dollars to achieve desired after-tax profit			
Guests served to achieve desired after-tax profit		Rounded up =	

c. Based on her calculations, Eunice doesn't think that she can attract as many guests as she needs to achieve her desired after-tax profit. Therefore, she has decided to make some changes to improve her situation. Due to these changes, she has been able to reduce her selling price by $1.00, decrease her variable cost percentage by 5 percentage points, and lower her fixed costs by $5,000. After these changes, what are Eunice's sales dollars and guests served to achieve her after-tax profit? Complete the following grid and calculations. (Spreadsheet hint: Use the ROUNDUP function for "Rounded up—Guests served to achieve desired after-tax profit.")

	Per Unit (Guest)	Percentage
SP		
VC		
CM		

Fixed costs	
Sales dollars to achieve desired after-tax profit	
Guests served to achieve desired after-tax profit	Rounded up =

7. Sinqobile runs a restaurant in an East Coast city that specializes in African-American cuisine. She has compiled her sales and cost data from last year, and she wants to develop a budget for this year. She has projected the following increases for this year:

Projected increases	
Meals served	4%
Selling price per meal	2%
Cost of food	5%
Cost of labor	10%
Other expenses	2%

Using this information, help Sinqobile complete her budget.

	Last Year	Budget
Meals served	122,000	
Selling price per meal	$12.50	

Last year's food cost per meal:	
Last year's food cost per meal + Estimated increase:	
Estimated cost of food this year:	

Last year's labor cost per meal:	
Last year's labor cost per meal + Estimated increase:	
Estimated cost of labor this year:	

	Last Year $	Last Year %	Budget $	Budget %
Revenue	$1,525,000.00			
Cost of food	$610,000.00			
Cost of labor	$488,000.00			
Other expenses	$245,760.00			
Total expenses	$1,343,760.00			
Profit	$181,240.00			

The owner of the restaurant has requested that Sinqobile make at least a 10 percent profit for this year. Based on her budget figures, is she likely to meet this goal? If not, what can she do to achieve a 10 percent profit?

8. Toni Lamazza is developing next year's foodservice budget for the Springdale school system, consisting of 17 different schools in a two-county area. Toni

knows the revenue she will get from the School Board but is not sure how much the Board will give her to pay for anticipated increases in employee benefits. Complete the chart below to help Toni determine the amount employee benefits can increase and still allow her to show a budget surplus.

		Budget with Employee Benefits Increases			
	Current Budget	5% Increase	10% Increase	15% Increase	20% Increase
Revenue	$7,000,000	$7,000,000	$7,000,000	$7,000,000	$7,000,000
Cost of food	$2,095,000	$2,095,000	$2,095,000	$2,095,000	$2,095,000
Cost of payroll	$3,700,000	$3,700,000	$3,700,000	$3,700,000	$3,700,000
Cost of employee benefits	$700,000				
Other expenses	$400,000	$400,000	$400,000	$400,000	$400,000
Total costs	$6,895,000				
Budget surplus/deficit	**$105,000**				

a. At what level of employee benefit cost increase will Toni have a break-even budget?

b. How much would her surplus/deficit be if benefits increase by 20 percent?

9. J. D. McAllister is really happy. He has just succeeded in securing a $2,000,000 bank loan with a variable percentage interest rate to build his dream restaurant. J. D.'s loan is for 25 years, and it carries an interest rate that is set at 7 percent for the first year (this year). Thus, his first year's monthly interest payments will be $14,136. An experienced restaurateur, J. D. has prepared the following annual budget for this year.

Annual Budget for This Year	
Total sales	$2,046,000.00
Variable costs	
Food	429,660.00
Beverage	257,796.00
Labor	572,880.00
Other variable costs	171,864.00
Total variable costs	1,432,200.00
Fixed costs (excluding loan repayment)	239,568.00
Loan repayment (7% interest)	169,632.00
Total fixed costs	409,200.00
Before tax profit	204,600.00
Taxes (40%)	81,840.00
After tax profit	122,760.00

J. D.'s interest rate will likely vary over the life of the loan because it is tied to the prime interest rate established by the Federal Reserve Board (part of the federal government). Assume that the Fed increases interest rates next year by half a percentage point, and thus J. D.'s interest rate moves to 7.5 percent. As a result, his new monthly loan repayment will be $14,780.

Also assume that his total variable cost percentage and fixed costs (excluding the increased loan repayment) will be the same next year, and his check average (selling price) will be $20 per cover. With the higher mortgage payment, what sales dollars and number of guests served will J. D. need to achieve next year to maintain the same number of after-tax profit dollars as he budgeted for this year?

	Per Unit (Guest)	Percentage
SP		
VC		
CM		

Fixed costs (excluding loan repayment)	
Loan repayment (7.5% interest)	
Total fixed costs	
Desired after-tax profit	
Tax rate	
Before-tax profit	
Sales dollars to achieve desired after-tax profit	
Guests served to achieve desired after-tax profit	

10. V. K. Sing is the manager of the Knight Kap restaurant. V. K. is preparing his budget for next year and would like to estimate his beverage revenue and expense. V. K. kept careful sales and cost records from last year and these are listed below.

	Guests Served	Selling Price	Beverage Cost
Beer	3,500	$7.75	$1.70
Wine	4,800	$11.00	$3.80
Spirits	2,500	$12.00	$1.80

V.K. has made the following assumptions about next year:

- The cost he will pay for alcoholic beverages will increase by 2 percent in all of the beverage categories.
- V.K. will increase the selling price of all alcoholic beverages by 5 percent.
- The number of guests served will increase by 2 percent for beer, 2 percent for wine, and 3 percent for spirits.

To help V.K. prepare next year's beverage revenue and expense budget, calculate the following for each of his three beverage types:

1. Forecasted total beverage revenue
2. Forecasted total beverage cost
3. Forecasted beverage cost percentage

Projected Changes (Percentage Increases or Decreases)			
	Guests Served	Selling Price	Beverage Cost
Beer			
Wine			
Spirits			

Forecasted Beverage Revenue and Expense Forecast: Knight Kap						
	Guests Served	Selling Price	Beverage Cost	Total Beverage Revenue	Total Beverage Cost	Beverage Cost %
Beer						
Wine						
Spirits						
Total						

11. Jerri Ellias is the owner of craft brewery. In a casual setting, her operation serves food, and a variety of specialty brewed beers. June is a busy month for her, and she has just completed it. Jerri uses a modified version of the Uniform System of Accounts for Restaurants (USAR) to prepare her budgets and income statements because she wants to identify her operation's total gross profit (see Chapter 9). Below are her original budget and her actual operating results for the month of June. Complete the worksheet for her by providing variance dollars and variance percentages for each budgeted item and then answer the questions that follow.

Jerri's Budget Versus Actual Results Comparison for June						
	Budget ($)	%	Actual ($)	%	Variance ($)	%
Sales						
Food	179,200	71.8%	193,251	68.4%		
Beverage	70,250	28.2%	89,250	31.6%		
Total sales	**249,450**	100.0%	**282,501**	100.0%		
Cost of sales						
Food	67,500	37.7%	68,925	35.7%		
Beverage	14,500	20.6%	20,415	22.9%		
Total cost of sales	**82,000**	32.9%	**89,340**	31.6%		
Gross profit						
Food	111,700	62.3%	124,326	64.3%		
Beverage	55,750	79.4%	68,835	77.1%		
Total gross profit	**167,450**	67.1%	**193,161**	68.4%		
Operating expenses						
Management & staff	38,750	15.5%	44,850	15.9%		
Employee benefits	9,500	3.8%	11,203	4.0%		
Direct operating expenses	26,500	10.6%	27,210	9.6%		
Music & entertainment	0	0.0%	0	0.0%		
Marketing	1,200	0.5%	150	0.1%		
Utility services	3,250	1.3%	3,289	1.2%		
Repairs & maintenance	3,500	1.4%	2,874	1.0%		
Administrative & general	24,750	9.9%	25,598	9.1%		
Occupancy	1,200	0.5%	1,200	0.4%		
Depreciation	2,200	0.9%	2,200	0.8%		
Total operating expenses	**110,850**	44.4%	**118,574**	42.0%		

Jerri's Budget Versus Actual Results Comparison for June						
	Budget ($)	**%**	**Actual ($)**	**%**	**Variance ($)**	**%**
Operating income	**56,600**	22.7%	**74,587**	26.4%		
Interest	7,100	2.8%	7,100	2.5%		
Income before income taxes	**49,500**	19.8%	**67,487**	23.9%		
Income taxes	12,625	5.1%	16,872	6.0%		
Net income	**36,875**	14.8%	**50,615**	17.9%		

a. What is the variance (dollar and percent) in Jerri's total cost of sales for the month of June?

b. What was the cause of the variance in Jerri's total cost of sales for the month of June?

c. What is the variance (dollar and percent) in Jerri's management and staff costs for the month of June?

d. To what would you attribute the variance in Jerri's management and staff costs for the month of June?

12. The Wheatfield Valley golf course has been owned by the Miley family for two generations. Currently it is managed by Cyrus Miley, a graduate of State University where he majored in hospitality management. Last year was a good one for the golf course. Now Cyrus is preparing next year's operating budget. He has gathered a great deal of information to help him prepare the best budget possible. After carefully analyzing that information, Cyrus predicts that next year the course will experience:

- A 5 percent increase in food sales
- A 3 percent increase in beverage sales
- No change in food or beverage product cost percentage
- Salaries and wages (management and staff) that will increase 4.5 percent
- Employee benefits that will increase 10 percent
- An increase of 2.5 percent in each Other Expense category
- No change in Occupancy costs and Depreciation & amortization.
- Interest payments of $1,000 per month
- Tax payments that are estimated to be 25 percent of Income Before Income Taxes

Calculate this year's operating percentages for the food and beverage department on the budget worksheet provided, and then using his assumptions about next year, create Cyrus' new operating budget in dollars and percentages.

The Wheatfield Valley Golf Course F&B Department Budget Worksheet for Next Year				
	This Year Actual $	%	Next Year Budget $	%
Sales				
Food	$169,250.00			
Beverage	$55,750.00			
Total sales				
Cost of sales				
Food	58,250.00			
Beverages	9,375.00			
Total cost of sales				
Labor				
Management	9,000.00			
Staff	27,250.00			
Employee benefits	6,200.00			
Total labor				
Prime cost				
Other controllable expenses				
Direct operating expenses	13,200.00			
Utilities	5,975.00			
General & administrative expenses	6,375.00			
Repairs & maintenance	1,750.00			
Total other controllable expenses				
Controllable income				
Non-controllable expenses				
Occupancy costs	15,000.00			
Equipment leases	0.00			
Depreciation & amortization	4,200.00			
Total non-controllable expenses				
Restaurant operating income				
Interest expense	12,000.00			
Income before income taxes				
Income taxes	$14,106.25			
Net income				

CHAPTER 11

Managing the Revenue Control System

OVERVIEW

This chapter will teach you the principles of revenue control, which includes protecting your sales revenue from external and internal threats of theft. This chapter will also teach you how to establish an effective revenue security system, which is a series of checks and balances that will help you protect your revenue from the time it is collected from your guests to the time it is deposited in your bank and appropriately spent.

Chapter Outline

- Revenue Security
- External Threats to Revenue Security
- Internal Threats to Revenue Security
- Developing the Revenue Security System
- Managing the Revenue Security System
- Technology Tools
- Apply What You Have Learned
- For Your Consideration
- Key Terms and Concepts
- Test Your Skills

LEARNING OUTCOMES

At the conclusion of this chapter, you will be able to:

- Identify internal and external threats to revenue.
- Create effective countermeasures to combat internal and external theft.
- Establish and monitor a complete and effective revenue security system.

REVENUE SECURITY

In Chapter 1, you learned that:

$$\text{Revenue} - \text{Expenses} = \text{Profit}$$

A close examination of that simple formula has led some foodservice operators to state that 50 percent of their time should be focused on managing revenue and 50 percent on managing expenses. This is actually quite reasonable because all of the cost control systems you develop will be of little use if you are unable to initially collect the revenue generated by your restaurant, have that revenue deposited into your bank account, and then spend it only on legitimate expenses.

The careful recording and reporting of revenue (sales) in a foodservice operation is important for a number of reasons. One of the most important is that in most states restaurant meal sales are subject to sales tax that must be collected by the foodservice operator and then paid to the appropriate taxing authority. The rules on sales tax collection and payment vary by state and local municipality, but must always be carefully followed.

Also, as previously addressed, revenue is the denominator for a number of important cost-related calculations (e.g., food cost percentage, labor cost percentage, and others), thus accuracy in determining actual revenue is critical. Finally, properly recording revenue is important because it allows operators to determine whether revenues are increasing or decreasing over time (see Chapter 2).

SALES VERSUS REVENUE

While the two terms are often used interchangeably, it is important to understand the subtle difference between sales and revenue. A sale is made when food and beverage products are delivered to a guest. Revenue (i.e., sales receipts) is generated when that guest pays for their purchases. Both the timing and method of payments available to guests can vary from one foodservice operation to another.

For example, in some foodservice operations payment for purchases is made prior to actual product delivery. In other operations, payment is collected at the time of product delivery. In table-service restaurants payment is not typically collected until guests have completed their meals, and in banquet and special event venues guests may be billed and payments collected well after guests have received their ordered products and services. These four various time frame/payment options are summarized in Figure 11.1, and the proper management of each option is essential.

FIGURE 11.1 Foodservice Guests Payment Options

Payment Made	Example(s)
1. Prior to guest receiving products	Guests orders placed via third-party delivery apps
2. At the time products are supplied to guests	Orders placed and received via drive-thru windows or picked up on-premise by guests
3. After guests have consumed their purchases	On-premise orders placed in table-service casual and fine-dining restaurants
4. Days or weeks after an event is held	Foodservice operations providing banquet services or direct billing options to selected guests

It is important to recognize that when restaurants utilize a **cash accounting** system, sales are recorded as "revenue" when payment for the sales have actually been received from guests.

When using an **accrual accounting** system, however, an operation's sales (revenues) are recorded when they are earned, not necessarily when cash is received. This means that if a product is delivered, the revenue associated with that sale is recorded at that time, even if the guest has not yet paid for their purchase.

FUN ON THE WEB!

Some foodservice industry experts recommend a cash accounting system be used for very small restaurants. They cite the simplicity of such a system as its major advantage.

Most restaurants, however, will find that the use of an accrual system, as recommended by the Uniform System of Accounts for Restaurants (USAR), provides better insight into an operation's financial performance and profitability.

To learn more about the advantages and disadvantages associated with each system, enter "Is cash or accrual the best accounting system for restaurants?" in your favorite search engine and review the results.

Regardless of *when* they pay for their purchases, in most foodservice operations guests can also choose *how* they make their payments. These most common of these are shown in Figure 11.2.

CASH

Most (but not all!) foodservice operations accept cash as a payment form. The number of guests using cash to pay for their purchases, however, has been steadily declining in recent years.[1]

Those foodservice operators who still encourage the use of cash as a method of payment cite the fact that, unlike payment cards, there are no transaction fees involved in processing a cash payment. Another advantage is that cash accepted as payment is immediately available for the operation's use. These pro-cash operators also point to enhanced guest privacy (because guests need not supply any personal or banking information when paying their bills).

FIGURE 11.2 Foodservice Guest Methods of Payment

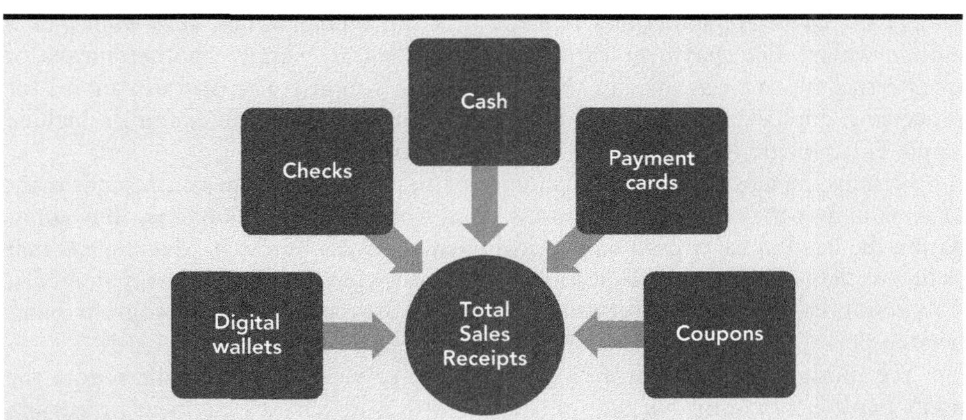

[1] https://www.restaurantbusinessonline.com/financing/pandemic-accelerated-americans-move-away-cash. Retrieved 7/25/2025.

CHECKS

While few foodservice operators still accept personal checks, checks are commonly accepted for payment of banquet services and for **house accounts** that allow for creditworthy companies and guests to pay for their purchases when they are billed for them, rather than when they actually make their purchases. This billing typically takes place after an event has concluded, and many foodservice operations bill house accounts monthly, rather than daily.

PAYMENT CARDS

Most (but not all!) foodservice operations will accept one or more types of electronic payment cards. The most popular of these options include:

- Travel and entertainment cards
- Credit cards
- Debit cards

Travel and entertainment (**T&E**) cards are a payment system by which the card issuer collects full payment from its card users on a monthly basis. These payment card companies do not typically charge interest on their user's purchases. Rather they rely on larger fees collected from the merchants (i.e., foodservice operators) who accept the cards to make their profits. Examples of T&E cards are American Express (Amex), Diners Club, and Carte Blanche. In some cases, the fees charged by T&E card issuers are very high, so some restaurateurs do not accept them as a payment form.

Also commonly known as "bank" cards, **credit cards** are simply a system by which banks loan money to the card's users as the users make purchases. The loans typically carry interest. Merchants, such as restaurants, accepting the cards for payment are charged a fee (by the banks) for the right to allow their customers to pay by credit card. Examples of popular credit cards are Visa, MasterCard, and Discover.

Debit cards are also an extremely popular form of guest payment. In this system, the funds needed to cover the card user's purchase are automatically transferred from the card user's bank account to the entity issuing the debit card, who then pays that amount to the operation accepting the card. As with T&E and credit cards, operators accepting debit cards are charged a fee for the right to do so. Debit cards look, and operate, in a manner similar to credit cards.

DIGITAL WALLETS

Increasingly, foodservice guests use a variety of **digital wallets** when making their purchases. These are sometimes referred to as an online wallet, an e-wallet, or a mobile wallet. Like payment cards, digital wallets are simply another means of transferring funds electronically, thus eliminating a foodservice operator's need for processing guests' cash or check payments. Currently popular examples include Apple Pay, Google Pay, Cash App Pay, PayPal, and Venmo.

Perhaps the most significant advantage of digital wallets from a customer's point of view is that the customer can access them from their mobile phone, thus eliminating the need to carry cash or physical payment cards. Another advantage is that online wallets often have built-in features such as loyalty card membership numbers. As a result, customers can access digital coupons or redeem loyalty program membership perks, again without having to carry a physical membership card.

Yet another advantage from a guest's point of view is that e-wallets store the user's bank information but do not share it with a merchant's payment processing system. E-wallets act as hubs that generate unique codes or payment codes that external systems can use only once, without accessing the guest's personal banking information. As a result, it eliminates the possibility of credit card fraud being initiated by a merchant's employees.

FUN ON THE WEB!

Perhaps the greatest reason for the increase in use of digital wallets is the ability of guests to use their cellphones to make purchases in person or online. This eliminates the need to fumble for cash or carry credit or debit cards.

There are, however, a variety of digital wallet types in use. These include mobile wallets, online wallets, open-wallets, closed-wallets, prepaid wallets, and bank-linked wallets. To better understand the differences between these e-wallet formats, enter "What are the different types of digital wallets in popular use today?" in your favorite search engine and view the results.

COUPONS

In most cases the use of a coupon is *not* considered a payment (revenue), but rather it is a discount (i.e., a reduction from revenue). Thus, for example, if a guest makes a $100 purchase and has a "$10 off" coupon, the foodservice operator will record revenue of $90.00 ($100 purchase − $10.00 coupon = $90.00 revenue), not $100.

Should they desire to do so, the total amount of the coupon values redeemed can be separately recorded as "Sales Discounts" or "Coupon Redemptions," both of which would reduce the overall reported revenue.

While not as common, if a coupon is issued by a manufacturer and the restaurant is reimbursed for the value of these redeemed coupons, the reimbursement amount is included in the restaurant's total taxable sales. Similarly, if a guest pays for a coupon initially (e.g., as part of a coupon booklet) a pro-rata share of the coupon booklet's cost, allocated to the specific coupon used, should be included as part of the operation's total taxable sales.

Regardless of when, and how, guests pay their bills, errors in revenue collection or security can be the result of simple employee mistakes or, in some cases, outright theft by either guests or employees. An important part of your job is to devise and manage a revenue security system that protects your operation's income, whether you receive it in the form of cash, checks, payment cards, digital wallets, coupons, or any other method of guest payment you accept.

In its simplest form, revenue control and security is a matter of matching products sold with funds received. Thus, an effective revenue security system ensures that the following five formulas reflect what happens in your foodservice operation:

1. Documented product requests = Product issues
2. Total product issues (by the kitchen or bar) = Total guest charges
3. Total guest charges = Total sales receipts
4. Total sales receipts = Total verified bank deposits
5. Total verified bank deposits = Funds available to pay legitimate expenses (i.e., **Accounts Payable**)

This chapter will address each of these important formulas in turn.

The potential for guest or employee theft or fraud exists in each of these five areas, so it is important for you to remain alert to irregularities that can help you know if theft or fraud is taking place.

To illustrate revenue-control-related problems that could occur, assume that you own a chain of 10 coffee/dessert shops. You call your units The Pie Parlor. You sell freshly baked gourmet pies of many varieties, all for the same price of $6 per slice and $45.00 per pie. A whole pie consists of eight slices.

In addition, you sell coffee for $5 per cup. Assume that Figure 11.3 details your sales record on a Monday for unit 6, which is one of your 10 stores. If you have in place a good system of controlling your revenue, you should have total receipts (revenue) of $4,000 for January 15 at this unit.

FIGURE 11.3 Sales Record

Unit Name: The Pie Parlor #6			Date: 1/15
Item	**Number Sold**	**Selling Price**	**Total Sales**
Apple pie			
Slices	60	$6.00	$360.00
Whole	11	$45.00	$495.00
Pumpkin pie			
Slices	40	$6.00	$240.00
Whole	14	$45.00	$630.00
Cherry pie			
Slices	75	$6.00	$450.00
Whole	5	$45.00	$225.00
Peach pie			
Slices	25	$6.00	$150.00
Whole	10	$45.00	$450.00
Coffee	200	$5.00	$1,000.00
Total			$4,000.00

Assume, however, that at the conclusion of the day you have only recorded $3,800 in actual revenue. You would be "short" $200 in revenue ($4,000 recorded revenue – $3,800 actual revenue = $200 shortage). In this case, a security problem exists; not perhaps, in the control of products (a cost control issue) but rather in the control of receipts (a revenue security issue).

If you were short of $200 in revenue per day for each of your 10 units and you were open 360 days per year, your company's revenue loss for the year would be a staggering $720,000 ($200 short per day × 10 units × 360 days = $720,000)!

There are several reasons why you might experience a revenue shortfall. One good way to view revenue-related concerns managers must address is to classify them as either:

1. External threats to revenue security

   or

2. Internal threats to revenue security

EXTERNAL THREATS TO REVENUE SECURITY

Both guests and employees can be threats to revenue. As a result, your facility could lose sales revenue because of guests defrauding your foodservice operation. This activity can take a variety of forms, and a very common one is for a guest to **walk**, or **skip** their bill. Guests are said to have walked, or skipped, when they consume products but then leave the foodservice operation without paying their bills. This type of theft is not generally present in, for example, a quick-service restaurant (QSR) where payment is collected before, or at the same time, the food is given to the guest.

In cases where a guest is in a busy table-service dining room, however, it is quite possible for one or more members of a dining party to slip outside while the server is busy with other guests. In any case, the loss of revenue to your business can be substantial if you do not take the necessary steps to reduce this type of theft. The fact is, in a busy restaurant, it can sometimes be relatively easy for a guest or an entire party to leave without properly settling their bill unless everyone on your staff is extremely vigilant. To help reduce this type of guest theft, implementation of the steps in Figure 11.4 is suggested.

It is, of course, also possible that a guest simply "forgets" to settle his or her bill and leaves the establishment without paying. However, if many of your guests walked their bills on your day of $4,000 sales in our Pie Parlor example, you would, of course, find that the products had indeed been consumed, but the receipts (revenue) in your cash drawer would not equal the sales value of the pie and coffee products issued by the kitchen to your wait staff, who then served the products to your guests.

In addition to skipping a bill, another form of external theft you must be aware of is that used by the quick-change artist. A **quick-change artist** is a guest who, having practiced the routine many times, attempts to confuse a cashier. In his or her confusion, the cashier gives the guest too much change. For example, a guest who should have received $5 in change may use a confusing routine to secure $15. To prevent this from happening, you must train your cashiers well and instruct them to notify management immediately if there is any suspicion of attempted fraud to your operation through quick-change routines by a guest.

Another form of external theft that you must guard against is that of a guest using counterfeit money. **Counterfeit money** is an imitation of currency intended to be passed off fraudulently as real money. Counterfeit money used by guests will not

FIGURE 11.4 **Steps to Reduce Guest Walks, or Skips**

1. If the custom of your restaurant is that guests order and consume their food prior to your receiving payment, instruct servers to present the bill for the food promptly when the guests have finished.
2. If your facility has a cashier in a central location in the dining area, have that cashier available and visible at all times.
3. If your facility operates in such a manner that each server collects for their own guest's charges, instruct the servers to return to the table promptly after presenting the guest's bill to secure a form of payment.
4. Train employees to be observant of exit doors near restrooms or other areas of the facility that may provide an unscrupulous guest the opportunity to exit the dining area without being easily seen.
5. If an employee sees a guest leave without paying the bill, management should be notified immediately.
6. Upon approaching a guest who has left without paying the bill, the manager should ask if the guest has inadvertently "forgotten" to pay. In most cases, the guest will then pay the bill.
7. Should a guest still refuse to pay, or flee the scene, the manager should note the following on an incident report:
   a. Number of guests involved
   b. Amount of the bill
   c. Physical description of the guest(s)
   d. Vehicle description if the guests flee in a car, as well as the license plate number if possible
   e. Time and date of the incident
   f. Name of the server(s) who actually served the guest
8. If the guest is successful in fleeing the scene, the police should be notified. In no case should your staff members or managers be instructed to attempt to physically detain the guest. The liability that could be involved should an employee or the guests be hurt in such an attempt is typically far greater than the value of a single skipped food and beverage bill.

be honored as legal currency by banks. Therefore, counterfeit money deposited by a restaurant manager will result in the ultimate loss of revenues if the perpetrator is not caught. You will have an exercise in the Test Your Skills section of this chapter that will help you learn how to identify counterfeit money.

THE ROLE OF MERCHANT SERVICE PROVIDERS

In our advanced technology society, fewer and fewer consumers carry significant amounts of cash. As a result, in most commercial foodservice operations, the majority of guests will use one or more of the noncash payment forms previously addressed in this chapter, but it is important to recognize that, when a foodservice operation agrees to accept a noncash form of payment, the operation will pay a fee for accepting that payment form.

If restaurant managers are to ensure that they collect all of the money they are due from the noncash payment forms they accept, they must effectively manage the **interface** (electronic connection) between the various noncash payment companies and their restaurants. That interface is typically provided by the restaurant's **merchant service provider (MSP)**. An MSP plays an important role as the restaurant's coordinator/manager of noncash payment form acceptance and funds collection.

It is important to understand that all noncash payment form companies, as well as MSPs who are assisting restaurants, only do so for a price. The noncash payment form issuers will charge the restaurant a fee for accepting the payment form and the MSP will charge a variety of fees, including initial setup fees, per transaction fees, programming fees, statement fees, and fees related to managing the payment form's interface. These interface-related fees may include equipment purchase charges (or leases) and connectivity fees.

When using the Uniform System of Accounts for Restaurants (USAR) these noncash payment-related expenses are not considered as a reduction of revenue, but rather they are recorded in the "Credit card commissions and fees" section of "General and Administrative Operating Expense" and thus are listed in the "Other Expense" category of the Income Statement (see Chapter 8).

In most cases, restaurants have little option but to accept the most popular forms of noncash payment. The amount charged by different MSPs, however, will vary and so will the amount of fees they charges. Overall, the MSP generally keeps 10 to 20 percent of all the payment card-related fees you will pay, and the noncash payment form issuing company (e.g., Visa, MasterCard, PayPal, and the like) receives the balance.

In most cases, the noncash payment company will establish an online account where you can view pending charges and, periodically, you will get a statement displaying the total transactions and the various fees you paid (because the fees are withdrawn by the MSP before your revenue is deposited in your bank account). Thus, for example, your operation may receive a total of $97.00 from a transaction in which a guest originally paid their $100.00 guest check using a noncash payment card. Such payment-related fees typically range from 2.0 to 4.0 percent of a transaction's total amount. Thus, an operation that charges a credit-card-paying guest $100 for items purchased by the guest will typically pay transaction fees ranging from $2.00 to $4.00 on that sale.

To help offset their noncash processing charges, some foodservice operations add a surcharge for those guests using a noncash payment form. Thus, for example, a guest who elects to use a credit card to pay a $100 guest check may be charged an additional 3.5 percent, bringing the guest's total to bill $103.50 ($\$100.00 + [\$100.00 \times 0.35] = \$103.50$). When this occurs the proper amount to be recorded as revenue would be $103.50, not $100.

Depending upon the specific noncash payment form used by a guest, it may take 1 to 7 days for these funds to actually show up in your restaurant's bank account. If a restaurant's point of sale (POS)/MSP interface is faulty or mismanaged, the result can be slow processing, increased errors, omissions, and even disappearing sales!

It is important to remember that maintaining the quality of relationship between your MSP and your restaurant's revenue is just as important as maintaining adequate control over the cash your restaurant accepts.

FUN ON THE WEB!

Each noncash payment form company or card issuer will charge a fee to the merchants accepting the payment form. These fees vary by company. To see an example of how the fees are assessed by one popular e-wallet company, enter "PayPal-Merchant fees" in your favorite search engine and view the results.

HERE'S HOW IT'S DONE 11.1

Because guests who pay their bills using a noncash method of payment often add a server's tip to the total they agree to pay, employers may have a decision to make. This is because the employer will be charged credit card processing fees on the guest's bill, as well as on the tip left for the server.

For example, if a guest's dinner bill is $100, and that guest decides to leave a $20 server tip, the server's employer will pay credit-card- and MSP-related fees based on a $120 transaction, not a $100 transaction. In most (but not all!) states, foodservice operations can legally deduct (from the gratuities to be paid to servers) the cost of processing the gratuity portion of a guest's payment. The employer *cannot* deduct from the server's gratuity the entire cost of processing the transaction.

Where allowed by law, calculating the amount of tipped income that can be legally withheld from tipped employees is an easy, three-step process.

Step 1: Identify the average cost of processing a noncash transaction

Payment card processing fees vary based on a number of things, including total number of transactions completed per month, transaction size, and even the way payment information is entered into the payment system. In most cases, it is acceptable to use the "average" percentage amount charged per transaction for purpose of calculating deductions for tipped income.

For example, if it costs, on average, 3.5 percent to process an operation's noncash payments and a customer leaves a $20 tip on a $100 bill, the total cost to process the transaction would be:

($100 bill + $20 gratuity) × 0.035 average processing fee = $4.20 fee

Step 2: Calculate the cost of processing the gratuity (only)

In this example, the server's tip is $20, thus the cost of processing the gratuity is

$20 gratuity × 0.035 average processing fee = $0.70 fee

Note that the employer *cannot* deduct the entire amount calculated in Step 1.

Step 3: Calculate the amount of gratuity to be paid to the employee

In this example, the employer could deduct a $0.70 processing fee from the server's tip, thus giving the server $20 − $0.70 = $19.30, rather than the full $20 tip amount.

Despite the ease with which it can be done, however, some foodservice operators elect *not* to reduce their employees' gratuities by the amount of processing fees paid, citing improved employee morale and enhanced serving staff recruitment abilities as the reasons for their decisions.

Unfortunately, the number of stolen and fraudulent payment cards in use today is very high. The federal government has passed laws prohibiting the fraudulent use of credit cards. Individuals who fraudulently use credit cards in interstate commerce to obtain goods or services of $1,000 or more in any given year could be subject to fines of up to $10,000 and prison terms of up to 10 years, yet payment card fraud continues.

In addition, issues of identity theft of customers are possible. The chance for a foodservice employee to defraud the customer through **credit card skimming** is always a concern. Skimming is the theft of payment card information used in an otherwise legitimate transaction. It is typically done by a dishonest employee, such as a server or bartender, who has easy access to a guest's payment card information. In the most common case, the employee secretly records the guest's name, credit card number, and the three- or four-digit security code number found on the back of the card. This information can be written down by the employee or photographed/scanned on a mobile device. The information is later used to make unauthorized purchases on that guest's payment card. One way to limit this activity is to provide handheld credit card scanners at the actual POS so that customers never lose sight of their cards.

Fortunately, payment card security has come a long way since the early introduction of these payment forms. Today, payment cards are issued with three-dimensional designs, magnetic strips, encoded numbers, card holder photos, and other features that reduce the chance of guest fraud as well as theft of the card user's personal information by unscrupulous employees. In addition, today's electronic payment card verification (interface) systems are fast, accurate, and designed to reduce the chances of income loss by your business.

FUN ON THE WEB!

Despite the best efforts of noncash payment card companies and those restaurants that accept these payment forms, increasing numbers of restaurant chains are encountering serious breaches of their online guests' payment card information. In one case, over 37 million card holders were affected![2] To see current examples of this problem, enter "restaurant credit card security breaches" in your favorite search engine and view the results.

To ensure that you can collect your money from the MSP that administers your account, your staff will be required to follow the specific procedures of the various payment card companies. If you agree to accept noncash payment forms, you should become very familiar with the procedures required by each card issuer/payment form manager and follow them carefully.

The Move to Off-Premise Dining

Payment card fees are not the only fees foodservice operators may have to pay. For example, fees charged to restaurants by third-party delivery apps such as Uber Etas, DoorDash, and Grubhub typically range from 15 to 30 percent of the order total (see Chapter 8).

While these fees are properly recorded as an "Other Expense," they have the indirect effect of reducing a foodservice operator's revenue. For example, if a restaurant guest orders $100.00 worth of food and it is delivered by a third-party delivery app that charges the operator 25 percent of an order's total, the operator will record revenue of $100.00 and delivery expense of $25.00, thus resulting in $75.00 available to cover the operator's expenses ($100.00 sale − $25.00 delivery fee = $75.00). This is the mathematical equivalent of achieving only $75.00 in revenue, not $100.00.

While each restaurant operator must determine the desirability of utilizing third-party delivery apps, it is also important to remember that such apps are subject to billing errors. In one widely reported case, a third-party delivery service charged a small bakery operator more than double what they should have: a 30 percent commission rate, when the bakery's contract clearly stated it should be 13 percent. The overcharges amounted to more than $100,000 over eight years![3]

The best operators carefully record and manage their actual revenue, but the move to off-premise dining means these operators must also carefully record, assess, and manage third-party delivery-related and other charges that have the mathematical effect of reducing their operations' revenue.

[2] https://www.eater.com/2018/4/3/17192264/panera-bread-data-credit-card-breach-37-million-consumers. Retrieved 2/6/2026.

[3] https://abc7news.com/post/doordash-mistakenly-overcharges-bay-area-family-bakery-8-years-shorts-100k/17134431. Retrieved 8/9/2025.

HERE'S HOW IT'S DONE 11.2

Just as foodservice managers must protect their current revenue from external threats, it is increasingly important to protect "future" revenue. One of the greatest threats to an operation's future revenue is guest comments about an operation that are posted online.

Study after study shows that when guests read negative comments about a foodservice operation posted on one of the popular social media restaurant review sites, these guests are less inclined to visit that operation.

Even the best-managed foodservice operations will encounter some negative reviews posted online, and operators should respond to these reviews promptly and professionally for several reasons:

- A boost in ranking or listing: Many data integration programs are developed to respond to and reward business-to-consumer interactions. When these programs detect that a business is regularly communicating with its customers, the business will be placed higher on search results pages.
- Showing the business cares: When customers complain, they typically want to know that someone hears them. A response (of almost any kind!) demonstrates that management does care about its guests' challenges and is interested in addressing them.
- Minimizing the damage: Once a negative review of an operation has been posted online, it will not, in most cases, be taken down at the request of the operator. However, when readers or viewers of a negative review immediately encounter a sincere and timely professional management response, the reputational damage done by that negative review is lessened.
- Demonstrating professionalism: A rapid response to negative reviews shows that a business is concerned about its customers, is responsive to them, and will respond quickly and professionally when issues arise. These are important traits all consumers look for when they buy products or services.

Foodservice operators today do not have the luxury of being able to ignore negative reviews in the hopes they will go away because, in nearly all cases, they will not.

Ignoring negative reviews increases their impact because a reply is not just a response to the person that posted the negative review. Instead, it is a message to all others who will read or see the review, including potential guests. A lack of response by a business tends to confirm the general position of the reviewer (that the business does not care about its customers!). So, how should foodservice operators respond to negative reviews that could threaten future income? Here are some "How To" guidelines for posting a professional response:

1. Start by thanking the reviewers for taking time to give feedback. (It is always best to start a response in a positive way.)
2. Apologize if the customer is correct about what happened.
3. Avoid arguing if the customer is incorrect about what happened. Instead, apologize and indicate that there was a misunderstanding.
4. Provide a very brief but clear and direct explanation of what caused the problem.
5. Assure the site's readers that specific actions have been taken to avoid a repeat of the problem.
6. Offer a direct line of communication between the business and the negative reviewer via e-mail, text, or phone to receive a personalized apology.
7. Emphasize any positive aspects of the post. For example, if part of the reviewer's comments were positive (e.g., if the reviewer says the server was "friendly" but the soup was "cold"), mention that it was good to learn the reviewer found the server to be "friendly."
8. Invite the customer to give the operation another chance to provide excellent service and products.
9. End the posting by again thanking the reviewer for feedback and helping to make the operation even better because of it.
10. Before hitting "Post" or "Send," double-check to ensure your response posting is short, well written, has been spell-checked, and is grammatically correct.

INTERNAL THREATS TO REVENUE SECURITY

In addition to guarding your revenue against external threats, it is also important to guard it against a number of internal threats. The greatest internal threat to an operation's revenue is its employees.

Most foodservice employees are honest, but some are not. In addition to protecting your revenue from unscrupulous guests, you must protect it from those few employees who may attempt to steal the revenue your facility has earned from sales. Cash is the most readily usable asset in a foodservice operation, and it is a major target for dishonest employees. In general, theft from service personnel is usually not a matter of removing large sums of cash at one time because this is too easy for management to detect. Rather, service personnel may use a variety of techniques to cheat an operation of a small amount of money at a time.

One of the most common theft techniques used by servers involves the omission of recording the guest's order. In this situation, the server does not record the sale in the operation's POS but charges the guest and pockets (keeps) the revenue from the sale. To reduce this possibility, managers must insist that all sales are recorded in the POS system.

The POS will then assign a unique **transaction number** to the sale. The transaction number created by a POS system is similar to the numbered **guest check** used prior to the introduction of POS systems. A sale's transaction number (like a numbered guest check) is a record of what the guest purchased and how much the guest was charged for the item(s) they chose.

The use of electronic guest checks produced by the POS is standard in the industry. A rule for all food and beverage production personnel is that no food or beverage item should be issued to a server unless the server has first recorded the sale in the POS system. This is so because modern POS systems assign unique transaction numbers to each guest check they create. This POS feature eliminates duplicate guest check fraud and is one of the main reasons even the smallest of foodservice operations should utilize an electronic POS system.

It is important to recognize that there are now a number of ways guest orders can be entered into the POS. In a bar, for example, a guest's order may be verbally communicated to a bartender who, before fulfilling the guest's order, would enter the order information in the POS, thus creating an individual transaction for that guest.

In a QSR, and in fast casual restaurants, a guest's order may be entered as it is received by the individual working in the drive-thru. In a fast casual restaurant, the sale may be entered before, or after, a guest actually receives their order.

In a table-service restaurant guest orders may be written on a note pad for later entry into one or more remote POS terminals (i.e., **user work stations**), or the orders may be immediately entered in the POS at the guest's table by servers using wireless hand-held POS terminals.

Increasingly, some QSR chains offer their guests the opportunity to enter their orders themselves using on-site kiosks. As well, the move to off-premise dining has resulted in more orders being directly entered by guests using a mobile ordering app.

Regardless of the order entry method used, the order information flow is from order taking and entry into the POS, to the interfaced **kitchen display system (KDS)** used by an operation's preparation staff to view orders, as shown in Figure 11.5.

FIGURE 11.5 Guest Order Information Flow. POS, point of sale; KDS, kitchen display system.

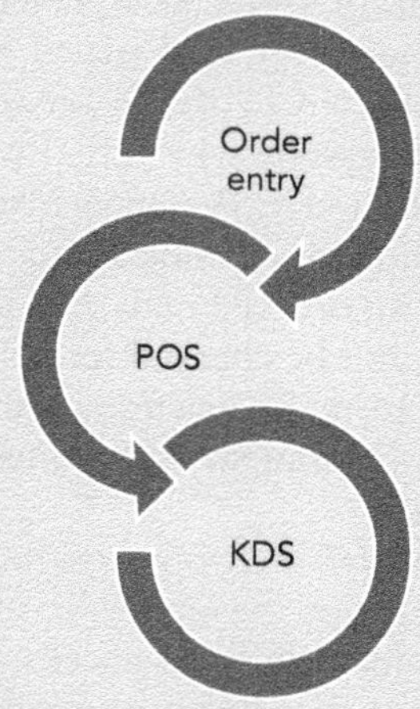

The Move to Off-Premise Dining

Historically, only a foodservice operation's employees were allowed to enter information into the operation's POS. Today, however, with the move to off-premise dining, more and more foodservice operations allow their guests to create orders that are immediately interfaced with their operations' POS systems.

Thus, for example, by using an ordering app. a guest may order a customized pizza (e.g., specifying options such as "extra cheese" or "extra pepperoni") by utilizing their mobile phone. In this example, that guest's order will be priced and immediately be entered into the operation's POS. In many cases, guests will also pay for their orders when the order is placed.

It is important to understand that while it is easy to replace an item and correct a mistake in order taking when a guest is in the dining room, mistakes made in orders received online are more problematic. On-premise orders can be corrected by replacing items, causing an increase in an operation's product costs, but no reduction in its revenue. In the typical case, however, improperly filled off-premises delivery orders can only be addressed by issuing a refund to the customer and the result is a reduction in revenue.

The error may actually be the fault of the guest or it may be the fault of the operation, but errors made in the fulfilling of online orders are common, and foodservice operators must design and manage the revenue control systems needed to minimize their occurrence and thus optimize their revenue.

Complete revenue control in a foodservice operation is a matter of developing the checks and balances necessary to ensure that the value of products sold and the amount of revenue received from their sale are equal to each other. For this reason, modern POS systems operate on a precheck/postcheck system for the recording of guest sales.

In a **precheck/postcheck** system kitchen and bar personnel are prohibited from issuing any products to guests without a prechecked and uniquely numbered transaction. Precheck means "before the check" and refers to a guest's order prior to payment for it.

For example, in a table-service restaurant, guest orders are pre-checked into the POS system. When the guest is ready to leave, the server or cashier retrieves the prechecked transaction and prepares a postcheck bill for the guest to pay. The postcheck bill should include the charges for all of the prechecked items ordered by the guest plus any service charges and taxes due.

The guest then pays the bill. In a precheck/postcheck system products ordered by the guest, prechecked by the server, and issued by the kitchen or bar should match the items and money collected by the server or cashier.

There can be a number of reasons why initial prechecked sales totals do not match the amount of money collected by the cashier. Many times, guests place orders and then change their minds, adding or subtracting from their original orders. In other cases, a guest may return an ordered item and, depending upon the reason for the item's return, the guest may not be charged for it. For these types of reasons, the question of who is authorized to make deletions (voids, returns, or reductions) to a prechecked guest check is a potential security concern to the restaurant manager and should be closely controlled.

A common method of employee theft involves failing to finalize a sale recorded on the precheck system and pocketing the money. To prevent such theft, management must have systems in place to identify open checks during and after each server's work shift. An **open check** is one that has been used to initially authorize product issues from the kitchen or bar (prechecked) but that has not been collected for (closed) and thus has not been added to the operation's sales total.

Unless all open checks are ultimately closed out and presented to guests for payment, the value of menu items issued will not equal the money collected for those items.

The totals of all cash and noncash transactions entered into the POS during a predetermined period are electronically tallied, so management can compare the sales recorded in the POS with the cash and other payment forms contained in the cash drawer of the POS. For that reason, it is important to properly count and close cash drawers at the end of each sales period.

For example, a cashier working a shift from 7:00 A.M. to 3:00 P.M. might have recorded \$1,000 in cash sales (including taxes) during that time period. If that were in fact the case, and if no errors in handling change occurred, the cash drawer should contain the \$1,000 in sales revenue in addition to the cash amount that was in the drawer at the beginning of the shift.

If the drawer contains less than \$1,000, it is said to be **short**; if it contains more than \$1,000, it is said to be **over**. Cashiers rarely steal large sums directly from the cash drawer because such theft is easily detected, but management must make it a policy that any cash shortages or overages will be investigated.

Some managers believe that only cash shortages, not overages, need to be monitored. This is not the case. Consistent cash shortages may be an indication of employee theft or carelessness and should, of course, be investigated. Cash overages, too, may be the result of very sophisticated theft by the cashier. Essentially, a cashier who is defrauding an operation because they know management does not investigate overages may remove \$18.00 from a cash drawer but falsify (reduce) sales records by \$20.00. The result is a \$2.00 cash "overage" even though the cashier stole \$18.00!

HERE'S HOW IT'S DONE 11.3

Restaurants use their POS systems to record their sales. Sales can be reported hourly, by shift, or by day, but in each case, management must decide when to start over again when recording sales. They start over by creating a Z report in their POS. To best understand Z reports, you must first understand X reports.

An **X report** details sales made during a specific time period such as an hour, shift, or day. Multiple X reports can be run during the designated time period. Thus, for example, a manager recording daily sales may elect to run an X report every 2 hours to continually monitor the day's sales. Thus, an X report gives a running total of sales made during a designated time period.

Managers start over in the recording of sales by running a Z report. A **Z report** is identical to an X report except that, when a Z report is run, the operation's sales figure is set back to zero to begin recording sales in a new time period. This is why some managers refer to Z reports as "closing" reports.

In most POS systems, managers can customize their Z reports, but all Z reports will detail an operation's sales and tax receipts for the defined sales period. The Z report will detail cash sales, tax collected, individual electronic sales (e.g., cash sales, Visa sales, Apple Pay, and MasterCard sales), gift card redemptions, and any other forms of payment accepted by the operation. Because it also identifies cash sales, it is the Z report that is used to ensure the amount of cash in a cashier's bank at the end of his or her shift is the amount that should be in the drawer.

When closing out a cashier drawer, managers use the following formula:

(Actual cash – Beginning cash) – Cash sales per the Z report = Cash overage (shortage)

To illustrate, if after carefully counting it, the amount of cash in the cashier bank at the end of the day is $2,965.15, and if the amount in the cashier bank at the start of the day was $500.00, and if the operation's Z report indicates that $2,451.82 is the amount of the operation's cash sales for the day, the over/short formula would be calculated as follows:

($2,965.15 – $500.00) – $2,451.82 = $13.33

Or, in this example,

$2,465.15 – $2,451.82 = $13.33

In this example, the cash drawer would be over by $13.33 and, by running the Z report, the POS would reset the sales figure to zero to be ready to begin calculating the next day's sales.

Even if you have controls that make internal theft difficult, the possibility of significant fraud still exists. Consequently, some companies protect themselves from employee dishonesty by bonding their employees. **Bonding** is simply a matter of management purchasing an insurance policy against the possibility that an employee(s) will steal.

Through the process of bonding an employer can be covered for the loss of money or other property sustained through dishonest acts of a bonded employee. Bonding can cover many types of acts, including larceny, theft, embezzlement, forgery, misappropriation, or other fraudulent or dishonest acts committed by an employee, alone or in collusion with other employees. Essentially, a business can select from several bonding options:

- Individual: covers one employee (e.g., a restaurant's bookkeeper or accountant)
- Position: covers all employees in a given position (e.g., all bartenders or all cashiers)
- Blanket: covers all employees

If an employee has been bonded and an operation can determine that they were indeed involved in the theft of a specific amount of money, the operation will be reimbursed for all or part of the loss by the bonding company. Although bonding will not completely eliminate theft, it is relatively inexpensive and well worth the cost to ensure that all employees who handle cash or other forms of operational revenue are bonded. Because the bonding company may require detailed background information on employees prior to agreeing to bond them, it is also an excellent pre-employment check to verify an employee's previous work history.

The scenarios discussed in this chapter do not list all possible methods of revenue loss, but it should be clear that you must have a complete revenue security system if you are to ensure that all products sold generate sales revenue that finds its way into your bank account. It is important to remember that even good revenue control systems present the opportunity for theft if management is not vigilant, or if two or more employees work together to defraud the operation.

Figures 11.6 and 11.7 outline common methods of theft and procedures managers can use to minimize theft by service employees.

FIGURE 11.6 Common Methods of Theft by Service Employees

- Omits recording the guest's order and keeps the money the guest pays.
- Voids a sale in the POS but keeps the money the guest paid.
- Enters another server's password in the POS and keeps the money.
- Fails to finalize a sale (keeps a check open) and keeps the money.
- Charges guests for items not purchased and then keeps the overcharge.
- Changes the totals on payment card charges after the guest has left.
- Enters additional payment card charges and pockets the cash difference.
- Incorrectly adds legitimate charges to create a higher than appropriate total, with the intent of keeping the overcharge.
- Purposely shortchanges guests when giving back change, with the intent of keeping the extra change.
- Charges higher than authorized prices for products or services, records the proper price, and then keeps the overcharge.
- Adds a coupon to the cash drawer and simultaneously removes sales revenue equal to the value of the coupon.
- Declares a transaction to be complimentary (comped) after the guest has paid the bill.
- Engages in collusion between two or more employees to defraud the operation.

FIGURE 11.7 Procedures to Minimize Theft by Service Employees

- Require a unique POS-produced transaction number to record each sale.
- Use a precheck/postcheck POS system.
- Do not allow employees to share their POS passwords and require employees to change their passwords frequently.
- Continually monitor and close out all open checks at the end of each sales period.
- Perform voids by supervisor only.
- Deny cashiers access to unredeemed cash value coupons.
- Require special authorization from management to "comp" guest purchases.
- Monitor cashier overages and shortages at each cashier shift change.
- Bond employees (acquire an insurance policy against employee theft).

FUN ON THE WEB!

A high-quality POS system always provides detailed reports. The sales report dashboard on a POS system provides an overview of all transactions completed during a selected time period. This includes net sales, service charges, tips, total guests served, table turn times, and a breakdown of all service types and payment methods. It should also provide detailed sales exception reports that allow users to quickly see an overview of all voids, discounts, and refunds.

It is important to recognize that, regardless of its sophistication, a POS system will not bring control to a foodservice operation. A high-quality POS system can, however, take good control systems that have been carefully designed by management and add to them speed, accuracy, and/or additional information.

Properly utilized, a POS system is of immense value. If, however, an operation has no formal revenue security plan, the POS system simply becomes a high-tech adding machine used primarily to sum guest purchases and little more. Properly selected and utilized, however, POS systems play a crucial role in the implementation of an operation's complete revenue security system.

To review the revenue control-related features of some currently popular POS systems enter "POS revenue control reports for restaurants" in your favorite search engine and view the results.

Consider the Costs

Donald Wright worked for 15 years as a snack bar cashier for the Sports Arena managed by Stanley Harper's company. Donald had twice won the company's "Employee of the Year" award, and Stanley Harper considered Donald a valued and trusted employee who had, on many occasions, performed above and beyond the call of duty.

Stanley was surprised when newly installed video surveillance equipment confirmed that Donald, despite rules against it, had, on several occasions, given free food and beverages to friends of his who had visited the arena to watch sporting events.

When confronted with the video evidence, Donald admitted that he had given away products in the past but promised never to do so again in the future and he also indicated that he really needed his job.

On the advice of the company's Human Resources Manager, Stanley is documenting, in writing, his decision on handling the situation.

1. Assume you are Stanley. What would you do about Donald?
2. Do you believe an employee caught defrauding their employer by taking products or money should ever be given a second chance? If so, under what circumstances?
3. What impact will Stanley's decision in this case have if, in the future, other employees are caught defrauding the business?

Experienced foodservice operators know they must be vigilant to protect against both internal external and internal threats to revenue. To show how that is done, we now turn our attention to the development of an effective revenue control system. It may not be possible to prevent all types of theft, but a good revenue control system can help determine if, in fact, a theft has occurred and also help you establish procedures to minimize revenue losses in your operation.

DEVELOPING THE REVENUE SECURITY SYSTEM

An effective revenue security system will help you verify:

1. Product issues
2. Guest charges
3. Sales receipts
4. Sales deposits
5. Accounts payable

You must consider each of these five areas when developing your total revenue security system. In many cases, each foodservice operation you manage will have a different manner of both selling products and accounting for revenue. It is useful, however, to view the revenue control system for any unit in terms of these five key areas and how they relate to each other.

In an ideal world, a product would be sold, its sale recorded, its selling price collected, the funds deposited in the foodservice operation's bank account, and the cost of providing the product would be paid for, all in a single step.

Rapid advances in the area of smart devices and digital wallets are making this a near reality for more foodservice operators each day. The following example from the grocery industry helps illustrate just how this works.

A grocery store customer uses their digital wallet when buying a frozen entrée dinner. The customer uses self-checkout and scans the product's bar code printed on the frozen entrée dinner. The following actions then take place:

- The amount the shopper owes the store is recorded in the store's POS system and the sale itself is assigned a tracking number (verification of product issued/sold).
- The sale amount is displayed for the shopper, who is asked to confirm its correctness; if it is correct, the shopper pays and a receipt is printed for the shopper (verification of guest charge).
- The store POS system records the amount of the sale as well as the form of payment used by this shopper (verification of sales receipt).
- The shopper's digital wallet electronically transfers funds from the shopper's account to that of the grocery store (verification of sales deposit).
- The funds on deposit in the grocery store's bank account are available to pay for the purchase of additional frozen entrée dinners (verification of accounts payable).

Of course, not all foodservice operations have the technological sophistication to duplicate the system described in our grocery store example. Foodservice operators should, however, adapt the technology they currently have available to the development of good revenue control systems.

In all cases, a foodservice manager should have a thorough understanding of how their revenue security system works and, thus, what is required to maintain it. To illustrate the five aspects (steps) required to help ensure revenue security, consider the situation of Faris, who operates the Caspian Grill, a Lebanese restaurant in New York City. His operation has a small cocktail area and 150 guest seats in the dining room. While at the present time on-premise guests still generate the majority of his revenue, an increasingly large percentage of his business now comes from carryout and off-premise delivery sales.

Revenue at Faris's exceeds $2.5 million per year. When he started the restaurant, he did not give a tremendous amount of thought to the design of his revenue control system because he was in the restaurant at all times. Due to his success, however, he spends more and more of his time developing a second restaurant and, thus, needs both the security of an adequate revenue security system and the ability to review it quickly to evaluate the sales levels in his original store. Thus, Faris has begun to develop a revenue security system, concentrating on the following formula:

Product issues = Guest charges = Sales receipts = Sales deposits
= Funds available for accounts payable

Faris knows that the first goal he must achieve is that of verifying his product issues.

STEP 1: VERIFICATION OF PRODUCT ISSUES

The key to verification of product issues in the revenue security system is to follow one basic rule: *No product shall be issued from the kitchen or bar unless a permanent record of the issue is made.*

In its simplest terms, this means that the kitchen or bar should not fill any server request for food or drink unless that request has been documented in this initial step of the revenue control system.

In some small restaurants, the server request for food or beverages might take the form of a single or multicopy written guest check, designed specifically for the purpose of revenue control. When this is the case, the top copy of this multicopy form would generally be sent to the kitchen or bar. The guest check, in this case, becomes the documented request for the food or beverage product.

This paper-only system can still work, but it is subject to a number of potential forms of abuse and fraud. For that reason, all operations should utilize an electronic POS system where the guest check consists of an electronic, rather than paper, record of product requests and issues. When this is the case, a guest's order information is viewed by the production staff on a computer terminal located in the kitchen or, in other cases, the POS system prints a hardcopy of the order for use by the production staff.

In either case, the software within the POS creates a permanent record (transaction number) for the product request. It is this record that authorizes the kitchen to prepare food or the bar to prepare a drink. If a foodservice operation elects to supply its employees with meals during work shifts, these meals, too, should be recorded in the POS system.

In the bar, this principle of verifying all product sales is especially important. Bartenders should be instructed never to issue a drink until that drink has first been recorded in the POS system. This should be the procedure, even if the bartender is busy and working alone. Even then care must be taken, because, for example, when a bartender is serving a guest they know is paying with cash, the bartender may take a guest's order for an $18 glass of wine but enter the sale into the POS as a $10 glass of wine, pocketing the $8 difference.

This rule regarding product issuing is important for two reasons. First, requiring a permanent documented order ensures that there is indeed a record of each product sale. Second, this record of product sales can be used to verify both proper inventory usage and product sales totals. Faris enforces this basic rule by requiring that no menu item be served from his kitchen or the bar without the sale first being entered into his POS system.

If his verification of product sales system is working correctly, Faris will find that the following formula should hold true:

$$\text{Documented product requests} = \text{Product issues}$$

Experienced foodservice operators know that it is possible for employees to issue products *without* a documented product request when:

1. Two or more employees work together to defraud the operation. Collusion of this type can be discovered when managers have a system in place to carefully count the number of items removed from inventory and then compare that number to the number of products that were actually issued.
2. A single employee (such as a bartender working alone) is responsible for both making and filling the product request. In this case also, managers will uncover this type of fraud when they carefully compare the number of items (or beverage servings) removed from inventory with the number of recorded product issues.

STEP 2: VERIFICATION OF GUEST CHARGES

When the production staff is required to prepare and distribute products only in response to a properly documented request, it is critical that those documented requests result in charges to the guest. It makes little sense to enforce a verification of product issues step without also requiring the service staff to ensure that guest charges match these requests.

This concept can be summarized as follows: *Product issues must equal guest charges*.

Guest charges are detailed on the guest bill that is generated by the POS. Figure 11.8 is a sample of a bill (guest charge) generated by a POS system. Historically a guest's bill in a restaurant simply listed the total amount owed, and for those guests choosing an electronic form of payment, it included a space to add a tip.

Today's guest receipts do those same things but often serve as a marketing tool as well as a key revenue control tool.

While each POS system will configure their guest charges summary somewhat differently, and most systems will allow operators to customize their billing formats to meet their own needs, Figure 11.8 is an example of a typical detailed guest receipt created each time a sale is made.

The shaded "Line number" column in Figure 11.8 has been added by the authors to aid in the understanding of each component in a typical POS system-generated guest receipt.

Lines 1–4 **Restaurant Information:** Typically shown in large letters at the top of a guest's bill, this section will normally include the operation's NAP (name, address, and phone number).

If the operation is part of a restaurant chain, the bill may also display a unique **store number** indicating the precise location of the operation. Store number is important when, for example, a district manager supervises several stores and wants to review the total revenue produced in their district, as well the revenue produced by each of their district's individual stores.

Line 5 **Date and Time of Sale:** This portion of the bill indicates the date and time the sale was entered into the POS system. In some POS systems the time of entry will be indicated to the whole minutes (i.e., 12:36 P.M., while in other POS systems this entry will also display time to the hundredth of a minute (i.e., 12:36:20 P.M.). This greater degree of accuracy is often important for foodservice operators who, for example, want to monitor their drive-thru times (i.e., the length of time elapsed from order entry to order delivery).

Line 6 **Employee Identification:** This portion of the bill indicates the individual counter worker, cashier, server, or bartender entering the sale. In some cases, such as that shown in Figure 11.8, the employee's name will be entered (i.e., Collins. A). In other operations those employees who are authorized to enter sales into the POS system will be identified by a unique number assigned to them by management. In all cases, however, it is important that management be able to identify the specific worker responsible for the accuracy of each guest's receipt.

Line 7 **Order ID:** This portion of the bill shows the individual transaction number, or order ID, assigned to this specific sale by the POS system. As previously addressed, a specific and unique order ID number will accompany each guest sale.

Line 8 **Sale Type:** In most cases, those foodservice operators who deliver their products to guests in a number of different ways will be interested in the proportion of sales each delivery method contributes. For example, the manager of a restaurant that has both a drive-thru window and an on-premise dining area would be interested in knowing what proportion of sales came from the drive-thru and what proportion was generated in the on-premise dining area.

In Figure 11.8 the entry for this order indicates that it is a carryout sale. Typical categories that can be assigned by a POS system for this purpose may include dine-in, carryout (pickup), drive-thru, and self or third-party delivery.

Lines 9–12 **Items Sold:** This is a list of the menu items actually ordered by the customer. This section of the guest bill would include:

- **a.** The number sold of a menu item (i.e., 2 iced teas)
- **b.** Menu item name (i.e., Shawarma Platter)
- **c.** Menu item **extended selling price** (e.g., $11.80 for two iced teas)

FIGURE 11.8 Typical Guest Receipt

Line 1		**Caspian Grill**	
2		2010 West 33rd. St	
3		New York, NY 10024	
4		Tel 212-412-61xx	
5	8/15/20xx	12:36:20 PM	
6	Cashier	Collins. A	
7	Order ID	204656	
8	Carryout		
9	1	Shawarma Platter	$43.90
10	1	Tabbouli	$14.95
11	1	Falafa	$9.95
12	2	Iced Tea	$11.80
13	**Order Subtotal**		$80.60
14	Sales Tax	8.875%	$7.15
15	**Order Total**		$87.75
16	Gratuity		$22.25
17	**Order Total (Before Convenience Fee)**		$110.00
18	Order Total w/3.5% credit card convenience fee		$113.85
19	Credit Card Sale	Visa xxxx xxxx xxxx 2525	
20	Reference ID	5228005422212	
21	Authorization ID	25219	
22	Time Order Closed	1:11:05 PM	
23	Caspian Grill Marketing Information		

An extended selling price is simple the price of a menu item multiplied by the number of times that menu item was purchased. The need for extended prices on a guest's bill can easily be understood if one considers the example of a table of six guests in a sports bar. Assume each of the guests at the table orders the same type of bottled beer, and over a period of several hours each guest consumes three beers. It makes little sense to list the purchase price of the beer 18 times on their bill. Thus, the POS system would be programmed to indicate 18 beers were sold and would then automatically display the extended price (18 beers sold × selling price of 1 beer = extended price).

If it is the preference of management, it is also possible to program the POS system to show the individual price of each menu item as well as its extended price. When only one menu item is sold, of course, its extended price will be equal to its menu price.

If a menu item has been modified by a guest, many POS systems will show the modification in this section of the guest's bill. Thus, for example, the guest in a burger restaurant may indicate the burger modifications chosen and the POS system will display special kitchen instructions such as, "No onions," "Well done," or "Double cheese"). Doing so can help ensure that those viewing the order on the KDS will prepare it properly.

If a **service charge** is mandated by the operation, then the POS will list the amount due in the "Item Sold" section of the bill. This is because the IRS considers service charges to be a taxable sale. Even if they are listed as on the guest receipt as a "Gratuity," the percentage that was charged and the dollar amount of the service charge must be listed.

Many foodservice operations mandate a service charge when, for example, the number of guests at a single table exceeds a predetermined number such as six or eight. Service charges are also typically imposed when a restaurant hosts a catered event or a banquet.

It is perfectly legal to charge a mandatory service fee, but in most jurisdictions the amount of the service fee must be announced to guests prior to the guests placing their orders.

Increasingly, some foodservice operations are adding additional fees and indicating that these funds are collected to offset specific expenses such as labor costs or employee benefits. Again, these are perfectly legal as long as they are clearly communicated to guests prior to the guests' placing their orders.

***Line 13* Order Subtotal:** This portion of the guest's bill shows a dollar amount "Sub Total," which is the sum of all charges incurred by the guest prior to the addition of any taxes or other charges due.

***Line 14* Sales Tax:** This portion of the bill indicates taxes due on the guest's purchase. Typically, the tax rate is shown (i.e., 8.875 percent in this example), as well as the amount of taxes due.

***Line 15* Order Total:** The "Order Total" portion of the bill indicates the sum of the cost of all items purchased (Order Subtotal) plus all sale taxes due.

***Line 16* Gratuity:** This section of the bill indicates the amount of gratuity the guest wants to leave for their server. From a revenue recording and control perspective this is an important row because gratuities are not considered sales. Rather they are considered gifts to employees, thus they are *not* included in revenue totals.

If a cash gratuity is left for a server, that does not affect their employer's revenue calculation, but it is mandatory for employees who receive more than $20.00 per month in tipped income to report the amount of their tips to their employers. Modern POS systems track and report the amount charged tips by shift and by server.

***Line 17* Order Total (Before Convenience Fee):** This line shows the total of the guest's bill before any extra charges for the use of a credit card as a payment form are added. It is not legal to charge a processing fee to guests who use a debit card.

Prior to 2013, restaurants in the United States were also prohibited from charging guests more when they used a credit card as their payment form. Since 2013, however, restaurants in most (but not all!) states are allowed to impose a surcharge for the use of a credit card. In the typical case this amounts to 3 to 4 percent of the guest's total bill amount.

In some POS systems Line 17 would be titled "Cash amount due" to indicate that this would be the amount to be paid if cash were used to pay the bill.

***Line 18* Order Total w/3.5 percent credit card convenience fee:** For those operators charging a fee for the use of a credit or debit card, the amount of the fee must be shown. This portion of the guest's bill will indicate the percentage amount to be charged, as well as the total dollar amount to be charged to the guest if a credit card is used as the guest's form of payment. It is important to note that these fees are considered revenue by taxing authorities and, if imposed, are subject to sales tax. Thus, in this example the guest would pay a convenience fee of 3.5 percent, which the restaurant would use to offset payment card charges and pay the sales taxes due on this additional revenue.

Line 19 **Credit Card (or other type) Sale:** This line is used to record the specific payment form used by guests. Typical categories include cash, debit card, various e-wallet forms and, as shown in this example, credit cards. From a revenue control perspective this is an important line because it informs operators about the amount of cash that should have been collected during a designated shift, and it also indicates their guests' various preferred payment forms.

In most cases the POS will list the specific type of payment form used and may indicate the last several digits of the credit or debit card if used (i.e., in this example: Visa xxxx xxxx xxxx 2525).

Line 20 **Reference ID:** This line, created by payment card issuers, is used to identify a specific transaction from start to finish (when using a payment card). It is this unique identifier that would be used by a foodservice operator, its MSP, and the guest's card issuer (bank) to record all of the parts of a transaction. Guests also use Reference IDs when they want to dispute the amount of a bill or get a refund.

Line 21 **Authorization ID:** This line is used to record the successful authorization of a payment by the card-issuing bank. Sometimes listed as an "Approval Code" this code verifies that the cardholder's account is in good standing and has adequate funds (debit card) or credit (credit card) to cover the amount of the bill.

To illustrate its use, assume that a guest offers a debit card as payment for their meal. A reference ID would have been established for this guest's transaction (Line 21) and an authorization ID would be created verifying that the guest, in fact, has sufficient funds in their account and the money can be transferred to the foodservice operator.

If there are insufficient funds in the guest's account then no authorization ID (approval) would be issued. Rather, a record would be made of the payment attempt, and the card issuer would send a "Card declined" code back to the foodservice operator. The guest would then be asked to provide an alternative, and acceptable, method of payment.

One good way to remember the difference between a reference ID and an authorization ID is that a reference ID is created by a card company, for example Visa or MasterCard, while an authorization ID is created by the bank or credit union actually issuing the card to the card holder, for example Bank of America, Wells Fargo, or the Navy Federal Credit Union).

Line 22 **Time Order Closed:** This portion of the bill documents the time a guest's payment was accepted and recorded in the POS. In many cases, managers of QSRs will be interested in the total length of time elapsing between order entry (Line 5) and order delivery and payment. Knowing this allows them to assess, for example, the length of time their average guest stays in an operation's drive-thru lanes.

The Move to Off-Premise Dining

The use of drive-thru lanes is becoming increasingly popular for many foodservice operations. One good way for QSRs to optimize their revenue is by providing fast and accurate service in their drive-thru lanes.

Average QSR drive-thru times vary by chain, but a recent study found an average of 5.5 minutes for the 10 major QSR chains studied, with Taco Bell being the fastest at about 4.3 minutes per order and Chick-fil-A filling its drive-thru orders in an average of about 8 minutes.[4]

A properly programmed POS system will inform managers about their average drive-thru order filling times, and careful examination of this data can help managers understand whether their delivery times are increasing, decreasing, or staying the same.

They can then take the steps necessary to make any changes or improvements they feel can optimize revenue by reducing drive-thru times while maintaining excellent guest service and order accuracy.

---

[4] https://www.washingtonpost.com/business/2025/03/28/wait-time-drive-through-restaurants. Retrieved 8/25/2025.

Line 23 **Marketing Information:** Foodservice operators increasingly use this optional portion of the bill. It may provide a QR code to access the operation's **guest rewards** program and sign for it, or to view a guest's current loyalty point balances. It also can be used to promote upcoming special events, or it may offer a "free" menu item on a future visit for those guests agreeing to take a survey or post a review. In other cases, it directs guests to a website where they can enter their e-mail and receive a few free products on a future return visit. This is one good way to help foodservice operator's increase the size of their e-mail address database.

Because the cost of implementing this marketing tool is so low, foodservice operators should take advantage of it in the ways they feel best benefit their operation's future revenue generation.

FUN ON THE WEB!

While the focus of this chapter is on controlling revenue that has already been generated, most foodservice operators would like to increase their profits by increasing the amount of revenue they generate in the future. As a result, they are interested in the latest marketing techniques and strategies designed to grow their businesses. The foodservice business is unique, and while there are many general marketing books available for sale, one of the best and most specifically targeted for foodservice operators is *Marketing in Foodservice Operations*, a part of the five-book John Wiley Foodservice Operations: The Essentials series.

In this book the authors explain how to identify a target market and create a marketing plan, as well as how menu prices impact an operation's marketing. They also describe how to use a foodservice operation's menu as a marketing tool and consider the importance of an active social media presence visible to the target market.

It includes practical discussions of digital foodservice marketing techniques, including social media plans and the increasing importance of mobile-friendly marketing strategies.

To view the book's Table of Contents, go to www.wiley.com, then enter "Marketing in Foodservice Operations" to see additional topics addressed by this targeted and very popular foodservice marketing tool.

While the specific format and wording used on an operation's guest bills will vary somewhat from that shown in Figure 11.8, when properly implemented, this second step of the revenue control system will result in the following formula:

Product issues = Guest charges

Faris has now implemented two key revenue control principles. The first one is that no product can be issued from the kitchen or bar unless the order is documented; the second one is that all product issues must be recorded as a guest charge.

With these two systems in place, Faris can deal with many problems. If, for example, a guest has "walked" his or her check, the operation would have a duplicate record of the transaction. The POS would have recorded which products were sold to this guest, which server sold them, and perhaps additional information, such as the time of the sale, the number of guests in the party, and, of course, the sales value of the products.

The POS system Faris utilizes also ensures that service personnel are not allowed to change the prices charged for items sold. This, of course, would be possible in an operation using manual guest checks.

To complete this aspect of his control system, Faris implements a strict policy regarding the documentation of employee meals. This has the added advantage of giving him a monthly total of the value of employee meals. Recall that he needs this figure to accurately compute his cost of food consumed (see Chapter 5).

Faris is now ready to address the next major component in his revenue security system. That component is the actual verification of total guest payments. These payments will represent Faris's actual sales receipts.

STEP 3: VERIFICATION OF SALES RECEIPTS

Sales receipts, as defined in this step, refer to actual revenue received by the cashier, server, bartender, or other designated personnel in payment for products served. In Faris's case, this means all sales revenue from his restaurant and lounge. The essential principle to recognize in this step is that: *Both the cashier(s) and a member of management must verify sales receipts.*

Verifying sales receipts is more than a simple matter of summing electronic charges and counting cash at the end of a shift. In fact, cash handling, although it is an important part of the total sales receipt reconciliation, is only one small part of the total sales receipt verification system.

To illustrate this, consider Figure 11.9. It is the Cashier Report that Faris uses to verify his total sales receipts. The form indicates that both a cashier and a supervisor make the sales receipt verification. Although this will not prevent possible collusion by this pair, it is extremely important that sales receipt verification be a two-person process.

FIGURE 11.9 Cashier Report

	Faris's		
			Performed By:
Date: 1/15/20xx			Cashier: Tammi F.
Shift: Dinner			Supervisor: Faris L.
Revenue per Guest Check			
Guest check totals		$7,500.00	
Service charges		$450.00	
Tax		$618.00	
Total guest check revenue			$8,568.00
Receipts			
Noncash payments			
Discover		$486.00	
Google Pay		$410.00	
MasterCard		$495.00	
Visa		$1,200.00	
Apple Pay		$975.00	
Total noncash receipts		$3,563.00	
Cash			
Twenties and larger	$4,840.00		
Tens	$1,480.00		
Fives	$240.00		
Ones	$196.00		
Change	$68.20		
Total cash	$6,824.20		
Less: Bank	$500.00		
Tip-outs	$1,320.00		
Net cash receipts		$5,004.20	
Net total receipts			$8,567.20
Variance check revenue to net receipts			$(0.80)

Faris wishes to ensure that the amount of cash in his cash drawer, when added to his noncash guest payments, matches the dollar amount that he has charged his guests (as shown on the shift or day's Z report). If this is so, he has accounted for all of his sales receipts, given that he has controlled both product issues and guest charges.

In most operations, individual guest charges are recorded only in the POS system. This is the case, for example, in a QSR or fast casual restaurant, where food purchases are totaled and paid for at the same time. It is also true in any operation using a precheck/postcheck POS system. In such instances, the POS system provides an accurate total of guest and other charges. Receipts collected should equal these charges. If Faris's revenue security system is working properly, he knows that the following formula should be in effect:

Total guest charges = Total sales receipts

Note that one function of a Cashier Report such as the one shown in Figure 11.9 is to require Faris's staff to reconcile total charges as recorded in the POS with actual revenue (net sales receipts). Cash overages, or shortages, are determined and, if they exceed predetermined allowable limits, are investigated by management.

Note that total POS recorded charges consist of all sales, service charges, and guest paid taxes. This is the total amount of money the operation should have received on this date.

Net total receipts refers to all noncash and cash payment forms used by Faris's customers. As you learned earlier in this chapter, when counting cash, Faris must subtract the value of his starting cash bank. In this example, he must also subtract any cash tip-outs, or gratuities, paid to his service personnel. Most often, guests choose the amount of gratuity to be added to their bills and the gratuities are paid out to the service personnel by management either in cash or in their paychecks because such tip income is taxable to the server.

Finally, note that Figure 11.9 has a space to indicate cash overages and shortages (POS recorded charges vs. Net sales receipts). Some foodservice operators investigate even very small amounts of cashier variance. Others routinely ignore minor variances. It will be up to you, as the manager, to determine the level of variance you are comfortable with as far as this reconciliation is concerned.

FUN ON THE WEB!

Noncash forms of payment have become so prevalent in foodservice operations that some owners have decided to accept only noncash payment forms from all of their guests. These operators cite the advantages of faster payment processing, decreased risk of theft and robbery, and reduced threats of internal pilferage.

Some states and cities, however, prohibit a business from refusing to accept cash. These include Colorado, Georgia, Michigan, New Jersey, New York City, Philadelphia, San Francisco, and the District of Columbia. In these jurisdictions lawmakers are concerned that cashless models may discriminate against those without bank accounts or credit cards, including the elderly, low-income communities, and people of color. As well, studies indicate millions of US households are unbanked and thus have no access to electronic payment forms.[5]

As fewer and fewer guests use cash as their preferred payment form, the noncash trend will likely become increasingly common and it will likely remain controversial. Thus, this situation bears continual monitoring. To do so, periodically enter "cashless restaurants" in your favorite search engine and review the current results.

[5] https://www.touchbistro.com/blog/are-cashless-restaurants-the-future-of-food-service. Retrieved 8/15/2025.

STEP 4: VERIFICATION OF SALES DEPOSITS

Many foodservice operations will make a sales deposit on a daily basis because keeping excessive amounts of cash on hand is not advisable. Those who accept only small amounts of cash payments will likely make their bank deposits less frequently.

It is strongly recommended that only management verify the actual deposit of daily sales revenue. A cashier or other clerical assistant may complete a **deposit slip** detailing the amount of cash and checks contained in the deposit and may even drop the deposit off at the bank, but management alone must bear the responsibility for monitoring the actual deposit of sales. This concept can be summarized as follows: *Management must personally verify all bank deposits.*

The verification of bank deposits is a two-step process:

Step 1: Verification that the deposit slip amount matches the sales amount shown in the POS.

Step 2: Verification that the deposit slip amount matches the bank's record of cash and checks actually received.

For example, if you deposit Thursday's sales on Thursday night, your bank's next day (Friday) verification of Thursday's deposit should match Thursday's cash and check receipts amount. If it does not, you have experienced some loss of revenue that has occurred *after* your cashier and a manager have reconciled sales receipts to guest charges.

Embezzlement is the term used to describe employee theft where the embezzler takes company funds they were entrusted to keep and diverts them to personal use. Embezzlement is a crime that often goes undetected for long periods of time because many times the embezzler is a trusted employee. Falsification of, or destroying, bank deposits is a common method of embezzlement. To prevent this activity, you should take the following steps to protect your deposits:

1. Make bank deposits of cash and checks daily if possible.
2. Ensure that the person preparing the deposit is not the one making the deposit—unless you or another manager completes both tasks. Also, ensure that the individual making the daily deposit is bonded.
3. Establish written policies for completing **bank reconciliations**, the regularly scheduled comparison of the business's deposit records with the bank's funds acceptance records. Payment card funds transfers to a business's bank account should be reconciled each time they occur. Today, in most cases, cash and noncash payment reconciliations can be accomplished on a daily basis via the use of online banking features.
4. Review and approve written bank statement reconciliations frequently and no less than once per month.
5. Change combinations on office safes periodically and share the combinations with the fewest employees possible.
6. Require that all cash-handling employees take regular and uninterrupted vacations so that another employee can assume their responsibilities and uncover any improper practices.
7. Employ an outside auditor to examine the accuracy of deposits on an annual basis.

If verification of sales deposits is done correctly and no embezzlement is occurring, the following formula should hold true:

Total sales receipts = Total verified bank deposits

STEP 5: VERIFICATION OF ACCOUNTS PAYABLE

It might seem that a foodservice operation's revenue is secure once it has been safely deposited in the operation's bank account. Unfortunately, that is not the case. One significant threat to revenue in every foodservice operation involves its accounts payable. Accounts payable (AP), as defined in this step, refers to the legitimate amount owed to suppliers and vendors for the purchase of products or services. The basic principle to be followed when verifying accounts payable is: *The individual authorizing the purchase should verify the legitimacy of the vendor's invoice before it is paid.*

Vendor and supplier payments are an often overlooked potential threat to the security of a restaurant's revenue. Of course, a restaurant should pay all of its valid expenses (i.e., its legitimate invoices for payment). However, both external vendors and the restaurant's employees can attempt to defraud a foodservice operation through the manipulation of accounts payable invoices.

For example, consider the case of Gabriela. One day, when paying her operation's bills, she encounters an invoice for fluorescent light bulbs. The invoice is for over $400 dollars, yet the invoice lists only two dozen bulbs as having been delivered. This is clearly a large overcharge. Gabriela is not familiar with this specific vendor, but the delivery slip included with the invoice was, in fact, signed (6 weeks ago) by her receiving clerk. Quite likely, in this case, Gabriela and her operation are the victims of a bogus invoice scam by the vendor and that scam threatens her facility's revenue.

Businesses, churches, and fraternal and charitable organizations are routinely targeted for invoice payment scams. The typical supplier scam involves goods or services that you would routinely order. These include items such as copier paper, maintenance supplies, equipment maintenance contracts, or classified advertising.

Dishonest suppliers can take advantage of weaknesses in an organization's purchasing procedures or of unsuspecting employees who may not be aware of their fraudulent practices. In addition, the supplies delivered by these bogus firms are most often highly overpriced and of poor quality.

In a revenue system that is working properly, the following formula should be in effect:

Total verified bank deposits = Funds available to pay legitimate expenses (Accounts Payable)

Funds available for accounts payable should, of course, only be used to pay legitimate expenses that result from a purchase that can be verified by authorized personnel within the foodservice operation.

MANAGING ACCOUNTS RECEIVABLE AND GIFT CARD REVENUE

In some situations, guests do not pay their bill upon finishing their meal, beverages, or reception, but instead are **direct billed** for their charges. This is often the case in hotel food and beverage operations as well as those restaurants with banquet facilities. When this is the case, creditworthy guests may be sent a bill for the value of the products they have consumed after their event is over. When this form of billing is utilized, it is important that the invoice accurately reflects all guest charges.

Consider, for example, the case of Ahmed, the manager of the Italian Garden restaurant. Ahmed has agreed to host a wedding banquet in his operation. The customer who arranged the event guarantees a count of 90 guests, but on the evening of the dinner, 100 guests actually attend. The customer should be billed for 100 guests if all were served, and future payment by the customer should reflect that.

In this manner, the principles of revenue control are still in place, that is, guest charges should equal revenue collected. Figure 11.10 is an example of a **banquet event order (BEO)** invoice that Ahmed would use to bill this customer.

FIGURE 11.10 Sample Banquet Event Order/Invoice

Caspian Grill **Date 1/15/20xx**

Banquet Event Order/Invoice

Day of Event: ____________ Date of Event: ____________

Time of Event: ________ to ________ Time Ready By: ____________

Type of Event: ____________ Location: ____________

Expected Count: ________ Guaranteed Count: ________ Final Count: ______

Organization Hosting Event: ____________

Organization Contact Person: ____________

Organization Address: ____________

Organization Telephone: ____________ E-mail: ____________

Price: ________ Tax: ________ Service Charge: ________ %

Deposit Amount: ____________ Deposit Received: ____________

Total Amount Due: ____________ Payment Due Date: ____________

Menu	Setup (Style of Room, A/V) Linen
Wines/Liquors	Decor/Flowers

Signature of Guest: ____________ Date:

Signature of Manager: ____________ Date:

Accounts receivable (AR) is the term used to refer to guest charges that have been billed to guests but that have not yet been collected. It's important to recognize that, when using the Uniform System of Accounts for Restaurants (USAR), sales that are to be direct billed are recorded as revenue on the date products and services are provided to guests, and not on the date on which payment is received.

Too high an AR amount owed to an operation should be avoided because the operation has already paid for the products consumed by guests and the labor to serve the products, but has not yet collected payment from the guest. In addition, collecting money after a guest has left your operation can be more difficult as time passes. For these reasons, you must be diligent in monitoring and promptly collecting all AR.

The increasingly popular sale of gift cards presents another revenue recording-related issue for foodservice operations. While direct billed products and services are counted as revenue when they are delivered, the revenue received from the sale of a gift card is not considered revenue until the card is redeemed.

For example, if a foodservice operation sells a $150 gift card, that $150 is immediately recorded as a "liability" on the operation's balance sheet. It is not considered revenue yet because the operation still "owes" the card's buyer $150 worth of products. That liability amount is then reduced by the amount of their purchase(s) when the card buyer redeems all or part of the card's value.

Thus, if the $150 gift card is presented and used to buy $50 worth of products, the operation would reduce its balance sheet liability by $50 and record the $50 charge as revenue. If any portion of the card remains unredeemed after a specific period of time (5 years at minimum in most cases), the operation may then record the unused balance as revenue and eliminate the card's remaining liability balance amount.

MANAGING THE REVENUE SECURITY SYSTEM

If he follows the steps outlined in the preceding pages, Faris will now have completed the development of his revenue security system. Its five key principles are:

1. No product shall be issued from the kitchen or bar unless a permanent record of the issue is made.
2. Product issues must equal guest charges.
3. Both the cashier and a member of management must verify sales receipts.
4. Management must personally verify all bank deposits.
5. The individual authorizing the purchase should verify the legitimacy of the vendor's invoice before it is paid.

In very small operations it is possible to develop and maintain a completely manual revenue control system using a simple **electronic cash register (ECR)** that provides only basic functions like sales tracking and cash management. In today's world, however, while it would be less expensive than investing in a modern POS system, such an approach is both wasteful of time and suspect in accuracy. The simple fact is that the amount of information you need to effectively operate your business grows constantly. Guest dining choices, vendor pricing, inventory levels, payroll statistics, and revenue control are simply a few of the issues involving a huge amount of data collection and manipulation of your business demands.

Fortunately, inexpensive POS systems that can help you easily and quickly assemble the data you need to make good management decisions are readily available. It is dangerous, however, for a foodservice manager to expect that a POS system will bring control to an operation. That happens rarely! A high-quality POS system will, however, take good control systems that have been carefully designed by management and add to them in terms of speed, accuracy, or additional information.

If you hope to improve your revenue security or any other cost control system in the operation, your POS system will be of immense value. If, however, your operation has no controls and you are not committed to the control process, your POS system will simply become a high-tech adding machine, used primarily to sum guest purchases and nothing more. Properly selected and utilized, however, POS systems play a crucial role in the management of a complete revenue security system.

Green and Growing!

In most foodservice operations, the use of sophisticated information technology systems continues to grow. Operators can purchase POS systems designed to gather and analyze more operational and customer-related data than ever before.

These same POS systems provide an excellent example of how companies seeking to reduce energy consumption and become more environmentally responsible can do so in every area of their operations. Considering the purchase of a "green" POS system means choosing energy-efficient systems designed for upgradability, expandability, and long life. By using energy-efficient POS equipment for longer periods of time, operating costs are reduced and landfill waste lessened. Green-oriented operators can also insist that POS system manufacturers do their part by utilizing sustainable shipment packaging materials. For example, by switching from Styrofoam packaging to recycled and reusable paper packaging, the average packaging footprint can be reduced by an estimated 50 percent.

Some recently introduced POS solutions consume less energy than a standard 100-watt light bulb! Efficiency advancements such as these help operators save money from high energy costs, while also helping to relieve the stress of excess energy consumption on the environment. When foodservice operators "go green" by purchasing refurbished computer equipment for use in their cost and revenue control systems, they may be able to save even more money. Both of these approaches to POS system implementation (choosing new or refurbished equipment) allow operators to aggressively pursue their profit goals in an environmentally friendly fashion.

Technology Tools

In this chapter, the principles of revenue control were introduced. In years past, the manual counting of money and balancing of cashier drawers was time-consuming and, in most cases, very tedious. With the increased use of noncash payment forms, the process has become easier, but its importance has not diminished.

Protecting sales revenue from external and internal threats of theft requires diligence and attention to detail. The POS software and specialized hardware now on the market that can help in this area include those that:

1. Maintain daily cash balances from all sources, including those of multiunit and international operations.
2. Reconcile inventory reductions with product issues from kitchen.
3. Reconcile product issues from kitchen with transaction or guest check totals.
4. Reconcile transaction totals with revenue totals.
5. Create over and short computations by server, shift, and day.
6. Balance daily bank deposits with daily sales revenue and identify variances.
7. Maintain accurate e-payment costs and bank deposit records
8. Maintain accurate accounts payable (AP) and accounts receivable (AR) records.
9. Maintain records related to the sale and redemption of gift cards.
10. Interface back office accounting systems with data compiled by the operation's POS system.
11. Interface budgeting software with revenue generation software.
12. Create monthly and annual income statements (profit and loss), balance sheets, and statements of cash flows.

It is important to note that interfacing (electronically connecting) the various software programs you select is critical. For example, a program that forecasts sales revenue and also supplies that data to the software program you are using to schedule labor hours needed will be more effective and helpful to you than one that does not directly connect to the scheduling feature.

In a similar manner, a software program that compares sales recorded on the POS with daily bank deposits is preferable to one that does not connect these two independent but correlated functions.

Apply What You Have Learned

"Are you sure?" asked Joey, as he looked up from the report he was reading.

"Positive," replied JeAnna.

Joey was the general manager of the Cattleman's Club restaurant and JeAnna held the position of kitchen manager.

"I left the restaurant last night just after 5:00 PM," said JeAnna, "I knew we were running that special on rib-eyes yesterday, so I wanted to make sure we had enough. I counted 33 steaks in the walk-in.

"Maybe you miscounted?" said Joey.

"No way," replied Jeanna, "I remember because we had three boxes on the shelf, and each box contains 10 steaks, and we also had three more individual steaks on the shelf, so that makes 33. It was an easy count and easy to remember. I also wrote it down."

"And you're saying that this morning when you counted we had five rib-eyes left?" said Joey.

"Right, we have five left in the walk-in. So that means we should have sold 28 rib-eyes last night" replied JeAnna.

"And how many did the POS say we sold last night?" asked Joey.

"The POS recorded 22 rib-eyes as sold last night," replied JeAnna.

"Then we have a real problem," said Joey.

"No doubt about that!" said JeAnna.

1. Given your knowledge of food and beverage cost control procedures, do you believe this loss of steaks was a result of poor product control procedures, poor revenue control procedures, or both?
2. What steps would you advise Joey and JeAnna to take to identify the source of this problem?
3. What steps would you advise Joey and JeAnna to take to prevent this type of problem from occurring in the future? Explain your answer.

For Your Consideration

1. As increased numbers of guests elect to use a noncash method of payment, the importance of cash control is being replaced, to varying degrees, by the importance of payment card control. While doing so remains controversial, do you see a time in the future when noncash payment forms will completely replace cash payments? Explain your answer.
2. One of the fastest growing career paths in the hospitality industry today is that of a data scientist. Data scientists, trained in the use of artificial intelligence, research design, statistics, computer coding, and sales data analysis, can help convert huge amounts of point-of-sale system and other data into usable and valuable management information. How do you see the skills of data scientists helping managers in the area of hospitality in which you most want to work?

Key Terms and Concepts

The following are terms and concepts addressed in the chapter that are important for you as a manager. To help you review, define the terms below:

Accounts payable (AP)
Automated clearing house (ACH)
Cash accounting (system)
Accrual accounting (system)
House account
Payment card
Travel and entertainment (T&E) card
Credit card
Debit card
E-payments
Walk/skip (a bill)
Quick-change artist
Counterfeit money
Interface
Merchant service provider (MSP)
Credit card skimming

Transaction number (POS)
Guest check
User work station
Kitchen display system (KDS)
Precheck/post-check system
Open check
Short (cash drawer)
Over (cash drawer)
X report
Z report
Bonding
Collusion
Store number
Extended selling price
Service charge
Guest rewards program
Deposit slip
Embezzlement
Bank reconciliation
Direct billed
Banquet event order (BEO)
Accounts receivable (AR)
Electronic cash register (ECR)

Test Your Skills

You may download the Excel spreadsheets for the Test Your Skills exercises from the student companion website at http://www.wiley.com/go/dopson/foodandbeveragecostcontrol8e. Complete the exercises by placing your answers in the shaded boxes and answering the questions as indicated.

1. Sue Pomplun is the owner of Susie's Hens and Fins, a newly opened restaurant in Wisconsin that is located on the shores of Lake Michigan. Because she used her own money to start her restaurant, and because her start-up funds were tight, Sue purchased a low-cost electronic cash register (ECR) system that she hopes to upgrade in the near future. For the time being, however, Sue must manually calculate her daily revenue based on information she receives from the ECR.

   The information below was provided by the ECR for June 15th. Complete the spreadsheet below to help Sue determine her actual revenue for June 15th.

ECR reading June 15 this year		
Susie's Hens and Fins	**Register Reading**	**Actual Revenue**
Food sales	$3,955.45	
Beverage sales	$1,006.10	
Sales tax	$298.44	
Payment card surcharges @ 3.5%	$87.55	
Gratuities	$1,235.00	
Service charges	$245.00	
Total day revenue		

2. Counterfeit money is a problem for all US businesses, including those in the hospitality industry. The US Department of Treasury has developed educational aids to assist managers who must train those who handle cash. To do so, they have compiled information that is critical in the detection of imitation currency and coins.

   Go to www.secretservice.gov to learn about how to detect counterfeit money. When you arrive at the site, enter "know your money" in the web site's search bar to learn how to detect this threat to revenue security. When you are done, prepare a training session appropriate for cashiers who may routinely be responsible for the detection of counterfeit money. Include a memo to your cashiers detailing what they should do if they suspect a bill is counterfeit.

3. Frank Ginzer is the owner of the Key Largo Bar and Grill. In the past, Frank has not reduced his tipped employees' wages by the amount of credit card processing fees he pays on the tips they receive from guests who pay their bills with a credit card.

   He is considering doing so in the future but wants to know what his actual savings amount would be on an average operating week. Help Frank find the answer to his question by completing the following spreadsheet he has begun and answer the question that follows.

Key Largo Credit Card Processing Fees			
March 1 to March 7	Charged Tips	Processing Fees @ 3.5%	Tips to Pay Out
Monday	$248.44	$8.70	
Tuesday			
Wednesday			
Thursday			
Friday			
Saturday			
Sunday	Closed	Closed	Closed
Total Week Charged Tips			

   How much money will Frank retain for his operation on this week if he reduces his tipped employees' pay 3.5 percent on those tips they received via credit card payments?

4. Victoria Font is the manager of a restaurant in a state that allows the restaurant's owner to deduct payment card processing fees from its servers' tipped income prior to paying the servers (in cash) the amount of tips they have earned each day. The average payment card processing fee at Victoria's restaurant is 3.25 percent of the total transaction amount.

   To keep proper records of the deductions her restaurant's owner is allowed, and to monitor the average tip amounts earned by each of her servers, Victoria has developed a daily Server Pay Out Report that she completes with information taken from her POS system. Help Victoria finish completing today's Server Pay Out Report and then answer the questions that follow.

Server Pay Out Report				Date	1-15-20xx
Server	Total Sales	Tips	Tip %	Processing Deduction	Amount to Pay Out
Seo-yun	$832.25	$123.84	14.9%		
Adnan	$781.56	$108.91			
Litzy	$1,254.36	$241.32			
Manon	$912.50	$171.45			
Noah	$425.63	$78.90			
Karis	$398.58	$89.00			
Total					

Which server achieved the highest tip percentage from the sales they made? What is the total amount Victoria's employer should deduct from server tips today because of the payment card processing fees the restaurant must pay?

5. Mary Margaret and Blue are the owners/operators of an extremely upscale bakery goods boutique, and they are interested in a complete asset control system that includes protection of both products and revenue. Identify two control devices/procedures that they could implement to help them control revenue security in the following areas and explain your reason for choosing each.

    a. Product issues

    b. Guest charges

    c. Sales receipts

    d. Sales deposits

    e. Accounts payable

6. Each of the following payment methods allows for potential employee and/or guest theft. Assume that Debbie operates a semiprivate country club where club members and the general public may purchase products in a variety of settings. Identify at least two potential methods of theft for each of the following, as well as a description of the specific procedures Debbie should implement to prevent such theft.

    a. Guest pays cashier directly.

    b. Guest pays server at the table.

    c. Guest is direct billed.

    Which system would you favor using?

7. Denise Cronin operates a quick-service sandwich restaurant in a busy section of a major downtown area. Last week, the POS system in Denise's operation reported the following guest charges, and Denise, upon verifying cash on hand at the end of each day, generated the following sales receipts. Determine Denise's daily and weekly overage and shortage amounts.

    What was the amount of over/short Denise's operation experienced in this week? Does Denise have a cash control problem? How often in a day do you believe Denise should balance her sales receipts with her guest charges? Why?

Day	Sales Receipts	Guest Charges	Over/(Short)
Monday	$3,587.74	$3,585.28	
Tuesday	$3,682.22	$3,693.35	
Wednesday	$3,120.35	$3,110.54	
Thursday	$2,985.01	$3,006.27	
Friday	$4,978.80	$4,981.50	
Saturday	$6,587.03	$6,588.82	
Sunday	$1,733.57	$1,747.93	
Total			

8. Allison Holmes has just been promoted to the job of Regional Beverage Manager at Appleboy's Restaurants. Her district includes six successful units. In one unit, Allison suspects that Ron, the restaurant manager, and Tony, the bar manager, have collaborated to defraud their restaurant by serving cash-paying guests at the bar, not ringing up the sales, and then splitting the revenue collected from those guests.

   Since the unit's cash drawers are always in balance, no one has previously investigated the possibility of employee theft in this unit. What are some indications of such fraudulent activity that would lend support to Allison's suspicions? How would you suggest Allison find out if she is actually correct?

9. Kathy, the general manager, was shocked to discover that the thief in her successful seafood restaurant was Dan, her own dining room supervisor. Dan had used his detailed knowledge of weaknesses in the POS data monitoring system Kathy had designed to void legitimate sales after the fact and thus defraud the restaurant out of an average of $800.00 in revenue per day, every day, for the past 6 months. When finally confronted, Dan confessed to the thefts, citing a personal gambling problem as the reason for his actions.

   The information below details Kathy's reported revenue, food cost, labor cost, other costs, and profit for the past 180 days. Calculate the potential financial performance in dollars and percentages her restaurant would have achieved had it not been for Dan's actions. (Spreadsheet hint: Food cost, Labor cost, and Other costs will be the same for Actual $ and Potential $.)

Kathy's Restaurant Performance				
Revenue Stolen				
	Actual %	Actual $	Potential $	Potential %
Revenue	100.0%	1,566,000		
Food cost	37.0%			
Labor cost	29.0%			
Other costs	25.5%			
Profit	8.5%			

   a. How much profit in dollars did Kathy lose because of Dan's fraudulent activities (difference between potential and actual)?

   b. How much profit percent did Kathy lose because of Dan's fraudulent activities (difference between potential and actual)?

   c. List three things that Kathy could have done to minimize the potential for Dan's illegal activity.

10. Kim owns a small bar in a suburban town and he caters to the "after five" business crowd. He has been experiencing his heaviest business on Wednesday, Thursday, and Friday evenings, so he has recently hired an extra bartender for those shifts. As part of his standard operating procedures, he requires all spills (errors in production or discarded drinks) to be recorded on a spillage report. His average drink price is $14.00 and his spillage has historically been running about 3 percent of sales.

   The information below details Kim's Spillage Report for last week. Calculate the dollar and percentage spilled to see how well Kim's bartenders are doing as compared to the historical average of 3 percent. Does Kim have any control problems? If so, what may be the contributing factors from a revenue security perspective and how might he correct them?

Kim's Spillage Report					
Day	Sales Receipts	# Spilled	Average Drink Price	$ Spilled	% of Sales Spilled
Monday	$1,563.00	6	$14.00		
Tuesday	$1,647.00	8	$14.00		
Wednesday	$1,936.00	20	$14.00		
Thursday	$2,045.00	23	$14.00		
Friday	$2,251.00	25	$14.00		
Saturday	$1,574.00	7	$14.00		
Sunday	$1,412.00	5	$14.00		
Total					

11. Erica is the Special Events Coordinator of a large hotel and she is reviewing her end of event reports after a wedding reception she has just managed. The bride and groom had planned on 300 guests, and they only wanted champagne and wine served. Based on historical data from similar receptions held at the hotel in the past 2 years, Erica estimated each guest would consume an average of three drinks and each bottle provided would yield five drinks. Based on that information, Erica had estimated the number of bottles she would need to serve during the reception.

The actual number of attendees at the reception was reported as 250 guests. The information below shows the beverage usage report that her banquet captain prepared at the end of the reception. That employee reported that a total of 28 bottles of champagne and 48 bottles of wine were served.

Calculate the difference in planned versus used beverages and the difference, if any, in number of guests served versus estimated. Are there any unusual discrepancies based on the number of actual reception attendees (250 guests)? Should Erica be concerned about these numbers? If so, what might Erica do to investigate these concerns?

Beverage Usage Report								
Beverage Selection	# of Guests Selecting	Average Drinks per Bottle	Estimated Bottles Used	Actual Bottles Used	Difference in Bottles Used	Difference in # of Drinks Served	Average Drinks per Guest	Difference in # of Guests Served
Champagne	100	5		28			3	
Wine	200	5		48			3	
Total		5					3	

12. Rami Kahn is the manager of a New York-style deli. At the end of the day, Rami and Samantha, his cashier for the second shift, are counting the cash drawer. Samantha began her shift with a $750.00 cash bank. Help Rami and Samantha complete the cashier reconciliation report he has begun and then answer the questions that follow.

Rami's Cashier Reconciliation Report					
Date	1-15-20xx			Prepared by	*Samantha*
Shift Cashier	*Samantha*			Verified by	*Rami*
Denomination	**Quantity on Hand**	**Value**	**Total on Hand**	**Amount per Z Report**	
Hundreds	4	$100.00		Cash sales	$1,403.62
Fifties	3	$50.00			
Twenties	56	$20.00		E-wallet sales	$135.21
Tens	28	$10.00			
Fives	15	$5.00		Credit and debit card sales	$789.96
Ones	55	$1.00			
Rolled quarters	1	$10.00		Less voids (enter as negative)	$−42.85
Rolled dimes	1	$5.00			
Rolled nickels	0	$20.00		Total Z sales/receipts	
Rolled pennies	1	$0.50			
Quarters	45	$0.25		**Total cash/noncash on hand**	
Dimes	32	$0.10		Less cash bank (enter as negative)	
Nickels	17	$0.05			
Pennies	56	$0.01		**Total shift receipts**	
Other coins/currency	0				
Total cash and coins				**Cash over/short**	
Noncash sales					
E-wallet	Receipts attached		$135.21		
Credit and debit cards	Receipts attached		$754.66		
Total cash/noncash on hand					

What is the amount of Rami's Cash Over/(Short) for this shift? What are likely causes of Rami's Cash Over/(Short) amount for this shift?

GLOSSARY

28 day-period. An accounting method that divides a year into 13 equal periods of 28 days each.

À la carte (menu). A menu format in which guests select individual menu items and each menu item is priced separately.

Acceptance hours. The hours of the day in which an operation is willing to accept food and beverage deliveries.

Accounting period. A period of time, that is, hour, day, week, or month, in which an operator wishes to analyze revenue and expenses.

Accounts payable. The legitimate amount owed to a vendor for products or services rendered.

Accounts receivable. The term used to refer to guest charges that have been billed to a guest but not yet collected.

Achievement budget. A forecast or estimate of projected revenue, expense, and profit for a short period such as a month or a week. It provides current operating information and, thus, assists in making current operational decisions.

Affordable Care Act. Formally known as The Patient Protection and Affordable Care Act; the federal law designed to expand health-care coverage for American citizens.

Aggregate statement. Summary of all details associated with the sales, costs, and profits of a foodservice establishment.

Alcoholic beverages. Those products that are meant for consumption as a beverage and that contain a significant amount of ethyl alcohol. They are classified as beer, wine, or spirits.

Ambience. The feeling or overall mood created in an operation by its decor, staff uniforms, music, and other factors that directly affect its atmosphere.

Annual budget. A forecast or estimate of projected revenue, expense, and profit for a period of one year.

AP. See ***As purchased.***

As needed (Just-in-time). A system of determining the purchase point by using sales forecasts and standardized recipes to decide how much of an item to place in inventory.

As purchased (AP). This term refers to the weight, amount, or count of a product as delivered to the foodservice operator.

Attainable product cost. That cost of goods consumed figure that should be achievable given the product sales mix of a particular operation.

Auditors. Those individuals responsible for reviewing and evaluating proper operational procedures.

Average (mean). The value arrived at by adding the quantities in a series and dividing the sum of the quantities by the number of items in the series.

Average sales per guest (check average). The mean amount of money spent per customer during a given financial accounting period. Often referred to as "check average."

Back of the house (BOH). The nonpublic kitchen production area of a foodservice establishment.

Baker's math. A product formula adjustment method that indicates the amount of an ingredient to use based on the ingredient's weight relative to 100% of the weight of the flour used in the formula.

Bank reconciliation. A comparison of a business's deposit and disbursements records with its bank's acceptance and disbursement records.

Beer. A fermented beverage made from grain and flavored with hops.

Beginning inventory. The dollar value of all products on hand at the beginning of the accounting period. This amount is determined by completing a physical inventory.

Beverage costs. The costs related to the sale of alcoholic beverages.

Beverage cost percentage. The ratio of an operation's beverage costs to its beverage sales. The formula used to calculate beverage cost percentage is Cost of beverage sold ÷ Beverage sales = Beverage cost percentage.

Bid sheet. A form used to compare prices among many vendors in order to select the best prices.

Bin card. An index card with both additions to and deletions from the inventory of a given product. To facilitate its use, the card is usually affixed to the shelf that holds the given item. Used in a perpetual inventory system.

BOGO. See ***Buy One Get One.***

Bonding. Purchasing an insurance policy to protect the operation in case of employee theft.

Bookkeeping. The process of recording and summarizing financial data.

Break-even point. The point at which an operation's expenses are exactly equal to its sales revenue.

Budget/Plan. A forecast or estimate of projected revenue, expense, and profit for a defined accounting period. Often referred to as a plan.

Bundling. The practice of selecting specific menu items and pricing them as a group, in such a manner that the single menu price of the group is lower than if the items comprising the group were purchased individually.

Business dining. Food is provided as a service to the company's employees either as a no-cost (to the employee) benefit or at a greatly reduced price.

Butcher's yield test. A procedure designed to identify losses due to the trimming of inedible fat and bones in meat as well as losses due to the meat's cooking or slicing.

Buy one get one (BOGO). A marketing technique in which buyers who purchase a single item receive a second and identical item for no additional charge.

Call liquors. Those spirits that are requested (called for) by a particular brand name.

Call-in. A system whereby employees who are off duty are required to check in with management on a daily basis to see if the volume is such that they may be needed.

Carryover. A menu item prepared for sale during a meal period but carried over for use in a different meal period.

Cellar temperature. The range of temperatures (approximately 50°F to 57°F (12°C to 14°C) highly recommended for the long-term storage of many types of bottled wines.

CEO (chief executive officer). The highest ranking leader/manager in an organization.

Certified Management Accountant (CMA). A professional designation and certification program designed for accounting and financial management professionals and administered by the Institute of Management Accountants, Inc.

Certified Public Accountant (CPA). A professional designation indicating the passing of a standardized examination (The Uniform Certified Public Accountant Examination) demonstrating the highest levels of competency in accounting.

Cherry pickers. A customer who buys only those items from a supplier that are the lowest in price among the supplier's competition.

Cloud storage. A form of data storage where the data is kept on a remote server rather than, or in addition to, on-site storage.

Cloud-based employee scheduling. A form of employee scheduling where work schedule information is digitally accessed by employees.

COLA. See ***Cost of living adjustment.***

Collusion. Secret cooperation between two or more individuals that is intended to defraud an operation.

Comp. Short for the word "complimentary," which refers to the practice of management giving a product to a guest without a charge. This can be done for a special customer or as a way of making amends for an operational error.

Contract price. A price mutually agreed upon by supplier and operator. This price is the amount to be paid for a product or products over a prescribed period of time.

Contribution margin. The profit or margin that remains after product cost is subtracted from a menu item's selling price.

Contribution margin for overall operation. The dollar amount that contributes to covering fixed costs and providing for a profit.

Contribution margin income statement. A financial summary that shows P&L items in terms of sales, variable costs, contribution margin, fixed costs, and profit.

Cost. See ***Expenses.***

Convenience foods (Ready foods). A menu item or ingredient that has been purchased preprepared to reduce on-site labor cost or to enhance product consistency.

Cost accounting. See ***Managerial accounting.***

Cost of beverage sold. The value of all beverage products sold as well as the costs of all beverages given away, wasted, or stolen in a defined accounting period.

Cost of food consumed. The actual dollar value of all food used, or consumed, by an operation in a defined time period.

Cost of food sold. The dollar amount of all food actually sold, thrown away, wasted, or stolen, plus or minus transfers from other units, minus employee meals.

Cost of living adjustment (COLA). A term used to describe the rationale for a raise in employee pay.

Cost of sales. The combined value of the food and beverage costs used to produce a known quantity of revenue in a defined accounting period.

Cost/volume/profit (CVP) analysis. Method that helps predict the sales dollars and volume required to achieve desired profit (or break-even) based on known costs.

Count (product size). Term used to designate size. Size as established by number of items per pound or number of items per container.

Counterfeit money. An imitation of currency intended to be passed off fraudulently as real money.

Cover. A foodservice term used to identify a single guest.

Craft beer. A custom-made beer traditionally produced in small batches.

Credit card. A card in a payment system by which a bank loans money to the card holder as the holder makes purchases using the card. The loans typically carry interest.

Credit card skimming. The theft of credit card information used in an otherwise legitimate transaction.

Credit memo. An addendum to the vendor's delivery slip (invoice) that reconciles differences between the delivery slip and the goods actually received.

Cycle menu. A menu that is in effect for a predetermined length of time, such as 7 days or 14 days.

Daily inventory sheet. A form that lists the items and ingredients in storage, the unit of purchase, and their par values. It also contains the following columns: on hand, special order, and order amount.

Daily menu. A menu that changes every day.

Debit card. A card in a payment system by which the funds needed to cover the user's purchases are automatically transferred from the user's bank account to the entity issuing the debit card.

Delivery invoice. A vendor's record of the amount due for items delivered to a customer.

Desired profit. The profit that an owner seeks to achieve on a predicted quantity of revenue.

Digital wallet. See ***E-wallet.***

Direct billed. In this situation, guests are not billed immediately upon consuming products and services, but are sent a bill for the value of the products and services after their visit or event is over.

Draft beer. The term used to identify beer products sold in a keg.

Dram shop laws. The term used to describe a series of legislative acts that, under certain conditions, hold businesses and, in some cases, individuals, personally responsible for the actions of guests who consume excessive amounts of alcoholic beverages. These "laws" shift the liability for acts committed by an individual under the influence of alcohol from that individual to the server or operation that supplied the intoxicating beverage.

Edible portion (EP). This term refers to the weight or count of a product after it has been trimmed, cooked, and portioned.

Edible portion (EP) cost. This term refers to the per-serving cost of a menu item.

Electronic funds transfer (EFT). Provides for the electronic payment and collection of money and information; EFT is safe, secure, efficient, and less expensive than paper check payments and collections.

Embezzlement. The term used to describe theft of a type where the money, although legally possessed by the embezzler, is diverted to the embezzler by his or her fraudulent action.

Employee meal cost. The value of a free or reduced-cost meal that is recorded as an employee benefit-related labor cost, not as a food-related cost.

Employee separation. An event that describes employees who quit, are dismissed, or in some other manner have their employment with an operation terminated.

Employment application. A formal document, completed by a candidate for employment, which lists the name, address, work experience, and related information of the candidate.

Empowerment. Giving employees the power to make decisions.

Empty for full system. The bartender is required to retain empty liquor bottles, and then each empty liquor bottle is replaced with a full one at the beginning of the next shift. The empty bottles are then either broken or disposed of, as local beverage law requires.

Ending inventory. The dollar value of all products on hand at the end of the accounting period. This amount is determined by completing a physical inventory.

Environmental sustainability. A variety of earth-friendly practices and policies designed to meet the needs of the present population without compromising the ability of future generations to meet their own needs.

EP. See ***Edible portion***.

Ethics. The choices of proper conduct made by an individual in his or her relationships with others.

Ethyl alcohol. The type of alcohol contained in intoxicating beverage products.

E-wallet (Digital wallet). A payment system that allows users to store their credit card and bank account information in a secure environment, thus eliminating the need to enter account information when making purchases.

Exempt employees. Salaried employees whose duties, responsibilities, and level of decisions make them "exempt" from the overtime provisions of the federal government's Fair Labor Standards Act (FLSA). Exempt employees are expected, by most organizations, to work whatever hours are necessary to accomplish the goals of the organization.

Expenses. The price paid to obtain the items required to operate the business. Often referred to as cost.

Extended price. The price per unit multiplied by the number of units. This refers to a total unit price on a delivery slip or invoice.

FIFO. See ***First-in, first-out***.

First-in, first-out (FIFO). Term used to describe a method of storage in which the operator intends to sell his or her oldest product before selling the most recently delivered product.

Fiscal year. Start and stop dates for a 365-day accounting period. This time period need not begin on January 1st and end on December 31st.

Fixed average. The average amount of sales or volume over a specific series or time period; for example, first month of the year or second week of the second month.

Fixed expense. An expense that remains constant despite increases or decreases in sales volume.

Fixed payroll. Those dollars spent on employees, such as managers, receiving clerks, and dietitians, whose presence is not generally directly dependent on the number of guests served.

Flow-through. The measure used to identify how well a foodservice operation converts additional sales dollars into additional restaurant operating income (ROI). The formula used to calculate flow-through is ROI this year − ROI last year/Sales this year − Sales last year = Flow-through

Food available for sale. The sum value of the beginning inventory of food and all food purchases made by an operation during a defined accounting period.

Food cost percentage. The portion of food sales that was spent on food expenses.

Food costs. The dollar costs associated with actually producing the menu item(s) a guest selects.

Food scoop. A portioning tool whose size is indicated by the number of level scoops required to fill a one-quart (32-ounce) container. Also commonly known as a "food" or "ice cream" scoop.

Franchisor. The entity responsible for selling and maintaining control over the franchise name. (Alternative spelling: franchiser.)

Free-pouring. Pouring liquor from a bottle without measuring the poured amount.

Freezer burn. A form of deterioration in product quality resulting from poorly wrapped or excessively old items stored at freezing temperatures.

Front of the house (FOH). The public (dining) areas of a foodservice establishment.

Full-time equivalent (FTE). One FTE is equivalent to one employee working full time. To calculate FTE, add the number of hours worked by part-time employees in a month and divide the total by 120.

Goal value analysis. A menu pricing and analysis system that compares the performance of individual menu items to the goals of a foodservice operation.

Goods available for sale. The sum of the beginning inventory and purchases during an accounting period. It represents the value of all food that was available for sale during the accounting period.

Gross profit. Sales minus cost of goods sold. Gross profit is listed on older editions of the Uniform System of Accounts for Restaurants (USAR) income (P&L) statement.

Guest check. A written record of what was purchased by the guest and how much the guest was charged for the item(s). Guest checks can be produced in hard copy or electronically.

Guest count. The number of individuals served in a defined time period at a scheduled event.

Guests served per labor dollar. A calculation used to measure worker productivity. The formula used to calculate guests served per labor dollar is Guests served ÷ Cost of labor = Guests served per labor dollar.

Guests served per labor hour. A calculation used to measure worker productivity. The formula used to calculate guests served per labor hour is Number of guests served ÷ Labor hours used = Guests served per labor hour.

Head size. The amount of space on the top of a glass of beer that is made up of foam. Thus, a glass of beer with 1 inch of foam on its top is said to have a 1-inch head.

Heating, ventilation, and air-conditioning (HVAC). Heating, ventilation, and air-conditioning.

House wines. The term used to indicate the type of wines to be served in the event a specific brand-name product is not requested by the guest.

Hosted bar. An arrangement in which no charge is made to guests for the individual drinks they consume at an event but, when the bar is closed, one total amount is charged to the event's host or sponsor. Also referred to as an "open" bar.

Hydrometer. An instrument used to measure the specific gravity of a liquid.

Ideal expense. Management's view of the correct or appropriate amount of expense necessary to generate a given quantity of sales.

Income before income taxes. A term utilized in the Uniform System of Accounts for Restaurants (USAR) indicating an operation's remaining revenue after subtracting all of its operating expenses but before subtracting the amounts of any income taxes that are due.

Income. See ***Net income.***

Income statement. See ***Profit and loss statement (P&L).***

Ingredient room. A storeroom or section of a storeroom where ingredients are weighed and measured according to standardized recipes and then delivered to the appropriate kitchen production area.

Interface. Electronic connection of two or more devices for the purpose of sharing data.

Inventory. All of a foodservice operation's currently stored food, beverage, and other products.

Inventory turnover. The number of times the total value of inventory has been purchased and replaced in an accounting period.

Inventory valuation sheet. A form that documents all inventory items, the quantity on hand, and the monetary value of each item.

Involuntary separation. Management causes the employee to separate from the organization (fires the employee).

Issuing. The process of transferring food or beverage products from storage to production areas for use in a foodservice operation.

Jigger. A small cup-like bar device used to measure predetermined quantities of alcoholic beverages. Jiggers usually are marked in ounces and portions of an ounce, for example, 1 ounce or 1.5 ounces.

Job description. A listing of the tasks to be performed in a particular position.

Job specification. A listing of the personal skills and characteristics needed to perform those tasks pertaining to a particular job description.

Keg beer. See ***Draft beer.***

Kilowatt hour (KWH). A measure of electrical usage.

KWH. See ***Kilowatt hour.***

Labor cost percentage. A ratio of labor costs to revenues. The formula used to calculate labor cost percentage is Cost of labor ÷ Total revenue = Labor cost percentage.

Labor costs. See ***Labor expense.***

Labor dollars per guest served. A calculation used to measure worker productivity. The formula used to calculate labor dollars per guest served is Cost of labor ÷ Guests served = Labor dollars per guest served.

Labor expense. All labor-related expenses (costs), including payroll, required to maintain a workforce in a foodservice operation.

LIFO (Last-in, first-out). Term used to describe a method of storage in which the operator intends to sell the most recently delivered products before selling older products.

Liquors. Fermented beverages that are distilled to increase their alcohol content. Also referred to as "spirits."

Long-range budget. A forecast or estimate of projected revenue, expense, and profit for a period of three to five years.

Loss leaders. Menu items that are priced very low for the purpose of drawing large numbers of customers to an operation.

Management by exception. If an expense is within an acceptable variation range, there is no need for management to intervene. Management takes corrective action only when operational results are outside (an exception to) the range of acceptability.

Managerial accounting. The process of documenting and analyzing sales, expenses, and profits. Sometimes referred to as cost accounting.

Matrix analysis. A method for comparisons between menu items. A matrix allows menu items to be placed into categories based on whether they are above or below overall menu item averages for characteristics such as food cost percentage, popularity, and contribution margin.

Menu engineering. A popular name for one variation of the contribution margin method of menu analysis and that classifies menu items as Stars, Puzzles, Plow Horses, or Dogs.

Menu specials. Menu items that will appear on the menu and be removed when they are either consumed or discontinued. These daily or weekly specials are an effort to provide variety, take advantage of low-cost raw ingredients, utilize carryover products, or test market potential of new menu items.

Merchant service provider (MSP). An MSP is the restaurant's coordinator/manager of payment card acceptance and funds collection, and it provides the electronic connection between payment card issuers and their merchants.

Minimum operating cost. Food cost % + Variable cost %, used in calculating minimum sales point.

Minimum sales point (MSP). The dollar sales volume in an operation that is required to justify staying open.

Minimum staff. The term used to designate the least number of employees, or payroll dollars, required to operate a facility or department within an operation.

Mixed drink. A drink made with one or more liquors (spirits) and other ingredients. Also referred to as a "cocktail."

Mixed expense. An expense that has properties of both a fixed and a variable expense.

Monetary unit. A specific currency denomination, for example US dollar, British pound, or Japanese yen.

MSP. See ***Minimum sales point.***

Mystery shopper. An individual employed by management to pose as an anonymous customer who, during an unannounced visit, observes an operation and reports the observations back to management. Also referred to as a "spotter."

Negligent hiring. Failure on the part of an employer to exercise reasonable care in the selection of employees.

Net income. The profit realized after all expenses and appropriate taxes for a business have been paid.

Noncash payment. A guest payment made using an electronic form of payment. Examples include debit card, credit card, and E-wallet payments.

Non-controllable expense. An expense that the foodservice manager can neither increase nor decrease.

Occupancy costs. Expenses related to occupying and paying for the physical facility that houses the foodservice unit.

OJT (On-the-job training). A method of training in which workers are training while they actually are performing their required tasks.

Onboarding. The procedures used to orient a new employee to the workplace.

On-call. A system whereby selected employees who are off duty can be contacted by management on short notice to cover for other employees who are absent or to come to work if customer demand suddenly increases.

On-the-floor. In the dining area.

Open bar (hosted bar). An arrangement in which no charge is made to guests for the individual drinks they consume at an event but, when the bar is closed, one total amount is charged to the event's host or sponsor. Sometimes referred to as a hosted bar.

Open check. A guest check that has been used to authorize product issues from the kitchen or bar in an operation, but that has not been collected for (closed) and thus the amount of the open check has not been added to the operation's sales total.

Operational efficiency ratio. A ratio used to compare an operation's attainable product costs to its actual product cost. The formula used to calculate an operational efficiency ratio is Actual product cost ÷ Attainable product cost = Operational efficiency ratio.

Orientation program. A program usually held during the first week of a new employee's job that provides information about important items such as dress code, disciplinary system, tip policy, lockers/security, sick leave policy, and retirement programs.

Other controllable expenses. Nonfood and beverage costs that can to some degree be directly influenced by management action. Examples include advertising expense, utility costs, and the costs of repair and maintenance of equipment.

Other expenses. The expenses of an operation that are neither food-, beverage-, or labor-related.

Over. Term used when the total amount of money in a cash drawer is more than the total amount of money that should be there based on sales receipts.

Overtime wages. Employee wages that, by law or policy, must be paid at a higher than normal rate.

Oxidation. A process that occurs when oxygen comes in contact with bottled wine, resulting in a deterioration of the wine product.

P&L. See ***Profit and loss statement***.

Padded inventory. The term used to describe the inappropriate activity of adding a value for nonexisting inventory items to the value of total inventory in an effort to understate actual costs.

Par level. A system of determining the point at which additional product should be ordered using management-established minimum and maximum allowable inventory levels for a given inventory item.

Payroll. Total wages and salaries paid by a foodservice operation to its employees.

Percent. The number "out of each hundred." Thus, 10 percent means 10 out of each 100. In this example, this would be computed by dividing the part (10) by the whole (100).

Percentage variance. The change in sales or an expense, expressed as a percentage, which results from comparing two operating periods.

Performance to budget. The percentage of the budget obtained and actually spent on expenses.

Perpetual inventory. An inventory control system in which additions to and deletions from total inventory are recorded as they occur.

Perpetual inventory card. A bin card that includes a product's price at the top of the card, allowing for continual tracking of the quantity of an item on hand and its value.

Physical inventory. An inventory control system in which an actual or physical count and valuation of all inventory on hand is taken at the close of each accounting period.

Plan. See ***Budget***.

Plate cost. The sum cost of preparing all of the items included in a single meal (or "plate") served to a guest and sold for one price.

PO. See ***Purchase order***.

Point of sales (POS) system. A system for controlling a hospitality operation's cash and product usage by using a computer processor and, depending on the size of the operation, additional computer hardware, interfaced communication devices, and software. Commonly referred to as a POS.

Popularity index. The percentage of total guests choosing a given menu item from a list of menu alternatives.

Portion cost. The product cost required to produce one serving of a menu item.

POS. See ***Point of sales***.

Potentially hazardous foods. Foods that must be carefully handled for time and temperature control to keep them safe to consume.

Precheck/postcheck system. In this system, a server electronically records (prechecks) an order as given to him or her by the guest. The products ordered by the guest, prechecked by the server, and issued by the kitchen or bar should match the items and money ultimately collected (postchecked) by the cashier.

Predicted number to be sold. The number of a given menu item that is likely to be sold if the total number of customers to be served is known.

Preemployment drug testing. A preemployment test used to determine if an applicant uses drugs. It is allowable in most states and can be a very effective tool for reducing insurance rates and potential employee liability issues.

Premium liquors. Expensive call liquors.

Price blending. The process of assigning prices based on product groupings of varying costs for the purpose of achieving predetermined overall cost objectives.

Price comparison sheet. A listing of multiple vendors' price bids on selected items that result in the selection of a vendor based on the best value offer.

Price spread. The difference in selling price between the lowest and highest priced menu item in the same menu category.

Price/value relationship. The guests' view of how much value they are receiving for the price they are paying.

Prime cost. An operation's total cost of sales added to its total labor cost.

Prix fixe (menu). A menu format in which guests select a preset grouping of menu items and then pay one set (fixed) price for all of the menu items included in the grouping.

Product mix. See ***Sales mix***.

Product specification (spec). A detailed description of an ingredient or menu item. Also referred to as a spec.

Product yield. The amount of a product remaining after cooking, trimming, portioning, or cleaning. Often expressed as a percentage of the product's as-purchased (AP) amount.

Productivity. The amount of work performed by a worker in a set amount of time.

Productivity ratio. A ratio formula calculated as total unit output divided by total unit input.

Productivity standard. Management's expectation of the desired productivity ratio of each employee. Also, management's view of what constitutes the appropriate productivity ratio of all employees in a given foodservice unit or units.

Profit. The dollars that remain after all expenses have been paid. Often referred to as net income.

Profit and loss statement. A detailed listing of revenue and expenses for a given accounting period. Also referred to as a P&L or an income statement.

Profit margin. This formula refers to net income divided by total revenues. Also referred to as return on sales.

Psychological testing. Preemployment testing that can include personality tests, tests designed to predict performance, or tests of mental ability.

Pull date (beer). Expiration date on beverage products, usually beers, after which they should not be sold.

Purchase order (PO). A listing of products requested by a purchasing agent. The purchase order lists various product information, including item and quantity ordered as well as price quoted by the vendor.

Purchase point. The point in time when an item held in inventory reaches a level that indicates it should be reordered.

Purchases. The sum cost of all food purchased during an accounting period. It is calculated by adding all properly tabulated invoices for the accounting period.

Quick-service restaurant (QSR). A quick-service restaurant offers a limited menu and is designed for the convenience of customers who want their food very fast. Sometimes referred to as "fast food."

Quick-change artist. A guest who, having practiced the routine many times, attempts to confuse a cashier so that the cashier, in his or her confusion, will give the guest too much change.

Recipe conversion factor (RCF). A mathematical formula that yields a number (factor) managers use to convert a recipe that produces a known yield to the same recipe producing a desired yield. The formula used to calculate a recipe conversion factor is Desired yield ÷ Current yield = Recipe conversion factor.

Recipe-ready. A recipe ingredient that has been cleaned, trimmed, cooked, and/or otherwise processed to the point it can be added as called for in a recipe.

Recodable electronic locks. A locking system that allows management to issue multiple keys and to identify precisely the time an issued key was used to access a lock, as well as to whom that key was issued.

Reference price. The price perceived by customers to be the normal and fair price for a product or service.

Refusal hours. Those hours of the day in which an operation refuses to accept food and beverage deliveries.

Reporting period. The process of reporting a time period for which records are being maintained. This may be of the same duration as an accounting period.

Requisition. When a food or beverage product is requested from storage by an employee for use in an operation.

Restaurant operating income. A term utilized in the Uniform System of Accounts for Restaurants (USAR) indicating an operation's remaining revenue after subtracting all operating expenses except those costs designated as corporate overhead, interest expense, and other income (expense).

Return on sales (ROS). Return on sales (ROS) percentage is calculated as Net income ÷ Total revenue. The ROS can also be stated in percentage or whole dollar terms. Often referred to as profit margin.

Revenue. The term used to indicate the dollars taken in by the business in a defined period of time. Often referred to as sales.

Revenue per available seat hour (RevPASH). The amount of hourly income generated by each available seat in a foodservice operation.

Rolling average. The average amount of sales or volume over a changing time period, for example the last 10 days or the last three weeks.

ROS. See ***Return on sales***.

Safety stock. These are additions to par stock, held as a hedge against the possibility of extra demand for a given product. This helps reduce the risk of being out of stock on a given item.

Salaried employee. An employee who receives the same income per week or month regardless of the number of hours worked.

Sales. See ***Revenue***.

Sales forecast. A prediction of the number of guests to be served and the revenues they will generate in a defined, future time period.

Sales history. A record of the sales achieved by an operation during a specifically identified time period.

Sales mix. The series of consumer-purchasing decisions that result in a specific food and beverage cost percentage. Sales mix can affect an operation's overall product cost percentage any time menu items have varying food and beverage cost percentages.

Sales per labor hour. A calculation used to measure worker productivity. The formula used to calculate sales per labor hour is Total sales ÷ Labor hours used = Sales per labor hour

Sales per seat. The total revenue generated by a facility divided by the number of available seats in its dining area.

Sales per square foot. A measure of an operation's revenue as it relates to the number of square feet the operation occupies. The formula used to calculate sales per square foot is Total sales ÷ Total square footage = Sales per square foot

Sales to date. The cumulative sales figures reported during a defined financial accounting period.

Sales variance. An increase or decrease from previously experienced or predicted sales levels.

Sales volume. The number of units sold.

Sarbanes–Oxley Act (SOX). Technically known as the Public Company Accounting Reform and Investor Protection Act, this law provides criminal penalties for those found to have committed accounting fraud. The SOX covers a whole range of corporate governance issues, including the regulation of those who are assigned the task of verifying a company's financial health.

Scoop (food). See ***Food scoop***.

Selling price. The total amount paid by guests for the purchase of a singly priced item.

Shelf life. The period of time an ingredient or menu item maintains its freshness, flavor, and quality.

Short. Term used when the total amount of money in a cash drawer is less than the total amount of money that should be there based on sales receipts.

Shorting. When the vendor is unable to deliver the quantity of item ordered for the appointed delivery date.

Skills tests. Preemployment tests such as data entry tests for office employees, computer application tests for those involved in using word processing or spreadsheet tools, or food production tasks, as in the case of cooks and chefs.

Skip. See ***Walk***.

SOP. See ***Standard operating procedure***.

Spec. See ***Product specification***.

Spirits. Fermented beverages that are distilled to increase the alcohol content of the product.

Split shift. An employee scheduling technique used to match individual employee work shifts with the peaks and valleys of customer demand.

Spotter. An individual employed by management for the purpose of inconspicuously observing bartenders and wait staff in order to detect any fraudulent or policy-violating behavior.

Standard labor cost. The labor cost needed to meet established productivity standards.

Standard menu. A printed and otherwise fixed menu that stays the same day after day.

Standard operating procedure (SOP). Term used to describe the way something should be done under normal business operating conditions.

Standardized recipe. The procedures to be used for consistently preparing and serving a specific menu item.

Standardized recipe cost sheet. A record of the ingredient costs required to produce a menu item sold by a foodservice operation.

Statement of income and expense. The accounting tool used to report an operation's revenue, expenses, and profit for a defined time period. Also known as the income statement, the profit and loss statement, and the P&L.

Supporting schedules. List of all detailed information associated with selected line items shown on the income statement.

Task training. The training undertaken to ensure that an employee has the skills to meet productivity goals.

Tenths system. A liquor inventory valuation system that requires the inventory taker to assign a product value based upon a visual inspection of a bottle's content. For example, a value of 10/10 is assigned for a full bottle, 5/10 is assigned to a half bottle, and so on.

Tip-on (menu). A small menu segment clipped to a larger and more permanent list of menu items.

Total bar system. An automated set of machines and procedures that combine sales information with product dispensing information to create a complete beverage revenue and production control program.

Transaction number (POS). The method by which a point of sale (POS) system identifies an individual guest order. Each guest order (guest check) is assigned a unique identifying (transaction) number.

Travel and entertainment (T&E) cards. Cards in a payment system by which the card issuer collects full payment from the card users on a monthly basis. These card companies do not typically assess their users' interest charges.

Two-key system. A control system to monitor access to storage areas.

Uniform system of accounts. A recommended and standardized (uniform) set of accounting procedures used for categorizing and reporting revenue and expense.

Uniform System of Accounts for Restaurants (USAR). The recommended and standardized (uniform) set of accounting procedures used for categorizing and reporting revenue and expense in the restaurant industry.

User work stations. A computer terminal in a point-of-sale system used only to record food and beverage orders.

Value pricing. The practice of reducing some or most prices on the menu in the belief that total guest counts will increase to the point that total sales revenue also increases.

Variable expense. An expense that generally increases as sales volume increases and decreases as sales volume decreases.

Variable payroll. Those dollars expended on employees whose presence is directly dependent on the number of guests served. These employees include, for example servers, bartenders, and dishwashers. As the number of guests served increases, the number of these individuals required to do their jobs also increases. As the number of guests served decreases, variable payroll should decrease.

Vintage. The specific year in which a wine's grapes were harvested and turned into wine.

Vintner. Wine producer.

Voluntary separation. When an employee makes the decision to leave the organization.

Walk/skip (the bill). A term used to describe a customer who has consumed a product, but leaves the foodservice operation without paying the bill for the product. Also known as a skip.

Waste percentage. This formula is defined as product loss divided by the product's as-purchased (AP) weight and refers to the amount of product lost in the preparation process.

Weighted average. An average that combines data on the number of guests served and how much each has spent during a given financial accounting period.

Well liquors. Those spirits that are served by an operation when the customer does not specify a particular brand name.

Wine. A fermented beverage made from grapes, fruits, or berries.

Wine list. A menu of an operation's wine offerings.

Working stock. The quantity of goods from inventory reasonably expected to be used between deliveries.

X report. A POS system report that identifies sales produced in a specific time period. An X report may be obtained multiple times in the same time period.

Yield percentage. This formula is defined as 1 − waste percentage and refers to the amount of product available for use by an operation after all preparation-related losses have been taken into account.

Yield test. A procedure used to determine actual edible portion (EP) ingredient costs. It is used to help establish actual per-serving costs on a product that will experience weight or volume loss in preparation.

Z report. A POS system report that identifies sales produced in a specific time period and that resets the recording system to zero in preparation for the next time period's sales data. A Z report may be obtained only once in a sales period.

APPENDIX A

Frequently Used Formulas for Managing Operations

CHAPTER 1: MANAGING REVENUE AND EXPENSE

$$\text{Revenue} - \text{Expenses} = \text{Profit}$$

$$\text{Revenue} - \text{Desired Profit} = \text{Ideal Expense}$$

$$\frac{\text{Part}}{\text{Whole}} = \text{Percent}$$

$$\frac{\text{Expense}}{\text{Revenue}} = \text{Expense \%}$$

$$\frac{\text{Profit}}{\text{Revenue}} = \text{Profit \%}$$

$$\frac{\text{Desired profit}}{\text{Revenue}} = \text{Desired Profit}$$

$$\text{Revenue} - (\text{Food cost} + \text{Beverage cost} + \text{Labor cost} + \text{Other expenses}) = \text{Profit}$$

$$\frac{\text{Food and beverage cost}}{\text{Revenue}} = \text{Food and beverage cost percentage.}$$

$$\frac{\text{Labor cost}}{\text{Revenue}} = \text{Labor cost \%}$$

$$\frac{\text{Other expenses}}{\text{Revenue}} = \text{Other expenses \%}$$

$$\frac{\text{Total expense}}{\text{Revenue}} = \text{Total expense \%}$$

$$\frac{\text{Profit}}{\text{Revenue}} = \text{Profit \%}$$

$$\frac{\text{Actual}}{\text{Budget}} = \text{\% of Budget}$$

CHAPTER 2: CREATING SALES FORECASTS

$$\frac{\text{Total sales}}{\text{Number of guests served}} = \text{Average sales per guest}$$

$$\text{Sales This Year} - \text{Sales Last Year} = \text{Variance}$$

$$\frac{\text{Sales this year} - \text{Sales last year}}{\text{Sales last year}} = \text{Percentage variance}$$

$$\frac{\text{Variance}}{\text{Sales Last Year}} = \text{Percentage Variance}$$

$$(\text{Sales This Year/Sales Last Year}) - 1 = \text{Percentage Variance}$$

$$\text{Sales Last Year} + (\text{Sales Last Year} \times \text{\% Increase Estimate}) = \text{Revenue Forecast}$$

$$\text{Sales Last Year} \times (1.00 + \text{\% Increase Estimate}) = \text{Revenue Forecast}$$

$$\text{Guest Count Last Year} + (\text{Guest Count Last Year} \times \text{\% Increase Estimate}) = \text{Guest Count Forecast}$$

$$\text{Guests Last Year} \times (1.00 + \text{\% Increase Estimate}) = \text{Guest Count Forecast}$$

$$\text{Last Year's Average Sales Per Guest} + \text{Estimated Increase in Sales Per Guest} = \text{Sales Per Guest Forecast}$$

$$\frac{\text{Revenue forecast}}{\text{Guest count forecast}} = \text{Average sales per guest forecast}$$

CHAPTER 3: PURCHASING AND RECEIVING

$$\frac{\text{Total Number of a Specific Menu Item Sold}}{\text{Total Number of All Menu Items Sold}} = \text{Popularity Index}$$

$$\text{Number of Guests Expected} \times \text{Item Popularity Index} = \text{Predicted Number of that Item to be Sold}$$

$$\frac{\text{Desired yield}}{\text{Current yield}} = \text{Recipe Conversion Factor (RCF)}$$

When using baker's math:

$$\frac{\text{Ingredient weight}}{\text{Flour weight}} \times 100\% = \text{Ingredient percentage}$$

To convert Celsius to Fahrenheit temperature

$$(\text{Celsius Temperature} \times 1.8) + 32 = \text{Fahrenheit Temperature}$$

$$\text{Unit Price} \times \text{Number of Units} = \text{Extended Price}$$

CHAPTER 4: MANAGING INVENTORY AND PRODUCTION

$$\text{Item Amount} \times \text{Item Value} = \text{Item Inventory Value}$$

$$\text{Par Value} - \text{On-Hand} + \text{Special Order} = \text{Order Amount}$$

$$\frac{\text{Total (Recipe) Cost}}{\text{Number of Portions}} = \text{Portion Cost}$$

$$\text{Prior-day carryover} + \text{Today's production} = \text{Today's sales forecast} \pm \text{Margin of error}$$

CHAPTER 5: MONITORING FOOD AND BEVERAGE PRODUCT COSTS

Beginning Inventory
+ Purchases
= Food Available for Sale
− Ending Inventory
= Cost of Food Consumed
− Employee Meals
= Cost of Food Sold

$$\frac{\text{Cost of food sold}}{\text{Food sales}} = \text{Food cost \%}$$

$$\frac{\text{Cost of food sold}}{\text{Total meals served}} = \text{Cost per meal}$$

Beginning Beverage Inventory
+ Purchases
= Beverages Available for Sale
− Ending Inventory
= Cost of Beverage Sold

$$\frac{\text{Cost of beverage sold}}{\text{Beverage sales}} = \text{Beverage cost \%}$$

Cost of Food Sold with Transfers:

Beginning Food Inventory
+ Purchases
= Food Available For Sale
− Ending Inventory
= Cost of Food Consumed
− Food Transfers From Kitchen
+ Beverage Transfers To Kitchen
− Employee Meals
= Cost of Food Sold

Cost of Beverage Sold with Transfers:

Beginning Beverage Inventory
+ Purchases
= Beverages Available For Sale
− Ending Inventory
= Cost of Beverage Consumed
+ Food Transfers From Kitchen
− Beverage Transfers To Kitchen
= Cost of Beverage Sold

$$\frac{\text{Cost in one product category}}{\text{Total cost in all Categories}} = \text{Proportion of total product cost}$$

$$\frac{\text{Purchases today}}{\text{Sales today}} = \text{Cost \% today (Six-column food cost \% estimate)}$$

$$\frac{\text{Purchases to date}}{\text{Sales to date}} = \text{Cost \% to date (Six-column food cost \% estimate)}$$

$$\text{Product as purchased (AP)} - \text{Losses due to preparation} = \text{Product edible portion (EP)}$$

$$\frac{\text{Product loss}}{\text{AP weight}} = \text{Waste \%}$$

$$\text{Waste \%} + \text{Yield \%} = 100\ \%$$

$$\text{Yield \%} = 1.00 - \text{Waste \%}$$

$$\frac{\text{EP required}}{\text{Product yield \%}} = \text{AP required}$$

$$\text{EP Required} = \text{AP Required} \times \text{Yield \%}$$

$$\frac{\text{EP weight}}{\text{AP weight}} = \text{Product yield \%}$$

$$\frac{\text{AP Price Per Pound}}{\text{Product Yield \%}} = \text{EP Cost (Per Pound)}$$

$$\frac{\text{Actual product cost}}{\text{Attainable product cost}} = \text{Operational efficiency ratio}$$

$$\frac{\text{Actual product cost} - \text{Attainable product cost}}{\text{Attainable product cost}} = \text{Cost percentage varience}$$

$$\frac{\text{Cost As Per Standardized Recipes}}{\text{Total Sales}} = \text{Attainable Food Cost \%}$$

$$\frac{\text{Item dollar sales}}{\text{Total dollar salees}} = \text{Item \% of total sales}$$

CHAPTER 6: MANAGING FOOD AND BEVERAGE PRICING

$$\text{Price} \times \text{Number sold} = \text{Total revenue}$$

$$\frac{\text{Cost of a specific food item sold}}{\text{Food sales of that item}} = \text{Food cost \% of that item}$$

$$\frac{\text{Cost of a specific food item sold}}{\text{Food cost \% of that item}} = \text{Food sales (selling price) of that item}$$

$$\frac{1.00}{\text{Desired product cost \%}} = \text{Pricing factor}$$

$$\text{Pricing Factor} \times \text{Item Cost} = \text{Item Menu (Selling) Price}$$

$$\text{Pricing Factor} \times \text{Plate Cost} = \text{Plate (Per Person) Selling Price}$$

$$\text{Selling Price} - \text{Product Cost} = \text{Contribution Margin}$$

$$\text{Product Cost} + \text{Contribution Margin Desired} = \text{Selling Price}$$

$$\frac{\text{Total buffet product cost}}{\text{Guests served}} = \text{Buffet product cost per guest}$$

CHAPTER 7: MANAGING THE COST OF LABOR

$$\frac{\text{Output}}{\text{Input}} = \text{Productivity ratio}$$

$$\frac{\text{Number of employees separated}}{\text{Number of employess in workforce}} = \text{Employee turnover rate}$$

$$\text{Involuntary employee turnover rate} = \frac{\text{Number of employees involuntarily separated}}{\text{Number of employees in workforce}}$$

$$\text{Voluntary employee turnover rate} = \frac{\text{Number of employees voluntarily separated}}{\text{Number of employees in workforce}}$$

$$\frac{\text{Cost of labor}}{\text{Total sales}} = \text{Labor cost \%}$$

$$\frac{\text{Total sales}}{\text{Labor hours used}} = \text{Sales per labor hour}$$

$$\frac{\text{Cost of labor}}{\text{Guests served}} = \text{Labor dollars per guest served}$$

$$\frac{\text{Guests served}}{\text{Cost of labor}} = \text{Guests served per labor dollar}$$

$$\frac{\text{Guests served}}{\text{Labor hours used}} = \text{Guests served per labor hour}$$

$$\frac{\text{Revenue}}{\text{Available seat hours}} = \text{Revenue per available seat hour (RevPASH)}$$

$$\frac{\text{Number of estimated guests served}}{\text{Guests served per labor dollar}} = \text{Estimated cost of labor}$$

$$\text{Forecasted Total Sales} \times \text{Labor Cost \% Standard} = \text{Cost of Labor Budget}$$

$$\frac{\text{Forecasted number of guests served}}{\text{Guests served per labor hour standard}} = \text{Labor hour budget}$$

$$\frac{\text{Actual amount}}{\text{Budgeted amount}} = \text{\% of budget}$$

CHAPTER 8: CONTROLLING OTHER EXPENSES

$$\frac{\text{Other expenses}}{\text{Total sales}} = \text{Other expenses cost\%}$$

$$\frac{\text{Other expense}}{\text{Number of guests served}} = \text{Other expense cost per guest}$$

CHAPTER 9: ANALYZING RESULTS USING THE INCOME STATEMENT

$$\frac{\text{Food Category Cost}}{\text{Total Food Sales}} = \text{Food Category Cost \%}$$

$$\text{Total Sales} - \text{Total Cost of Sales} = \text{Gross Profit}$$

$$\frac{\text{Cost of food consumed}}{\text{Average inventory value}} = \text{Food inventory turnover}$$

$$\frac{\text{Beginning inventory value} + \text{Ending inventory value}}{2} = \text{Average inventory value}$$

$$\frac{\text{Totallabor cost}}{\text{Total sales}} = \text{Labor cost \%}$$

$$\text{This Year's Sales} \times \text{Last Year's Adjusted Labor Cost \%} = \text{This Year's Projected Labor Cost}$$

$$\frac{\text{Net income}}{\text{Total sales}} = \text{Profit margin (Return on Sales (ROS))}$$

$$\text{NetIncome This Year} - \text{NetIncome Last Year} = \text{NetIncome Variance}$$

$$\frac{\text{Net income this period} - \text{Net income last period}}{\text{Netincome last period}} = \text{Percentage variance}$$

$$\frac{\text{Variance}}{\text{Net income last year}} = \text{Percentage variance}$$

$$\left[\frac{\text{Net income this year}}{\text{Net income last year}} - 1.00\right] = \text{Percentage variance}$$

CHAPTER 10: PLANNING FOR PROFIT

$$\text{Selling Price} - \text{Item Cost} = \text{Contribution Margin Per Menu Item}$$

$$\text{Total Sales} - \text{Total Item (Product) Costs} = \text{Total Contribution Margin}$$

$$\frac{\text{Total contribution margin}}{\text{Number of items sold}} = \text{Average contribution margin per item}$$

$$\frac{\text{Total sales}}{\text{Total number of items sold}} = \text{Weighted average selling price}$$

$$\frac{\text{Total cost}}{\text{Total number of items sold}} = \text{Weighted average item cost}$$

$$\frac{\text{Total contribution margin (CM)}}{\text{Total number of items sold}} = \text{Weighted average item CM}$$

$$\frac{\text{Total food cost}}{\text{Total sales}} = \text{Weighted average food cost}$$

$$A \times B \times C \times D = \text{Goal Value}$$

where

$A = .00 - \text{Food Cost \%}$
$B = \text{Item Popularity}$
$C = \text{Selling Price}$
$D = 1.00 - (\text{Variable Cost \%} + \text{Food Cost \%})$

$$\text{Total Sales} - \text{Variable Costs} = \text{Contribution Margin for Overall Operation}$$

$$\frac{\text{Total sales}}{\text{Guest served (Units)}} = \text{Selling Price Per Unit}$$

$$\frac{\text{Variable Costs}}{\text{Guest served (Units)}} = \text{Variable Costs Per Unit}$$

$$\text{(Per Unit) Selling Price} - \text{Variable Cost Per Unit} = \text{Contribution Margin Per Unit}$$

$$\frac{\text{Variable Cost Per Unit}}{\text{Selling Price Per Unit}} = \text{Variable Cost \%}$$

$$\frac{\text{Variable Contribution Margin Per Unit}}{\text{Selling Price Per Unit}} = \text{Contribution Margin \%}$$

$$\text{Selling Price \%} - \text{Variable Cost \%} = \text{Contribution Margin \%}$$

or

$$100\% - \text{Variable Cost \%} = \text{Contribution Margin \%}$$

$$\frac{\text{Fixed costs}}{\text{Contribution margin \%}} = \text{Break-even point in sales}$$

$$\frac{\text{Fixed costs}}{\text{Contribution margin per unit (per guest)}} = \text{Break-even point in guests (covers) served}$$

$$\frac{\text{Fixed costs} + \text{Before-tax profit}}{\text{Contribution margin \%}} = \text{Sales dollars to achieve desired after-tax profit}$$

$$\frac{\text{After-tax profit}}{(1.00 - \text{Tax rate})} = \text{Before-tax profit}$$

$$\frac{\text{Fixed costs} + \text{Before-tax profit}}{\text{Contribution Margin Per Unit (Per Guest)}} = \text{Guests to be Served to Achieve Desired After-Tax Profit}$$

$$\text{Food Cost \%} + \text{Variable cost \%} = \text{Minimum Operating Cost}$$

$$\frac{\text{Minimum labor cost}}{1.00 - \text{Minimum operating cost}} = \text{MSP}$$

or

$$\frac{\text{Minimum labor cost}}{1.00 - (\text{Food cost \%} + \text{Variable cost \%})} = \text{MSP}$$

(MSP = minimum sales point)

$$\text{Budgeted Revenue} - \text{Budgeted Expense} = \text{Budgeted Profit}$$

$$\frac{\text{Total sales}}{\text{Available seats}} = \text{Sales per seat}$$

$$\frac{\text{Total sales}}{\text{Total square footage}} = \text{Sales per square foot}$$

CHAPTER 11: MANAGING THE REVENUE CONTROL SYSTEM

Documented Product Requests = Product Issues

Total Product Issues (by the kitchen or bar) = Total Guest Charges

Total Guest Charges = Total Sales Receipts

Total Sales Receipts = Total Verified Bank Deposits

Total Verified Bank Deposits = Funds Available To Pay Legitimate Expenses (AP)

(Actual cash – Beginning cash) – Cash sales per the Z report = Cash overage (shortage)

Product Issues = Guest Charges = Sales Receipts = Sales Deposits = Funds Available for Accounts Payable

INDEX

A

B

C

D

E

F

G

H

I

J

K

L

M

N

O

P

Q

R

S

T

U

V

W

X

Y

Z